Lecture Notes in Computer Science 12095

More information about this series at http://www.springer.com/series/7407

Alexander Kononov · Michael Khachay ·
Valery A Kalyagin · Panos Pardalos (Eds.)

Mathematical Optimization Theory and Operations Research

19th International Conference, MOTOR 2020
Novosibirsk, Russia, July 6–10, 2020
Proceedings

 Springer

Editors
Alexander Kononov (iD)
Sobolev Institute of Mathematics SB RAS
Novosibirsk, Russia

Valery A Kalyagin (iD)
National Research University Higher School
of Economics
Nizhny Novgorod, Russia

Michael Khachay (iD)
Krasovsky Institute of Mathematics
and Mechanics
Yekaterinburg, Russia

Panos Pardalos (iD)
University of Florida
Gainesville, FL, USA

ISSN 0302-9743   ISSN 1611-3349 (electronic)
Lecture Notes in Computer Science
ISBN 978-3-030-49987-7   ISBN 978-3-030-49988-4 (eBook)
https://doi.org/10.1007/978-3-030-49988-4

LNCS Sublibrary: SL1 – Theoretical Computer Science and General Issues

This Springer imprint is published by the registered company Springer Nature Switzerland AG
The registered company address is: Gewerbestrasse 11, 6330 Cham, Switzerland

Preface

This volume contains the refereed proceedings of the 19th International Conference on Mathematical Optimization Theory and Operations Research (MOTOR 2020)[1] held on July 6–10, 2020, near Novosibirsk, Russia.

MOTOR 2020 was the second joint scientific event[2] unifying a number of well-known international and Russian conferences held in Ural, Siberia, and the Far East for a long time:

- The Baikal International Triennial School Seminar on Methods of Optimization and Their Applications (BITSS MOPT), established in 1969 by academician N. N. Moiseev, was the 17th event[3] in this series, held in 2017 in Buryatia
- The All-Russian Conference on Mathematical Programming and Applications (MPA), established in 1972 by academician I. I. Eremin, was the 15th conference[4] in this series, held in 2015 near Ekaterinburg
- The International Conference on Discrete Optimization and Operations Research (DOOR) was organized nine times since 1996, and the last event[5] was held in 2016 in Vladivostok
- The International Conference on Optimization Problems and Their Applications (OPTA) was organized regularly in Omsk since 1997, and the 7th event[6] was held in 2018

As per tradition, the main conference scope included, but was not limited to, mathematical programming, bi-level and global optimization, integer programming and combinatorial optimization, approximation algorithms with theoretical guarantees and approximation schemes, heuristics and meta-heuristics, game theory, optimization in machine learning and data analysis, and valuable practical applications in operations research and economics.

In response to the call for papers, MOTOR 2020 received 175 submissions. Out of 102 full papers considered for reviewing (73 abstracts and short communications were excluded because of formal reasons) only 31 papers were selected by the Program Committee (PC) for publication in this volume. Each submission was reviewed by at least three PC members or invited reviewers, experts in their fields, in order to supply

---

[1] http://math.nsc.ru/conference/motor/2020/.

[2] http://motor2019.uran.ru.

[3] http://isem.irk.ru/conferences/mopt2017/en/index.html.

[4] http://mpa.imm.uran.ru/96/en.

[5] http://www.math.nsc.ru/conference/door/2016/.

[6] http://opta18.oscsbras.ru/en/.

detailed and helpful comments. In addition, the PC recommended to include 33 papers in the supplementary volume after their presentation and discussion during the conference and subsequent revision with respect to the reviewers' comments.

The conference featured nine invited lectures:

- Prof. Aida Abiad (Eindhoven University of Technology, The Netherlands, and Ghent University, Belgium), "Graph invariants and their application to the graph isomorphism problem"
- Prof. Evripidis Bampis (Sorbonne Université, France), "Multistage Optimization Problems"
- Prof. Bo Chen (University of Warwick, UK), "Capacity Auctions: VCG Mechanism vs. Submodularity"
- Prof. Sergei Chubanov (Bosch Research, Germany), "Convex geometry in the context of artificial intelligence"
- Prof. Igor Konnov (Kazan Federal University, Russia) "Equilibrium Formulations of Relative Optimization Problems"
- Prof. Alexander Kostochka (University of Illinois at Chicago, USA), "Long cycles in graph and hypergraphs"
- Prof. Panos Pardalos (University Florida, USA), "Inverse Combinatorial Optimization Problems"
- Prof. Soumyendu Raha (Indian Institute of Science, Bangalore, India) "Optimal complexity Matrix Multiplication like computation structures on network-on-chip architecture subject to conflicts of resource allocation constraints"
- Prof. Yakov Zinder (University of Technology Sydney, Australia), "Two-stage Scheduling Models with Limited Storage"

The following five tutorials were given by outstanding scientists:

- Prof. Alexander Grigoriev (Maastricht University, The Netherlands), "Evolution of sailor and surgical knots"
- Prof. Michael Khachay (Krasovsky Institute of Mathematics and Mechanics, Russia), "Metrics of a fixed doubling dimension: an efficient approximation of combinatorial problems"
- Prof. Vladimir Mazalov (Institute of Applied Mathematical Research, Russia), "Game Theory and Social Networks"
- Dr. Andrey Melnikov (Sobolev Institute of Mathematics, Russia), "Practice of using the Gurobi optimizer"
- Prof. Konstantin Vorontsov (Institute of Physics and Technology, Russia), "A survey of machine learning problems from optimization point of view".

We thank the authors for their submissions, members of the PC, and external reviewers for their efforts in providing exhaustive reviews. We thank our sponsors and partners: Mathematical Center in Akademgorodok, Russian Foundation for Basic Research, Sobolev Institute of Mathematics, Novosibirsk State University, Krasovsky Institute of Mathematics and Mechanics, Higher School of Economics, and Melentiev

Energy Systems Institute. We are grateful to Alfred Hofmann, Aliaksandr Birukou, Anna Kramer, and colleagues from Springer LNCS and CCIS editorial boards for their kind and helpful support.

July 2020

Alexander Kononov
Michael Khachay
Valery Kalyagin
Panos Pardalos

Organization

Program Committee Chairs

Vladimir Beresnev	Sobolev Institute of Mathematics, Russia
Alexander Kononov	Sobolev Institute of Mathematics, Russia
Michael Khachay	Krasovsky institute of Mathematics and Mechanics, Russia
Valery Kalyagin	Higher School of Economics, Russia
Panos Pardalos	University of Florida, USA

Program Committee

Edilkhan Amirgaliev	Suleyman Demirel University, Kazakhstan
Anatoly Antipin	Dorodnicyn Computing Centre, FRC, CSC, RAS, Russia
Evripidis Bampis	Sorbonne Université, France
Olga Battaïa	ISAE-Supaero, France
René van Bevern	Novosibirsk State University, Russia
Oleg Burdakov	Linköping University, Sweden
Maxim Buzdalov	ITMO University, Russia
Igor Bykadorov	Sobolev Institute of Mathematics, Russia
Tatjana Davidović	Mathematical Institute SANU, Serbia
Stephan Dempe	Freiberg University, Germany
Anton Eremeev	Sobolev Institute of Mathematics, Russia
Adil Erzin	Novosibirsk State University, Russia
Stefka Fidanova	Institute of Information and Communication Technologies, Bulgaria
Alexander Filatov	Far Eastern Federal University, Russia
Fedor Fomin	University of Bergen, Norway
Edward Gimadi	Sobolev Institute of Mathematics, Russia
Alexander Grigoriev	Maastricht University, The Netherlands
Evgeny Gurevsky	University of Nantes, France
Klaus Jansen	Kiel University, Germany
Vadim Kartak	Ufa State Aviation Technical University, Russia
Alexander Kazakov	Matrosov Institute of System Dynamics and Control Theory, Russia
Lev Kazakovtsev	Siberian State Aerospace University, Russia
Oleg Khamisov	Melentiev Energy Systems Institute, Russia
Andrey Kibzun	Moscow Aviation Institute, Russia
Donghyun (David) Kim	Kennesaw State University, USA
Yury Kochetov	Sobolev Institute of Mathematics, Russia
Igor Konnov	Kazan Federal University, Russia

Vadim Levit	Ariel University, Israel
Bertrand M. T. Lin	National Chiao Tung University, Taiwan
Vittorio Maniezzo	University of Bologna, Italy
Vladimir Mazalov	Institute of Applied Mathematical Research, Russia
Nenad Mladenović	Khalifa University, UAE
Rolf Niedermeier	Technical University of Berlin, Germany
Yury Nikulin	University of Turku, Finland
Evgeni Nurminski	Far Eastern Federal University, Russia
Leon Petrosyan	Saint Petersburg State University, Russia
Petros Petrosyan	Yerevan State University, Armenia
Alex Petunin	Ural Federal University, Russia
Sergey Polyakovskiy	Deakin University, Australia
Leonid Popov	Krasovsky Institute of Mathematics and Mechanics, Russia
Mikhail Posypkin	Dorodnicyn Computing Centre, Russia
Artem Pyatkin	Sobolev Institute of Mathematics, Russia
Soumyendu Raha	Indian Institute of Science, India
Yaroslav Sergeyev	University of Calabria, Italy
Sergey Sevastyanov	Sobolev Institute of Mathematics, Russia
Natalia Shakhlevich	University of Leeds, UK
Alexander Shananin	Moscow Institute of Physics and Technology, Russia
Angelo Sifaleras	University of Macedonia, Greece
Vladimir Skarin	Krasovsky Institute of Mathematics and Mechanics, Russia
Petro Stetsyuk	Glushkov Institute of Cybernetics, Ukraine
Alexander Strekalovsky	Matrosov Institute for System Dynamics and Control Theory, Russia
Vitaly Strusevich	University of Greenwich, UK
Maxim Sviridenko	Yahoo! Labs, USA
Tatiana Tchemisova	University of Aveiro, Portugal

Additional Reviewers

Agafonov, Evgeny	Borisovsky, Pavel	Filchenkov, Andrey
Alazmi, Hakem	Bulanova, Nina	Filippova, Tatiana
Alekseeva, Ekaterina	Buzhinsky, Igor	Forghani, Majid
Almeida, Ricardo	Bykadorov, Igor	Gluschenko, Konstantin
Antipin, Anatoly	Camacho, David	Gomoyunov, Mikhail
Antipov, Denis	Chandramouli, Shyam	Grage, Kilian
Ayzenberg, Natalya	Chentsov, Alexander	Gromova, Ekaterina
Benmansour, Rachid	Chicano, Francisco	Gruzdeva, Tatiana
Bentert, Matthias	Chivilikhin, Daniil	Gubar, Elena
Berikov, Vladimir	Dang, Duc-Cuong	Gudyma, Mikhail
Berndt, Sebastian	Davydov, Ivan	Gusev, Mikhail
Boehmer, Niclas	Dobrynin, Andrey	Himmel, Anne-Sophie

Jaksic Kruger, Tatjana
Karavaev, Vitalii
Khandeev, Vladimir
Khoroshilova, Elena
Khutoretskii, Alexander
Kobylkin, Konstantin
Kononova, Polina
Kovalenko, Yulia
Krylatov, Alexander
Kumacheva, Suriya
Latyshev, Aleksei
Lempert, Anna
Lengler, Johannes
Levanova, Tatyana
Lezhnina, Elena
Liberti, Leo
Maack, Marten
Martins, Natália
Masich, Igor
Melnikov, Andrey
Mironovich, Vladimir

Molter, Hendrik
Muravyov, Sergey
Naumova, Nataliya
Neznakhina, Ekaterina
Ogorodnikov, Yuri
Olenev, Nicholas
Orlov, Andrei
Parilina, Elena
Plotnikov, Roman
Plyasunov, Alexander
Polyakova, Anastasiya
Prolubnikov, Alexander
Renken, Malte
Rettieva, Anna
Sedakov, Artem
Servakh, Vladimir
Shary, Sergey
Shenmaier, Vladimir
Shkaberina, Guzel
Sidorov, Alexander
Simanchev, Ruslan

Sleptchenko, Andrei
Smetannikov, Ivan
Sopov, Evgenii
Stanimirovic, Zorica
Stashkov, Dmitry
Stupina, Alena
Suslov, Nikita
Suvorov, Dmitrii
Tsoy, Yury
Urazova, Inna
Vasilyev, Igor
Vasin, Alexander
Yanovskaya, Elena
Zabashta, Alexey
Zabudsky, Gennady
Zalyubovskiy Vyacheslav
Zaozerskaya, Lidia
Zenkevich, Anatolii
Zeufack, Vannel
Zschoche, Philipp

Industry Section Chairs

Vasilyev Igor

Matrosov Institute for System Dynamics and Control
Theory, Russia

Alexander Kurochkin

Sobolev Institute of Mathematics, Russia

Organizing Committee

Yury Kochetov   Sobolev Institute of Mathematics, Russia
Nina Kochetova   Sobolev Institute of Mathematics, Russia
Polina Kononova   Sobolev Institute of Mathematics, Russia
Timur Medvedev   Higher School of Economics, Russia
Tatyana Gruzdeva   Matrosov Institute for System Dynamics and Control
Theory, Russia

Ivan Davydov   Sobolev Institute of Mathematics, Russia
Adil Erzin   Sobolev Institute of Mathematics, Russia
Sergey Lavlinsky   Sobolev Institute of Mathematics, Russia

Organizers

Sobolev Institute of Mathematics, Russia
Novosibirsk State University, Russia
Krasovsky Institute of Mathematics and Mechanics, Russia
Higher School of Economics, Russia
Melentiev Energy Systems Institute, Russia

Sponsors

Mathematical Center in Akademgorodok, Russia
Russian Foundation for Basic Research

Abstracts of Invited Talks

On Graph Invariants and Their Application to the Graph Isomorphism Problem

Aida Abiad (ORCID)

Eindhoven University of Technology and Ghent University, The Netherlands
aida.abiad@ugent.be

Abstract. Graphs invariants have proven to be one of the most important and fruitful concepts in modern Combinatorics and Theoretical Computer Science. Besides being a fascinating study subject for their own sake, they play an important role in the famous graph isomorphism problem. Their success serves as a natural motivation for the following natural question: what are the graph properties that can be deduced from a certain graph invariant? In this talk we will give an overview and will report on recent results concerning two graph invariants: the status sequence of a graph and the graph spectrum.

Keywords: Graph isomorphism problem · Graph invariant · Graph spectrum

Multistage Optimization Problems

Evripidis Bampis

Sorbonne Universite, Paris, France
bampis@gmail.com

Abstract. Many systems have to be maintained while the underlying constraints, costs and/or profits change over time. Although the state of a system may evolve during time, a non-negligible transition cost is incurred for transitioning from one state to another. In order to model such situations, Gupta et al. (ICALP 2014) and Eisenstat et al. (ICALP 2014) introduced a multistage model where the input is a sequence of instances (one for each time step), and the goal is to find a sequence of solutions (one for each time step) that are both (i) near optimal for each time step and (ii) as stable as possible. In this talk, we will give a survey of recent results in algorithmic multistage optimization, both in the offline and the online contexts and we will discuss connections with other models taking into account the evolution of data during time.

Keywords: Multistage optimization · Approximation algorithms ·
Online algorithms

Capacity Auctions: VCG Mechanism vs. Submodularity

Bo Chen (ID)

University of Warwick, UK
Bo.Chen@wbs.ac.uk

Abstract. We study a form of capacity mechanism that combines capacity and supply auctions. We characterize how participants bid in this auction and show that, on a pay-as-bid basis, an equilibrium behavior gives Vickrey-Clarke-Groves (VCG) profits and achieves efficient outcomes when there is submodularity, which is in stark contrast with what in the existing literature—at equilibrium VCG payments achieve truthful bids and efficiency. We also provide some necessary and sufficient conditions for submodularity.

Keywords: Capacity mechanism · Supply auction · Submodularity

Convex Geometry in the Context of Artificial Intelligence

Sergei Chubanov

Bosch Research, Germany
Sergei.Chubanov@de.bosch.com

Abstract. Applications of convex optimization in machine learning include support vector machines, polyhedral classifiers, deduction of disjunctive and conjunctive normal forms, time-series clustering, image segmentation, different models based on information theory, e.g., those involving Shannon entropy and Kullback-Leibler divergence. Virtually the whole spectrum of standard methods of convex optimization such as the gradient descent, the Frank-Wolfe algorithm, and interior-point methods is used for training deep neural networks. At the same time, some new results in the area of linear programming and convex optimization indicate that there are methodologies beyond the classical approaches which can lead to substantially more efficient machine learning algorithms and better interpretable machine learning models. So in this lecture we will address recent developments in convex optimization and convex analysis, in particular in the context of machine learning.

Keywords: Machine learning · Polyhedral theory · Information theory

Equilibrium Formulations of Relative Optimization Problems

Igor Konnov (iD)

Kazan Federal University, Russia
konn-igor@yandex.ru

Abstract: We consider relative or subjective optimization problems where the goal function and feasible set are dependent of the current state of the system under consideration. We propose equilibrium formulations of the corresponding problems that lead to general (quasi-)equilibrium problems. We propose to apply a regularized version of the penalty method for the general quasi-equilibrium problem, which enables us to establish existence results under weak coercivity conditions and replace the quasi-equilibrium problem with a sequence of the usual equilibrium problems. We describe several examples of applications and show that the subjective approach can be extended to non-cooperative game problems.

Keywords: Quasiequilibrium problems · Penalty method · Weak coercivity conditions

Long Cycles in Graph and Hypergraphs

Alexander Kostochka[iD]

University of Illinois at Chicago, USA
kostochk@illinois.edu

Abstract. Finding long cycles in graphs is an NP-hard problem. Cycles in hypergraphs can be defined in several natural ways. Since finding long cycles in hypergraphs is hard for all kinds of cycles, it makes sense to consider approximate algorithms and extremal problems on long cycles in hypergraphs. We discuss several such extremal problems, recent progress on them and possible algorithms based on the proofs.

Keywords: Long cycles · Approximate algorithm · NP-hard problem

Inverse Combinatorial Optimization Problems

Panos M. Pardalos ⓘ

University Florida, USA
pardalos@ise.ufl.edu

Abstract: Given an optimization problem and a feasible solution to it, the corresponding inverse optimization problem is to find a minimal adjustment of the cost vector under some norm such that the given solution becomes optimum. Inverse optimization problems have been applied in diverse areas, ranging from geophysical sciences, traffic networks, communication networks, facility location problems, finance, electricity markets, and medical decision-making. It has been studied in various optimization frameworks including linear programming, combinatorial optimization, conic, integer and mixed-integer programming, variational inequalities, and countably infinite linear problems and robust optimization. In this talk, we mainly concentrate on inverse combinatorial optimization problems (ICOP). We will introduce some classes of ICOP as well as general methods to solve them. Some open problems are proposed. We also discuss some generalized inverse optimization problems. We introduce inverse optimization problems on spanning trees and mainly concentrate on the inverse max+sum spanning tree problems in which the original problem aims to minimize the sum of a maximum weight and a sum cost of a spanning tree.

Keywords: Inverse optimization problem · Spanning tree · Optimization framework

Optimal Complexity Matrix Multiplication Like Computation Structures on Network-on-Chip Architecture Subject to Conflicts of Resource Allocation Constraints

Soumyendu Raha(iD)

Indian Institute of Science, Bangalore, India
raha@iisc.ac.in

Abstract. Massively parallel many-core coarse-grain reconfigurable system-on-chip (SoC) solutions are being increasingly deployed for solving compute intensive problems in an energy constrained environment. In this talk we will present how computational structures similar to matrix multiplication, as in all pairs shortest path algorithm, convolution neural networks, digital filtering, etc. can be optimally (in terms of power, performance, and resource utilization) constructed, and realized as efficient datapaths on a matrix of compute units refered to as Hypercells. Several of such Hypercells interconnected over a network-on-chip (NoC) make up the massively parallel many-core runtime reconfigurable SoC. Deadlock free data communication on the network-on-chip (NoC) is provisioned and scheduled in a way such that the optimality of the computation structures is preserved both at the level of individual Hypercells, and at the level of the overall many-core SoC.

Keywords: Matrix multiplication · Hypercells · System-on-chip

Two-Stage Scheduling Models with Limited Storage

Yakov Zinder

University of Technology, Sydney, Australia
Yakov.Zinder@uts.edu.au

Abstract. Publications on the two-stage scheduling systems such as two-machine flow shops, job shops and open shops, single machine with coupled tasks, and their various generalisations constitute a significant part of the scheduling literature. Many of these publications consider a limited storage (buffer) or an additional limited resource. The majority of publications on scheduling with a buffer consider the buffer as storage that limits the number of jobs that have completed their first operation and are waiting for the commencement of the second one. The majority of scheduling models with an additional resource assume that the resource is used only during the processing on a machine. Despite numerous possible applications that include, for example, supply chains, multimedia systems and data gathering networks, much less research has been done on the models where the resource (storage space, buffer) is allocated to a job from the start of its first operation till the end of its second operation and where the storage requirement varies from job to job. The talk presents a survey of recent publications on this type of scheduling problems, which includes NP-hardness proofs, particular cases amenable for polynomial-time algorithms, polynomial approximation schemes, and integer programming based algorithms, including for example Lagrangian relaxation.

Keywords: Scheduling with limited storage · Approximation algorithms · NP-hard problem

Abstracts of Tutorials

Evolution of Sailor and Surgical Knots

Alexander Grigoriev ⓘ

Maastricht University, Netherlands
a.grigoriev@maastrichtuniversity.nl

Abstract. This is a survey of recent developments in the computational knot theory. We start with retrospective of well-known results and techniques bringing topological knot theory and graph theory together: knots equivalence, Reidemeister moves, knot diagrams and knot polynomials. Then, we briefly address the complexity of the unknotting problem. We illustrate the difficulty of unknotting on small and insightful examples of knots. The easy unknotting cases, e.g., knot diagrams of treewidth 2, are addressed in details. We wrap up the tutorial posing numerous open questions and introducing new research directions.

Keywords: Knot theory · Unknotting problem · Graph theory

Metrics of a Fixed Doubling Dimension: An Efficient Approximation of Combinatorial Problems

Mikhail Khachay [iD]

Krasovsky Institute of Mathematics and Mechanics, Ekaterinburg, Russia
mkhachay@imm.uran.ru

Abstract. For decades, for many well-known combinatorial optimization problems, the approximability results in the class of algorithms with theoretical performance guarantees have had the quite similar nature. For instance, the classic Traveling Salesman Problem (TSP) is strongly NP-hard both in general and even in very specific settings, e.g. in the Euclidean plane. The problem is hardly approximable in general setting, it is APX-complete for an arbitrary metric, whilst, the problem admits polynomial time approximation schemes (PTAS) in the Euclidean space of an arbitrary fixed dimension. Recent results in the field of the analysis of finite metric spaces shed a light to the design of approximation schemes for a wide family of metric settings of these problems. In this tutorial, we give a short overview of such an approach leading to the PTAS for the metric TSP in metric space of any fixed doubling dimension.

Keywords: Metric combinatorial problem · APX-completness · PTAS

Game Theory and Social Networks

Vladimir V. Mazalov 🆔

Institute of Applied Mathematical Research, Petrozavodsk, Russia
vmazalov@krc.karelia.ru

Abstract. Social networks represent a new phenomenon of our life. The growing popularity of social networks in the Web dates back to 1995 when American portal Classmates.com was launched. This project facilitated the soon appearance of online social networks (SixDegrees, LiveJournal, LinkedIn, MySpace, Facebook, Twitter, YouTube, and others) in the early 2000s. In Russia, the most popular networks are VKontakte and Odnoklassniki. Social networks are visualized using social graphs. Graph theory provides main analysis tools for social networks. In particular, by calculating centrality measures for nodes and edges one may detect active participants (members) of a social network. We use for the analysis of social networks game-theoretic approach. We propose a new concept of the betweenness centrality for weighted graphs using the methods of cooperative game theory. The characteristic function is determined by special way for different coalitions (subsets of the graph). The betweenness centrality is determined as the Myerson value. The results of computer simulations for some examples of networks, in particular, for the popular social network "VKontakte", as well as the comparing with the PageRank method are presented. Then we apply game-theoretic methods for community detection in networks. Finally, for approaches based on potential games we suggest a very efficient computational scheme using Gibbs sampling.

Keywords: Online social networks · Cooperative game theory · Social graphs

Practice of Using the Gurobi Optimizer

Andrey Melnikov [ID]

Sobolev Institute of Mathematics, Russia
a.a.melnikov@hotmail.com

Abstract. The general-purpose optimization software has never been more powerful than today. It is universal, customizable, controllable, and enables a user with a variety of tools and features to get better solutions in a shorter time. In this tutorial, we will overview the ingredients of one of the fastest MIP solvers, the Gurobi optimizer, that are relevant in academic studies. The key topics would be the internal organization of the solver, its tuning when solving LPs and MIPs, the most newly added features, and other selected ones.

Keywords: Optimization models · Mixed-integer programming · Software

A Survey of Machine Learning Problems from Optimization Point of View

Konstantin Vorontsov[ID]

Moscow Institute of Physics and Technology, Russia
kvorontsov@hse.ru

Abstract. In recent years, machine learning problems have become increasingly diverse and even exotic. We can no longer say that machine learning is basically classification, clustering, regression, and density estimation from empirical data. New types of machine learning, such as transfer, self-supervised, adversarial, privileged, meta, one-shot, few-shot, positive-unlabeled, and others, are expanding the boundaries of AI applications. Despite their diversity, each of them remains an optimization problem for the sum of a large number of terms. In this tutorial, you will learn how to set the meaning of a machine learning task by changing the construction of the terms, whether it be a parametric model of data, loss function, or regularizer.

Keywords: Machine learning · AI applications · Optimization

Contents

Game Theory

Scheduling Problem

Heuristics and Metaheuristics

Operational Research Applications

Session 1: Invited Talks

Global and Local Search Methods for D.C. Constrained Problems

Alexander S. Strekalovsky(✉) (ID)

Matrosov Institute for System Dynamics and Control Theory of SB of RAS,
Irkutsk, Russia
strekal@icc.ru

Abstract. This paper addresses the general optimization problem $(\mathcal{P})$ with equality and inequality constraints and the cost function given by d.c. functions. We reduce the problem to a penalized problem $(\mathcal{P}_\sigma)$ without constraints with the help of the Exact Penalization Theory. Further, we show that the reduced problem is also a d.c. minimization problem. This property allows us to prove the Global Optimality Conditions (GOCs), which reduce the study of the penalized problem to an investigation of a family of linearized (convex) problems tractable with the help of standard convex optimization methods and software.

In addition, we propose a new Local Search Scheme (LSS1) which produces a sequence of vectors converging to a so-called critical point. On the other hand, the vector satisfying the GOCs turns out to be also a critical point.

On the basis of the GOCs for finding a global solution to $(\mathcal{P}_\sigma)$, we develop a Global Search Scheme, including the LSS1 with an update of the penalty parameter, and a special stopping criteria allowing detection of a feasible vector in the original problem $(\mathcal{P})$, and, consequently, a global solution to the original Problem $(\mathcal{P})$.

Keywords: Difference of convex functions · Equality and inequality constraints · Exact penalty · Linearized problem · Local search · Critical vector · Global search

1 Introduction

According to the opinions of the well-known specialists, the optimization problems can be separated into two different classes: convex and nonconvex [1, 4, 6, 10–13, 16–19, 30].

At present, the convex problems look as commonly solved by standard classical methods and software [11, 12, 16]. Meanwhile, the nonconvex problems from different fields of applications generated so many various difficulties (when it comes to finding a global solution) that the majority of researchers prefer to use the simplest approaches, such as "branch and bounds", which lead, in general, to "the dimension curse" and do not allow finding a global solution. As a result, they only derive "suboptimal points". Even such approaches as a "bioiniciated

© Springer Nature Switzerland AG 2020
A. Kononov et al. (Eds.): MOTOR 2020, LNCS 12095, pp. 3–18, 2020.
https://doi.org/10.1007/978-3-030-49988-4_1

method", a genetic algorithm, the neuron's sets, become more and more popular without any mathematical substantiation. In contrast to this, we prefer to follow another way, which might be viewed as a more "mathematical" one.

Below, we consider a rather general optimization problem when the data is given by d.c. functions. As well-known, the linear space of d.c. functions merits to be used for rather large fields of problems, since any continuous function over a compact can be approximated by a d.c. function in the topology of uniformed convergence [6, 10–13, 16–19, 30]. Besides, we attack d.c. problems with the solid theoretical foundation of the Global Optimality Conditions (GOCs), the Exact Penalization Theory (EPT), the Local and Global Search convergence theory and our algorithmic and computational experience [1, 2, 4–10, 14, 19–23, 25–29].

The most advantageous property of the developed methodology with respect to the other approaches [1, 2, 4–6, 13, 15] is the employing all the achievements of the Theory and Methods of the Modern Optimization [1–6, 9–18, 30].

In particular, in the interior of the Local Search Schemes (LSS) for solving the convex linearized problems one can use any relevant method of unconstrained and constrained optimization and the corresponding computational software (CPLEX, PATH, XpressMP, KNITRO etc.) [7, 8, 19, 21, 25].

Moreover, since the LSS and the linearized problems are used in global search procedures based on GOCs, one can conclude that, first, the Modern Optimization Theory and Methods provides the bricks with which one construct the new numerical Global Search methods for nonconvex optimization.

Hence, it is clear, that this new methodology is completely different from the B&B and its satellites, and off-causes from the "'bioniciated'" ideology which has no relations to modern optimization methods.

As result, here we propose a new approach, which we present step by step.

After the statement of Problem $(\mathcal{P})$ and the auxiliary Problem $(\mathcal{P}_\sigma)$, we show in Sect. 2 that the objective function $F_\sigma(\cdot)$ of the penalized problem is also a d.c. function, which allows us to present the GOCs in Theorem 1 and to comment the first properties of these new tools of Optimization.

In Sect. 4, we discuss the Local Search for Problem $(\mathcal{P}_\sigma)$ and develop the Local Search Method (LSS1) with an update of the penalty parameter value.

Furthermore, we study the convergence properties of LSS1 (Proposition 1 and 2 and Theorem 2). Theorem 2 is the principal result of Sec. 4: the limit point x_* of the sequence $\{x^s\}$, produced by the LSS1, is a critical point, i.e. the solution to the linearized problem $(\mathcal{P}_* L_*)$ (linearized just at the point x_*). Moreover, this point, according to Theorem 3, turns out to be a KKT-point for the original Problem $(\mathcal{P})$ with the Lagrange multipliers provided by the auxiliary linearized problem (23).

In Sect. 5, we present the Global Search Scheme (GSS) constructed on the base of the GOCs (Theorem 4). The GSS includes the LSS1 (for instance), a GSS for Problem $(\mathcal{P}_\sigma)$ and, of course, a penalty parameter update procedure, the convergence properties of which was presented in [29].

The new stopping criteria at the final stage of the GSS1 allows us to obtain a solution to the original problem $(\mathcal{P})$.

2   Problem Statements and Exact Penalty

Consider the problem with constraints:

$$(\mathcal{P}): \qquad \left. \begin{array}{ll} f_0(x) := g_0(x) - h_0(x) \downarrow \min_x, & x \in S, \\ f_i(x) := g_i(x) - h_i(x) \leq 0, & i \in I = \{1, \ldots, m\}, \\ f_i(x) := g_i(x) - h_i(x) = 0, & i \in \mathcal{E} = \{m+1, \ldots, l\}; \end{array} \right\}$$

where the functions $g_i(\cdot)$, $h_i(\cdot)$ are convex and sufficiently smooth on $I\!\!R^n$, so that the functions $f_i(\cdot)$, $i \in \{0\} \cup I \cup \mathcal{E}$, are the d.c. functions [6,10,13,19,30]. Assume that the set $S \subset I\!\!R^n$ is convex and compact.

Further, suppose that the feasible set $\mathcal{F}$,

$$\mathcal{F} := \{x \in S \mid f_i(x) \leq 0,\ i \in I,\ f_i(x) = 0,\ i \in \mathcal{E}\},$$

of Problem $(\mathcal{P})$, and the set $Sol(\mathcal{P})$ of global solutions to Problem $(\mathcal{P})$,

$$Sol(\mathcal{P}) := \{x \in \mathcal{F} \mid f_0(x) = \mathcal{V}(\mathcal{P}) := \inf(f_0, \mathcal{F}) := \inf_x \{f_0(x) \mid x \in \mathcal{F}\}$$

are non-empty, $\mathcal{F} \neq \emptyset \neq Sol(\mathcal{P})$.

In addition, the optimal value $\mathcal{V}(\mathcal{P})$ of Problem $(\mathcal{P})$ is supposed to be finite:

$$(\mathcal{A}_f): \qquad \mathcal{V}(\mathcal{P}) := \inf(f_0, \mathcal{F}) = \inf_x \{f_0(x) \mid x \in \mathcal{F}\} > -\infty. \tag{1}$$

Along with Problem $(\mathcal{P})$, we consider the auxiliary (penalized) problem

$$(\mathcal{P}_\sigma): \qquad F_\sigma(x) := f_0(x) + \sigma W(x) \downarrow \min_x, \qquad x \in S, \tag{2}$$

$$W(x) := \max\{0, f_1(x), \ldots, f_m(x)\} + \sum_{j \in \mathcal{E}} |f_j(x)| \tag{3}$$

without constraints, where the penalty function $W(x)$ has a mixed form, so that the inequality constraints are penalized by the term defined with the help of l_∞-norm, meanwhile for the equality constraints one employs the l_1-norm penalization.

As known, the penalization aims at the replacing Problem $(\mathcal{P})$ with constraints by Problem $(\mathcal{P}_\sigma)$ without them, with the hope that $(\mathcal{P}_\sigma)$ might be easier to solve than $(\mathcal{P})$, due to the strong nonconvexity of $(\mathcal{P})$. For instance, if $z \in Sol(\mathcal{P}_\sigma)$, and, in addition, z is feasible for Problem $(\mathcal{P})$, i.e. $z \in \mathcal{F}$, then z turns out to be a global solution to $(\mathcal{P})$: $z \in Sol(\mathcal{P})$ [1,2,4,5,9,15,16].

Therefore, the key feature of the Exact Penalization Theory (EPT) is the existence of a threshold value $\sigma_* \geq 0$ of the penalty parameter $\sigma \geq 0$ for which $Sol(\mathcal{P}_\sigma) \subset Sol(\mathcal{P}) \ \forall \sigma \geq \sigma_*$. The latter means that for $\sigma \geq \sigma_*$ Problems $(\mathcal{P})$ and $(\mathcal{P}_\sigma)$ are equivalent in the sense that $Sol(\mathcal{P}) = Sol(\mathcal{P}_\sigma)$ (see [4], [12, Chapt. VII, Lemma 1.2.1]).

It is worth noting that under various constraint qualification (CQ) conditions (MFCQ, etc.), the error bound properties, etc., one can prove the existence of the exact penalty threshold $\sigma_* \geq 0$ for a local and global solutions, as well. [1,4,5,9,14].

Assume henceforth that some regularity conditions for the existence of an exact penalty threshold $\sigma_* \geq 0$ are fulfilled.

However, it can be readily seen that Problem $(\mathcal{P}_\sigma)$ stays to be nonconvex and, moreover, preserves the d.c. structure, so that the idea of solving $(\mathcal{P}_\sigma)$ by the classical (standard) optimization method (SQP, CGM, TRM, IPM, etc.) is not sufficiently substantiated.

In order to easily assimilate this idea, let us show that the cost function $F_\sigma(\cdot)$ of Problem $(\mathcal{P}_\sigma)$ is a d.c. function. Consequently, Problem $(\mathcal{P}_\sigma)$ turns out to be nonconvex. Indeed, since $|f_i(x)| = 2\max\{g_i(x), h_i(x)\} - [g_i(x) + h_i(x)]$, it can be readily seen that

$$F_\sigma(x) \stackrel{\triangle}{=} f_0(x) + \sigma \max\{0, f_i(x), \ i \in I\} + \sigma \sum_{i \in \mathcal{E}} |f_i(x)| = G_\sigma(x) - H_\sigma(x), \quad (4)$$

$$H_\sigma(x) := h_0(x) + \sigma \left[\sum_{i \in I} h_i(x) + \sum_{j \in \mathcal{E}} (g_j(x) + h_j(x)) \right], \quad (5)$$

$$G_\sigma(x) := F_\sigma(x) + H_\sigma(x) = g_0(x) + 2\sigma \sum_{i \in \mathcal{E}} \max\{g_i(x); h_i(x)\}$$

$$+ \sigma \max \left\{ \sum_{j \in I} h_j(x); \left[g_i(x) + \sum_{\substack{j \neq i \\ j \in I}} h_j(x) \right], \ i \in I \right\}. \quad (6)$$

It is clear, that $G_\sigma(\cdot)$ and $H_\sigma(\cdot)$ are both convex functions [11,12,17,18], so that $F_\sigma(\cdot)$ is a d.c. function, as claimed.

3   Global Optimality Conditions

It is easy to see that for a feasible (in $(\mathcal{P})$) point $z \in S$ we have $W(z) = 0$, and therefore, with $\zeta := f_0(z)$, we obtain

$$F_\sigma(z) \stackrel{\triangle}{=} f_0(z) + \sigma W(z) = f_0(z) = \zeta. \quad (7)$$

The following result was proved in [19,23,24,26,27].

Theorem 1. *Let a feasible vector* $z \in \mathcal{F}$, $\zeta := f_0(z)$, *be a solution to Problem* $(\mathcal{P})$ *and* $\sigma \geq \sigma_* > 0$, *where* $\sigma_* \geq 0$ *is a threshold value of the penalty parameter, such that* $Sol(\mathcal{P}) = Sol(\mathcal{P}_\sigma) \ \forall \sigma \geq \sigma_*$.

Then, for every pair $(y, \beta) \in \mathbb{R}^n \times \mathbb{R}$ *satisfying the equation*

$$H_\sigma(y) = \beta - \zeta, \quad (8)$$

the following inequality holds

$$G_\sigma(x) - \beta \geq \langle \nabla H_\sigma(y), x - y \rangle \quad \forall x \in S. \quad (9)$$

Remark 1. It can be readily seen that Theorem 1 reduces the solution of the nonconvex Problem $(\mathcal{P}_\sigma)$ to a study of the family of the convex (linearized) problems as follows

$$(\mathcal{P}_\sigma L(y))\colon \qquad \Phi_{\sigma y}(x) := G_\sigma(x) - \langle \nabla H_\sigma(y), x \rangle \downarrow \min_x, \quad x \in S, \qquad (10)$$

depending on the pairs $(y, \beta) \in I\!\!R^{n+1}$, fulfilling the Eq. (8).

Moreover, the linearization is carried out with respect to the "united" nonconvexity of Problem $(\mathcal{P}_\sigma)$, which is accumulated by the function $H_\sigma(\cdot)$ (see $(\mathcal{P})$, $(\mathcal{P}_\sigma)$–(2) and (5)).

Remark 2. If we can produce a triple $(y, \beta, u) \in I\!\!R^{n+1} \times S, \quad H_\sigma(y) = \beta - \zeta$, $u \in S$, which brakes down the principal inequality (9) of Theorem 1, i.e.

$$0 > G_\sigma(u) - \beta - \langle \nabla H_\sigma(y), u - y \rangle,$$

then, with the help the Eq. (8) and the convexity of the function $H_\sigma(\cdot)$, we derive

$$0 > G_\sigma(u) - \beta - H_\sigma(u) + H_\sigma(y) = F_\sigma(u) - \zeta,$$

or, $F_\sigma(z) > F_\sigma(u), \quad u \in S, \quad z \in \mathcal{F}$. Hence, the vector z is not a solution to $(\mathcal{P}_\sigma)$.

If, in addition, u is feasible in $(\mathcal{P})$, $z, u \in \mathcal{F}$, $W(u) = 0 = W(z)$, we obtain $f_0(z) = F_\sigma(z) > F_\sigma(u) = f_0(u)$, so that $z \notin Sol(\mathcal{P})$ and the point u is better than $z \in \mathcal{F}$.

It means that the conditions (8)–(9) of Theorem 1 possess the classical constructive (algorithmic) property: once the conditions are violated, one can find a feasible (in $(\mathcal{P})$) vector u, which is better than the point $z \in \mathcal{F}$ in question.

Employing the property of the Global Optimality Conditions [7,8,19–29], we developed numerical methods of the local and global search in Problem $(\mathcal{P}_\sigma)$, which will be used below for finding a global solution to Problem $(\mathcal{P})$.

4   Local Search in Problem $(\mathcal{P}_\sigma)$

Below we develop the standard scheme [15,19,22,25,28] of consecutive solution of the convex (linearized) problems of type $(\mathcal{P}_\sigma L(y))$–(10), where y is replaced by the current iterate $x^s \in S$, $s = 0, 1, 2, \dots$.

Our principal objective here is the investigation of the convergence of the local search and the properties of a cluster point. For the sake of simplicity, in this section, we use the following notations: $H_s(\cdot) := H_{\sigma_s}(\cdot)$, $G_s(\cdot) := G_{\sigma_s}(\cdot)$.

Suppose, that a starting point $x_0 \in S$ and a current iterate $x^s \in S$ are given. Besides, let a value of the penalty parameter $\sigma_s > 0$ be also given. Then one can find the next iterate $x^{s+1} \in S$, as satisfying the inequality as follows

$$\Phi_s(x^{s+1}) \overset{\triangle}{=} G_s(x^{s+1}) - \langle \nabla H_s(x^s), x^{s+1} \rangle - \delta_s \leq$$
$$\leq \inf_x \{ \Phi_s(x) = G_s(x) - \langle \nabla H_s(x^s), x \rangle \mid x \in S \} =: V_s, \qquad (11)$$

where $\delta_s > 0$, $s = 0, 1, 2, \ldots$, is the accuracy of a solution to the following linearized problem $((\mathcal{P}_s L_s) := (\mathcal{P}_{\sigma_s} L_s))$

$$(\mathcal{P}_s L_s): \qquad \Phi_s(x) := G_s(x) - \langle \nabla H_s(x^s), x \rangle \downarrow \min_x, \quad x \in S. \qquad (12)$$

Assume, henceforth, the following condition on a number sequence $\{\delta_s\}$

$$\delta_s > 0, \; s = 0, 1, 2, \ldots, \; \sum_{s=0}^{\infty} \delta_s < +\infty. \qquad (13)$$

Furthermore, in virtue of the assumption $(\mathcal{A}_f)$–(1), the goal function $F_\sigma(x)$ of Problem $(\mathcal{P}_\sigma)$ satisfies the similar condition

$$\inf_x \{F_\sigma(x) \mid x \in S\} > -\infty, \qquad (1')$$

due to the obvious inequalities: $\sigma \geq 0, W(x) \geq 0, f_0(x) \leq F_\sigma(x) \overset{\triangle}{=} f_0(x) + \sigma W(x)$.

Now we are ready to present a Local Search Scheme (LSS) for Problem $(\mathcal{P}_\sigma)$, which is based on the idea of consecutive solution of linearized problems [15, 22, 25] but with a penalty parameter update procedure [28].

Let there be given a starting point $x_0 \in S$, an initial value $\sigma^0 > 0$ of the penalty parameter $\sigma \geq 0$ along with two parameters $\eta_1 \in]0, 1[$, $\eta_2 \in [2, 10]$ of the LSS. Then the first LSS can be described as follows.

Local Search Scheme 1 (LSS1)

Step 0. Set $s := 0$, $x^s := x_0$, $\sigma_s := \sigma^0$.
Step 1. Solve the subproblem $(\mathcal{P}_s L_s)$–(12) to get $x(\sigma_s) \in \delta_s - Sol(\mathcal{P}_s L_s)$.
Step 2. If $W(x(\sigma_s)) = 0$, then set $\sigma_+ := \sigma_s$, $x(\sigma_+) := x(\sigma_s)$, and go to Step 6.
Step 3. $(W(x(\sigma_s)) > 0)$ If the inequality

$$\Phi_s(x^s) - \Phi_s(x(\sigma_s)) \geq \eta_1 \sigma_s [W(x^s) - W(x(\sigma_s))] \qquad (14)$$

holds, then set $\sigma_+ := \sigma_s$, $x(\sigma_+) := x(\sigma_s)$, and go to Step 6.
Step 4. (Else) Increase $\sigma_s > 0$, so that $\sigma_+ := \eta_2 \sigma_s$ with $\eta_2 \in [2, 10]$, and solve the linearized problem

$$(\mathcal{P}_+ L_+): \qquad \Phi_+(x) := G_+(x) - \langle \nabla H_+(x(\sigma_s)), x \rangle \downarrow \min_x, \quad x \in S, \qquad (15)$$

with $G_+ := G_{\sigma_+}$, $H_+ := H_{\sigma_+}$, to get $x(\sigma_+) \in S$, such that $x(\sigma_+) \in \delta_s - Sol(\mathcal{P}_+ L_+)$.
Step 5. Set $x(\sigma_s) := x(\sigma_+)$, $\sigma_s := \sigma_+$ and return to Step 3.
Step 6. Set $\sigma_{s+1} := \sigma_+$, and $x^{s+1} := x(\sigma_+) \in S$, $s := s + 1$, and loop to Step 1.

It is worth noting that the above scheme is not yet to become a proper algorithm, since it is not clear what stopping criteria can be used here.

In order to handle these issue, let us consider the first convergence features of the scheme [22, 25, 28].

Introduce now the following assumption

$$(\mathcal{A}_W): \qquad \begin{array}{ll} (a): & \displaystyle\sum_{s=0}^{\infty} \xi_s \overset{\triangle}{=} \sum_{s=0}^{\infty} (\sigma_{s+1} - \sigma_s) W(x^s) < +\infty \\[3mm] (b): & \xi_s := (\sigma_{s+1} - \sigma_s) W(x^s) \geq 0, \ s = 0, 1, 2, \ldots \end{array} \right\} \qquad (16)$$

Note that it follows from $(\mathcal{A}_W)$–(16) (b) that $\sigma_{s+1} \geq \sigma_s > 0$.

Furthermore, it can be readily seen [25, 28] that

$$F_{s+1}(x^{s+1}) \leq F_{(}x^s) + \xi_s, \ s = 0, 1, 2, \ldots \qquad (17)$$

and, due to $(\mathcal{A}_W)$ the number sequence $\{F_s(x^s)\}$ produced by the LSS1, turns out to be almost decreasing (see (16)–(17)), and, therefore, converging [25, 28].

Proposition 1. *Let the assumption $(\mathcal{A}_f)$–(1) and $(\mathcal{A}_W)$–(16) be fulfilled.*

Then, the sequence $\{x^k\} \subset S$ produced by the LSS1 satisfies the following conditions.

The number sequences $\{F_s(x^s)\}$ and $\{\Delta\Phi_{s+1}\}$, where $\Delta\Phi_{s+1} := \Phi_{s+1}(x^s) - \Phi_{s+1}(x^{s+1}) \overset{\triangle}{=} G_{s+1}(x^s) - G_{s+1}(x^{s+1}) + \langle \nabla H_{s+1}(x^s), x^{s+1} - x^s \rangle$, converge, so that

$$\begin{array}{ll} (a): & \displaystyle\lim_{s \to \infty} F_s(x^s) =: F_* > -\infty; \\[2mm] (b): & \displaystyle\lim_{s \to \infty} \Delta\Phi_{s+1} = 0. \end{array} \right\} \qquad (18)$$

$$\#$$

Furthermore, it can be readily seen that, under the following assumption

$$(\mathcal{A}_{str}): \qquad \begin{array}{l} \text{At least one of the functions } h_i, \ i \in \mathcal{I} \cup \mathcal{E} \cup \{0\}, \\ g_j, \ j \in \mathcal{E} \text{ is strongly convex;} \end{array} \right\} \qquad (19)$$

the functions $H_\sigma(\cdot)$, $H_s(\cdot)$, $H_{s+1}(\cdot)$ turn out to be strongly convex.

Moreover, it is well-known [10–12] that in a DC decomposition $f(x) = g(x) - h(x)$ of a DC function $f(\cdot)$, the convex components $g(\cdot)$ and $h(\cdot)$ can be always chosen to be strongly convex [6, 10–13, 19, 30].

Proposition 2. *[25, 28] Let the assumption $(\mathcal{A}_{str})$–(19) be satisfied. Then, the sequence $\{x^s\}$, produced by the LSS1, is the Cauchy sequence, i.e*

$$\lim_{s \to \infty} = ||x^s - x^{s+1}|| = 0. \qquad (20)$$

$$\#$$

To handle other properties of the sequence $\{x^s\}$, suppose now that the following assumption holds

$$(\mathcal{A}_\sigma): \qquad \exists \sigma_{up} \in \mathbb{R}: \ \sigma_{up} \geq \sigma_s, \ s = 0, 1, 2, \ldots \qquad (21)$$

Clearly, from the practical view-point $(\mathcal{A}_\sigma)$–(21) looks too natural, and, on the other hand, it is related to existence of the threshold value $\sigma_* \geq 0$ of the penalty parameter $\sigma \geq 0$, i.e. to the Exact Penalty Theory [1,2,4,5,9,14,24].

Moreover, combining (16$'$) and (21) we derive that there exists $\sigma_* > 0$, such that $\lim\limits_{s \to \infty} \sigma_s = \sigma_*$ and one immediately gets the next result as was in [25,28].

Theorem 2. *Let the assumptions* $(\mathcal{A}_W)$–(16), $(\mathcal{A}_f)$–(1), $(\mathcal{A}_{str})$–(19) *and* $(\mathcal{A}_\sigma)$–(21) *be fulfilled. Then, a limit point* x_* *of the sequence* $\{x^s\}$, *produced by the LSS1, is a solution to the following convex problem*

$$(\mathcal{P}_*L_*): \qquad \Phi_*(x) := G_*(x) - \langle \nabla H_*(x_*), x \rangle \downarrow \min_x, \ x \in S, \qquad (22)$$

where $G_*(x) := G_{\sigma_*}(x) = g_0(x) + \sigma_* G_W(x)$, $H_*(x) := H_{\sigma_*}(x) = h_0(x) + \sigma_* H_W(x)$, $W(x) = G_W(x) - H_W(x)$, $\sigma_* = \lim\limits_{s \to \infty} \sigma_s$.

Remark 3. According to Theorem 2, any cluster point x_* of the sequence $\{x^s\}$, generated by LSS1, is a solution to the linearized Problem $(\mathcal{P}_*L_*)$ (linearized just by $\nabla H_*(x_*)$ at the point x_* with the limit value $\sigma_* = \lim\limits_{s \to \infty} \sigma_s$ of the corresponding number sequence $\{\sigma_s\}$ of the penalty parameter). The point x_* is now said to be a critical in Problem $(\mathcal{P}_\sigma)$ (with respect to the LSS1).

However, it is not the only remarkable property of this point. It can be shown, in addition, that a critical point x_* turns out to be a KKT-vector for the original Problem $(\mathcal{P})$ with the Lagrange multipliers provided by an auxiliary convex optimization problem [22–25,28].

Indeed, with the help of the explicit form (5), (6) of the functions $G_\sigma(\cdot)$ and $H_\sigma(\cdot)$, it can be readily seen that the linearized Problem $(\mathcal{P}_*L_*)$ in not smooth, since the function $G_*(\cdot) = G_{\sigma_*}(\cdot)$ is not differentiable, while the entries of $(\mathcal{P})$ are supposed to be smooth. To avoid the difficulties generated by non-smoothness, let us apply Lemma 4.1 from [23] (see also [10,16]), which states the equivalence of Problem $(\mathcal{P}_*L_*)$–(22) and the following problem with the supplementary parameters $(\gamma, t_{m+1}, ..., t_l)$:

$$\left.\begin{array}{l} g_0(x) - \langle \nabla H_*(x_*), x \rangle + \sigma_* \gamma + 2\sigma_* \sum\limits_{j \in \mathcal{E}} t_j \downarrow \min\limits_{(x,\gamma,t)}, \ x \in S, \\[2mm] g_i(x) + \sum\limits_{\substack{j \neq i \\ i \in \mathcal{I}}} h_j(x) \leq \gamma, \ i \in \mathcal{I}, \ \sum\limits_{j \in \mathcal{I}} h_j(x) \leq \gamma, \ \gamma \in \mathbb{R}; \\[2mm] g_p(x) \leq t_p, \ h_p(x) \leq t_p, \ p \in \mathcal{E}, \ t = (t_{m+1}, \ldots, t_l) \in \mathbb{R}^{l-m} \end{array}\right\} \quad (23)$$

One can easily see that the problem (23) is a convex optimization problem with the variables $(x, \gamma, t) \in \mathbb{R}^{n+1} \times \mathbb{R}^{l-m}$. Unlike Problem $(\mathcal{P}_*L_*)$, since the entries of Problem $(\mathcal{P})$ are smooth, the problem (23) stays smooth too. On the other hand, comparing with Problem $(\mathcal{P})$, (23) has only $(m+1)+2(l-m)$ inequalities, whereas $(\mathcal{P})$ includes equality and inequality constraints.

Moreover, it can be readily seen that in (23) the Slater's condition holds. Hence, we have $\mu_0 = 1$ in the corresponding to (23) Lagrange function.

In addition, the KKT-condition for (23) becomes necessary and sufficient for a triple (x_*, γ_*, t_*) to belong to $Sol(23)$, since x_* is the solution to $(\mathcal{P}_* L_*)$ and (23) is a convex optimization problem.

Constructing the corresponding Lagrange function [24, 25, 27, 28] and employing the KKT-system for $(x_*, \gamma_*, t_*) \in Sol$ (23) with

$$\left. \begin{array}{l} \gamma_* = \max\{\sum_{j \in \mathcal{I}} h_j(x_*); [g_i(x_*) + \sum_{j \in \mathcal{I}} h_j(x_*)], i \in \mathcal{I}\}, \\ t_{p*} = \max\{g_p(x_*); h_p(x_*), \ p \in \mathcal{E}\} \end{array} \right\} \tag{24}$$

we obtain, in particular, the following relations (see [25, 26, 28])

$$\mu_{m+1} + \sum_{i \in \mathcal{I}} \mu_i = \sigma_*, \ \eta_p + \nu_p = 2\sigma_*, \ p \in \mathcal{E}, \tag{25}$$

between the Lagrange multipliers (μ, η, ν) and the penalty parameter $\sigma_* > 0$.

Furthermore, from the equality $0_n = \nabla_x \mathcal{L}(x_*, \gamma_*, t_*)$ one can derive the principal KKT-equation

$$0_n = \nabla f_0(x_*) + \sum_{i \in \mathcal{I}} \mu_i \nabla f_i(x_*) + \sum_{p \in \mathcal{E}} (\eta_p - \sigma_*) \nabla f_p(x_*), \tag{26}$$

where $\nabla f_i(x_*) = \nabla g_i(x_*) - \nabla h_i(x_*), \ i \in \{0\} \cup \mathcal{I} \cup \mathcal{E}$ (see [23–25, 27, 28]).

Clearly, (26) is the principal equation of the KKT system for the original Problem $(\mathcal{P})$ at the critical point $x_* \in S$ with the Lagrange multipliers $\lambda_0 = 1$ and $\lambda_i, \ i \in \mathcal{I} \cup \mathcal{E}$, satisfying the condition

$$\lambda_i = \mu_i \geq 0, \ i \in \mathcal{I}, \ \lambda_p = \eta_p - \sigma_*, \ \eta_p \geq 0, \ p \in \mathcal{E}. \tag{27}$$

Moreover, it can be readily shown that, if the critical vector x_* is feasible in Problem $(\mathcal{P})$, i.e. $W(x_*) = 0$, then the complementarity conditions in Problem $(\mathcal{P})$ are also fulfilled with $\lambda_i \geq 0, \ i \in \mathcal{I}$, defined in (27) [25, 28].

Besides, it turns out that the feasibility of the cluster point x_* can be proved under now natural condition of Proposition 1.

Proposition 3. *Let the assumption* $(\mathcal{A}_W)$–*(16) and the following condition hold*

$(\mathcal{A}_\sigma 2)$: $\qquad$ *If* $W(x^s) > 0$, *then* $\sigma_{s+1} \geq \sigma_s + \text{æ}, \ \text{æ} > 0.$ $\qquad$ (28)

Then, the limit point x_* *of the sequence* $\{x^s\}$, *produced by the LS Scheme 1, is feasible in Problem* $(\mathcal{P})$, *i.e.* $W(x_*) = 0 = \lim_{s \to \infty} W(x^s)$.

However, we should not forget that we just proved the following result.

Theorem 3. *The limit point* x_* *of the sequence* $\{x^s\}$, *produced by the LSS1, turns out to be a KKT-vector for the original Problem* $(\mathcal{P})$ *with the Lagrange multipliers* $\lambda_0 = 1, \ \lambda_i \geq 0, \ i \in \mathcal{I}, \ \lambda_j \in \mathbb{R}, \ j \in \mathcal{E}$, *completely defined by the Lagrange multipliers* $(\mu, \eta, \nu) \in \mathbb{R}_+^m \times \mathbb{R}^{2(l-m)}$ *of the problem (23) and the limit penalty parameter* $\sigma_* > 0$, *all satisfying (25), (27).*

Remark 4. Thus, it is clear, that x_* is considerably stronger and better than the usual stationary (KKT) point provided by classical optimization methods [16] in Problem $(\mathcal{P})$.

Let us add a few words about the stopping criteria that play an important role in computational solutions, simulations, and experiments.

Besides, it would be reasonable to propose a few combinations of different stopping criteria for the LSS1 and the Global Search in $(\mathcal{P}_\sigma)$ on the basis of Theorems 2, 3 and Propositions 1–3. Let us begin with the well-known and obvious conditions as follows $(x_+ := x(\sigma_+))$, see Steps 2 and 4 of the LSS1):

$$W(x_+) = 0 \quad (\text{or} \quad W(x_+) \leq \tau), \tag{29}$$

$$||x_+ - x^s|| \leq \tau. \tag{30}$$

For (29), it is quite clear, that the fulfillness of (29) means only that x_+ is feasible in Problem $(\mathcal{P})$ and nothing else. Meanwhile, the inequality (30) implies that the process of convergence: $||x^{s+1} - x^s|| \downarrow 0$, described in Theorem 2, might be closed to or situated at the terminal stage.

On the other hand, Proposition 1 suggests that the inequality

$$\Phi_s(x^s) - \Phi_s(x^{s+1}) \leq \frac{\tau}{2} \tag{31}$$

may also be considered as a stopping criterion for the LSS1.

Indeed, due to the principal inequality (11) and (31) we have

$$\Phi_s(x^s) \overset{\triangle}{=} G_s(x^s) - \langle \nabla H_s(x^s), x^s \rangle \leq \frac{\tau}{2} + \Phi_s(x^{s+1}) \leq \mathcal{V}_s + \delta_s + \frac{\tau}{2},$$

where $\mathcal{V}_s = \inf_x \{\Phi_s(x) \mid x \in S\}$ is the optimal value of the problem $(\mathcal{P}_s L_s)$.

Whence, by choosing $\delta_s \leq \frac{\tau}{2}$, we obtain $\Phi_s(x^s) \leq \mathcal{V}_s + \tau$. It means that x^s is an τ-solution to $(\mathcal{P}_s L_s)$, which is rather satisfactory for any local search method, in particular, in Problem $(\mathcal{P}_{\sigma_s}) = (\mathcal{P}_s)$.

Hence, the inequality (31) together with the inequality $\sigma_s \leq \frac{\tau}{2}$ can be used as a reasonable stopping criterion.

Suppose, in addition, that the assumption $(\mathcal{A}_{str})$–(19) holds, then the function $H_s(\cdot)$ is strongly convex, i.e. with some $\rho_s > 0$ one has $\forall x, y \in \mathbb{R}^n$

$$H_s(x) - H_s(y) \geq \langle \nabla H_s(y), x - y \rangle + \frac{\rho_s}{2} ||x - y||^2. \tag{32}$$

Then, with the help of the inequalities (11) and (32), we derive

$$\frac{\rho_s}{2} ||x - y||^2 \leq \Phi_s(x^s) - \Phi_s(x^{s+1}) + \frac{\rho_s}{2} ||x^s - x^{s+1}||^2 + \delta_s \leq \\ \leq F_s(x^s) - F_s(x^{s+1}) + \delta_s, \tag{33}$$

from which the two inequalities follow:

$$||x^s - x^{s+1}|| \leq \sqrt{\frac{2}{\rho_s}} \left[F_s(x^s) - F_s(x^{s+1}) + \delta_s \right]^{1/2}; \tag{33a}$$

$$-\frac{\delta_s}{2} \leq \Phi_s(x^s) - \Phi_s(x^{s+1}) + \frac{\delta_s}{2} \leq F_s(x^s) - F_s(x^{s+1}). \tag{33b}$$

Note that (33b) was obtained under the condition that $\frac{\delta_s}{2} \leq \frac{\rho_s}{2}||x^s - x^{s+1}||$. From (33a) it can be readily seen that the stopping criterion

$$|F_s(x^s) - F_s(x^{s+1})| \leq \frac{\tau}{2}, \quad \delta_s \leq \frac{\tau}{2} \tag{34}$$

is more suitable and practical than, for example, the inequality $||x^s - x^{s+1}|| \leq \frac{\tau}{2}$.

Indeed, suppose, for instance, we have $|F_s(x^s) - F_s(x^{s+1})| \leq 2 \cdot 10^{-4}$, $\delta_s \leq 2 \cdot 10^{-4}$. Then, from (33a) with $\rho_s = 2$ it follows that $||x^s - x^{s+1}|| \leq 2 \cdot 10^{-2}$, which can be insufficient and unsatisfactory even for $\tau = 10^{-3}$. Moreover, (33b) means that the numbers $\Delta\Phi_s \stackrel{\triangle}{=} \Phi_s(x^s) - \Phi_s(x^{s+1})$ and $[F_s(x^s) - F_s(x^{s+1})]$ are approximately of the same order when $\delta_s \geq 0$ is sufficiently small. The latter result suggests that instead of the criterion (34) we can use the inequality (31) together with $\sigma_s \leq \frac{\tau}{2}$.

It is worth nothing that the latter criterion is easier to verify, since we compute these numbers at every iteration of the LSS1 by solving the linearized Problems (P_sL_s).

5   Global Search Scheme for Problem (P)

Here, we return to the problem of global search in Problem (P) with the equality and inequality constraints. To this end, we will use Problem (P_σ) (without constraints). At present, there exists only one known tool for escaping a stationary point (provided by classical optimization methods [16]) and improving the value of a cost functional. This tool is the Global Optimality Conditions (GOCs) for Problem (P_σ), in particular, Theorem 1. In addition, there exist two supplementary theorems, one of which yields the theoretical foundation for escaping from a local pitfall and is presented below (see [23, 24, 26, 27]).

Theorem 4. *Assume that a feasible in Problem (P) point z is not a ε-solution to (P), i.e. $\inf(f_0, \mathcal{F}) + \varepsilon = \mathcal{V}(P) + \varepsilon < \zeta := f_0(z)$. In addition, let a vector $v \in \mathbb{R}^n$ satisfy the following inequality $f_0(v) > \zeta - \varepsilon$.*

Then, for any penalty parameter $\sigma > 0$ there exists a tuple (y, β, u), $(y, \beta) \in \mathbb{R}^{n+1}$, $u \in \mathcal{F}$, such that the following conditions hold

$$\left.\begin{array}{ll}(a) \ H_\sigma(y) = \beta - \zeta + \varepsilon; & (b) \ G_\sigma(y) \leq \beta, \\ (c) \ G_\sigma(u) - \beta < \langle \nabla H_\sigma(y), u - y \rangle.\end{array}\right\} \tag{35}$$

Remark 5. It is worth noting now that if a point $z \in S$ satisfies the GOCs, then it turns out to be a critical vector in Problem (P_σ) (see Theorem 2). To show this, it is sufficient to set in the conditions (8)–(9) of Theorem 1 $y := z$. Hence, on account of the convergence property of a sequence $\{x^s\}$ generated by the LSS1 of Sect. 4 (see Theorem 2), the notion of the critical point of Problem (P_σ) seems to be the principal, essential and fundamental feature in DC optimization.

On the basis of such theoretical foundation and on account of our computational experience [7,8,19–26,29], we are now able to develop the following Global Search Scheme (GSS) for Problem $(\mathcal{P}_\sigma)$, where $\sigma > 0$ is fixed.

Suppose, we have a current iterate $z^k \in S$, $\zeta_k := F_\sigma(z^k)$.

(a) Choose a number β, such that

$$\beta_- := \inf(G_\sigma(\cdot), S) \leq \beta \leq \beta_+ := \sup(G_\sigma(\cdot), S). \tag{36}$$

Then, for the level surface $Y_k = Y(\zeta_k, \beta) = \{y \in \mathbb{R}^n : H_\sigma(y) = \beta - \zeta_k\}$ of the convex function $H_\sigma(\cdot)$, we construct a finite approximation

$$\mathcal{A}(\zeta_k, \beta) = \{y^1, ..., y^N \mid H_\sigma(y^i) = \beta - \zeta_k, \ G_\sigma(y^i) \leq \beta, \ i = 1, ..., N\}.$$

(b) For every $y^i \in \mathcal{A}_k(\beta) := \mathcal{A}(\zeta_k, \beta)$, solve the linearized problem as follows

$$(\mathcal{P}_\sigma L_i): \qquad G_\sigma(x) - \langle \nabla H_\sigma(y^i), x \rangle \downarrow \min_x, \quad x \in S. \tag{37}$$

Let $\bar{u}^i \in S$ be an approximate solution to $(\mathcal{P}_\sigma L_i)$.

(c) By starting at the point $\bar{u}^i \in S$, find a critical (to the LSS1, for example) vector u^i, i.e. which is an approximate solution to the linearized problem

$$(\mathcal{P}_\sigma L(u^i)): \qquad G_\sigma(x) - \langle \nabla H_\sigma(u^i), x \rangle \downarrow \min_x, \quad x \in S, \tag{38}$$

(linearized at the point u^i just).

(d) Furthermore, solve "the level problem"

$$(Lev\, \mathcal{P}_\sigma): \qquad \langle \nabla H_\sigma(v), u^i - v \rangle \uparrow \max_v, \ H_\sigma(v) = \beta - \zeta_k. \tag{39}$$

Let w^i be an approximate solution to Problem $(Lev\, \mathcal{P}_\sigma)$.

(e) Compute the number $\eta_k(\beta) = \eta(\zeta_k, \beta) := \eta_0(\zeta_k, \beta) - \beta$,

$$\eta_0(\zeta, \beta) = G_\sigma(u^j) - \langle \nabla H_\sigma(w^j), u^j - w^j \rangle :=$$
$$= \min_{1 \leq i \leq N} \{G_\sigma(u^i) - \langle \nabla H_\sigma(w^i), u^i - w^i \rangle\}.$$

(f) If $\eta_k(\beta) < 0$, then the vector $u^j \in S$ is better, than the point z^k in question, due to the convexity of $H_\sigma(\cdot)$, so that $F_\sigma(u^j) < F_\sigma(z^k)$. Hence, in this case, one can go to the next iteration, i.e. $k := k + 1$, $z^{k+1} := u^j$.

(g) When $\eta(\zeta_k, \beta) \geq 0$, we have to change the value β for $\bar{\beta} := \beta + \Delta\beta \in [\beta_-, \beta_+]$ with the help of some one-dimensional search methods.   #

By developing this general GSS for finding a global solution to $(\mathcal{P}_\sigma)$, we proposed an algorithmic form of the GSS and proved the convergence of the sequence $\{z^k\}$ produced by the method [29].

Now it is necessary to do the next step and to advance a new method for finding a global solution to original Problem $(\mathcal{P})$, which was our principal objective from the very beginning.

It is clear that this algorithmic procedure should be a mature scheme including, without doubts,

(A) a local search method (LSM) (say, LSS1);
(B) a global search scheme (say, the GSS1 from [29]) aiming at a global solution to the auxiliary Problem $(\mathcal{P}_\sigma)$ along with
(C) an update procedure for the penalty parameter providing a feasible (for the original Problem $(\mathcal{P})$) iterate z^k with a corresponding value of $\sigma_k > 0$.

For the sake of simplicity, let us use the following notations: $F_k(\cdot) := F_{\sigma_k}$, $G_k := G_{\sigma_k}$, $H_k := H_{\sigma_k}$. In addition, let there be given a starting vector $x_0 \in S$ and a starting penalty parameter value $\sigma^0 > 0$ (say, $\sigma^0 = 1$), and, besides, the number sequences $\{\omega_k\}, \{\tau_k\}, \{\delta_k\}, \{\varepsilon_k\}$ such that $\omega_k, \tau_k, \delta_k, \varepsilon_k > 0$, $k = 0, 1, 2, \ldots, \omega_k \downarrow 0, \tau_k \downarrow 0, \delta_k \downarrow 0, \varepsilon_k \downarrow 0 \ (k \to \infty)$ (see [19,21,25,26,29]).

Finally, suppose that numbers $\gamma_1 \in]0, 1[$ and $\gamma_2 \in [2, 10]$ (the parameters of the GSS) are also given. Then the Global Search Procedure for the original Problem $(\mathcal{P})$ can be described as follows.

Global Search Scheme 1 (GSS1) for $(\mathcal{P})$

Step 0. Set $k := 0$, $x^k := x_0$, $\sigma_k := \sigma^0$.
Step 1. By starting at $x^k \in S$ and with the help of a LSM (for example, LSS1) for Problem $(\mathcal{P}_\sigma)$, produce a τ_k-critical point $z^k \in S$, $\zeta_k := F_k(z^k) \le F_k(x^k)$ (for instance, satisfying the following inequality with $\Phi_k(x) = G_k(x) - \langle \nabla H_k(z^k), x \rangle$

$$\Phi_k(z^k) - \tau_k \le \inf_x \{\Phi_k(x) | \ x \in S\}). \tag{40}$$

Step 2. If $W(z^k) > \omega_k$, then increase σ_k up to σ_+ to fulfill the inequality

$$\Phi_k(x^k) - \Phi_k(x(\sigma_+)) \ge \gamma_1 \sigma_+ [W(z^k) - W(x(\sigma_+))], \tag{41}$$

where $x(\sigma_+) \in \delta_k\text{-}Sol(\mathcal{P}_+ L_k); (\mathcal{P}_+ L_k) := (\mathcal{P}_{\sigma_+} L(z^k)), G_+ := G_{\sigma_+}, H_+ := H_{\sigma_+}$,

$$(\mathcal{P}_+ L_k): \qquad G_+(x) - \langle \nabla H_+(z^k), x \rangle \downarrow \min_x, \ x \in S. \tag{42}$$

Set $z^k := x(\sigma_+)$, $\sigma_k := \sigma_+, G_k := G_+, H_k := H_+$.
Step 3. Choose a number $\beta \in [\beta_k^-, \beta_k^+]$. (One can set $\beta_0 := G_k(z^k)$), $\beta_k^- := \inf\{G_k(x) | \ x \in S\}, \beta_k^+ = \sup\{G_k(x) | \ x \in S\}$.
Step 4. Construct an approximation

$$\mathcal{A}_k(\beta) := \{y^1, \ldots, y^{N_k} | \ H_k(y^i) = \beta - \zeta_k, \ i = 1, \ldots, N_k, \ N_k = N_k(\beta)\},$$

and, according to Theorem 3, form a collection of indexes I_k defined as follows $I_k = I_k(\beta) = \{i \in \{1, \ldots, N_k\} | \ G_k(y^i) \le \beta\}$.
Step 5. If $I_k = \emptyset$, set $\beta := \beta + \Delta\beta \in [\beta_k^-, \beta_k^+]$ and loop to Step 4.
Step 6. For every $i \in I_k$ find a global $2\delta_k$-solution $\overline{u}^i \in S$ to the linearized convex problem $(\mathcal{P}_k L(y^i))$, and, after that, starting at $\overline{u}^i \in S$, use a LSM for $(\mathcal{P}_\sigma)$ to produce a $2\tau_k$-critical vector $u^i \in S$, so that

$$G_k(u^i) - \langle \nabla H_k(u^i), u^i \rangle - 2\tau_k \le \inf_x \{G_k(x) - \langle \nabla H_k(u^i), x \rangle | \ x \in S\}. \tag{43}$$

Step 7. Further, for every $i \in I_k$ find a global $2\delta_k$-solution v^i: $H_k(v^i) = \beta - \zeta_k$, to the level problem (see the GOCs (8)–(9))

$$(Lev\ \mathcal{P}_k): \qquad \langle \nabla H_k(v), u^i - v \rangle \uparrow \max_v, \quad H_k(v) = \beta - \zeta_k. \qquad (44)$$

Note that for a quadratic function $H_k(\cdot)$, this problem can be solved manually.

Step 8. Compute the number $\eta_k(\beta) := \eta_k^0(\beta) - \beta$, where

$$\eta_k^0(\beta) := G_k(u^j) - \langle \nabla H_k(v^i), u^j - v^j \rangle := \\ = \min_{i \in I_k} \{ \langle G_k(u^i) - \nabla H_k(v^i), u^i - v^i \rangle \}. \qquad (45)$$

Step 9. If $\eta_k(\beta) < 0$, then set $x^{k+1} := u^j$, $\sigma_{k+1} := \sigma_k$, $k := k + 1$ and loop to Step 1.

Step 10. (Else) Set $\beta := \beta + \Delta\beta \in [\beta_-, \beta_+]$, and go to Step 3.

Step 11. If $\eta_k(\beta) \geq 0 \; \forall \beta \in [\beta_-, \beta_+]$ (i.e. one-dimensional search on β is terminated) and, in addition

$$\left. \begin{array}{l} W(z^k) \leq \omega_* \text{ (that means } z^k \text{ is approximately feasible in } (\mathcal{P})) \\ \text{and } \delta_k \leq \delta_*, \ \tau_k \leq \tau_*, \ \varepsilon_k \leq \varepsilon_*, \end{array} \right\} \qquad (46)$$

where $\omega_* > 0$, $\delta_* > 0$, $\tau_* > 0$, $\varepsilon_* > 0$ are the fixed accuracies of corresponding computations, then STOP.

Step 12. (Else) Set $x^{k+1} := z^k$ (and perhaps $\sigma_{k+1} := \gamma_2 \sigma_k$), $k := k + 1$, and loop to Step 1.

It is clear that due to the stopping criteria (46) at Step 11 we obtain a feasible in Problem $(\mathcal{P})$ point z^k ($W(z^k) \approx 0$) and, therefore, z^k turns out to be a solution to $(\mathcal{P})$, since, due to Steps 3–10 (see [29]), $z^k \in Sol(\mathcal{P}_\sigma)$ under the corresponding assumptions on the approximation $\mathcal{A}_k(\beta)$ (Step 4) (see [19, 29]).

References

1. Burke, J.: An exact penalization viewpoint of constrained optimization. SIAM J. Control Optim. **29**, 968–998 (1991)
2. Byrd, R., Lopez-Calva, G., Nocedal, J.: A line search exact penalty method using steering rules. Math. Progr. Ser. A **133**, 39–73 (2012)
3. De Cosmis, S., De Leone, R.: The use of Grossone in mathematical programming and operations research. Appl. Math. Comput. **218**(16), 8029–8038 (2012)
4. Demyanov, V.F.: Extremum's conditions and variational calculus. High school edition, Moscow (2005). (In Russian)
5. Di Pillo, G., Lucidi, S., Rinaldi, F.: An approach to constrained global optimization based on exact penalty functions. J. Global Optim. **54**, 251–260 (2012)
6. Floudas, C.A., Pardalos, P.M. (eds.): Frontiers in Global Optimization. Kluwer Academic Publishers, Dordrecht (2004)
7. Gaudioso, M., Gruzdeva, T.V., Strekalovsky, A.S.: On numerical solving the spherical separability problem. J. Global Optim. **66**(1), 21–34 (2015). https://doi.org/10.1007/s10898-015-0319-y

8. Gruzdeva, T.V., Strekalovskiy, A.S.: On solving the sum-of-ratios problem. Appl. Math. Comput. **318**, 260–269 (2018)
9. Han, S., Mangasarian, O.: Exact penalty functions in nonlinear programming. Math. Progr. **17**, 251–269 (1979)
10. Hiriart-Urruty, J.B.: Generalized differentiability, duality and optimization for problems dealing with difference of convex finctions. In: Ponstein, J. (ed.) Convexity and Duality in Optimization. LNEMS, vol. 256, pp. 37–69. Springer, Heidelberg (1985). https://doi.org/10.1007/978-3-642-45610-7_3
11. Hiriart-Urruty, J.-B.: Optimisation et Analyse Convex. Presses Universitaires de France, Paris (1998)
12. Hiriart-Urruty, J.-B., Lemaréchal, C.: Convex Analysis and Minimization Algorithms. Springer, Heidelberg (1993). https://doi.org/10.1007/978-3-662-02796-7
13. Horst, R., Tuy, H.: Global Optimization: Deterministic Approaches. Springer, Heidelberg (1996). https://doi.org/10.1007/978-3-662-03199-5
14. Kruger, A., Minchenko, L., Outrata, J.: On relaxing the Mangasarian-Fromovitz constraint qualiffcation. Positivity **18**, 171–189 (2014)
15. Le Thi, H.A., Huynh, V.N., Dinh, T.P.: DC programming and DCA for general DC programs. In: van Do, T., Thi, H.A.L., Nguyen, N.T. (eds.) Advanced Computational Methods for Knowledge Engineering. AISC, vol. 282, pp. 15–35. Springer, Cham (2014). https://doi.org/10.1007/978-3-319-06569-4_2
16. Nocedal, J., Wright, S.J.: Numerical Optimization. Springer, New York (2006). https://doi.org/10.1007/978-0-387-40065-5
17. Rockafellar, R.T.: Convex Analysis. Princeton University Press, Princeton (1970)
18. Rockafellar, R.T., Wets, R.J.-B.: Variational Analysis. Springer, New York (1998). https://doi.org/10.1007/978-3-642-02431-3
19. Strekalovsky, A.S.: Elements of Nonconvex Optimization. Nauka, Novosibirsk (2003)
20. Strekalovsky, A.S.: Global optimality conditions for optimal control problems with functions of A.D. Alexandrov. J. Optim. Theory Appl. **159**, 297–321 (2013)
21. Strekalovsky, A.S.: On solving optimization problems with hidden nonconvex structures. In: Rassias, T.M., Floudas, C.A., Butenko, S. (eds.) Optimization in Science and Engineering, pp. 465–502. Springer, New York (2014). https://doi.org/10.1007/978-1-4939-0808-0_23
22. Strekalovsky, A.S.: On local search in D.C. optimization problems. Appl. Math. Comput. **255**, 73–83 (2015)
23. Strekalovsky, A.S.: Global optimality conditions in nonconvex optimization. J. Optim. Theory Appl. **173**(3), 770–792 (2017). https://doi.org/10.1007/s10957-016-0998-7
24. Strekalovsky, A.S.: Global optimality conditions and exact penalization. Optim. Lett. **13**(3), 597–615 (2017). https://doi.org/10.1007/s11590-017-1214-x
25. Strekalovsky, A.S., Minarchenko, I.M.: A local search method for optimization problem with D.C. inequality constraints. Appl. Math. Model. **58**, 229–244 (2018)
26. Strekalovskiy, A.: Nonconvex optimization: from global optimality conditions to numerical methods. In: AIP Conference Proceedings, LeGO 2018, 2070, Leiden, Netherlands (2019)
27. Strekalovsky, A.S.: New global optimality conditions in a problem with D.C. constraints. Trudy Inst. Mat. i Mekh. UrO RAN **25**, 245–261 (2019). (in Russian)
28. Strekalovsky, A.S.: Local search for nonsmooth DC optimization with DC equality and inequality constraints. In: Bagirov, A.M., Gaudioso, M., Karmitsa, N., Mäkelä, M.M., Taheri, S. (eds.) Numerical Nonsmooth Optimization, pp. 229–261. Springer, Cham (2020). https://doi.org/10.1007/978-3-030-34910-3_7

29. Strekalovsky, A.S.: On a global search in D.C. optimization problems. In: Jaćimović, M., Khachay, M., Malkova, V., Posypkin, M. (eds.) OPTIMA 2019. CCIS, vol. 1145, pp. 222–236. Springer, Cham (2020). https://doi.org/10.1007/978-3-030-38603-0_17
30. Tuy, H.: DC optimization: theory, methods and algorithms. In: Horst, R., Pardalos, P.M. (eds.) Handbook of Global Optimization, pp. 149–216. Kluwer Academic Publisher, Dordrecht (1995)

Dual Newton's Methods for Linear Second-Order Cone Programming

Vitaly Zhadan[1,2][⊠] [iD]

[1] Dorodnicyn Computing Centre, FRC "Computer Science and Control" of RAS,
40, Vavilova str., Moscow 119333, Russia
zhadan@ccas.ru
[2] Moscow Institute of Physics and Technology (State Research University),
9 Institutskiy per., Dolgoprudny, Moscow Region 141701, Russia

Abstract. The linear second-order cone programming problem is considered. For its solution, two dual Newton's methods are proposed. These methods are constructed with the help of optimality conditions. The nonlinear system of equations, obtained from the optimality conditions and depended only from dual variables, is solved by the Newton method. Under the assumption that there exist strictly complementary solutions of both primal and dual problems the local convergence of the methods with super-linear rate is proved.

Keywords: Linear second-order cone programming · Dual Newton's method · Local convergence · Super-linear rate of convergence

1 Introduction

The second-order cone programming problem (SOCP) is one of the main programs in cone programming. The linear SOCP is a problem in which the linear objective function is minimized on the intersection of a linear manifold with a second-order cone (the Lorentz cone) (see [1]). Many optimization problems, including combinatorial problems, can be reduced to the SOCP programs [1–3].

Today, there are some numerical methods for solving SOCP programs. From these methods, the primal-dual path-following methods are the most known [4,5]. In [6] the dual barrier-projection methods have been proposed for SOCP programs. These methods are generalizations of the corresponding methods for linear programming [7]. The primal Newton's method for SOCP have been considered in [8]. Both dual barrier-projection methods and the primal Newton's method had been worked out with the help of optimality conditions.

In dual methods the dual variables depending on primal variable are defined. As a result, the system of nonlinear equations with respect to dual variables, including a slack dual variable, is derived. In [6] this derived system of nonlinear equations is solved by the fix point method. The proposed in [6] dual methods are of the affine-scaling type. In present paper unlike to [6] the Newton method

© Springer Nature Switzerland AG 2020
A. Kononov et al. (Eds.): MOTOR 2020, LNCS 12095, pp. 19–32, 2020.
https://doi.org/10.1007/978-3-030-49988-4_2

is used for solving the derived system of nonlinear equations. Under assumption that the solutions of primal and dual problems are strictly complementary dual Newton's methods converge locally to these solutions with super-linear rate.

The paper is organized as follows. In Sect. 1, we formulate the SOCP program. Section 2 is principal in the paper. In this section the dual iterative methods for SOCP programs, based on the Newton method, are constructed. In Sect. 3 we show that in the case of non-degenerate problem these dual methods are well-posed. Finally, in Sect. 4, the local convergence of the methods is proved.

In what follows, the identity matrix of order s is denoted by I_s. The symbol 0_s is used for denoting the zero s-dimensional vector, and the symbol 0_{sk} is used for denoting $s \times k$ zero matrix. By Diag (x) is denoted the diagonal matrix with a vector x at its diagonal. Similarly, a block diagonal matrix with diagonal blocks $M_1, \ldots, M_k$ is denoted by DIAG $(M_1, \ldots, M_k)$.

2   The Linear Second-Order Cone Programming Problem

Let $\mathcal{K} \subset \mathbb{R}^n$ denote a closed convex pointed cone with the nonempty interior. This cone induces in $\mathbb{R}^n$ a partial order, that is: $x_1 \succeq_K x_2$, if $x_1 - x_2 \in K$. The linear cone programming problem is

$$\min \langle c, x \rangle, \quad \mathcal{A}x = b, \quad x \in \mathcal{K}, \tag{1}$$

where $\mathcal{A}$ is a $m \times n$ matrix, and $c = [c^1; \ldots; c^n] \in \mathbb{R}^n$, $b = [b^1; \ldots; b^m] \in \mathbb{R}^m$. The semicolons between vectors or components of a vector denote that these vectors or components are placed one under another. The angle brackets denotes the usual Euclidean scalar product.

The linear SOCP program is a special case of the problem (1). Let $c_i \in \mathbb{R}^{n_i}$, $1 \leq i \leq r$. Let also matrices A_i have dimensions $m \times n_i$, $1 \leq i \leq r$. Consider the problem

$$\min \sum_{i=1}^r \langle c_i, x_i \rangle,$$
$$\sum_{i=1}^r A_i x_i = b, \quad x_1 \succeq_{K_2^{n_1}} 0_{n_1}, \quad \ldots, \quad x_r \succeq_{K_2^{n_r}} 0_{n_r}. \tag{2}$$

Here $K_2^{n_i}$ is the second order cone (the Lorentz cone) in $\mathbb{R}^{n_i}$, defined as

$$K_2^{n_i} = \left\{ [x^0; \bar{x}] \in \mathbb{R} \times \mathbb{R}^{n_i - 1} : x^0 \geq \|\bar{x}\| \right\}, \quad 1 \leq i \leq r,$$

where $\|\cdot\|$ is the Euclidean norm. The cone $K_2^{n_i}$ is self-dual, that is $(K_2^{n_i})^* = K_2^{n_i}$.

The following problem is dual to (2)

$$\max \langle b, u \rangle,$$
$$A_i^T u + y_i = c_i, \quad 1 \leq i \leq r; \quad y_1 \succeq_{K_2^{n_1}} 0_{n_1}, \quad \ldots, \quad y_r \succeq_{K_2^{n_r}} 0_{n_r}, \tag{3}$$

in which $u \in \mathbb{R}^m$.

Denote $n = n_1 + \cdots + n_r$. If set $c = [c_1; \ldots; c_r]$, $x = [x_1; \ldots; x_r]$, $y = [y_1; \ldots; y_r]$ and $\mathcal{A} = [A_1, \ldots A_r]$, $\mathcal{K} = K_2^{n_1} \times \cdots \times K_2^{n_r}$, then the problem (2) can be written in the form of (1). The cone $\mathcal{K}$ is self-dual. We assume that both problems (2) and (3) have solutions, and that rows of the matrix $\mathcal{A}$ are linear independent. We assume also that $r > 1$.

Let $y(u) = c - \mathcal{A}^T u$. By

$$\mathcal{F}_P = \{x \in \mathcal{K} : \ \mathcal{A}x = b\}, \quad \mathcal{F}_D = \{[u, y] \in \mathbb{R}^m \times \mathcal{K} : \ y = y(u)\}$$

we will denote the feasible sets in problems (2) and (3), respectively. By $\mathcal{F}_{D,u}$ we will denote the projection of the set $\mathcal{F}_D$ onto the space $\mathbb{R}^m$, i.e. the set $\mathcal{F}_{D,u} = \{u \in \mathbb{R}^m : \ y(u) \in \mathcal{K}\}$.

If x and $[u, y]$ are solutions of problems (2) and (3), then they satisfy to the following system of equalities

$$\langle x, y \rangle = 0, \qquad \mathcal{A}x = b, \qquad y = c - \mathcal{A}^T u, \tag{4}$$

and to inclusions: $x \in \mathcal{K}$, $y \in \mathcal{K}$. Taking into account these inclusions, the equality $\langle x, y \rangle = 0$ from (4) can be replaced by n other equalities

$$x_i \circ y_i = 0_{n_i}, \quad 1 \le i \le r, \tag{5}$$

where the product between vectors $x_i \in \mathbb{R}^{n_i}$ and $y_i \in \mathbb{R}^{n_i}$ is defined by the following way $x_i \circ y_i = \left[x_i^T y_i; \ x_i^0 \bar{y} + y_i^0 \bar{x}_i\right]$. By introducing the matrix

$$\mathrm{Arr}\,(x_i) = \begin{bmatrix} x_i^0 & \bar{x}_i^T \\ \bar{x}_i & x_i^0 I_{n-1} \end{bmatrix},$$

the product $x_i \circ y_i$ can be represented as $x_i \circ y_i = \mathrm{Arr}\,(x_i)\, y_i = \mathrm{Arr}\,(y_i)\, x_i$.

Compose the block-diagonal matrix $\mathbf{Arr}(y) = \mathrm{DIAG}\,[\mathrm{Arr}\,(y_1), \ \ldots, \ \mathrm{Arr}\,(y_r)]$. Then equalities (4) can be rewritten as

$$\mathbf{Arr}\,(y)\, x = \mathbf{Arr}\,(x)\, y = 0_n, \qquad \mathcal{A}x = b, \qquad y = c - \mathcal{A}^T u, \tag{6}$$

where, recall, $x \in \mathcal{K}$, $y \in \mathcal{K}$.

3   The Dual Newton's Methods

Consider the iterative dual methods for solving problems (2) and (3). These methods are analogs of the primal method proposed in [8]. In dual methods the dependence $x(u)$ or more general dependence $x(u, y)$ are used to derive from (6) the system of nonlinear equations depending on only dual variables.

In order to obtain $x(u)$ we multiply the second equality from (6) by the matrix $\mathcal{A}^T$ and sum it with the first equality (6). As a result, we get the equation with respect to x:

$$\Phi(y)x = \mathcal{A}^T b, \tag{7}$$

where by $\Phi(y)$ is denoted the matrix: $\Phi(y) = \mathcal{A}^T \mathcal{A} + \mathbf{Arr}\,(y)$. The matrix $\Phi(y)$ is symmetric of order n. If $\Phi(y)$ is nonsingular, then, solving the Eq. (7), we obtain

$$x = x(y) = \Phi^{-1}(y)\mathcal{A}^T b.$$

Taking $y = y(u)$, we conclude that in fact the matrix $\Phi(y)$ depends on u.

Substituting the founded $x(u) = x(y(u))$ into the second equation from (6), we get the system of nonlinear equations with respect to u, namely,

$$\left[I_m - \mathcal{A}\mathbf{\Phi}^{-1}(y(u))\mathcal{A}^T \right] b = 0_m. \tag{8}$$

The system (8) consists of m equations. The number of unknowns is also equal to m.

Applying the Newton method to solve (8), we obtain the iterative process

$$u_{k+1} = u_k - \mathbf{G}^{-1}(u_k)\left(\mathcal{A}x_k - b \right). \tag{9}$$

Here $x_k = x(u_k)$ and $\mathbf{G}(u) = \frac{d}{du}\mathcal{A}x(u) = \mathcal{A}x_u(u)$.

Treating (7) as the identity with respect to u, we obtain after differentiating

$$\mathbf{Arr}_u(y(u))\,x(u) + \mathbf{\Phi}(y(u))\,x_u(u) = 0_{nm}.$$

If $\mathbf{\Phi}(y(u))$ is a nonsingular matrix, then

$$x_u(u) = -\mathbf{\Phi}^{-1}(y(u))\,\mathbf{Arr}_u(y(u))\,x(u). \tag{10}$$

Since $y(u) = c - \mathcal{A}^T u$, we get $\mathbf{Arr}_u(y(u)) = -\mathbf{Arr}_y(y)\mathcal{A}^T$.

Proposition 1. *For any $x \in \mathbb{R}^n$ the equality*

$$\mathbf{Arr}_u(y(u))\,x = -\mathbf{Arr}\,(x)\mathcal{A}^T \tag{11}$$

holds.

Proof. Let us take the product z_i of any matrix $\mathrm{Arr}(y_i)$ on the vector x_i and differentiate each row of z_i by y_i separately. First of all, the "null" row is the following: $z_i^0 = \sum_{j=0}^{n-1} x_i^j y_i^j$. Therefore,

$$\frac{d}{dy} z_i^0 = \left[x_i^0;\ x_i^1;\ \ldots;\ x_i^{n-1} \right]. \tag{12}$$

Further, for any consequent j-th row: $z_i^j = y_i^j x_i^0 + y_i^0 x_i^j$. Hence

$$\frac{d}{dy} z_i^j = \left[x_i^j;\ 0;\ \ldots;0;\ x_i^0;\ 0;\ \ldots,0 \right], \quad 1 \le j \le n-1. \tag{13}$$

From (12) and (13) we derive that $\mathbf{Arr}_y(y)x = \mathbf{Arr}(x)$. Hence, the equality (11) takes place. $\square$

According to Proposition 1 and to (10) $\mathbf{G}(u) = \mathcal{A}\mathbf{\Phi}^{-1}(u)\mathbf{Arr}(x(u))\mathcal{A}^T$. Thus, the iterative method (9) can be written in the following form

$$u_{k+1} = u_k - \left[\mathcal{A}\mathbf{\Phi}^{-1}(y_k)\mathbf{Arr}(x_k)\mathcal{A}^T \right]^{-1} \left(\mathcal{A}x_k - b \right), \tag{14}$$

where $x_k = x(u_k)$, $y_k = y(u_k)$.

It is possible to consider the more general with respect to (9) iterative process. In this process both variables u and y are updated at each iteration. In order to construct the method we add to the right side of Eq. (7) the second equality from (6), multiplied by some parameter $\tau > 0$. As a result, we obtain instead of (7) the system of equations

$$\mathbf{\Phi}(y)x = \mathcal{A}^T b + \tau \left(y + \mathcal{A}^T u - c \right) \tag{15}$$

with the solution $x(u, y) = \mathbf{\Phi}^{-1}(y)f(u, y)$, where $f(u, y) = \mathcal{A}^T b + \tau \left(y + \mathcal{A}^T u - c \right).$

Substituting $x(u, y)$ in first and second equalities from (6), we obtain the system of $n + m$ equations

$$\mathcal{A}\mathbf{\Phi}^{-1}(y)f(u, y) - b = 0_m, \qquad \mathbf{Arr}(y)\mathbf{\Phi}^{-1}(y)f(u, y) = 0_n. \tag{16}$$

Denote $w = [u; y]$ and $\mathbf{\Psi}(w) = \left[\mathbf{\Psi}^{(1)}(w); \mathbf{\Psi}^{(2)}(w) \right]$, where

$$\mathbf{\Psi}^{(1)}(w) = \mathcal{A}\mathbf{\Phi}^{-1}(y)f(u, y) - b, \quad \mathbf{\Psi}^{(2)}(w) = \mathbf{Arr}(y)\mathbf{\Phi}^{-1}(y)f(u, y).$$

Lemma 1. *Let the point $w = [u; y] \in \mathcal{F}_D$ be such that the matrix $\mathbf{\Phi}(y)$ is nonsingular. Then the matrix $\mathbf{\Psi}_w(w)$ has the form*

$$\mathbf{\Psi}_w(w) = \begin{bmatrix} \tau\mathcal{A}\mathbf{\Phi}^{-1}\mathcal{A}^T & \mathcal{A}\mathbf{\Phi}^{-1}\left[\tau I_n - \mathbf{Arr}(x(w))\right] \\ \tau\mathbf{Arr}(y)\mathbf{\Phi}^{-1}\mathcal{A}^T & \left[I_n - \mathbf{Arr}(y)\mathbf{\Phi}^{-1}\right]\mathbf{Arr}(x(w)) + \tau\mathbf{Arr}(y)\mathbf{\Phi}^{-1} \end{bmatrix}. \tag{17}$$

where $\mathbf{\Phi}^{-1} = \mathbf{\Phi}^{-1}(y)$.

Proof. Differentiating $\mathbf{\Psi}^{(1)}$, we obtain: $\mathbf{\Psi}_u^{(1)}(w) = \mathcal{A}x_u(w)$, $\mathbf{\Psi}_y^{(1)}(w) = \mathcal{A}x_y(w)$. Moreover,

$$\mathbf{\Psi}_u^{(2)}(w) = \mathbf{Arr}(y)\, x_u(w), \quad \mathbf{\Psi}_y^{(2)}(w) = \mathbf{Arr}(x(w)) + \mathbf{Arr}(y)\, x_y(w).$$

Because of (15), the function $x(w)$ is satisfied to the identity

$$\left[\mathcal{A}^T\mathcal{A} + \mathbf{Arr}(y)\right]x(w) \equiv \mathcal{A}^T b + \tau \left(y + \mathcal{A}^T u - c\right). \tag{18}$$

After differentiation (18) by u we obtain

$$\left[\mathcal{A}^T\mathcal{A} + \mathbf{Arr}(y)\right]x_u(w) = \tau\mathcal{A}^T. \tag{19}$$

Respectfully, after differentiation (18) by y we derive the equality $\mathcal{A}^T\mathcal{A}x_y(w) + \frac{\partial}{\partial y}\mathbf{Arr}(y)x(w) = \tau I_n$ or

$$\mathcal{A}^T\mathcal{A}x_y(w) + \mathbf{Arr}(x(w)) + \mathbf{Arr}(y)\, x_y(w) = \tau I_n. \tag{20}$$

Equalities (19) and (20) can be written as

$$\mathbf{\Phi}(y)x_u(w) = \tau\mathcal{A}^T, \quad \mathbf{\Phi}(y)x_y(w) + \mathbf{Arr}(x(w)) = \tau I_n.$$

If $\mathbf{\Phi}(y)$ is a nonsingular matrix, we derive from here that

$$x_u(w) = \tau\mathbf{\Phi}^{-1}(y)\mathcal{A}^T, \quad x_y(w) = \mathbf{\Phi}^{-1}(y)\left[\tau I_n - \mathbf{Arr}(x(w))\right].$$

Thus, the matrix $\mathbf{\Psi}_w(w)$ has the form (17). $\square$

If the matrix $\mathbf{\Psi}_w(w)$ is nonsingular for all points w in some neighbourhood of the solution w_* of the problem (3), then it is possible to apply the Newton method for solving the system of nonlinear equations (16). We obtain the dual iterative method

$$w_{k+1} = w_k - \mathbf{\Psi}_w^{-1}(w_k)\mathbf{\Psi}(w_k). \tag{21}$$

The point w_0 must be taken from some vicinity of the solution w_*.

Denote $\mathbf{\Gamma}(y) = \mathcal{A}\mathbf{\Phi}^{-1}(y)\mathcal{A}^T$, $\mathbf{K}_1(y) = \mathcal{A}\,\mathbf{\Phi}^{-1}(y)$, $\mathbf{K}_2(y) = \mathbf{Arr}(y)\mathbf{\Phi}^{-1}(y)$. Then the matrix (17) can be written as

$$\mathbf{\Psi}_w(w) = \begin{bmatrix} \tau\mathbf{\Gamma}(y) & \mathbf{K}_1(y)\left[\tau I_n - \mathbf{Arr}(x(w))\right] \\ \tau\mathbf{Arr}(y)\mathbf{K}_1^T(y) & \left[I_n - \mathbf{K}_2(y)\right]\mathbf{Arr}(x(w)) + \tau\mathbf{K}_2(y) \end{bmatrix}.$$

Let the point $w \in \mathcal{F}_D$ be such that the matrix $\mathbf{\Psi}_w(w)$ is nonsingular. In this case the matrix $\mathbf{\Gamma}(y)$ is also nonsingular. Denote $\mathbf{K}_3(y) = \mathbf{K}_1^T(y)\mathbf{\Gamma}^{-1}\mathbf{K}_1(y)$ and

$$\mathbf{\Omega}(w) = \left[I_n - \mathbf{K}_2(y)\right]\mathbf{Arr}(x(w)) + \tau\mathbf{K}_2(y) - \mathbf{Arr}(y)\mathbf{K}_3(y)\left[\tau I_n - \mathbf{Arr}(x(w))\right].$$

Then, by the Frobenius formula, we obtain

$$\mathbf{\Psi}_w^{-1}(w) = \begin{bmatrix} \mathbf{P}_1 & \mathbf{P}_2 \\ \mathbf{P}_3 & \mathbf{P}_4 \end{bmatrix},$$

where $\mathbf{P}_4 = \mathbf{\Omega}^{-1}$ and

$$\mathbf{P}_1 = \tau^{-1}\mathbf{\Gamma}^{-1} + \tau^{-1}\mathbf{\Gamma}^{-1}\mathbf{K}_1(y)\left[\tau I_n - \mathbf{Arr}(x(w))\right]\mathbf{\Omega}^{-1}\mathbf{Arr}(y)\mathbf{K}_1^T(y)\mathbf{\Gamma}^{-1},$$

$$\mathbf{P}_2 = -\tau^{-1}\mathbf{\Gamma}^{-1}\mathbf{K}_1(y)\left[\tau I_n - \mathbf{Arr}(x(w))\right]\mathbf{\Omega}^{-1}, \quad \mathbf{P}_3 = -\mathbf{\Omega}^{-1}\mathbf{Arr}(y)\mathbf{K}_1^T(y)\mathbf{\Gamma}^{-1}.$$

At last, denote $\mathbf{W} = \mathbf{\Gamma}^{-1}\mathbf{K}_1(y)\left[\tau I_n - \mathbf{Arr}(x(w))\right]\mathbf{\Omega}^{-1}$. It follows from previous formulas that $\mathbf{P}_2 = -\tau^{-1}\mathbf{W}$ and $\mathbf{P}_1 = \tau^{-1}\left[I_m - \mathbf{W}\mathbf{Arr}(y)\mathbf{K}_1^T(y)\right]\mathbf{\Gamma}^{-1}$.

Therefore, the formulas (21) for updating the point $[u_k; y_k]$ are following:

$$u_{k+1} = u_k - \tau^{-1}\left[\left(I_m - \mathbf{W}\mathbf{Arr}(y)\mathbf{K}_1^T(y)\right)\mathbf{\Gamma}^{-1}\left(\mathcal{A}x_k - b\right) + \mathbf{W}\mathbf{Arr}(y_k)x_k\right],$$

$$y_{k+1} = y_k + \mathbf{\Omega}^{-1}\left[\mathbf{Arr}(y)\mathbf{K}_1^T(y)\mathbf{\Gamma}^{-1}\left(\mathcal{A}x_k - b\right) - \mathbf{Arr}(y_k)x_k\right],$$

where $x_k = x(w_k) = \mathbf{\Phi}^{-1}(y_k)f(w_k)$.

4   Non-degeneracy in the Dual Problem

Let us show that the matrix $\mathbf{\Phi}(y)$ is nonsingular, if the point $[u,y] \in \mathcal{F}_D$ is non-degenerate.

Definition 1. [1]. *The point $[u,y] \in \mathcal{F}_D$ is called non-degenerate if $\mathcal{T}_{\mathcal{K}}(y) + \mathcal{R}(\mathcal{A}^T) = \mathbb{R}^n$, where $\mathcal{T}_{\mathcal{K}}(y)$ is the tangent space to the cone $\mathcal{K}$ at the point y and $\mathcal{R}(\mathcal{A}^T)$ is the image of the matrix $\mathcal{A}^T$.*

Let $[u,y] \in \mathcal{F}_D$, and let the vector $y \in \mathcal{K}$ be partitioned onto three blocks of components: $y = [y_F; y_I; y_N]$. We assume for definiteness that these blocks are consisted from components y_i ordered in the following way: $y_F = [y_1; \ldots; y_{r_F}]$, $y_I = [y_{r_F+1}; \ldots; y_{r_F+r_I}]$, $y_N = [y_{r_F+r_I+1}; \ldots; y_{r_F+r_B+r_N}]$. Recall that $r = r_F + r_I + r_N$.

The partition of the vector y induces the partition of the index set $J^r = [1:r]$ onto three subsets:

$$J_F^r(y) = [1, \ldots, r_F], \quad J_I^r(y) = [r_F + 1, \ldots, r_F + r_I], \quad J_N^r(y) = [r_F + r_I + 1, \ldots, r].$$

If $i \in J_F^r(y)$, then $y_i \neq 0_{n_i}$ and $y_i \in \partial K_2^{n_i}$, where $\partial K_2^{n_i}$ is the boundary of the cone $K_2^{n_i}$. If $i \in J_I^r(y)$, then $y_i = 0_{n_i}$. At last, in the case, where $i \in J_N^r(y)$, the inclusion $y_i \in \text{int}K_2^{n_i}$ holds. According to the partition of the vector y onto three blocks of components we partition also the matrix $\mathcal{A} = [\mathcal{A}_F, \mathcal{A}_I, \mathcal{A}_N]$ and the vector $c = [c_F; c_I; c_N]$.

For any nonzero component $y_i \in \mathbb{R}^{n_i}$, $i \in J^r$, the following spectral decomposition

$$y_i = \theta_{i,1} \mathbf{d}_{i,1} + \theta_{i,n_i} \mathbf{d}_{i,n_i} \tag{22}$$

takes place [1]. Here the pair of vectors

$$\mathbf{d}_{i,1} = \frac{1}{\sqrt{2}} \left[1; \frac{\bar{y}_i}{\|\bar{y}_i\|} \right], \qquad \mathbf{d}_{i,n_i} = \frac{1}{\sqrt{2}} \left[1; -\frac{\bar{y}_i}{\|\bar{y}_i\|} \right],$$

is a Jordan frame. The coefficients $\theta_{i,1}$ and θ_{i,n_i} in (22) are following:

$$\theta_{i,1} = \frac{1}{\sqrt{2}} \left(y_i^0 + \|\bar{y}_i\| \right), \qquad \theta_{i,n_i} = \frac{1}{\sqrt{2}} \left(y_i^0 - \|\bar{y}_i\| \right).$$

Both vectors $\mathbf{d}_{i,1}$ and $\mathbf{d}_{i,n_i}$ are orthogonal each to other and their lengths equal to one.

If $y_i \in K_2^{n_i}$, then $\theta_{i,1} \geq 0$ and $\theta_{i,n_i} \geq 0$. In the case, where $y_i \neq 0_{n_i}$ and $y_i \in \partial K_2^{n_i}$, the equality $y_i^0 = \|\bar{y}_i\|$ holds. Hence, only the first coefficient $\theta_{i,1} = \sqrt{2} y_i^0 = \sqrt{2} \|\bar{y}_i\|$ differs from zero.

Let us assume that $y_i \in K_2^{n_i}$ and $y_i \neq 0_{n_i}$. The matrix $\mathrm{Arr}\,(y_i)$ is symmetric. Denote by H_i the orthogonal matrix with columns being eigenvectors of $\mathrm{Arr}\,(y_i)$. The vectors $\mathbf{d}_{i,1}$ and $\mathbf{d}_{i,n_i}$ are among eigenvectors of $\mathrm{Arr}\,(y_i)$. The matrix H_i can be taken in the following form

$$H_i = [\mathbf{d}_{i,1}, h_{i,2}, \ldots, h_{i,n_i-2}, \mathbf{d}_{i,n_i}].$$

The eigenvectors $h_{i,2}, \ldots h_{i,n_i-2}$ are arbitrary vectors from the subspace

$$\mathbb{R}_0^{n_i} = \left\{ z = [z^0; \bar{z}] \in \mathbb{R}^{n_i} : z^0 = 0 \right\}.$$

All these vectors have the unit length and are orthogonal each to others. Moreover, they are orthogonal to the vectors $\mathbf{d}_{i,1}$ and $\mathbf{d}_{i,n_i}$.

Eigenvalues $y_i^0 + \|\bar{y}_i\|$ and $y_i^0 - \|\bar{y}_i\|$ correspond to the eigenvectors $\mathbf{d}_{i,1}$ and $\mathbf{d}_{i,n_i}$, respectively. The eigenvalue y_i^0 has the multiplicity $n_i - 2$ and corresponds to eigenvectors $h_{i,2}, \ldots h_{i,n_i-2}$. Denoting by Σ_i the diagonal matrix

$$\Sigma_i = \mathrm{Diag} \left(\sqrt{2}\theta_{i,1}, y_i^0, \ldots, y_i^0, \sqrt{2}\theta_{i,n_i} \right),$$

we have $\mathrm{Arr}\,(y_i) = H_i \Sigma_i H_i^T$.

If $i \in J_I^r(y)$, then $y_i = 0_{n_i}$. In this case the identity matrix I_{n_i} can be taken as the orthogonal matrix H_i. It is evident that $\Sigma_i = 0_{n_i n_i}$ for this $\mathrm{Arr}\,(y_i)$.

Introduce into consideration the block-diagonal matrices

$$\mathbf{H}_F = \mathrm{DIAG}\,[H_1, \ldots, H_{r_F}], \quad \mathbf{H}_I = \mathrm{DIAG}\,[H_{r_F+1}, \ldots, H_{r_F+r_I}],$$

The matrices $\mathbf{H}_F$ and $\mathbf{H}_I$ are orthogonal. In the same way we combine the diagonal matrices Σ_i:

$$\Sigma_F = \mathrm{DIAG}\,[\Sigma_1, \ldots, \Sigma_{r_F}], \quad \Sigma_I = \mathrm{DIAG}\,[\Sigma_{r_F+1}, \ldots, \Sigma_{r_F+r_I}],$$

Let $\mathcal{A}_F^{\mathbf{H}} = \mathcal{A}_F \mathbf{H}_F$, and let $\tilde{\mathcal{A}}_F^{\mathbf{H}}$ be the matrix $\mathcal{A}_F^{\mathbf{H}}$, from which all columns are removed, except the columns being the first columns of matrices $A_i H_i$, $i \in J_F^r(y)$. The matrix $\tilde{\mathcal{A}}_F^{\mathbf{H}}$ has the dimension $m \times r_F$. Denote $\mathcal{A}_{FI}^{\mathbf{H}} = \left[\tilde{\mathcal{A}}_F^{\mathbf{H}}, \mathcal{A}_I^{\mathbf{H}} \right]$, where $\mathcal{A}_I^{\mathbf{H}} = \mathcal{A}_I \mathbf{H}_I$. The following criterion of the non-degeneracy of the point $[u, y]$ is valid [1].

Proposition 2. *The point $[u, y] \in \mathcal{F}_D$ is non-degenerate if and only if columns of the matrix $\mathcal{A}_{FI}^H$ are linear independent.*

It follows from Proposition 2, that at a non-degenerate point $[u, y]$ the inequality $r_F + n_I \leq m$ takes place, where $n_I = \sum_{i \in J_I^r(y)} n_i$.

Proposition 3. *[6] Let the point $[u, y] \in \mathcal{F}_D$ be non-degenerate. Then the matrix $\mathbf{\Phi}(y)$ is nonsingular.*

We call the dual problem (3) *non-degenerate*, if all points $[u, y] \in \mathcal{F}_D$ are non-degenerate. Below we suppose that the problem (3) *is non-degenerate*.

5   Local Convergence of the Dual Methods

Let x_* and $[u_*, y_*]$ be the solutions of problems (2) and (3), respectively. We assume without loss of generality that for the vector x_* the following partition onto three blocks of components $x_* = [x_{*,F}, x_{*,I}, x_{*,N}]$ holds. Moreover, the number of component in blocks $x_{*,F}$, $x_{*,I}$ and $x_{*,N}$ is equal to r_F, r_I and r_N, respectively. Each component $x_{*,i}$ from the block $x_{*,F}$ belongs to the boundary of the cone $K_2^{n_i}$. Each component $x_{*,i}$ from the block $x_{*,I}$ is an interior point of $K_2^{n_i}$. All $x_{*,i}$ from the block $x_{*,N}$ are zero vectors.

Besides, let for the vector y_* the decomposition onto block of components $y_* = [y_{*,F}, y_{*,I}, y_{*,N}]$ take place. Moreover, the number of components in blocks is equal to $\bar{r}_F$, $\bar{r}_I$ and $\bar{r}_N$, respectively. But unlike to x_*, components $y_{*,i}$ from the block $y_{*,I}$ are zero vectors. On the contrary, $y_{*,i}$ from the block $y_{*,N}$ is an interior point of the cone $K_2^{n_i}$.

According to (5) the following *complementary condition* $x_{*,i} \circ y_{*,i} = 0$, $1 \leq i \leq r$, holds. The strict complementary condition means that additionally $x_{*,i} + y_{*,i} \in \text{int } K_2^{n_i}$. In this case $\bar{r}_F = r_F$, $\bar{r}_I = r_I$ and $\bar{r}_N = r_N$. Furthermore, the matrices $\text{Arr}(x_{*,i})$ and $\text{Arr}(y_{*,i})$ commute between themselves. The following decompositions

$$\text{Arr}(x_{*,i}) = H_i \Lambda_i H_i^T, \qquad \text{Arr}(y_{*,i}) = H_i \Sigma_i H_i^T, \qquad (23)$$

take place. Here H_i is an orthogonal matrix, and Λ_i and Σ_i are diagonal matrices with eigenvalues of $\text{Arr}(x_{*,i})$ and $\text{Arr}(y_{*,i})$ at their diagonals, respectively. Below we set $r_{FI} = r_F + r_I$ and $J_F^r = [1 : r_F]$, $J_I^r = [r_F + 1 : r_{FI}]$, $J_N^r = [r_{FI} + 1 : r]$, $J_{FI}^r = J_F^r \cup J_I^r$.

Similar to (22) for $x_{*,i}$ the spectral decomposition

$$x_{*,i} = \eta_{i,1} \mathbf{e}_{i,1} + \eta_{i,n_i} \mathbf{e}_{i,n_i} \qquad (24)$$

holds, where

$$\mathbf{e}_{i,1} = \frac{1}{\sqrt{2}} \left[1; \frac{\bar{x}_{*,i}}{\|\bar{x}_{*,i}\|} \right], \qquad \mathbf{e}_{i,n_i} = \frac{1}{\sqrt{2}} \left[1; -\frac{\bar{x}_{*,i}}{\|\bar{x}_{*,i}\|} \right]$$

are frame vectors. The coefficients $\eta_{i,1}$ and η_{i,n_i} in (24) are following:

$$\eta_{i,1} = \frac{1}{\sqrt{2}} \left(x_{*,i}^0 + \|\bar{x}_{*,i}\| \right), \qquad \eta_{i,n_i} = \frac{1}{\sqrt{2}} \left(x_{*,i}^0 - \|\bar{x}_{*,i}\| \right).$$

Both $\mathbf{e}_{i,1}$ and $\mathbf{e}_{i,n_i}$ are unit vectors and orthogonal each to other.

The orthogonal matrix H_i in (23) has the form

$$H_i = [\mathbf{e}_{i,1}, h_{i,2}, \dots, h_{i,n_i-2}, \mathbf{e}_{i,n_i}], \tag{25}$$

where $h_{i,2}, \dots, h_{i,n_i-2}$ are unit vectors from the subspace $\mathbb{R}_0^{n_i}$. The matrix $\Lambda_i = \mathrm{Diag}\left(\sqrt{2}\eta_{i,1}, x_{*,i}^0, \dots, x_{*,i}^0, \sqrt{2}\eta_{i,n_i}\right)$ is diagonal with eigenvalues of $\mathrm{Arr}(x_{*,i})$ on its diagonal, $i \in J_{FI}^r$. Remark, that for $i \in J_F^r$ the last eigenvalue is zero, that is $\Lambda_i = \mathrm{Diag}\left(2x_{*,i}^0, x_{*,i}^0, \dots, x_{*,i}^0, 0\right)$.

At solutions x_* and y_* according to the complementary condition the vector $\mathbf{e}_{i,n_i}$ must be collinear to the vector $\mathbf{d}_{i,1}$ from the spectral decomposition (22) for $y_{*,i}$. Hence, the orthogonal matrix (25) can be used also in the spectral decomposition of the matrix $\mathrm{Arr}(y_{*,i})$, $i \in J_{FI}^r$, i.e. $\mathrm{Arr}(y_{*,i}) = H_i \Sigma_i H_i^T$, where Σ_i is a diagonal matrix with the vector of eigenvalues of the matrix $\mathrm{Arr}(y_{*,i})$ at its diagonal. For $i \in J_F^r$ the matrix Σ_i has the form $\Sigma_i = \mathrm{Diag}\left(0, y_{*,i}^0, \dots, y_{*,i}^0, 2y_{*,i}^0\right)$. The matrix Λ_i is zero for $i \in J_N^r$, and, vice verse, the matrix Σ_i is zero, when $i \in J_I^r$.

In addition, let the orthogonal matrix H_i for $i \in J_N^r$ be defined by the matrix $\mathrm{Arr}(y_{*,i})$, that is $\mathrm{Arr}(y_{*,i}) = H_i \Sigma_i H_i^T$. Then $H_i = [\mathbf{d}_{i,1}, h_{i,2}, \dots, h_{i,n_i-2}, \mathbf{d}_{i,n_i}]$ and $\Sigma_i = \mathrm{Diag}\left(y_{*,i}^0 + \|\bar{y}_{*,i}\|, y_{*,i}^0, \dots, y_{*,i}^0, y_{*,i}^0 - \|\bar{y}_{*,i}\|\right)$. Moreover, Λ_i is a zero matrix for $i \in J_N^r$.

Let $\mathbf{\Lambda} = \mathrm{DIAG}\left(\mathbf{\Lambda}_F, \mathbf{\Lambda}_I, \mathbf{\Lambda}_N\right)$, $\mathbf{\Sigma} = \mathrm{DIAG}\left(\mathbf{\Sigma}_F, \mathbf{\Sigma}_I, \mathbf{\Sigma}_N\right)$, where

$$\mathbf{\Lambda}_F = \mathrm{DIAG}\left(\Lambda_1, \dots, \Lambda_{r_F}\right), \quad \mathbf{\Sigma}_F = \mathrm{DIAG}\left(\Sigma_1, \dots, \Sigma_{r_F}\right)$$

and

$$\mathbf{\Lambda}_I = \mathrm{DIAG}\left(\Lambda_{r_F+1}, \dots, \Lambda_{r_{FI}}\right), \quad \mathbf{\Sigma}_I = \mathrm{DIAG}\left(\Sigma_{r_F+1}, \dots, \Sigma_{r_{FI}}\right),$$
$$\mathbf{\Lambda}_N = \mathrm{DIAG}\left(\Lambda_{r_{FI}+1}, \dots, \Lambda_r\right), \quad \mathbf{\Sigma}_N = \mathrm{DIAG}\left(\Sigma_{r_{FI}+1}, \dots, \Sigma_r\right).$$

Set also $\mathbf{H} = \mathrm{DIAG}\left(H_1, \dots, H_r\right)$ and denote: $\mathcal{A}^{\mathbf{H}_F} = \mathcal{A}\mathbf{H}_F$, $\mathcal{A}^{\mathbf{H}_I} = \mathcal{A}\mathbf{H}_I$, $\mathcal{A}^{\mathbf{H}_N} = \mathcal{A}\mathbf{H}_N$, $\mathcal{A}^{\mathbf{H}} = \mathcal{A}\mathbf{H}$. For $\mathcal{A}^{\mathbf{H}}$ the decomposition $\mathcal{A}^{\mathbf{H}} = \left[\mathcal{A}^{\mathbf{H}_F}, \mathcal{A}^{\mathbf{H}_I}, \mathcal{A}^{\mathbf{H}_N}\right]$ is valid. With the introduced notations the matrix $\mathbf{G}(u_*)$ can be submitted in the form

$$\mathbf{G}(u_*) = \mathcal{A}^{\mathbf{H}} \mathbf{\Phi}^{-\mathbf{H}}(y_*) \mathbf{\Lambda}(x_*)(\mathcal{A}^{\mathbf{H}})^T, \quad \mathbf{\Phi}^{-\mathbf{H}}(y_*) = \left(\mathbf{\Phi}^{\mathbf{H}}(y_*)\right)^{-1}, \tag{26}$$

where $y_* = y(u_*)$ and $\mathbf{\Phi}^{\mathbf{H}}(y_*) = \mathbf{H}^T \mathbf{\Phi}(y_*)\mathbf{H}$.

We have by aforesaid

$$\mathbf{\Phi}^{\mathbf{H}}(y_*) = \begin{bmatrix} \left(\mathcal{A}_F^{\mathbf{H}}\right)^T \mathcal{A}_F^{\mathbf{H}} + \mathbf{\Sigma}_F & \left(\mathcal{A}_F^{\mathbf{H}}\right)^T \mathcal{A}_I^{\mathbf{H}} & \left(\mathcal{A}_F^{\mathbf{H}}\right)^T \mathcal{A}_N^{\mathbf{H}} \\ \left(\mathcal{A}_I^{\mathbf{H}}\right)^T \mathcal{A}_F^{\mathbf{H}} & \left(\mathcal{A}_I^{\mathbf{H}}\right)^T \mathcal{A}_I^{\mathbf{H}} & \left(\mathcal{A}_I^{\mathbf{H}}\right)^T \mathcal{A}_N^{\mathbf{H}} \\ \left(\mathcal{A}_N^{\mathbf{H}}\right)^T \mathcal{A}_F^{\mathbf{H}} & \left(\mathcal{A}_N^{\mathbf{H}}\right)^T \mathcal{A}_I^{\mathbf{H}} & \left(\mathcal{A}_N^{\mathbf{H}}\right)^T \mathcal{A}_N^{\mathbf{H}} + \mathbf{\Sigma}_N \end{bmatrix}.$$

All diagonal entrees of the matrix $\mathbf{\Sigma}_N$ are strictly positive. The diagonal matrix $\mathbf{\Sigma}_F$ is such that there are r_F zero entrees at its diagonal. All these zero entrees are first diagonal elements of the matrices Σ_i, $i \in J_F^r$.

Compute $\mathbf{\Phi}^{-\mathbf{H}}(y_*)$. For this purpose we firstly rearrange rows and columns of the matrix. Suppose that first columns of the matrices $\mathcal{A}_i^H$, $i \in J_F^r$, are removed from $\mathcal{A}_i^H$, and the separate sub-matrix $\tilde{\mathcal{A}}_F^H$ is composed from these first columns. The dimension of $\tilde{\mathcal{A}}_F^H$ is $m \times r_F$. Denote by $\hat{\mathcal{A}}_F^H$ the sub-matrix of the matrix $\mathcal{A}_F^H$ composed from the rest columns of $\mathcal{A}_F^H$. Add the sub-matrix $\tilde{\mathcal{A}}_F^H$ to the matrix $\mathcal{A}_I$, putting it before $\mathcal{A}_I$. The resulting $m \times (r_F + n_I)$ matrix denote by $\mathcal{A}_{FI}^{\mathbf{H}}$. Moreover, denote by $\hat{\mathbf{\Sigma}}_F$ the diagonal sub-matrix of the matrix $\mathbf{\Sigma}_F$, from which first diagonal entrees of the matrices Σ_i, $i \in J_F^r$, are eliminated. Let $\mathbf{\Pi}$ be a permutation matrix, realizing the mentioned changes of rows and columns of $\mathbf{\Phi}^{\mathrm{H}}(y_*)$. Then the matrix $\mathbf{\Phi}^{\mathrm{H}}(y_*)$ can be written in the form

$$\mathbf{\Phi}^{\mathrm{H}}(y_*) = \mathbf{\Pi} \begin{bmatrix} \left(\hat{A}_F^{\mathrm{H}}\right)^T \hat{A}_F^{\mathrm{H}} + \hat{\mathbf{\Sigma}}_F & \left(\hat{A}_F^{\mathrm{H}}\right)^T A_{FI}^{\mathrm{H}} & \left(\hat{A}_F^{\mathrm{H}}\right)^T A_N^{\mathrm{H}} \\ (A_{FI}^{\mathrm{H}})^T \hat{A}_F^{\mathrm{H}} & (A_{FI}^{\mathrm{H}})^T A_{FI}^{\mathrm{H}} & (A_{FI}^{\mathrm{H}})^T A_N^{\mathrm{H}} \\ (A_N^{\mathrm{H}})^T \hat{A}_F^{\mathrm{H}} & (A_N^{\mathrm{H}})^T A_{FI}^{\mathrm{H}} & (A_N^{\mathrm{H}})^T A_N^{\mathrm{H}} + \mathbf{\Sigma}_N \end{bmatrix} \mathbf{\Pi}^{\mathrm{T}}. \quad (27)$$

Partition the matrix (27) onto four blocks:

$$\mathbf{\Phi}^{\mathrm{H}}(y_*) = \mathbf{\Pi} \begin{bmatrix} \mathcal{W}_{11} & \mathcal{W}_{12} \\ \mathcal{W}_{12}^T & \mathcal{W}_{22} \end{bmatrix} \mathbf{\Pi}^{\mathrm{T}},$$

where

$$\mathcal{W}_{11} = \begin{bmatrix} \left(\hat{A}_F^{\mathrm{H}}\right)^T \hat{A}_F^{\mathrm{H}} + \hat{\mathbf{\Sigma}}_F & \left(\hat{A}_F^{\mathrm{H}}\right)^T A_{FI}^{\mathrm{H}} \\ (A_{FI}^{\mathrm{H}})^T \hat{A}_F^{\mathrm{H}} & (A_{FI}^{\mathrm{H}})^T A_{FI}^{\mathrm{H}} \end{bmatrix}, \quad \mathcal{W}_{12} = \begin{bmatrix} \left(\hat{A}_F^{\mathrm{H}}\right)^T A_N^{\mathrm{H}} \\ (A_{FI}^{\mathrm{H}})^T A_N^{\mathrm{H}} \end{bmatrix}$$

and $\mathcal{W}_{22} = \mathbf{\Sigma}_N + \left(A_N^{\mathrm{H}}\right)^T A_N^{\mathrm{H}}$.

If the non-degeneracy condition holds at the point $[u_*, y_*]$, then according to Proposition 3 the matrix $\mathbf{\Phi}^{\mathrm{H}}(y_*)$ is positive definite. Therefore, the diagonal blocks $\mathcal{W}_{11}$ and $\mathcal{W}_{22}$ are also positive definite matrices.

Using the Frobenius formula, we obtain

$$\mathbf{\Phi}^{-\mathrm{H}}(y_*) = \mathbf{\Pi} \begin{bmatrix} \mathcal{V}_{11} & \mathcal{V}_{12} \\ \mathcal{V}_{12}^T & \mathcal{V}_{22} \end{bmatrix} \mathbf{\Pi}^{\mathrm{T}},$$

where

$$\mathcal{V}_{11} = \mathcal{W}_{11}^{-1} + \mathcal{W}_{11}^{-1} \mathcal{W}_{12} \mathcal{Z}^{-1} \mathcal{W}_{12}^T \mathcal{W}_{11}^{-1}, \quad \mathcal{V}_{12} = -\mathcal{W}_{11}^{-1} \mathcal{W}_{12} \mathcal{Z}^{-1}, \quad \mathcal{V}_{22} = \mathcal{Z}^{-1} \quad (28)$$

and $\mathcal{Z} = \mathcal{W}_{22} - \mathcal{W}_{12}^T \mathcal{W}_{11}^{-1} \mathcal{W}_{12}$.

Firstly, compute the matrix $\mathcal{W}_{11}^{-1}$. According to Proposition 2 the matrix $\left(A_{FI}^{\mathrm{H}}\right)^T A_{FI}^{\mathrm{H}}$ at the non-degenerate point $[u_*, y_*]$ is nonsingular. Denote

$$\mathcal{Y} = \left(\hat{A}_F^{\mathrm{H}}\right)^T \hat{A}_F^{\mathrm{H}} + \hat{\mathbf{\Sigma}}_F - \left(\hat{A}_F^{\mathrm{H}}\right)^T A_{FI}^{\mathrm{H}} \left[\left(A_{FI}^{\mathrm{H}}\right)^T A_{FI}^{\mathrm{H}}\right]^{-1} \left(A_{FI}^{\mathrm{H}}\right)^T \hat{A}_F^{\mathrm{H}}. \quad (29)$$

Denote also $\mathcal{P} = A_{FI}^{\mathrm{H}} \left[\left(A_{FI}^{\mathrm{H}}\right)^T A_{FI}^{\mathrm{H}}\right]^{-1} \left(A_{FI}^{\mathrm{H}}\right)^T$. The matrix $\mathcal{P}$ is an orthogonal projector onto the linear sub-space $\mathcal{L}$, generated by columns of the matrix A_{FI}^{H}. The matrix $\mathcal{P}_\perp = I - \mathcal{P}$ projects onto the orthogonal complement $\mathcal{L}^\perp$ to the sub-space $\mathcal{L}$. By (29)

$$\mathcal{Y} = \hat{\mathbf{\Sigma}}_F + \left(\hat{A}_F^{\mathrm{H}}\right)^T \mathcal{P}_\perp \hat{A}_F^{\mathrm{H}}. \quad (30)$$

Let $\mathcal{E} = \left[(A_{FI}^{\mathrm{H}})^T A_{FI}^{\mathrm{H}}\right]^{-1}$, $\mathcal{S} = \hat{A}_F^{\mathrm{H}} \mathcal{Y}^{-1} \left(\hat{A}_F^{\mathrm{H}}\right)^T$. With the help of the Frobenius formula, we obtain

$$\mathcal{W}_{11}^{-1} = \begin{bmatrix} \mathcal{Y}^{-1} & -\mathcal{Y}^{-1} \left(\hat{A}_F^{\mathrm{H}}\right)^T A_{FI}^{\mathrm{H}} \mathcal{E} \\ -\mathcal{E}(A_{FI}^{\mathrm{H}})^T \hat{A}_F^{\mathrm{H}} \mathcal{Y}^{-1} & \mathcal{E} + \mathcal{E}(A_{FI}^{\mathrm{H}})^T \mathcal{S} A_{FI}^{\mathrm{H}} \mathcal{E} \end{bmatrix}, \quad (31)$$

The matrix $\mathcal{P}_\perp$ is idempotent, that is $\mathcal{P}_\perp = \mathcal{P}_\perp^2$. Using the Sherman-Morrison-Woodbury formula, we derive from (30)

$$\mathcal{Y}^{-1} = \hat{\Sigma}_F^{-1} - \hat{\Sigma}_F^{-1} \left(\hat{\mathcal{A}}_F^{\mathbf{H}}\right)^T \mathcal{P}_\perp \left[I_m + \mathcal{P}_\perp \hat{\mathcal{A}}_F^{\mathbf{H}} \hat{\Sigma}_F^{-1} \left(\hat{\mathcal{A}}_F^{\mathbf{H}}\right)^T \mathcal{P}_\perp\right]^{-1} \mathcal{P}_\perp \hat{\mathcal{A}}_F^{\mathbf{H}} \hat{\Sigma}_F^{-1}. \quad (32)$$

Introduce the additional notation $\hat{\mathcal{A}}_{FI}^{\mathbf{H}} = \left[\hat{\mathcal{A}}_F^{\mathbf{H}}, \mathcal{A}_{FI}^{\mathbf{H}}\right]$. Then the matrix $\mathcal{Z}$ can be written in the form

$$\mathcal{Z} = \Sigma_N + \left(\mathcal{A}_N^{\mathbf{H}}\right)^T \left[I_m - \hat{\mathcal{A}}_{FI}^{\mathcal{H}} \mathcal{W}_{11}^{-1} \left(\hat{\mathcal{A}}_{FI}^{\mathcal{H}}\right)^T\right] \mathcal{A}_N^{\mathbf{H}}. \quad (33)$$

It can be seen from (33) that the matrix $\mathcal{Z}$ is a Schur complement of the positive definite matrix $\mathcal{W}_{11}$ at (27). Therefore, $\mathcal{Z}$ is a positive definite matrix too.

Proposition 4. Let $\hat{S} = \mathcal{P} + \mathcal{P}_\perp S \mathcal{P}_\perp$. Then $\hat{\mathcal{A}}_{FI}^{\mathcal{H}} \mathcal{W}_{11}^{-1} \left(\hat{\mathcal{A}}_{FI}^{\mathcal{H}}\right)^T = \hat{S}$.

Proof. This equality can be obtained by direct calculations.   □

Corollary 1. *According to* (33) $\mathcal{Z} = \Sigma_N + \left(\mathcal{A}_N^{\mathbf{H}}\right)^T \left(I_m - \hat{S}\right) \mathcal{A}_N^{\mathbf{H}}$. *Since* Σ_N *is a positive definite diagonal matrix, we obtain by the Sherman–Morrison–Woodbury formula*

$$\mathcal{Z}^{-1} = \Sigma_N^{-1} - \Sigma_N^{-1} \left(\mathcal{A}_N^{\mathbf{H}}\right)^T \left(I - \hat{S}\right)^{1/2}$$
$$\cdot \left[I + \left(I - \hat{S}\right)^{1/2} \mathcal{A}_N^{\mathbf{H}} \Sigma_N^{-1} \left(\mathcal{A}_N^{\mathbf{H}}\right)^T \left(I - \hat{S}\right)^{1/2}\right]^{-1} \left(I - \hat{S}\right)^{1/2} \mathcal{A}_N^{\mathbf{H}} \Sigma_N^{-1} \quad (34)$$

Using the matrices (31) *and* (34), *it is possible by* (28) *compute the matrix* $\Phi^{-\mathbf{H}}(y_*)$.

Below, we will need in the definition of non-degeneracy of a point $x \in \mathcal{F}_P$ in the primal problem (2).

Definition 2. [1]. *The point* $x \in \mathcal{F}_P$ *is called non-degenerate, if* $\mathcal{T}_{\mathcal{K}}(x) + \mathcal{N}(\mathcal{A}) = \mathbb{R}^n$, *where* $\mathcal{T}_{\mathcal{K}}(x)$ *is a tangent space to the cone* $\mathcal{K}$ *at the point* x, *and* $\mathcal{N}(\mathcal{A})$ *is a null-space of the matrix* $\mathcal{A}$.

Denote by H_i^L, $i \in J_F^r$, the left $n_i \times (n_i - 1)$ sub-matrix of the matrix H_i. In other words, H_i^L is the matrix H_i, from which the last column $\mathbf{e}_{i,n_i}$ is removed. Denote also by $A_i^{H_L} = A_i H_i^L$. Compose from $A_i^{H_L}$, $i \in J_F^r$, the matrix $\mathcal{A}_F^{\mathbf{H}_L} = \left[A_1^{H_L}, \dots, A_{r_F}^{H_L}\right]$ with the dimension $m \times (n_F - r_F)$, where $n_F = \sum_{i \in J_F^r} n_i$. Introduce additionally the matrix $\mathcal{A}_{FI}^{\mathbf{H}_L} = \left[\mathcal{A}_F^{\mathbf{H}_L}, \mathcal{A}_I^{\mathbf{H}}\right]$. The following criterion of non-degeneracy of the point $x \in \mathcal{F}_P$ is valid.

Proposition 5. [1]. *The point* $x = [x_F; x_I; x_N]$ *is non-degenerate if and only if rows of the matrix* $\mathcal{A}_{FI}^{\mathbf{H}_L}$ *are linear independent.*

Lemma 2. *Let* $x_* \in \mathcal{F}_P$ *and* $[u_*, y_*] \in \mathcal{F}_D$ *be non-degenerate solutions of problems* (2) *and* (3), *respectively. Let also the solutions* x_* *and* y_* *be strictly complementary. Then the matrix* $\mathbf{G}(u_*) = \mathcal{A} \Phi^{-1}(u_*) \mathbf{Arr}(x(u_*)) \mathcal{A}^T$ *is nonsingular.*

Proof. Since $\mathcal{A}x_* = b$ and $\mathbf{Arr}\,(y_*)x_* = 0_n$, we have $\left[\mathbf{Arr}\,(y_*) + \mathcal{A}^T\mathcal{A}\right]x_* = \mathcal{A}^T b$. Hence, $x(u_*) = x_*$. Moreover, the expression (26) for $\mathbf{G}\,(u_*)$ takes place.

Let show, that the homogeneous system of linear equations

$$\mathbf{G}\,(u_*)z = 0_m \tag{35}$$

has only zero solution $z = 0_m$. It follows from here that the matrix $\mathbf{G}\,(u_*)$ is nonsingular.

Denote $\hat{\mathcal{A}}^{\mathbf{H}} = \mathcal{A}^{\mathbf{H}}\mathbf{\Pi}$. By (26) $\mathbf{G}\,(u_*) = \hat{\mathcal{A}}^{\mathbf{H}}\,\hat{\mathbf{\Phi}}^{-\mathbf{H}}(y_*)\hat{\mathbf{\Lambda}}(x_*)\left(\hat{\mathcal{A}}^{\mathbf{H}}\right)^T$, where $\hat{\mathcal{A}}^{\mathbf{H}} = \left[\hat{\mathcal{A}}_{FI}^{\mathbf{H}}, \hat{\mathcal{A}}_N^{\mathbf{H}}\right]$ and $\hat{\mathbf{\Phi}}^{-\mathbf{H}}(y_*) = \mathbf{\Pi}^T\mathbf{\Phi}^{-\mathbf{H}}(y_*)\mathbf{\Pi}$. The block-diagonal matrix $\hat{\mathbf{\Lambda}} = \hat{\mathbf{\Lambda}}(x_*)$ is obtained from the matrix $\mathbf{\Lambda}$ by rearrangement of rows and columns with the help of the permutation matrix $\mathbf{\Pi}$, that is $\hat{\mathbf{\Lambda}} = \mathbf{\Pi}\mathbf{\Lambda}\mathbf{\Pi}^T$. This matrix can be written also in the form $\hat{\mathbf{\Lambda}} = \mathrm{DIAG}\left[\hat{\mathbf{\Lambda}}_{FI}, \mathbf{\Lambda}_N\right]$.

The right lower block $\mathbf{\Lambda}_N$ is a zero matrix, therefore

$$\mathbf{G}\,(u_*) = \hat{\mathcal{A}}_{FI}^{\mathbf{H}}\mathcal{V}_{11}\hat{\mathbf{\Lambda}}_{FI}\left(\hat{\mathcal{A}}_{FI}^{\mathbf{H}}\right)^T + \hat{\mathcal{A}}_N^{\mathbf{H}}\mathcal{V}_{12}^T\hat{\mathbf{\Lambda}}_{FI}\left(\hat{\mathcal{A}}_{FI}^{\mathbf{H}}\right)^T.$$

Substituting $\mathcal{V}_{11}$ and $\mathcal{V}_{12}$, we derive that $\mathbf{G}\,(u_*)$ is the matrix of the following form

$$\mathbf{G}\,(u_*) = \hat{\mathcal{A}}_{FI}^{\mathbf{H}}\mathcal{W}_{11}^{-1}\hat{\mathbf{\Lambda}}_{FI}\left(\hat{\mathcal{A}}_{FI}^{\mathbf{H}}\right)^T$$
$$+ \left[\hat{\mathcal{A}}_{FI}^{\mathbf{H}}\mathcal{W}_{11}^{-1}\left(\hat{\mathcal{A}}_{FI}^{\mathbf{H}}\right)^T - I_m\right]\mathcal{A}_N^{\mathbf{H}}\mathcal{Z}^{-1}\left(\mathcal{A}_N^{\mathbf{H}}\right)^T\hat{\mathcal{A}}_{FI}^{\mathbf{H}}\mathcal{W}_{11}^{-1}\hat{\mathbf{\Lambda}}_{FI}\left(\hat{\mathcal{A}}_{FI}^{\mathbf{H}}\right)^T.$$

Denote $\mathcal{U} = \mathcal{A}_N^{\mathbf{H}}\mathcal{Z}^{-1}\left(\mathcal{A}_N^{\mathbf{H}}\right)^T$. Then, using Proposition 4, we come to conclusion that

$$\mathbf{G}\,(u_*) = \left[I_m - \left(I_m - \hat{\mathcal{S}}\right)\mathcal{U}\right]\hat{\mathcal{A}}_{FI}^{\mathbf{H}}\mathcal{W}_{11}^{-1}\hat{\mathbf{\Lambda}}_{FI}\left(\hat{\mathcal{A}}_{FI}^{\mathbf{H}}\right)^T.$$

We multiply the equality (35) from the left on the matrix $\left(\mathcal{A}_{FI}^{\mathbf{H}}\right)^T$. Since $I_m - \hat{\mathcal{S}} = \mathcal{P}_\perp - \mathcal{P}_\perp\mathcal{S}\mathcal{P}_\perp$, we derive that $\left(\mathcal{A}_{FI}^{\mathbf{H}}\right)^T\left(I_m - \hat{\mathcal{S}}\right) = 0$. Thus, we have

$$\left(\mathcal{A}_{FI}^{\mathbf{H}}\right)^T\hat{\mathcal{A}}_{FI}^{\mathbf{H}}\mathcal{W}_{11}^{-1}\hat{\mathbf{\Lambda}}_{FI}\left(\hat{\mathcal{A}}_{FI}^{\mathbf{H}}\right)^T z = 0_{l_1}, \tag{36}$$

where $l_1 = r_F + n_I$.

Assume that $z \neq 0_m$ and consider separately two possibilities.

1) $\left(\mathcal{A}_{FI}^{\mathbf{H}}\right)^T z \neq 0_{l_1}$. In this case, taking into account the expression (31) for the matrix $\mathcal{W}_{11}^{-1}$, we get

$$\left(\mathcal{A}_{FI}^{\mathcal{H}}\right)^T\hat{\mathcal{A}}_{FI}^{\mathbf{H}}\mathcal{W}_{11}^{-1} = \left(\mathcal{A}_{FI}^{\mathcal{H}}\right)^T\left[\mathcal{P}_\perp\hat{\mathcal{A}}_F^{\mathbf{H}}\mathcal{Y}^{-1}, (I_m - \mathcal{P}_\perp\mathcal{S})\mathcal{A}_{FI}^{\mathbf{H}}\mathcal{E}\right] = [0_{l_1 l_2}, I_{l_1}],$$

where $l_2 = n_F - r_F$. Hence, the equation (36) is reduced to the following one: $\Lambda_{FI}\left(\hat{\mathcal{A}}_{FI}^{\mathbf{H}}\right)^T z = 0_{l_1}$, where Λ_{FI} is a right lower diagonal $l_1 \times l_1$ sub-matrix of the matrix $\hat{\mathbf{\Lambda}}_{FI}$. Because all diagonal entrees of the matrix Λ_{FI} are positive numbers, this equality does not fulfilled, when $z \neq 0_m$.

2) $\left(\mathcal{A}_{FI}^{\mathbf{H}}\right)^T z = 0_{l_1}$. Under this assumption $z \in \mathcal{L}^\perp$, therefore,

$$\hat{\mathcal{A}}_{FI}^{\mathbf{H}}\mathcal{W}_{11}^{-1}\hat{\mathbf{\Lambda}}_{FI}\left(\hat{\mathcal{A}}_{FI}^{\mathbf{H}}\right)^T z = \mathcal{P}_\perp\hat{\mathcal{A}}_F^{\mathbf{H}}\mathcal{Y}^{-1}\hat{\mathbf{\Lambda}}_F\left(\hat{\mathcal{A}}_F^{\mathbf{H}}\right)^T\mathcal{P}_\perp z. \tag{37}$$

By (32) $\mathcal{P}_\perp \hat{\mathcal{A}}_F^H \mathcal{Y}^{-1} = \left(I_m + \hat{\mathcal{C}}\right)^{-1} \mathcal{P}_\perp \hat{\mathcal{A}}_F^H \hat{\Sigma}_F^{-1}$, where $\hat{\mathcal{C}} = \mathcal{P}_\perp \hat{\mathcal{A}}_F^H \hat{\Sigma}_F^{-1} (\hat{\mathcal{A}}_F^H)^T \mathcal{P}_\perp$. It follows from here and (37) that

$$\hat{\mathcal{A}}_{FI}^H \mathcal{W}_{11}^{-1} \hat{\Lambda}_{FI} \left(\hat{\mathcal{A}}_{FI}^H\right)^T z = \left(I_m + \hat{\mathcal{C}}\right)^{-1} \mathcal{P}_\perp \hat{\mathcal{A}}_F^H \hat{\Sigma}_F^{-1} \hat{\Lambda}_F \left(\hat{\mathcal{A}}_F^H\right)^T \mathcal{P}_\perp z = 0.$$

But the matrix $\left(I_m + \hat{\mathcal{C}}\right)^{-1}$ is positive definite. Therefore, this equality is possible only when $\mathcal{P}_\perp \hat{\mathcal{A}}_F^H \hat{\Sigma}_F^{-1} \hat{\Lambda}_F \left(\hat{\mathcal{A}}_F^H\right)^T \mathcal{P}_\perp z = 0$.

All diagonal entrees of the diagonal matrix $\Sigma_F^{-1} \hat{\Lambda}_F$ are positive except of r_F diagonal entrees equal to zero. All these zero entrees correspond to the last diagonal entrees of matrices Λ_i in the decomposition $\mathrm{Arr}(x_{*,i}) = H_i \Lambda_i H_i^T$, $i \in J_F^r$. Denote by $\mathcal{A}_F^{H_L}$ the sub-matrix of the matrix $\mathcal{A}_F^H$, from which the last columns of the matrices A_i^H are removed.

Let $p = \hat{\Sigma}_F^{-1} \hat{\Lambda}_F \left(\hat{\mathcal{A}}_F^H\right)^T \mathcal{P}_\perp z$. The vector p is non-zero. Really, otherwise, because of $\left(\hat{\mathcal{A}}_{FI}^H\right)^T z = 0$, the rows of the matrix $\left[\mathcal{A}^{H_L}, \mathcal{A}_{FI}^H\right]$ are linear dependent. This contradicts to non-degeneracy of the point x_*.

By the same reason the equality $\mathcal{P}_\perp \hat{\mathcal{A}}_F^H p = 0$ is also impossible, since in the opposite case we have contradiction with Proposition 5. $\square$

Lemma 3. *Let assumptions of Lemma 2 hold. Then the matrix $\Psi_w(w_*)$ is nonsingular, where $w_* = [u_*; y_*]$ is the solution of problem (3).*

Proof. Multiplying the right column of the matrix $\Psi_w(w_*)$ from the right by $\mathcal{A}^T$ and subtracting this column from the left column, we obtain the matrix

$$\begin{bmatrix} \mathcal{A}\Phi^{-1}\mathrm{Arr}(x_*)\mathcal{A}^T & \mathcal{A}\Phi^{-1}\left[\tau I_n - \mathrm{Arr}(x(w))\right] \\ \left[\mathrm{Arr}(y_*)\Phi^{-1} - I_n\right]\mathrm{Arr}(x_*)\mathcal{A}^T & \left[I_n - \mathrm{Arr}(y_*)\Phi^{-1}\right]\mathrm{Arr}(x_*) + \tau\mathrm{Arr}(y)\Phi^{-1} \end{bmatrix},$$
(38)

where $x_* = x(w_*)$ is the solution of the primal problem (2), and $\Phi^{-1} = \Phi^{-1}(y_*)$.

Further, we multiply the first row of the matrix (38) from the left by the matrix $\mathcal{A}^T$ and sum it with the second row. Since $\left[\mathcal{A}^T\mathcal{A} + \mathrm{Arr}(y_*)\right]\Phi^{-1}(y_*) = I_n$, we amount to the matrix

$$\begin{bmatrix} \mathcal{A}\Phi^{-1}(y_*)\mathrm{Arr}(x_*)\mathcal{A}^T & \mathcal{A}\Phi^{-1}(y)\left[\tau I_n - \mathrm{Arr}(x(w))\right] \\ 0_{nm} & \tau I_n \end{bmatrix}.$$
(39)

By Lemma 2 the left upper sub-matrix $\mathcal{A}\Phi^{-1}(y_*)\mathrm{Arr}(x_*)\mathcal{A}^T$ is non-singular. Under $\tau > 0$ the right lower sub-matrix of the matrix (39) is also non-singular. Therefore, the matrix (38) is non-singular too. $\square$

Remark 1. If the point $[u_*, y_*] \in \mathcal{F}_D$ is non-degenerate, then due to continuity the points $[u, y]$ in some vicinity of $[u_*, y_*]$ are also non-degenerate. Thus, the algorithmic mappings in methods (14) and (21) are completely defined in some domain containing points u_* and w_*, respectively.

Theorem 1. *Let all conditions of Lemma 2 be valid. Then the iterative methods (14) and (21) converge locally to the solutions u_* and w_* with super-linear rate.*

Proof. The proof follows from well-known results concerning the Newton method and from Lemmas 2, 3. $\square$

6   Conclusion

We have proposed two variants of the dual Mewton's method for solving linear second order cone programming problems. Both variants of the method converge locally with the super-linear rate. From theoretical point of view dual methods are preferable in compare with primal methods, when the number of equalities in the primal problem is not large.

References

1. Alizadeh, F., Goldfarb, D.: Second-order cone programming. Math. Program. Ser. B. **95**, 3–51 (2003)
2. Lobo, M.S., Vandenberghe, L., Boyd, S., Lebret, H.: Applications of second order cone programming. Linear Algebra Appl. **284**, 193–228 (1998)
3. Anjos, M.F., Lasserre, J.B. (eds.): Handbook of Semidefinite, Cone and Polynomial Optimization: Theory, Algorithms, Software and Applications, p. 915. Springer, New York (2011). https://doi.org/10.1007/978-1-4614-0769-0
4. Nesterov, Y.E., Todd, M.J.: Primal-dual interior-point methods for self-scaled cones. SIAM. J. Optim. **8**, 324–364 (1998)
5. Monteiro, R.D.C., Tsuchiya, T.: Polynomial convergence of primal-dual algorithms for second-order cone program based on the MZ-family of directions. Math. Program. **88**(1), 61–83 (2000)
6. Zhadan, V.: Dual multiplicative-barrier methods for linear second-order cone programming. In: Jaćimović, M., Khachay, M., Malkova, V., Posypkin, M. (eds.) OPTIMA 2019. CCIS, vol. 1145, pp. 295–310. Springer, Cham (2020). https://doi.org/10.1007/978-3-030-38603-0_22
7. Evtushenko, Y.G., Zhadan, V.G.: Dual barrier-projection and barrier-newton methods for linear programming. Comp. Maths. Math. Phys. **36**(7), 847–859 (1996)
8. Zhadan, V.G.: Primal Newton method for the linear cone programming problem. Comput. Mathe. Mathe. Physics. **58**(2), 207–214 (2018)

Discrete Optimization

On Symmetry Groups of Some Quadratic Programming Problems

Anton V. Eremeev[1]([✉])[iD] and Alexander S. Yurkov[2][iD]

[1] Sobolev Institute of Mathematics, Omsk, Russia
eremeev@ofim.oscsbras.ru
[2] Institute of Radiophysics and Physical Electronics Omsk Scientific Center SB RAS,
Omsk, Russia
fitec@mail.ru

Abstract. Solution and analysis of mathematical programming problems may be simplified when these problems are symmetric under appropriate linear transformations. In particular, a knowledge of the symmetries may help reduce the problem dimension, cut the search space by linear cuts or obtain new local optima from the ones previously found. While the previous studies of symmetries in the mathematical programming usually dealt with permutations of coordinates of the solutions space, the present paper considers a larger group of invertible linear transformations. We study a special case of the quadratic programming problem, where the objective function and constraints are given by quadratic forms, and the sum of all matrices of quadratic forms, involved in the constraints, is a positive definite matrix. In this setting, it is sufficient to consider only orthogonal transformations of the solution space. In this group of orthogonal transformations, we describe the structure of the subgroup which gives the symmetries of the problem. Besides that, a method for finding such symmetries is outlined, and illustrated in two simple examples.

Keywords: Non-convex programming · Orthogonal transformation · Symmetry group · Lie group

1 Introduction

Solution and analysis of mathematical programming problems may be simplified when these problems are symmetric under appropriate linear transformations. In particular, a knowledge of the symmetries may help reduce the problem dimension, cut the search space by symmetry-breaking linear cuts or obtain new local optima from the ones previously found. These methods are applicable in the case of a continuous solutions domain [3,6,8] as well as in the integer programming [1,2,7,11,16] and in the mixed integer programming [10,12]. While most of the applications of symmetries are aimed at speeding up the exact optimization algorithms, yet in some cases the knowledge of symmetries may also be useful in designing evolutionary algorithms [13] and other heuristics.

© Springer Nature Switzerland AG 2020
A. Kononov et al. (Eds.): MOTOR 2020, LNCS 12095, pp. 35–48, 2020.
https://doi.org/10.1007/978-3-030-49988-4_3

In the present paper, we study the case of continuous solutions domain. While the previous studies of symmetries in mathematical programming usually dealt with permutations of coordinates of the solutions space [7,8,10], the present paper considers a larger group of invertible linear transformations. We study the special case of quadratically-constrained quadratic programming problem in $\mathbb{R}^N$, where the objective function and the constraints are given by quadratic forms, A, and $B_1, \ldots, B_M$ respectively:

$$\begin{cases} x^T A x \to \max, \\ x^T B_1 x \leq 1, \\ \ldots \\ x^T B_M x \leq 1, \end{cases} \tag{1}$$

where x is an N-component column vector of variables, and the superscript T denotes matrix transposition. In what follows, without loss of generality we assume that $N \times N$ matrices $A, B_i, i = 1, \ldots, M$ are symmetric (note that any matrix can be decomposed into a sum of symmetric matrix S and skew-symmetric matrix C, and the quadratic form $x^T C x$ identically equals zero). A more substantial assumption that we will make in this paper is that $B_\Sigma := \sum_{i=1}^{M} B_i$ is a positive definite matrix. An example of application of quadratic programming problems with such a property in radiophysics may be found e.g. in [4].

The results of this paper may also be used for finding symmetries if some of the problem constraints have the inequality $\leq$, some have the inequality $\geq$ and some have the equality sign. We will consider only the inequalities $\leq$ for the notational simplicity. The obtained results may also be applied in semidefinite relaxation methods, see e.g. [15]. Note that in [15] the well-known Maximum Cut problem (which is NP-hard) is reduced to the problem considered here.

By a symmetry of problem (1) we mean a set of linear transformations

$$x \to y = Px, \tag{2}$$

defined by a non-degenerate matrix P such that the problem (1), expressed in terms of the transformed space (i.e., through the vector columns y), coincides with the original problem. That is, in terms of the vectors y our optimization problem again has the form

$$\begin{cases} y^T A y \to \max, \\ y^T B_1 y \leq 1, \\ \ldots \\ y^T B_M y \leq 1, \end{cases} \tag{3}$$

with the same matrix A and the same set of matrices $\{B_i : i = 1, \ldots, M\}$. We emphasize that, in the set of constraints, matrices B_i may be numbered arbitrarily, which, obviously, does not change the problem. The transformations given by the matrices P obviously form a group, which we denote by $\mathcal{G}$. The goal of the paper is to analyse group $\mathcal{G}$ and propose an algorithm for finding it.

In some cases, it may also be of interest to find the symmetry group of the set of constraints only. Obviously, this is not much different from the search for symmetry group $\mathcal{G}$ of the problem; one just needs to exclude matrix A from the consideration (i.e. formally assume that A is a zero matrix). Furthermore, the set of symmetries of the constraints is not larger than the set of all invertible linear transformations, bijectively mapping the feasibility domain of the problem $\mathcal{D} := \{x \in \mathbb{R}^N : x^T B_i x \leq 1, i = 1, \ldots, M\}$ onto itself. Therefore, the symmetry group of the set of constraints is a subgroup in the symmetry group of invertible linear transformations of $\mathcal{D}$.

The structure of the paper is the following. In Sect. 2, it is shown that the group of linear symmetries of the problem is a subgroup of orthogonal transformations. Also, the structure of the group of symmetries and the corresponding Lie algebra are discussed. In Sect. 3, a general algorithm for finding the symmetries is proposed, and in Sect. 4 it is illustrated in two simple examples. A discussion of obtained results and conclusions are provided in Sects. 5 and 6.

2 Structure of the Symmetry Group

Invariance of the problem under transformation P implies that

$$P^T A P = A, \qquad P^T B_i P = \sum_{j=1}^{M} L_{ij} B_j, \ j = 1, \ldots, M, \tag{4}$$

where L_{ij} are the elements of a permutation matrix, i.e. matrix $L = (L_{ij})$ has a single "1" in each column and in each row, other elements of L are zeros.

If (4) holds, then the invariance condition of the matrix B_Σ is satisfied:

$$P^T B_\Sigma P = B_\Sigma. \tag{5}$$

Naturally, the converse is not true in the general case, but at least we can say that the desired group $\mathcal{G}$ is a subgroup of the invariance group of B_Σ. This matrix may be represented as a congruent transformation of a diagonal matrix:

$$B_\Sigma = S^T D S, \tag{6}$$

where D is a diagonal matrix, which can have only "0", "1", or "−1" on its main diagonal. Essentially, we are talking about reducing the quadratic form corresponding to matrix B_Σ to its canonical form. So matrix S can be constructively obtained, for example, by the finite Lagrange method ([9], Ch. 5).

Now, if we restrict ourselves to the special case where matrix B_Σ is positive definite (it occurs, for example, in the radiophysical problem of optimizing the excitation of antenna arrays [4]), then D will be the unit matrix and it may be omitted in (6). Condition (5) then turns into

$$P^T S^T S P = S^T S \tag{7}$$

or

$$(SPS^{-1})^T (SPS^{-1}) = E, \tag{8}$$

where E is a unit matrix. This means that matrix

$$Q := SPS^{-1} \tag{9}$$

is in the group of orthogonal transformations $O(N)$ (see e.g. [17]). So we proved

Proposition 1. *If B_Σ is positive definite then group $\mathcal{G}$ is isomorphic to some subgroup of $O(N)$ and this isomorphism is given by Eq. (9).*

Since $P = S^{-1}QS$ by (9), so application of (4) gives

$$(S^{-1}QS)^T A (S^{-1}QS) = A,$$
$$(S^{-1}QS)^T B_i (S^{-1}QS) = \sum_{j=1}^{N} L_{ij} B_j, \ i = 1, \dots, M, \tag{10}$$

and after a simple transformation we have

$$Q^T \tilde{A} Q = \tilde{A}, \quad Q^T \tilde{B}_i Q = \sum_{i=1}^{N} L_{ij} \tilde{B}_j, \ i = 1, \dots, M, \tag{11}$$

where

$$\tilde{A} = \left(S^{-1}\right)^T A S^{-1}, \quad \tilde{B}_i = \left(S^{-1}\right)^T B_i S^{-1}, \ i = 1, \dots, M. \tag{12}$$

So using isomorphism (9) we can substitute Eqs. (4) by the similar Eqs. (11), but with the matrix substitution

$$A \to \tilde{A}, \quad B_i \to \tilde{B}_i, \ i = 1, \dots, M. \tag{13}$$

and substituting P by the orthogonal matrix Q. These equations are significantly simpler, since in this case condition (11) may be formulated linearly in Q:

$$\tilde{A} Q = Q \tilde{A}, \quad \tilde{B}_i Q = Q \sum_{j=1}^{M} L_{ij} \tilde{B}_j, \ i = 1, \dots, M. \tag{14}$$

If one finds all suitable orthogonal mappings Q, then it will be easy to restore the corresponding matrices P. Assuming all this, we omit the tildes above matrices A and B_i further in order to simplify the notation.

It is well-known that the orthogonal group $O(N)$ consists of two connected components, for one of them the determinant of the matrix equals 1, for the other it equals -1 (see e.g. [17]). The first component is a subgroup of $O(N)$, denoted by $SO(N)$ and also called the *rotation group*, due to the fact that in dimensions 2 and 3, its elements are the usual rotations around a point or a line, respectively. The second component does not constitute a subgroup of $O(N)$, since it does not contain the identity element. Matrices from the second component can be represented, for example, in the following form: $\mathrm{diag}\{-1, 1 \dots 1\}Q$, where $Q \in SO(N)$, so between these components there is a one-to-one correspondence (which is not an isomorphism in the group-theoretical sense, since it does not preserve the group operations). The required matrices Q can belong to both the first component and the second.

The standard facts of topological groups theory (see e.g. [17], Ch. 1) imply the following properties of symmetry group $\mathcal{G}$, endowed with the standard topology of $\mathbb{R}^{N^2}$, applicable to the space of $(N \times N)$-matrices. As any topological group, $\mathcal{G}$ consists of connected components (in the topological sense), only one of which, hereafter denoted as $\mathcal{G}_1$, contains the identity element. This $\mathcal{G}_1$ is invariant subgroup of $\mathcal{G}$, see Theorem 1 [17], and called *the continuous subgroup of symmetries* in what follows. The remaining connected components (not being subgroups) can be considered as products of the elements of the group outside $\mathcal{G}_1$ and the elements of $\mathcal{G}_1$ i.e. the cosets of $\mathcal{G}_1$. These cosets make up a discrete group. Given that $\mathcal{G}_1$ is an invariant subgroup, multiplication of the cosets of this discrete group is determined naturally, and the discrete group is a factor group $\mathcal{G}/\mathcal{G}_1$. These cosets can be identified by indicating one (any) representative of a coset.

Naturally, degenerate cases are possible. First, when $\mathcal{G}_1$ degenerates into the identity element, the entire symmetry group $\mathcal{G}$ is a purely discrete group (each coset consists of one element). Secondly, there may be no other elements of discrete symmetry but only the continuous subgroup of symmetries $\mathcal{G}_1$. And finally, the entire symmetry group $\mathcal{G}$ may consist of only the identity element.

3   Finding the Symmetry Group

Due to the observations from Sect. 2, the search for all appropriate symmetry transformations Q may be divided into two parts: the search in the first component of $O(N)$ (i.e., in subgroup $SO(N)$) and the search in the second component where the determinant of orthogonal matrices equals -1. Initially we restrict ourselves to the first subset. A generalization to the whole group $O(N)$ will be done by analogous consideration of the second subset while searching for discrete symmetries. The only difference will be that in the second case, it will be necessary to replace $Q \to \operatorname{diag}\{-1, 1, \ldots, 1\}Q$.

3.1   Continuous Subgroup of Symmetry

First, we consider the continuous subgroup of symmetry $\mathcal{G}_1$. Nontrivial permutations of matrices B_i can not result from transformations which belong to $\mathcal{G}_1$, since it is impossible to continuously move from the identical transformation (which implies that matrices B_i are not permuted) to any transformation Q yielding a non-trivial permutation of matrices B_i. Note that any such Q has a neighborhood of transformations which do not yield the trivial permutation of the matrices B_i. So the invariance conditions must hold:

$$Q^T A Q = A, \qquad Q^T B_i Q = B_i, \ i = 1, \ldots, M. \tag{15}$$

For orthogonal transformations Q, this is equivalent to commutativity:

$$AQ = QA, \qquad B_i Q = QB_i, \quad i = 1, \ldots, M. \tag{16}$$

Proposition 2 may be considered a "folklore" fact of matrix analysis and therefore provided without a proof here.

Proposition 2. *Any matrix $Q \in SO(N)$ can be represented as a matrix exponential function of a skew-symmetric matrix. The converse is also true: the exponential function of any skew-symmetric matrix is an orthogonal matrix.*

So with some skew-symmetric matrix X we have $Q = e^X$. The set of skew-symmetric matrices X make up the Lie algebra corresponding to this Lie group [17]. (The Lie algebra corresponding to $SO(N)$ is usually denoted by $so(N)$.) Any Lie algebra is also a linear space, any of its elements can be expressed by means of basis elements, called generators. Thus, any element of the Lie algebra can be represented as:

$$X = \sum_n a_n G_n, \tag{17}$$

where a_n are real numbers, G_n are the generators. The space of skew-symmetric matrices has a dimension $N(N-1)/2$, and there will be as many coefficients a_n and as many generators. As generators, one can choose matrices containing one unit element above the main diagonal (the rest are zeros), then the skew-symmetry uniquely determines the remaining matrix elements of these generators. So, any element Q of $SO(N)$ can be represented as:

$$Q = e^{\sum_n a_n G_n}. \tag{18}$$

Since the desired continuous subgroup of symmetry $\mathcal{G}_1$ is a subgroup of $SO(N)$, so representation (18) is also valid for it, but, generally speaking, the parameters a_n are not independent now. Thus, the search for this subgroup essentially reduces to finding the restrictions on parameters a_n.

It is quite obvious that in order for commutativity conditions (16) to be satisfied, it suffices that the following conditions hold true:

$$\begin{cases} B_i \left(\sum_n a_n G_n \right) = \left(\sum_n a_n G_n \right) B_i, \\[2mm] A \left(\sum_n a_n G_n \right) = \left(\sum_n a_n G_n \right) A. \end{cases} \tag{19}$$

It means that if matrix X commutes with all matrices B_i and with matrix A, then X lies in Lie algebra of $\mathcal{G}_1$. Indeed, expanding the exponential function in a power series, we see that if the matrices A and B_i commute with the argument of this function, then they commute with the exponential function itself.

Note that condition (19), generally speaking, is not necessary to fulfill (16). However, the continuous subgroup of symmetry, as a connected Lie group, is completely determined by its Lie algebra, so it is completely determined by the restrictive relations for elements of the Lie algebra[1]. Thus, in search for the continuous subgroup of symmetry, (16) may be replaced with (19).

Equations (19) are a system of linear algebraic equations that determine parameters a_n. This system is homogeneous, so it has a continuum of nonzero solutions. Note that there is always a trivial zero solution to the system of equations (19) corresponding to an identity matrix Q. Some of parameters a_n remain "free" (these will be the parameters of the desired subgroup), and the rest of a_n may be linearly expressed through the "free" ones. The solution to this system of equations (19) can be obtained constructively by the Gauss method.

The condition of problem invariance under transformation Q turnes into

$$Q = e^{\sum_n a_n \hat{G}_n}, \tag{20}$$

where the sum goes over the "free" parameters a_n, and the new generators denoted by $\hat{G}_n$ are linear combinations of the former generators G_n. The set of all Q matrices satisfying (20) is parameterized by a finite set of real parameters a_n. Note that this set of matrices is not necessarily isomorphic to a Euclidean space, since more than one set of parameters a_n can correspond to the same Q.

Let us show that the set of matrices defined by formula (20) is a group. To this end, it is sufficient to prove that this matrix set $\hat{A} = \{\hat{X}, \hat{X} = \sum_n a_n \hat{G}_n\}$ is a Lie algebra. For a matrix algebra to be a Lie algebra, it is necessary and sufficient to be closed relative to the calculation of the commutator, i.e. $\hat{A}$ is Lie algebra if and only if for any $\hat{X}_i, \hat{X}_j \in \hat{A}$ a commutator

$$[\hat{X}_i, \hat{X}_j] = \hat{X}_i \hat{X}_j - \hat{X}_j \hat{X}_i, \tag{21}$$

is also an element of $\hat{A}$. This is easily verified in our case. Indeed, since all $\hat{X}$ lie in $so(N)$, their commutators also lie in $so(N)$. Therefore, for them to lie not only in $so(N)$, but also in $\hat{A}$, that is, for this algebra to be a Lie algebra, it is sufficient that these commutators satisfy the same restrictive conditions that distinguish set $\hat{A}$ from $so(N)$. The restrictive conditions (19) mean that all $\hat{X}$ commute with all matrices B_i and with matrix A. But then all the products of such $\hat{X}$ also commute with all matrices B_i and with matrix A. And then the commutator $[\hat{X}_j, \hat{X}_j]$, which is a product difference, satisfies the same restrictive conditions. Thus, the set of matrices $\hat{A}$ is a Lie algebra, and therefore the set of matrices defined by formula (20) is a Lie group.

Now let us prove that the set of matrices defined by formula (20) is *the whole* continuous subgroup of symmetries $\mathcal{G}_1$. We will show that a converse leads to a contradiction. Indeed, the converce assumption implies that in the algebra of the continuous group of symmetry there is at least one more generator G_{extra}

---

[1] For abstract groups, such a unique connection exists only in the case of simply connected groups; otherwise, an abstract exponent cannot be uniquely determined. But in our case of a matrix group, the matrix exponent is uniquely determined.

(with its own coefficient, let it be b), linear independent from generators $\hat{G}_n$. But then there is a one-dimensional subgroup of $\mathcal{G}_1$ produced by the element $Q = e^{bG_{extra}}$. If we substitute this Q into the invariance condition (16), differentiate with respect to b and set $b = 0$, then it turns out that G_{extra} satisfies exactly the same condition, which distinguishes the set of matrices $\hat{A}$ from the entire Lie algebra of group $SO(N)$. So this additional generator lies in the linear hull of the generators $\hat{G}_n$. Which is a contradiction. So we have proved the following

Theorem 1. *The continuous subgroup of symmetries $\mathcal{G}_1$ consists of orthogonal transformations with matrices expressed by the matrix exponential function $e^{\sum_n a_n \hat{G}_n}$, where a_n are any real-valued parameters, and all $\hat{G}_n$ make up a basis of the space of solutions to the system of linear equations (19) in the linear space of the $(N \times N)$ skew-symmetric matrices.*

3.2 Discrete Group of Symmetry

In the case of discrete symmetry, nontrivial permutations of matrices B_i are possible. Therefore, the condition (16) is replaced by the following:

$$AQ = QA, \qquad B_iQ = Q\sum_{j=1}^{N} L_{ij}B_j, \quad i = 1, \ldots, M. \tag{22}$$

There are $M!$ permutation matrices L and they can be enumerated for small problems. Then we can assume that in (22) L_{nm} are known. (Note that if we generalize Problem (1) so that some of the constraints have inequalities $\leq$, some have inequalities $\geq$, and some have equalities, then the permutations in each of these three subgroups should be considered.) Furthermore, iterating over all possible matrices L, one can solve Eqs. (22) with respect to Q. But it must be taken into account that matrix Q lies in $SO(N)$, otherwise Eq. (22) is not valid. To this end, one can represent Q as a matrix exponential function (18) and solve the equation for $N(N-1)/2$ parameters a_n as variables. The same should be done with matrix $Q\mathrm{diag}\{-1,1,\ldots,1\}$. The resulting equations will involve exponential functions, so for their solution in each particular case, it is necessary to develop a special numerical method. Alternatively, one can solve Eqs. (22) for matrix Q as a variable, conditioned that $QQ^T = E$.

4 Illustrative Examples

4.1 Example with Trivial Continuous Subgroup of Symmetries

Let us apply the obtained results to a quadratic programming problem with $N = M = 2$, defined by the following matrices (see Fig. 1)

$$A = \begin{pmatrix} 1.0 & 0.0 \\ 0.0 & 0.8 \end{pmatrix}, \tag{23}$$

$$B_1 = \begin{pmatrix} 0.5 & 2.0 \\ 2.0 & 0.5 \end{pmatrix}, \; B_2 = \begin{pmatrix} 0.5 & -2.0 \\ -2.0 & 0.5 \end{pmatrix}. \tag{24}$$

In this example, B_Σ is the identity matrix, and so $S = E$, therefore transformation (13) is not necessary. The feasibility area corresponding to matrices B_1 and B_2 is shown in Fig. 1. Its rotational symmetry properties (as well as the symmetry properties of the problem which involves matrix A) are obvious from geometric considerations: the symmetry group of the domain $\mathcal{D}$ consists of the identical transformation (the identity matrix), rotations of 90°, 180° and 270° (the latter is also the inverse element to the rotation of 90°). In total, there are four elements of the group.

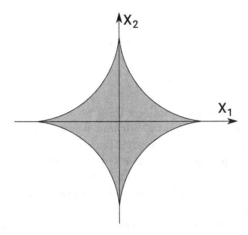

Fig. 1. Feasibility domain defined by matrices B_1 and B_2 in Subsect. 4.1.

For the symmetry group of the problem, 90 and 270° rotations disappear, the two other elements of the group remain. It is also clear that there will be four local optima, two of which are global.

Let us now verify that the results described above give the same result. Firstly, in this two-dimensional case there is only one generator:

$$G = \begin{pmatrix} 0 & 1 \\ -1 & 0 \end{pmatrix}. \tag{25}$$

Accordingly, there is only one coefficient a. The generator G does not commute with any of the matrices written above. Therefore, the system of equations (19) has only one zero solution corresponding to an identity matrix E. The continuous subgroup of symmetry in this example degenerates into a trivial subgroup of one identity element.

To find a discrete symmetry by direct calculations, we note that

$$e^{aG} = \begin{pmatrix} \cos a & \sin a \\ -\sin a & \cos a \end{pmatrix}, \tag{26}$$

$$B_1 e^{aG} = 0.5 \cos a\, E - 2 \sin a\, D + 0.5 \sin a\, G + 2 \cos a\, H,$$

$$e^{aG} B_1 = 0.5 \cos a\, E + 2 \sin a\, D + 0.5 \sin a\, G + 2 \cos a\, H,$$

$$B_2 e^{aG} = 0.5 \cos a\, E + 2 \sin a\, D + 0.5 \sin a\, G - 2 \cos a\, H,$$

$$e^{aG} B_2 = 0.5 \cos a\, E - 2 \sin a\, D + 0.5 \sin a\, G - 2 \cos a\, H, \tag{27}$$

$$A e^{aG} = 0.9 \cos a\, E + 0.1 \cos a\, D + 0.9 \sin a\, G + 0.1 \sin a\, H,$$

$$e^{aG} A = 0.9 \cos a\, E + 0.1 \cos a\, D + 0.9 \sin a\, G - 0.1 \sin a\, H.$$

where

$$D = \begin{pmatrix} 1 & 0 \\ 0 & -1 \end{pmatrix},$$

$$H = \begin{pmatrix} 0 & 1 \\ 1 & 0 \end{pmatrix}. \tag{28}$$

Substituting this all into the equations from Sect. 3, we obtain the following. When considering the symmetry of $\mathcal{D}$ without permutations of matrices B_i, we obtain the equation $\sin a = 0$, and with permutations, the equation $\cos a = 0$. The first one corresponds to the identical transformation and a rotation of $180°$ $(a = 0, \pi)$. The second one corresponds to rotations of 90 and 270° $(a = \pi/2, 3\pi/2)$. Thus, a formal application of the above formulas agrees with the geometric considerations.

If we additionally require the symmetry of the objective function, then in both cases (with the permutation and without it) the second equation $\sin a = 0$ will appear, excluding rotations of 90 and 270°. Finally, to obtain *all* symmetries of the problem, one has to solve Eqs. (22) for the matrix $\mathrm{diag}\{-1,1\}e^{aG}$ and join resulting symmetries with the rotations found before.

4.2   Example with Non-trivial Continuous Subgroup of Symmetries

As a second example, now with a continuous symmetry, we can take a problem with $N = 3, M = 2$ defined by the following matrices

$$A = \mathrm{diag}\{1, 1, 1\},$$

$$B_1 = \mathrm{diag}\{2, 2, 0\}, \tag{29}$$

$$B_2 = \mathrm{diag}\{-1, -1, 1\}.$$

In this example, the objective function is obviously invariant under any transformations from $SO(3)$, so the symmetry of the problem coincides with the symmetry of $\mathcal{D}$. Again, transformation (13) is not necessary here, since B_Σ is the identity matrix.

In this example, we will choose the generators in the following form:

$$G_1 = \begin{pmatrix} 0 & 0 & 0 \\ 0 & 0 & -1 \\ 0 & 1 & 0 \end{pmatrix}, \; G_2 = \begin{pmatrix} 0 & 0 & 1 \\ 0 & 0 & 0 \\ -1 & 0 & 0 \end{pmatrix}, \; G_3 = \begin{pmatrix} 0 & -1 & 0 \\ 1 & 0 & 0 \\ 0 & 0 & 0 \end{pmatrix}. \tag{30}$$

Substituting this into (19) we see that $a_1 = a_2 = 0$, and the arbitrary parameter is a_3. Thus, the continuous symmetry subgroup is described by the following one-parameter matrix family:

$$e^{a_3 G_3} = \begin{pmatrix} \cos a_3 & -\sin a_3 & 0 \\ \sin a_3 & \cos a_3 & 0 \\ 0 & 0 & 1 \end{pmatrix}. \tag{31}$$

To find the discrete symmetry in this particular case, it is more convenient to represent matrix Q not in the exponential form (18) but rather through the Euler parameters α, β and γ, as a product of three exponential functions:

$$Q = e^{\alpha G_3} e^{\beta G_1} e^{\gamma G_3}. \tag{32}$$

Now we substitute (32) into Eq. (14), which may be written as

$$e^{-\gamma G_3} e^{-\beta G_1} e^{-\alpha G_3} B_i e^{\alpha G_3} e^{\beta G_1} e^{\gamma G_3} = \sum_{j=1}^{M} L_{ij} B_j, i = 1, \ldots, M. \tag{33}$$

Note that $\exp(\alpha G_3)$ commutes with both matrices B_1, B_2, and therefore the left factor cancels out. The last factor also cancels out after multiplying the equations on the left and on the right side by the similar exponential functions. So the defining Eq. (14) reduces to

$$e^{-\beta G_1} B_i e^{\beta G_1} = \sum_{j=1}^{M} L_{ij} B_m, \; i = 1, \ldots, M. \tag{34}$$

We have two options for permutations: one trivial and one non-trivial. Accordingly, two options are obtained. The first:

$$\begin{cases} e^{-\beta G_1} B_1 e^{\beta G_1} = B_1, \\ e^{-\beta G_1} B_2 e^{\beta G_1} = B_2. \end{cases} \tag{35}$$

the second:

$$\begin{cases} e^{-\beta G_1} B_1 e^{\beta G_1} = B_2, \\ e^{-\beta G_1} B_2 e^{\beta G_1} = B_1. \end{cases} \tag{36}$$

We note that due to the equality $B_2 = E - B_1$, in both cases the second equation can be reduced to the first one and vice versa. So from two equations it is enough to solve only one. By direct calculations we obtain the following:

$$e^{\beta G_1} = \begin{pmatrix} 1 & 0 & 0 \\ 0 & c & -s \\ 0 & s & c \end{pmatrix}, \tag{37}$$

where for simplicity of notation we denote $s := \sin \beta$, $c = \cos \beta$. Further direct calculations give

$$e^{-\beta G_1} B_1 e^{-\beta G_1} = 2 \begin{pmatrix} 1 & 0 & 0 \\ 0 & c^2 & -cs \\ 0 & -cs & s^2 \end{pmatrix}. \tag{38}$$

In the case of the trivial permutation, this reduces to a system of equations that has two obvious solutions: $c = \pm 1$, $s = 0$. This results in two options for Q:

$$e^{\alpha G_3} \, \mathrm{diag}\{1, 1, 1\} \, e^{\gamma G_3},$$

$$e^{\alpha G_3} \, \mathrm{diag}\{1, -1, -1\} \, e^{\gamma G_3}. \tag{39}$$

Obviously, the first matrix belongs to a continuous subgroup of symmetry, it does not need to be taken into account, since such matrices are already taken into account above. The second matrix, however, does not belong to the continuous subgroup[2]. As a representative of this component, we can take the above expression, written for $\alpha = \gamma = 0$, i.e. just $\mathrm{diag}\{1, -1, -1\}$.

In the second case, where the permutation of matrices B_i is non-trivial, the system of equations obviously has no solutions.

Thus, the subgroup of orthogonal symmetries with determinant 1 in this example consists of two connected components. The first one is described by the matrix family (31), parametrized by one real parameter (rotation angle). The second one is described by the same matrices, but multiplied by $\mathrm{diag}\{1, -1, -1\}$.

To obtain the whole group $\mathcal{G}$, one has to solve the equations from Sect. 3 for the matrix $\mathrm{diag}\{-1, 1, 1\}Q$ and join the resulting symmetries Q to the subgroup of orthogonal symmetries with determinant 1 which we found above.

5   Discussion

As a "brute force" approach to finding all symmetries of the problem, one can formulate a non-linear optimization problem in $\mathbb{R}^{N^2}$:

$$\min \left(||AQ - QA|| + \sum_{i=1}^{M} ||B_i Q - Q C_i(L)|| \; : \; QQ^T = E \right)$$

where Q is a matrix of variables, the matrices $C_i(L)$ are defined by L as $C_i(L) := \sum_{j=1}^{M} L_{ij} B_j$, and $||\cdot||$ denotes any matrix norm. A set of optimal solutions (with zero objective value) gives the set of orthogonal symmetry transformations. The union of $M!$ such sets, taken over all permutation matrices L, makes up the whole group $\mathcal{G}$. In the case of trivial continuous subgroup of symmetry, each of the $M!$ problems has a discrete set of optimal solutions, which, in principle, may be found e.g. by a multi-start of a gradient descent method.

There are other options to find group $\mathcal{G}$ using non-linear programming. For example, one can similarly formulate a minimization problem with respect to

---

[2] This is because Q_{33} is -1, rather than 1 as in the continuous subgroup.

the elements of matrix P. Moreover, there is no need to impose the condition $\det(P) \neq 0$, since it follows from (5) that the square of this determinant is equal to 1. Analysis of the properties and methods of solution of these non-linear optimization problems is beyond the scope of the paper.

In applications of quadratic programming, it is not necessary to find all symmetries of a problem to improve performance of solution algorithms, such as the branch and cut method. If some valid cuts are known already for the problem instance, then each linear symmetry of a problem may be used to double the set of valid cuts. Even if there were no cuts known before, then any symmetry $P \in \mathcal{G}$ which maps a hemi-space $\{x : a^T x \geq 0\}$ into the hemi-space $\{x : a^T x \leq 0\}$ with some $a \in \mathbb{R}^N$ then the constraint $a^T x \geq 0$ may be added to the set of problem constraints as a valid cut.

If $\mathcal{G}$ has a non-trivial continuous subgroup so large that any element of $\mathcal{D}$ may be mapped onto some hyper-plane in $\mathbb{R}^N$ by a corresponding $P \in \mathcal{G}$, then the problem dimension may be decreased by one, see e.g. the problem from Subsection 4.2, where any vector may be rotated by mapping (31) with an appropriate angle a_3 into the subspace $\{x : x_1 = 0\}$. In this respect, it would be appropriate to study the following hypothesis: *Problem* (1) *may be reduced to a problem of the same form in solutions space* $\mathbb{R}^{N-K}$, *where K is the size of the basis mentioned in Theorem 1.*

In local search, the problem symmetries may be used to identify equivalence classes of local optima (consisting of local optima, identical up to a symmetry transformation) since obviously, local optima are mapped to local optima under invertible linear symmetries of the problem. In the multi-start procedure, a smaller number of visited equivalence classes, compared to the number of visited local optima, should tighten estimates of the total local optima number [5,14].

6 Conclusions

The results obtained in this paper further extend the applicability of the approach to improving algorithms performance in the mathematical programming, employing symmetries of the problem. The authors are not aware of other works on problem symmetries, based on the theory of Lie groups and Lie algebras. It is expected that the proposed approach may be extended to other types of problems in the mathematical programming. In particular, it would be interesting to try extending the analysis to the general case of problem (1) without the assumption of positive-definiteness of the sum of matrices of quadratic forms. It is challenging in this case that instead of the group of orthogonal transformations $O(n)$ one would have to consider the more general pseudo-orthogonal group $O(p, q)$. Technical development of the outlined method for finding problem symmetries is also a subject of further research.

Acknowledgments. The authors thank V.M. Gichev for helpful comments on the preliminary version of the manuscript. The work on Sects. 2 and 3 was funded in accordance with the state task of the Omsk Scientific Center SB RAS (project AAAA-A19-119052890058-2). The rest of the work was funded by the program of fundamental scientific research of the SB RAS, I.5.1., project 0314-2019-0019.

References

1. Bödi, R., Herr, K., Joswig, M.: Algorithms for highly symmetric linear and integer programs. Math. Program. **137**, 65–90 (2013)
2. Chervyakov, O.: Affine symmetries of the polyhedron of independence system with uhit shift. Discretnyi Analiz i Issledovanie Operacii **2**(2), 82–96 (1999). (in Russian)
3. Costa, A., Hansen, P., Liberti, L.: On the impact of symmetry-breaking constraints on spatial branch-and-bound for circle packing in a square. Discrete Appl. Math. **161**(1), 96–106 (2013)
4. Eremeev, A.V., Tyunin, N.N., Yurkov, A.S.: Non-convex quadratic programming problems in short wave antenna array optimization. In: Khachay, M., Kochetov, Y., Pardalos, P. (eds.) MOTOR 2019. LNCS, vol. 11548, pp. 34–45. Springer, Cham (2019). https://doi.org/10.1007/978-3-030-22629-9_3
5. Garnier, J., Kallel, L.: How to Detect All Maxima of a Function, pp. 343–370. Springer, Heidelberg (2001). https://doi.org/10.1007/978-3-662-04448-3_17
6. Gatermann, K., Parrilo, P.A.: Symmetry groups, semidefinite programs, and sums of squares. J. Pure Appl. Algebra **192**(1), 95–128 (2004). https://doi.org/10.1016/j.jpaa.2003.12.011
7. Kolokolov, A.A., Orlovskaya, T.G., Rybalka, M.F.: Analysis of integer programming algorithms with l-partition and unimodular transformations. Autom. Remote Control **73**(2), 369–380 (2012)
8. Kouyialis, G., Wang, X., Misener, R.: Symmetry detection for quadratic optimization using binary layered graphs. Processes **7**(11) (2019). https://doi.org/10.3390/pr7110838
9. Lancaster, P., Tismenetsky, M.: The Theory of Matrices. Academic Press, Cambridge (1985)
10. Liberti, L.: Reformulations in mathematical programming: automatic symmetry detection and exploitation. Math. Program. **131** (2010). https://doi.org/10.1007/s10107-010-0351-0
11. Margot, F.: Symmetry in integer linear programming. In: Jünger, M., et al. (eds.) 50 Years of Integer Programming 1958-2008, pp. 647–686. Springer, Heidelberg (2010). https://doi.org/10.1007/978-3-540-68279-0_17
12. Pfetsch, M.E., Rehn, T.: A computational comparison of symmetry handling methods for mixed integer programs. Math. Program. Comput. **11**(1), 37–93 (2018). https://doi.org/10.1007/s12532-018-0140-y
13. Prugel-Bennett, A.: Symmetry breaking in population-based optimization. Trans. Evol. Comp **8**(1), 63–79 (2004). https://doi.org/10.1109/TEVC.2003.819419
14. Reeves, C., Eremeev, A.: Statistical analysis of local search landscapes. J. Oper. Res. Soc. **55**(7), 687–693 (2004)
15. Shor, N.Z.: Semidefinite Programming Bounds for Extremal Graph Problem, pp. 265–298. Springer, Boston (1998). https://doi.org/10.1007/978-1-4757-6015-6_8
16. Simanchev, R.: Linear symmetries of matchings polyhedron and graph automorphisms. Vestnik Omskogo Universiteta **1**, 18–20 (1996). (in Russian)
17. Zhelobenko, D.P.: Compact Lie Groups and their Representations, Translations of mathematical monographs, vol. 40. Providence, AMS (1973)

An Extension of the Das and Mathieu QPTAS to the Case of Polylog Capacity Constrained CVRP in Metric Spaces of a Fixed Doubling Dimension

Michael Khachay[1,2,3]([✉]) [iD], Yuri Ogorodnikov[1,2] [iD], and Daniel Khachay[1] [iD]

[1] Krasovsky Institute of Mathematics and Mechanics, Ekaterinburg, Russia
{mkhachay,yogorodnikov}@imm.uran.ru, daniil.khachay@gmail.com
[2] Ural Federal University, Ekaterinburg, Russia
[3] Omsk State Technical University, Omsk, Russia

Abstract. The Capacitated Vehicle Routing Problem (CVRP) is the well-known combinatorial optimization problem having numerous practically important applications. CVRP is strongly NP-hard (even on the Euclidean plane), hard to approximate in the general case and APX-complete for an arbitrary metric. Meanwhile, for the geometric settings of the problem, there are known a number of quasi-polynomial and even polynomial time approximation schemes. Among these results, the well-known QPTAS proposed by A. Das and C. Mathieu appears to be the most general. In this paper, we propose the first extension of this scheme to a more wide class of metric spaces. Actually, we show that the metric CVRP has a QPTAS any time when the problem is set up in the metric space of any fixed doubling dimension $d > 1$ and the capacity does not exceed polylog (n).

1 Introduction

The Capacitated Vehicle Routing Problem (CVRP) is one of the widely known and actively studied combinatorial problems with numerous important applications in operations research [30]. To the best of our knowledge, the problem was introduced by G. Dantzig and J. Ramser in their seminal paper [10], which provided the first mathematical model of gasoline distribution over the network of gas stations.

Since then, the field of the algorithmic design for the CVRP is developed in a number of research directions. The first direction is based on a reduction of the problem in question to some appropriate mixed-integer program and finding an optimal solution of this program using some of the well-known branch-and-price methods [12]. Recently, a significant success was achieved both in the development of such algorithms and computational hardware [16,27]. Unfortunately, due to the known strongly NP-hardness of the CVRP, instances of this problem that are managed to be solved efficiently within this framework still remain quite modest.

© Springer Nature Switzerland AG 2020
A. Kononov et al. (Eds.): MOTOR 2020, LNCS 12095, pp. 49–68, 2020.
https://doi.org/10.1007/978-3-030-49988-4_4

A wide range of modern heuristic algorithms and metaheuristics makes up the basis of the second research direction. To date, the most significant numerical results were obtained for local search algorithms [6], Tabu search [28], Variable Neighborhood Search (VNS) [13], machine learning [24], evolutionary [31], and bio-inspired algorithms [25], as well as their various combinations [9]. Often heuristic algorithms demonstrate remarkable performance, yielding close to optimal or even optimal solutions for CVRP instances of extremely large size. Nevertheless, an absence of any theoretical guarantees implies additional computational expenses related to numerical performance evaluation and possible tuning of their internal parameters during the transition to any novel class of instances.

The arguments above confirm the relevance of the third direction related to the design of approximation algorithms with theoretical performance guarantees. It is known that CVRP is NP-hard in the strong sense, enclosing the classic Traveling Salesman Problem (TSP), and remains intractable even on the Euclidean plane [26]. The problem is hard to approximate in the general case (provided $P \neq NP$), APX-complete for an arbitrary metric [5,15] even for an arbitrary fixed capacity $q \geq 3$.

Related Work. In the field of approximation algorithms with theoretical bounds, the most significant results were achieved for the settings of CVRP in finite-dimensional Euclidean spaces. All of them date back to the celebrated papers by M. Haimovich and A. Rinnooy Kan [15] and S. Arora [4]. At the moment, the most general result for the CVRP on the Euclidean plane is the Quasi-Polynomial-Time Approximation Scheme (QPTAS) proposed by A. Das and C. Mathieu [11]. For the planar CVRP with restricted capacity growth, there are known a number of Polynomial-Time Approximation Schemes (PTAS), among them, the algorithm [2] appears to be state-of-the-art. This PTAS allows to find an $(1 - \varepsilon)$-approximate solution of the problem in polynomial time provided $q \leq 2^{log^{\delta(\varepsilon)} n}$. The approach proposed in [15] was extended to several modifications of the problem including the CVRP settings in Euclidean spaces of an arbitrary fixed dimension [17,22,23], additional time windows constraints [19,20], and heterogeneity of demand [21].

Thus, until now, the class of metric CVRP instances approximable by PTAS or QPTAS was exhausted by the Euclidean settings of the problem except maybe some special cases investigated in [8,18]. For a long time, the similar theoretic gap remained for a the very close Traveling Salesman Problem, until the pioneering papers by K. Talwar [29], and Y. Bartal et al. [7] providing an opportunity for the extension of famous Arora's PTAS [3] to the universe of metric spaces of a fixed doubling dimension. In this paper, we try to bridge a similar gap for the metric Capacitated Vehicle Routing Problem.

Our Contribution. The contribution of this paper is twofold.

(i) we show that the approach proposed by Das and Mathieu for the efficient approximation of the Euclidean CVRP can be extended to a significantly wider class of metric CVRP settings. We prove that this framework combined with recent approximation results obtained for the metric TSP, for any given $\varepsilon > 0$, provides a $(1 + O(\varepsilon))$-approximate solution for the CVRP formulated in a metric space of an arbitrary fixed doubling dimension $d > 1$.

(ii) nevertheless, broadly speaking, the approximation scheme obtained by the straightforward application of the Das and Mathieu framework is no longer a QPTAS in general metric space of a fixed doubling dimension, even for an arbitrary fixed capacity $q > 2$. Therefore, in this paper, we introduce a refinement of their algorithm by replacing the stage of exhaustive search with our internal dynamic program, such that the resulting scheme becomes a QPTAS again, at least for $q = \operatorname{polylog} n$.

The rest of the paper is structured as follows. In Sect. 2, we recall the statement of the metric CVRP. Then, in Sect. 3 we overview some basic notation regarding the metrics of a fixed doubling dimension. Main results of the paper are presented in Sect. 4 and Sect. 5. In particular, Sect. 4 deals with approximation properties of the proposed scheme, whilst, in Sect. 5, we present an upper bound of its running time. Finally, Sect. 6 summarizes the work and provides a short overview of some questions that still remain open.

2   Problem Statement

The Capacitated Vehicle Routing Problem (CVRP) can be formulated informally as follows. We are given by a set of customers X, each of them has a unit demand on some homogeneous commodity. All the customer's demand should be serviced by identical vehicles of a fixed capacity q located initially at the given depot y. The problem is to construct a minimum cost family of cyclic routes servicing the total customer demand, each of them departs from and arrives at the depot y and satisfies the capacity constraint.

For the sake of convenience, we give a mathematical statement of a slightly more general problem, where each customer is free to have a non-unit integer demand, which can be split between several routes. In the literature, this problem is referred to as the Capacitated Vehicle Routing Problem with Splittable Demand (CVRP-SD).

An instance of the CVRP-SD is given by a complete weighted graph $G = (X \cup \{y\}, E, D, w)$ and a natural number q. Here, $X = \{x_1, \ldots, x_n\}$ is a set of customers, y is a depot, the non-negative weighting function $D : X \to \mathbb{Z}_+$ specifies customer demand, the symmetric weighting function $w : E \to \mathbb{R}_+$, to any couple of nodes $\{u, v\} \subset X \cup \{y\}$, assigns the transportation cost $w(u, v)$ related to the direct transition along the edge $\{u, v\} \in E$, and q is an upper vehicle capacity bound.

A *route* is an ordered pair $\mathcal{R} = (\pi, S_{\mathcal{R}})$, such that $\pi = y, x_{i_1}, \ldots, x_{i_t}, y$ is a cycle in the graph G and the function $S_{\mathcal{R}} \colon X \to \mathbb{Z}_+$ defines a distribution of the serviced customer demand. For the route $\mathcal{R}$, its *cost* $w(\mathcal{R})$ is defined as follows

$$w(\mathcal{R}) = w(y, x_{i_1}) + w(x_{i_1}, x_{i_2}) + \cdots + w(x_{i_{t-1}}, x_{i_t}) + w(x_{i_t}, y).$$

The route $\mathcal{R}$ is called *feasible*, if

$$S_{\mathcal{R}}(x) \begin{cases} \leq D(x) & \text{for any } x \in \{x_{i_1}, \ldots, x_{i_t}\}, \\ = 0, & \text{otherwise} \end{cases} \quad \text{and} \quad \sum_{x \in X} S_{\mathcal{R}}(x) \leq q.$$

The goal is to construct the cheapest family $\mathfrak{S}$ of feasible routes, which services the total customer demand

$$w(\mathfrak{S}) \equiv \sum_{\mathcal{R} \in \mathfrak{S}} w(\mathcal{R}) \to \min$$

$$s.t. \sum_{\mathcal{R} \in \mathfrak{S}} S_{\mathcal{R}}(x) = D(x) \quad (x \in X). \tag{1}$$

Obviously, a statement of the classical CVRP can be obtained by restriction of the above setting with the additional constraint $D(x) \equiv 1$.

If the function w satisfies the triangle inequality, i.e. $w(v_1, v_2) \leq w(v_1, v_3) + w(v_3, v_2)$ holds for any subset $\{v_1, v_2, v_3\} \subset X \cup \{y\}$, the instance of CVRP is called *metric*. In this case, nodes of the graph G are called *points*, $w(u, v)$ is referred to as a *distance* between the points u and v, and the cost $w(\mathcal{R})$ of an arbitrary route $\mathcal{R}$ is called its *length*.

In this paper, we consider the metric CVRP restricted as follows:

(i) the ordered pair (Z, ρ), where $Z = X \cup \{y\}$ and the metric $\rho|_E \equiv w$, is a finite metric space of some fixed doubling dimension $d > 1$;
(ii) the vehicle capacity bound q does not exceed polylog n.

Hereinafter, we do not distinguish the weight function w and the corresponding metric ρ and use the notation $\mathrm{CVRP}(Z, w, q)$ and $\mathrm{CVRP}^*(Z, w, q)$ for the instance specified by the graph $G = (X \cup \{y\}, E, w)$ and capacity q and its optimum value, respectively[1].

3   Metric Spaces of a Fixed Doubling Dimension

For the subsequent constructions, we need to recall some definitions and preliminary technical results.

Suppose we are given by some metric space (Z, ρ). For any $z_0 \in Z$ and a number $R \geq 0$, the set $B(z_0, R) = \{z \in Z \colon \rho(z_0, z) \leq R\}$ is called a *metric ball* of a radius R centered at the point $z_0 \in Z$.

---

[1] And the notation $\mathrm{CVRP\text{-}SD}(Z, D, w, q)$ and $\mathrm{CVRP\text{-}SD}^*(Z, D, w, q)$ for the case of CVRP-SD as well.

Definition 1 (see, e.g [1]). *For a number $d > 1$, the space (Z, ρ) is referred to as a metric space of the fixed dimension d, if, for an arbitrary $z_0 \in Z$ and $R > 0$, there exist points $z_1, \ldots, z_M \in Z$, such that*

$$B(z_0, R) \subseteq \bigcup_{j=1}^{M} B(z_j, R/2) \text{ and } M \leq 2^d.$$

It is easy to verify that, for any $d \geq 1$ and $p \geq 1$, the space l_p^d is a metric space of doubling dimension $O(d)$. On the other hand, there are known many metrics of a fixed dimension that appear to be very far from the finite-dimensional numeric spaces (see, e.g. [14]).

Next, let $Z' \subset Z$ be an arbitrary nonempty subspace of the space Z (of doubling dimension d). By $\Delta = \Delta_\rho(Z') = \sup\{\rho(u, v) : u, v \in Z'\}$ and $\alpha = \alpha_\rho(Z') = \inf\{\rho(u, v) : \{u, v\} \subset Z'\}$ we denote an upper and a lower bounds for the distances between the distinct points in Z', respectively.

Lemma 1 ([29]). *Let $0 < \alpha \leq \Delta < \infty$. Then, the subspace Z' is finite and*

$$|Z'| \leq \left(\frac{2\Delta}{\alpha}\right)^d.$$

In this paper, we restrict ourselves to finite metric spaces induced by complete weighted graphs $G = (Z, E, w)$. Let, further, $U \subset Z$ be an arbitrary nonempty node subset of the graph G, $\mathrm{MST}(U)$ be the minimum spanning tree for the induced subgraph $G\langle U \rangle$, and $R = R(U)$ be a radius of the minimal ball (centered at some point $z \in Z$) enclosing the subset U.

Lemma 2.
$$w(\mathrm{MST}(U)) \leq 12R \cdot |U|^{1-1/d}. \tag{2}$$

4   Extended Das and Mathieu Approximation Scheme

In this section, we show that the well-known QPTAS proposed by A. Das and C. Mathieu [11] for the Euclidean CVRP can be extended to the case of metric spaces of any fixed doubling dimension $d > 1$. Supplementing the main idea of their scheme with the technical results underlying the recent PTAS of Y. Bartal et al. [7] for the metric TSP formulated in such spaces, we propose an algorithm that, for an arbitrary $0 < \varepsilon < 1/8$ finds a $(1 + O(\varepsilon))$-approximate solution of the CVRP in a metric space of any doubling dimension $d > 1$. On the other hand, we show that the resulting algorithm, generally speaking, ceases to be a QPTAS, even for a fixed capacity q. Further, in Subsect. 4.5, we propose a novel version of Das and Mathieu scheme and show that its complexity bound is quasi polynomial, provided $q = O(\mathrm{polylog}\, n)$.

Similarly to the original scheme, our algorithm consists of several consecutive stages, as follows:

(i) **Preliminary processing and accuracy driven rounding.** At this stage, given by $\varepsilon > 0$, to the instance in question, we assign an auxiliary instance of more simple structure, called *rounded*, such that an arbitrary $(1 + \varepsilon)$-approximate solution of this instance can be transformed efficiently to the appropriate $(1 + O(\varepsilon))$-approximate solution of the initial problem.

(ii) **Randomized hierarchical clustering.** Given by values of random parameters, at this stage, we construct a number of mutually nested partitions of the set $X \cup \{y\}$. Then, in each cluster located at any level of the resulting hierarchy, we point out some number of special points (we call them *portals*). Following to the approach proposed in [29], we show that, for any rounded instance, there exist $(1 + \varepsilon)$-approximate solutions, each their route crosses any cluster at most r times (for some number r, which will be defined later) and at portals exclusively. Such routes are referred to as *net-respecting and r-light* (see, e.g. [7]).

(iii) **Dynamic Programming and Iterated Tour Partition.** At this stage, following [11], we allow some routes of the constructed solutions (we call them *relaxed*) slightly violate the capacity constraint. Then, to obtain a required feasible approximate solution,

  (a) we apply dynamic programming to find a relaxed net-respecting and r-light solution minimizing some especially penalized objective function
  (b) applying a randomized rank procedure for the demand covered by the routes of the solution obtained, we ensure that each route covers at most q demand units of the highest rank; following to [11], we call such units *black*
  (c) all other units (we call them *red*) are excluded from these routes and covered separately, by the additional routes constructed using the well-known Iterated Tour Partition (ITP) heuristic [15]
  (d) thus, we obtain two partial solutions $\mathfrak{S}_{black}$ and $\mathfrak{S}_{red}$, such that their combination is a feasible solution to the problem in question.

  Finally, we show that the expected cost of this combined solution over random clustering and demand ranking fulfils the following equation

$$\mathbb{E}(w(\mathfrak{S}_{black}) + w(\mathfrak{S}_{red})) = (1 + O(\varepsilon)) \cdot \text{CVRP}^*(Z, w, q).$$

(iv) **Derandomization.** Relying on the arguments from [11] and [29], we show that the proposed algorithm admits polynomial time derandomization.

4.1 Accuracy Driven Rounding

This stage dates back to the classic PTAS proposed by S. Arora for the Euclidean TSP [3]. As above, let $\Delta = \Delta_w(Z) = \max\{w(u, v) \colon u, v \in Z = X \cup \{y\}\}$ be the diameter of the set Z. Without loss of generality, we assume that $\Delta = n/\varepsilon$. Indeed, otherwise, to the initial instance $\text{CVRP}(Z, E, w)$, we can easily assign an equivalent (in terms of optimality sets) scaled instance $\text{CVRP}(Z, E, w')$ with the following weighting function: $w'(u, v) = w(u, v) \cdot \frac{n}{\varepsilon \cdot \Delta}$.

We define the desired *rounded* instance in terms of metric nets.

Definition 2. *A subset $N \subseteq Z$ is called a δ-net in the metric space (Z, ρ) for some given $\delta > 0$, if the following conditions holds*

(i) for any $u \in Z$, there exists $v = v(u) \in N$, such that $\rho(u, v) \leq \delta$;
(ii) for an arbitrary distinct points $v_1, v_2 \in N$, the distance $\rho(v_1, v_2) > \delta$.

Let $N_1' = \{\xi_1, \dots, \xi_J\}$ be an arbitrary 1-net of the set X. Assuming $N_1 = N_1' \cup \{y\}$, to the initial instance CVRP(Z, w, q), we assign the rounded one CVRP-SD(N_1, D, w_1, q) as follows:

(i) breaking tights arbitrarily, we define a mapping $\xi \colon X \to N_1'$ such that $w(x, \xi(x)) \leq 1$ holds for any $x \in X$;
(ii) to any node $\xi_j \in N_1'$, we assign the accumulated customer demand $D(\xi_j) = |\xi^{-1}(\xi_j)|$;
(iii) as new weighting function w_1, we take a restriction $w|_{N_1}$ of the function w to the set $N_1 \subset Z$.

Lemma 3 establishes a close relation between optimum values of the initial and rounded instances.

Lemma 3.

$$\text{CVRP}^*(Z, w, q) - 2n \leq \text{CVRP-SD}^*(N_1, D, w_1, q) \leq \text{CVRP}^*(Z, w, q) + 2n.$$

Notice that the procedures required for construction of the net N_1' as well as the ones for assigning to the initial CVRP(Z, w, q) its rounded instance CVRP-SD(N_1, D, w_1, q) and finally the reconstruction of solution $\mathfrak{S}$ according to solution $\bar{\mathfrak{S}}$ could be done in polynomial time.

As a simple corollary, we show that an arbitrary approximate solution of CVRP-SD(N_1, D, w_1, q) corresponds to the suitable approximate solution of the initial CVRP(Z, w, q).

Corollary 1. *For any $(1 + \varepsilon)$-approximate solution of* CVRP-SD(N_1, D, w_1, q) *can be transformed efficiently to an appropriate $(1 + O(\varepsilon))$-approximate solution of CVRP(Z, w, q).*

Thus, in the sequel, without loss of generality, we assume that we a given by a rounded instance.

4.2   Randomized Hierarchical Clustering

Following to [5], we fix a number $s \geq 6$ and put $L = \lceil \log_s \Delta_w(Z) \rceil = O(\log n - \log \varepsilon)$. Then, for each $l = 0, 1, \dots, L+1$, we fix an arbitrary s^{L-l}-net $N(l)$ of the set Z. Without loss of generality, assume that $N(l) \subset N(l+1)$ for any $0 \leq l \leq L$. Notice, that the net $N(L+1) = Z$, whilst the net $N(0)$ is a singleton.

In the following, we construct a randomized hierarchical clustering of Z by induction on level $l = 0, \dots, L+1$ as proposed in the paper [29].

We start with level $l = 0$, where we have a single cluster C_1^0. Further, let $Z = C_1^l \cup C_2^l \dots \cup C_K^l$ be a clustering at the level $l < L$. To proceed with the clustering at level $l + 1$, we partition each cluster C_j^l separately, applying the following simple procedure

(i) pick a random permutation σ of the $s^{L-(l+1)}$-net $N(l+1) = \{h_1, \ldots, h_{t_{l+1}}\}$;

(ii) to an arbitrary $h_{\sigma(i)} \in N(l+1)$, assign a number μ picked uniformly at random in $[1, 2)$;

(iii) define a subset C_{ji}^{l+1} by the formula

$$C_{ji}^{l+1} = B\left(h_{\sigma(i)}, \mu \cdot s^{L-(l+1)}\right) \cap C_j^l \setminus \bigcup_{k=1}^{i-1} C_{jk}^{l+1};$$

(iv) construct a partition of the cluster C_j^l from all non-empty subsets C_{ji}^{l+1}.

where μ is an arbitrary value of uniform distribution on $[1, 2)$. Finally, we obtain the resulting clustering of the set Z at level $l + 1$ by combining individual partitions for all clusters C_j^l.

By construction, at level $L + 1$, all the clusters are singletons, while, at level $l = 0$, we have the only cluster C_1^0. Thus, the total number of clusters is at most $(n + 1) \cdot (L + 1) = O(n(\log n - \log \varepsilon))$.

For the further constructions, we need to introduce a special type of routes.

Definition 3. *A route* $\mathcal{R} = (\pi, S_{\mathcal{R}})$ *is called net-respecting relatively to a given hierarchy* $N(l), l = 0, 1, \ldots, L+1$ *and value* $\varepsilon > 0$*, if, for an arbitrary edge* $\{u, v\}$ *of the cycle* π*, both its endpoints belong to* $N(l)$*, such that*

$$s^{L-l} \leq \varepsilon \cdot \rho(u, v) < s^{L-l+1}.$$

We say that a route $\mathcal{R} = (\pi, S_{\mathcal{R}})$ crosses the boundary of some cluster C_j^l at level $l > 0$, if π contains an edge $\{u, v\}$, such that $|\{u, v\} \cap C_j^l| = 1$. In the following, we introduce a special type of the net-respecting routes, each of them is restricted to cross the boundary of any cluster not too often and at *portals* exclusively.

Definition 4. *Let M be some degree of s, for which*

$$\frac{M}{s} \leq \frac{dL}{\varepsilon} < M. \tag{3}$$

We call a portal an arbitrary point from $C_j^l \cap N(l + \log_s M)$.

Applying Lemma 1 we obtain the following upper bound for the number m of portals of any cluster C_j^l.

$$m \leq \left(2\frac{4s^{L-l}}{s^{L-l}/M}\right)^d = (8M)^d = O\left(\left(\frac{d \cdot (\log n - \log \varepsilon)}{\varepsilon}\right)^d\right). \tag{4}$$

Definition 5. *A route $\mathcal{R}$ crossing the boundary of any cluster C_j^l at most r times, is called r-light.*

The main result of Subsect. 4.2 is the following Structure Theorem.

Theorem 1. *Let $r = m$ and $d > 1$. For any fixed $\varepsilon \in (0, 1/8)$ and an arbitrary feasible solution $\mathfrak{S}$ of $\mathrm{CVRP}(Z, E, w)$, there exists an appropriate feasible solution $\tilde{\mathfrak{S}}$ consisting of net-respecting and r-light routes, such that*

$$\mathbb{E}(w(\tilde{\mathfrak{S}})) = (1 + O(\varepsilon))w(\mathfrak{S}),$$

where the expectation is made over the random hierarchical clustering.

As it follows from Theorem 1, any time, when we need to find an approximate solution of the initial problem, we can restrict ourselves to the solutions consisting of net-respecting and r-light routes exclusively. In the sequel, we call such solutions net-respecting and r-light as well.

4.3 Demand Ranking and Relaxed Solutions

The aforementioned approach relying upon the minimization of total transportation cost in the class of net-respecting and r-light solutions yields a number of seminal approximation results for intractable routing problems, including the well-known Arora's PTAS for the Euclidean TSP [3] and its extension to metric spaces of a fixed doubling dimension [7]. Unfortunately, it is well-known that, for the CVRP, this approach results in tremendously time expensive algorithms. In this subsection, following to the main idea of the paper [11], we outline another approach that leads us to really efficient approximation algorithms based on a concept of *relaxed solutions*.

We start with some necessary definitions and notation. Consider a net-respecting route $\mathcal{R}$ that enters and leaves the cluster C_j^l (located at some level $l > 0$) at portals p^{in} and p^{out} respectively. We call an arbitrary maximal by inclusion fragment

$$\sigma = p^{in}, x_{i_1}, \dots, x_{i_k}, p^{out}, \tag{5}$$

which entirely belongs to the cluster C_j^l, a *crossing segment* of the route $\mathcal{R}$ with respect to the cluster C_j^l (or just a *segment*).

Definition 6. *Let $\Lambda = \lceil \log_{1+\varepsilon/(L+1)}(q\varepsilon) + 1/\varepsilon \rceil$. Numbers t_i, $i = \overline{1, \Lambda}$ are called rounding thresholds for covered customer demand, if*

$$t_i = \begin{cases} i & \text{for all } i = 1, \dots, \lfloor 1/\varepsilon \rfloor \\ t_{i-1}(1 + \varepsilon/(L+1)) & \text{otherwise.} \end{cases}$$

Next, we proceed with *ranking* of customer demand. We assume that each unit of the demand has an integer *rank* from the range $0, 1 \dots, L + 1$. Each customer can have demand units of different ranks. An arbitrary demand unit can be either *active* or *non-active* depending on its rank and level of the considered enclosing cluster. Namely, a demand unit of rank $\mathbf{r}$ is called *active* with respect to any enclosing cluster located at level $l > \mathbf{r}$ (otherwise, this unit is called inactive). By convention, demand units of rank 0 are active at any level.

A segment σ is called *rounded* inside the cluster C_j^l, if it covers exactly t active demand units for some threshold t. Otherwise, σ is called *unrounded*.

Definition 7. *A set of tours $\mathfrak{S}$ is called a relaxed solution if it covers the total customer demand and there exists an assignment of ranks for all demand units, such that*

(i) each route $\mathcal{R} \in \mathfrak{S}$ covers at most q units of the rank 0;
(ii) if a route $\mathcal{R}$ covers exactly t units of active demand at level l, then at level $l+1$, it covers at most $t(1 + \varepsilon/(L+1))$ such units;
(iii) for any route $\mathcal{R} \in \mathfrak{S}$, if the number of its segments crossing some cluster C exceeds $\gamma = \left(\frac{L+1}{\varepsilon}\right)^{2d}$, then all these segments are rounded. Otherwise, all of them are unrounded.

In the following, we call any relaxed solution $\mathfrak{S}$ that is also net-respecting and r-light a *structured solution*. Such solutions are essential point of our approach. Given a random hierarchical clustering, we find a structural solution minimizing the following auxiliary objective function

$$F(\mathfrak{S}) = \sum_{\mathcal{R} \in \mathfrak{S}} w(\mathcal{R}) + \frac{\varepsilon}{L+1} \sum_{\mathcal{R} \in \mathfrak{S}} \sum_{l=1}^{L+1} c(\mathcal{R}, l) \cdot s^{L-l}, \qquad (6)$$

where, for any route $\mathcal{R} \in \mathfrak{S}$, $c(\mathcal{R}, l)$ is the number of crossings the boundaries of all clusters at level l.

Notice that with respect to feasible solutions the initial objective function $w(\mathfrak{S})$ and the introduced above function F behave quite similarly.

Theorem 2. *The hypothesis of Theorem 1 implies*

$$\mathbb{E}(F(\tilde{\mathfrak{S}})) = (1 + O(\varepsilon))w(\mathfrak{S}),$$

where the expectation is made over the random hierarchical clustering.

Let, further, for a given random clustering, $\mathfrak{S}_{DP}$ be a minimizer of the function F in the class of structured solutions[2]. To address the possible infeasibility of $\mathfrak{S}_{DP}$, we introduce a random ranking of the customer demand by Algorithm 1.

Given by a demand ranking, we color each demand unit of the rank 0 in black and all other units in red. After that, we transform the solution $\mathfrak{S}_{DP}$ to the partial solution $\mathfrak{S}_{black}$ by exclusion all the red units. Then, we employ the ITP heuristic to find an approximate CVRP solution $\mathfrak{S}_{red}$ that covers the remaining red demand. Obtain upper bounds for $\mathbb{E}(w(\mathfrak{S}_{black}))$ and $\mathbb{E}(w(\mathfrak{S}_{red}))$ individually. Indeed, by definition of the function F, for any fixed hierarchical clustering,

$$w(\mathfrak{S}_{black}) \leq F(\mathfrak{S}_{black}) \leq F(\mathfrak{S}_{DP}) \leq F(\tilde{\mathfrak{S}}^*),$$

where $\tilde{\mathfrak{S}}^*$ is the net-respecting and r-light feasible solution associated with an arbitrary optimal solution $\mathfrak{S}^*$ of the initial problem, whose existence is guaranteed by Theorem 1. The right-most inequality is valid, since $\tilde{\mathfrak{S}}^*$ is a structured solution, by Lemma 5 from [11]. Then, by Theorem 2, we obtain

$$\mathbb{E}(w(\mathfrak{S}_{black})) \leq \mathbb{E}(F(\tilde{\mathfrak{S}}^*)) = (1+O(\varepsilon))w(\mathfrak{S}^*) = (1+O(\varepsilon))\text{CVRP}^*(Z, w, q), \quad (7)$$

---

[2] In Sect. 4.4, we provide a dynamic programming algorithm, which finds such a solution for any given random clustering.

Algorithm 1. Demand Ranking Algorithm

Input: a structured solution $\mathfrak{S}_{DP}$ with respect to some random hierarchical clustering
Output: ranking of all units of demand
1: initialize rank of each unit of demand by 0
2: **for all** level l from $L+1$ to 0 **do**
3:   **for all** cluster C_j^l crossed by more than γ segments **do**
4:     **for all** segment σ crossing the cluster C_j^l **do**
5:       Let **a** be the number of active demand units covered by the segment σ and t be the largest threshold, such that $t \leq \mathbf{a}$.
6:       Pick an active demand unit at random and $\mathbf{a} - t - 1$ consecutive units (wrapping around to the start of σ if necessary) and assign to them the rank l.
7:     **end for**
8:   **end for**
9: **end for**

---

where the expectation is taken over random clustering. The latter upper bound is given by Lemma 4.

Lemma 4. *For an arbitrary clustering and the expected value of $w(\mathfrak{S}_{red})$ over random ranking of the demand, the following equation*

$$\mathbb{E}(w(\mathfrak{S}_{red})) = O(\varepsilon) \cdot (F(\mathfrak{S}_{DP}) + \mathrm{CVRP}^*(Z, w, q)) \qquad (8)$$

is valid.

Finally, relying on Eq. (7), Lemma 4, and Theorem 2, we easily obtain the main result of this subsection.

Theorem 3. *Let an instance of the CVRP be given in a metric space of a fixed dimension $d > 1$ and $r = m$. Then, for any $\varepsilon \in (0, 1/8)$, Das and Mathieu randomized scheme provides an approximate solution $\mathfrak{S}_{black} \cup \mathfrak{S}_{red}$, such that*

$$\mathbb{E}(w(\mathfrak{S}_{black}) + w(\mathfrak{S}_{red})) = (1 + O(\varepsilon))\mathrm{CVRP}^*(Z, w, q),$$

where the expectation is taken over random clustering and ranking of the demand.

The obtained results shed new light on the approximation of the Capacitated Vehicle Routing Problem formulated in metric spaces of a fixed dimension. Actually, Theorem 3 implies that any structured solution $\mathfrak{S}_{DP}$ minimizing the auxiliary objective function F can be transformed into a required approximate solution of the given problem. Furthermore, as it follows from the seminal paper [15], such post-processing can be carried out in polynomial time. In the sequel, we develop an efficient procedure for finding such structured solutions.

4.4   Baseline Dynamic Programming

In this section, we present a short overview of our adaptation of the initial Das and Mathieu dynamic programming algorithm to the case of metric spaces of a fixed doubling dimension.

We start with some necessary notation. We encode an arbitrary crossing segment (5) by a tuple $(p^{in}, p^{out}, \mathbf{s}, \mathbf{d})$, where $\mathbf{s}$ is the amount of covered active demand units and $\mathbf{d}$ indicates whether this segment should visit the depot y.

Given by a fixed hierarchical clustering, we index entries of the lookup table of our dynamic program by couples $(C, \mathfrak{C})$, where C is a cluster and $\mathfrak{C}$ is a *configuration* defining behavior patterns for all segments crossing the boundary of the cluster C. Depending on the number of segments described, we distinguish two kinds of configurations, *unrounded* and *rounded*.

An *unrounded configuration* is just a finite sequence

$$((p^{in}_\nu, p^{out}_\nu, \mathbf{s}_\nu, \mathbf{d}_\nu): \nu = \overline{1, k_u})$$

of at most γ tuples, each of them represents a single unrounded crossing segment. On the other hand, a *rounded configuration* is set of ordered pairs

$$\{(s_\nu, m_\nu): \nu = \overline{1, k_r}\}, \ s_{\nu_1} \neq s_{\nu_2},$$

each of them defines a common behavior pattern $s_\nu = (p^{in}_\nu, p^{out}_\nu, t_\nu, \mathbf{d}_\nu)$ for exactly m_ν rounded segments. Namely, all such segments should enter and leave the cluster in portals p^{in}_ν and p^{out}_ν respectively, cover t_ν units of active demand exactly (for some threshold t_ν), and visit the depot according to the value of $\mathbf{d}_\nu$.

To define the concept of a *feasible* lookup table entry, we need some technical notation. A family Σ of segments crossing the boundary of some cluster C augmented by a number of routes enclosed within this cluster is called a *partial relaxed solution* for the cluster C, if this family covers all the customer demand in this cluster and fulfills conditions (i), (ii), and (iii) enlisted in Definition 7 (with respect to this cluster).

Definition 8. *An entry $(C, \mathfrak{C})$ is called feasible, if there exists a partial relaxed solution $\Sigma = \Sigma(C)$, such that*

(i) if $\mathfrak{C}$ is unrounded, then all the segments of Σ are unrounded and are too, s.t. there exists a one-to-one correspondence between them and the entries of the configuration $\mathfrak{C}$;

(ii) otherwise, if $\mathfrak{C}$ is rounded, then the family Σ is partitioned into k_r subfamilies, such that the ν-th subfamily consists of m_ν rounded crossing segments sharing the same behavior pattern s_ν.

As usual, the lookup table entries are computed bottom-up. The base case corresponds to the level $L+1$, where all the clusters are singletons. Thus, all the entries can be computed trivially.

To proceed with the recurrence, assume that all the entries for the levels $l+1, \ldots, L+1$ are calculated. Fix an arbitrary cluster C^l_j and try to compute the entry $(C^l_j, \mathfrak{C})$ for some configuration $\mathfrak{C}$. By the given clustering, we have a partition $C^l_j = C^{l+1}_{j1} \cup \ldots \cup C^{l+1}_{jK}$ for some $K = 2^{O(d)}$. Guided by the approach proposed in [11], to compute the entry $(C^l_j, \mathfrak{C})$, it is necessary to employ the two-stage exhaustive search as follows:

Stage (i) to enumerate all the combinations

$$((C_{j1}^{l+1}, \mathfrak{C}_1), \dots, (C_{jK}^{l+1}, \mathfrak{C}_K)) \tag{9}$$

of the computed already entries induced by the child subclusters;
Stage (ii) for any given combination (9), enumerate all the ways to stitch child configurations to fulfill the initial configuration $\mathfrak{C}$. Any time when such a stitching is possible, the record value of the function F is updated.

Thus, the entry $(C_j^l, \mathfrak{C})$ is filled by the resulting record value, if such a value was updated at least once. Otherwise, the entry is set to be infeasible and excluded from the consideration. To obtain the desired structured solution $\mathfrak{S}_{DP}$ minimizing the objective function F, it is sufficient to compute the only entry $(C_1^0, \mathfrak{C})$ at level 0 for the empty configuration $\mathfrak{C}$.

The point is that although, for the finite dimensional Euclidean spaces considered by Das and Mathieu, Stage (ii) can be calculated efficiently, in metric spaces even of a fixed doubling dimension, its running time is no longer quasi-polynomial.

Indeed, at Stage (ii), the calculations are specified in terms of *concatenation profiles* and *interface vectors*. A concatenation profile defines the stitching order for any single segment crossing the boundary of the cluster C_j^l (or a route contained in it).

Namely, a finite sequence of tuples $\varphi = ((p_k^{in}, p_k^{out}, x_k, \mathbf{d}_k): k = \overline{1, \theta_\varphi})$ is called a concatenation profile, if, for each j-th tuple,

(i) p_k^{in} and p_k^{out} are some child portals
(ii) x_k is either a threshold or a natural number from $[1, \gamma]$
(iii) $\mathbf{d}_k$ indicates whether depot should be visited.

In turn, each entry of an interface vector specifies the number of times when some concatenation profile is used during the stitching procedure. By definition, an interface vector has the form $\mathfrak{I} = (n_1, \dots, n_{|\Phi|})$, where $n_i \in [0, n \cdot r]$ and Φ is the number of all possible concatenation profiles. Since, by construction, $|\Phi| = (\log n)^{\Omega(r)}$, the number of distinct interface vectors enumerated at Stage (ii) is at least

$$(nr)^{|\Phi|} = (nr)^{(\log n)^{\Omega(r)}}. \tag{10}$$

Evidently, the lower bound (10) is not quasi-polynomial for an arbitrarily slowly increasing function $r = r(n)$. Therefore, we cannot claim that the aforementioned algorithm retains quasi-polynomial running time bound in metric spaces of a fixed doubling dimension, even for any fixed $q > 2$, since at the moment no structure theorems are known for such spaces, proved for a constant r (see, e.g. [7]).

In the following subsection, we propose our modification of this scheme, where, at Stage (ii) of the recursive step, the exhaustive search for the optimal interface vector is replaced with an internal dynamic program, such that the resulting scheme becomes QPTAS again, at least for $q = \text{polylog } n$.

4.5   Our Improvement

For the sake of brevity, we restrict ourselves on the special case, where the cluster C_j^l contains no depots and all the configurations $\mathfrak{C} = \{(s_i, m_i) : i = \overline{1, k_{\mathfrak{C}}}\}$ and $\mathfrak{C}_u = \{(s_v^u, m_v^u) : v = \overline{1, k_u}\}$ for $u \in \{1, \dots, K\}$ are rounded[3].

Then, to match the configuration $\mathfrak{C}$ with child configurations $\mathfrak{C}_1, \dots, \mathfrak{C}_K$, we need to assign to each s_i a sequence $\Phi_i = (\varphi_{i,1}, \dots, \varphi_{i,m_i})$ of not necessary distinct concatenation profiles, such that

(i)   each profile $\varphi_{i,j}$ consists of tuples s_v^u only;
(ii) any tuple s_v^u is contained in the profiles m_v^u times in total;
(iii) for any tuple $s_i = (p_i^{in}, p_i^{out}, t_i, \mathbf{d}_i)$, the following inequality

$$t_i \le D_{\varphi_{i,j}} < t_i \left(1 + \frac{\varepsilon}{L+1}\right)$$

holds, where $D_{\varphi_{i,j}}$ is the total active demand covered by the profile $\varphi_{i,j}$;
(iv) their total cost

$$\sum_{i=1}^{k_{\mathfrak{C}}} \sum_{j=1}^{m_i} \text{cost}(\varphi_{i,j}) \to \min, \tag{11}$$

such that, for any profile $\varphi = ((p_k^{in}, p_k^{out}, x_k, \mathbf{d}_k) : k = \overline{1, \theta})$,

$$\text{cost}(\varphi) = \sum_{k=1}^{\theta-1} \tilde{\rho}(p_k^{out}, p_{k+1}^{in}) + \frac{2\theta \cdot \varepsilon}{L+1} \cdot s^{L-l-1}, \tag{12}$$

where

$$\tilde{\rho}(p_k^{out}, p_{k+1}^{in}) = \begin{cases} \rho(p_k^{out}, p_{k+1}^{in}), & \text{if } p_k^{out} \text{ and } p_{k+1}^{in} \text{ satisfy Definition 3,} \\ +\infty, & \text{otherwise.} \end{cases}$$

Notice, that criterion (11) and the reduced costs (12) of concatenation profiles can be obtained straightforwardly from the auxiliary objective function (6). Indeed, for any given configuration $\mathfrak{C}$ and child configurations $\mathfrak{C}_1 \dots, \mathfrak{C}_K$, thanks to condition (ii), the total cost of all child subsegments is constant and does not depend on profiles $\varphi_{i,j}$. Therefore, we exclude it from (11) and (12).

Further, notice that each concatenation profile $\varphi_{i,j}$ to be constructed can have its own size $\theta_{\varphi_{i,j}}$ fulfilling the condition $\theta_{\varphi_{i,j}} \le K \cdot r$, since the resulting solution is r-light. To ensure that each profile has the same size $\bar{r} = K \cdot r$, we *pad* it by enough copies of the dummy tuple σ_0. Further, we introduce the set

$$\bar{S} = \{\sigma_0\} \cup \bigcup_{u=1}^{K} \{s_1^u, \dots, s_{k_u}^u\} = \{\sigma_0, \sigma_1, \dots, \sigma_{\mathcal{K}}\}, \quad \mathcal{K} = \sum_{u=1}^{K} k_u$$

---

[3] The general case can be treated similarly, we postpone its consideration to the forthcoming paper.

containing all the tuples s_v^u from the child configurations augmented by the dummy tuple σ_0,

By a *resource matrix*, we call any three-dimensional matrix A of size $[k_{\mathfrak{C}} \times (\mathcal{K} + 1) \times \bar{r}]$, whose entry $a_{i,\nu}^p$ specifies how many times the tuple σ_ν is used in concatenation profiles Φ_i at position p. For any fixed i, we call the submatrix $A_i = \|a_{i,\nu}^p\|$, where $p = \overline{1,\bar{r}}$ and $\nu = \overline{0,\mathcal{K}}$, the *$i$-th resource row*.

Our Algorithm 2 comes as a replacement of Stage (ii) of the initial Das and Mathieu scheme. Skipping the rigorous definition of such a compatibility, we illustrate it by the simple example (see Example 1). Its main idea is based on the construction of a minimum cost family of concatenation profiles Φ_i *compatible* with any given resource row A_i.

Example 1. A family of concatenation profiles Φ_i compatible with the resource row A_i for $\mathcal{K} = 3$, $m_i = 5$, and $\bar{S} = \{\sigma_0, \dots, \sigma_3\}$

$$
A_i
\begin{array}{c|cc|c}
 {}_\sigma\!\diagdown\!{}^p & 1 & 2 & \dots \\
\hline
0 & 0 & 1 & \\
\hline
1 & 2 & 1 & \\
\hline
2 & 3 & 1 & \\
\hline
3 & 0 & 2 & \\
\end{array}
\qquad
\Phi_i =
\begin{bmatrix}
(\sigma_1, \sigma_2, \dots) \\
(\sigma_1, \sigma_3, \dots) \\
(\sigma_2, \sigma_1, \dots) \\
(\sigma_2, \sigma_3, \dots) \\
(\sigma_2, \sigma_0)
\end{bmatrix}
$$

Algorithm 2. Our 'Stage (ii)'

Input: a parent cluster C_j^l with associated configuration $\mathfrak{C}$ and the child DP table entries $(C_1^{l+1}, \mathfrak{C}_1), \dots, (C_K^{l+1}, \mathfrak{C}_K)$

Output: the minimum value of the objective function F for the given configurations $\mathfrak{C}, \mathfrak{C}_1, \dots, \mathfrak{C}_K$

1: **for each** resource matrix A **do**
2:     check the validity of the feasibility constraints $\sum_{\nu=0}^{\mathcal{K}} a_{i,\nu}^p = m_i$, $(p = \overline{1,\bar{r}}, i = \overline{1,k_{\mathfrak{C}}})$ and $m_i t_i \le \sum_{p=1}^{\bar{r}} \sum_{\nu=1}^{\mathcal{K}} a_{i,\nu}^p \cdot t_\nu < m_i t_i \left(1 + \frac{\varepsilon}{L+1}\right)$, $(i = \overline{1,k_{\mathfrak{C}}})$
3:     **if** the matrix A is feasible **then**
4:         **for each** $i \in \{1, \dots, k_{\mathfrak{C}}\}$ **do**
5:             employ the Internal Dynamic Program (Algorithm 3) to obtain the minimum cost family Φ_i of m_i concatenation profiles compatible with A_i (or show that it is impossible)
6:         **end for**
7:         **if** all Φ_i are constructed **then**
8:             sum up their costs and update the record
9:         **end if**
10:    **end if**
11: **end for**

Internal Dynamic Programming Algorithm. The goal of the algorithm is to construct a family of the minimum total cost (induced by the objective function F), which consists of m_i segments crossing the boundary of the cluster C_j^l, each of them corresponds to the behavior pattern s_i. Every such a segment is stitched from the child subsegments (defined by the patterns $\sigma_\nu \in \bar{S}$) in accordance to some concatenation profile $\varphi_{i,j} \in \Phi_i$. For the sake of simplicity, in the sequel, we do not distinguish such segments and the concatenation profiles that specify them and call the desired family Φ_i as well.

We construct the desired family Φ_i by recursion on the position p in concatenation profiles. Each entry of the internal dynamic programming lookup table is indexed by a couple (p, H_p), where $p = 1, \ldots, \bar{r}$ indicates the current position, and the matrix $H_p = \|h_{\nu,c}^p\|$, $\nu = \overline{0, \mathcal{K}}$, $c = \overline{0, q}$ specifies terminal constraints on a family $\Phi_i^{(p)}$ of m_i partial concatenation profiles of length p.

Actually, each entry $h_{\nu,c}^p$ of the matrix H_p denotes the number of such profiles (in this family), that cover exactly c units of active demand in total and have the same tuple σ_ν at position p. A matrix H_p is called *compatible* with the p-th column of a resource A_i, if $\sum_{c=0}^q h_{\nu,c}^p = a_{i,\nu}^p$ is valid for any $\nu = \overline{0, \mathcal{K}}$. In addition, $H_{\bar{r}}$ is compatible if and only if, for any ν, $h_{\nu,c}^{\bar{r}} > 0$ implies $c \in [t_i, t_i(1+\varepsilon/(L+1)))$.

Notice, that for any given resource row A_i, the sum of terms penalizing for crossings all the boundaries of the child subclusters (at level $l + 1$) is fixed and does not depend on Φ_i. Therefore, we can restrict ourselves to the minimization of the stitching costs for child subsegments only.

Thus, we define our reduced internal objective function $\tilde{F}$ as follows. Let $\Phi_i^{(p)}$ be a family of partial concatenation profiles $\varphi_{i,1}^{(p)}, \ldots, \varphi_{i,m_i}^{(p)}$, each of them consists of p tuples. Then, $\bar{F}(\Phi_i^{(p)}) = \sum_{j=1}^{m_i} \overline{\text{cost}}(\varphi_{i,j}^{(p)})$, where, for any partial profile $\varphi^{(p)} = (\sigma_{i_1}, \ldots, \sigma_{i_p})$, its reduced cost is defined by

$$\overline{\text{cost}}(\varphi^{(p)}) = \sum_{k=1}^{p-1} \text{conn}(\sigma_{i_k}, \sigma_{i_{k+1}}) = \sum_{k=1}^{p-1} \tilde{\rho}\left(p^{out}(\sigma_{i_k}), p^{in}(\sigma_{i_{k+1}})\right).$$

Further, the Bellman function $\bar{D}$ takes the form

$$\bar{D}(p, H_p) = \min\{\bar{F}(\Phi_i^{(p)}) \colon \Phi_i^{(p)} \text{ satisfies the constraints imposed by the matrix } H_p\}.$$

Thus, to define the Bellman equation, we introduce a special kind of matrices, establishing relationships between any pair of consecutive entries $(p - 1, H_{p-1})$ and (p, H_p). We call a three-dimensional matrix $X = \|x_{\nu_1,\nu_2}^c\|$, $c = \overline{0, q}$, $\nu_1, \nu_2 = \overline{1, \mathcal{K}}$ a *transition matrix* for some entries $(p - 1, H_{p-1})$ and (p, H_p), if x_{ν_1,ν_2}^c coincides with the number of partial concatenation profiles, that cover exactly c units of active demand in total, and have the same tuples σ_{ν_1} and σ_{ν_2} at positions $p - 1$ and p, respectively. By construction, any transition matrix satisfies the following evident constraints

$$\sum_{\nu_1=1}^{\mathcal{K}} x_{\nu_1,\nu_2}^c = h_{\nu_2,c}^p, \ (\nu_2 = \overline{1, \mathcal{K}}), \quad \sum_{\nu_1=0}^{\mathcal{K}} x_{\nu_1,0}^c = h_{0,c}^p. \tag{13}$$

Algorithm 3. Internal Dynamic Program

Input: a resource row A_i.
Output: a family Φ_i of concatenation profiles compatible with A_i and minimizing the function $\bar{F}$.

1: base case: the only feasible entry $(1, H_1)$, where $h^1_{\nu,c} = \begin{cases} a^1_{i,\nu}, & \text{if } \nu > 0, \ c = t_\nu \\ 0, & \text{otherwise} \end{cases}$

   and $\bar{D}(1, H_1) = 0$
2: recursive step: assume that all feasible entries are computed for any $p' < p$
3: **for each** H_p compatible with the p-th column of the resource row A_i **do**
4:   apply the Bellman equation to compute an entry (p, H_p)

$$\bar{D}(p, H_p) = \min_{X = \|x^c_{\nu_1,\nu_2}\|} \left\{ \bar{D}(p-1, H_{p-1}(X)) + \sum_{\nu_1=1}^{\mathcal{K}} \sum_{\nu_2=1}^{\mathcal{K}} \sum_{c=0}^{q} x^c_{\nu_1,\nu_2} \operatorname{conn}(\sigma_{\nu_1}, \sigma_{\nu_2}) \right\},$$
(14)

   where the minimization is carried out over feasible entries $(p-1, H_{p-1}(X))$ only. If at least one such an entry is found, then the result is stored in (p, H_p)
5: **end for**
6: **if** there are no feasible entries $(\bar{r}, H_{\bar{r}})$ or $\inf\{\bar{D}(\bar{r}, H_{\bar{r}})\} = \infty$ **then**
7:   output 'no profile families compatible with A_i'.
8: **else**
9:   the cost of the desired family Φ_i is contained within the entry

$$(\bar{r}, H^*_{\bar{r}}) = \arg\min\{\bar{D}(\bar{r}, H_{\bar{r}})\}.$$
(15)

10:   output the optimal solution Φ_i, which can be obtained from (15) by backtracking.
11: **end if**

---

5 Complexity Bounds

In this section, we find an upper bound for the time complexity of the proposed scheme. First of all, we evaluate the maximum size of the lookup table for the master (Das and Mathieu) dynamic program. The total amount of all clusters is at most $O(n \log n)$. Then, to each cluster, we have at most $(2m^2 q)^\gamma$ and $(n \cdot r)^{2m^2 L \log q}$ options to assign an unrounded and a rounded configuration respectively. Therefore, an upper bound for the size of this lookup table is

$$O(n \log n)\mathfrak{C}_{max}, \text{ where } \mathfrak{C}_{max} = (n \cdot r)^{2m^2 L \log q} + (2m^2 q)^\gamma = O\left((n \cdot r)^{2m^2 L \log q}\right).$$

Next, consider the complexity of computing an arbitrary entry $(C, \mathfrak{C})$ of this table. In order to proceed, we enumerate all possible combinations (9), which are exactly $(\mathfrak{C}_{max})^K$, and apply Algorithm 2 to any such a combination.

In turn, Algorithm 2 enumerates all the possible resource matrices and, for any such a matrix A, it applies Algorithm 3 to each its resource row A_i. Therefore, its complexity is determined by the running time of Algorithm 3 multiplied by the factor $(n \cdot r)^{k_{\mathfrak{C}}(\mathcal{K}+1)\bar{r}} \cdot k_{\mathfrak{C}}$ which is polylog $n \cdot (n \cdot r)^{O(m^4 L^2 \log^2 q)}$ for any fixed d.

Finally, the complexity of Algorithm 3 is determined by the number of entries in the lookup table of the internal dynamic program and the upper running time bound for computation of any such an entry, i.e. $\bar{r} \cdot (n \cdot r)^{\mathcal{K}q} \times (n \cdot r)^{\mathcal{K}^2 q} = (n \cdot r)^{O(\mathcal{K}^2 q)}$, since $\bar{r} = K \cdot r = 2^{O(d)} \cdot r$. Further, combining all the terms, we obtain the desired time complexity bound

$$\text{poly}(n) \cdot ((n \cdot r)^{2m^2 L \log q}) 2^{O(d)} \cdot (n \cdot r)^{O(m^4 L^2 \log^2 q + \mathcal{K}^2 q)} = \text{poly}(n) \cdot n^{O(m^4 L^2 q \log^2 q)}.$$

where $m = r = O\left(\left(\frac{d \cdot (\log n - \log \varepsilon)}{\varepsilon} \right)^d \right)$ and $L = O(\log n - \log \varepsilon)$.

Applying the techniques proposed in [11] and [29], we can derandomize our scheme in polynomial time.

Theorem 4. *For the CVRP in a metric space of an arbitrary doubling dimension $d > 1$, an $(1 + O(\varepsilon))$-approximate solution can be found by the randomized approximation algorithm within time $\text{poly}(n) \cdot n^{O(m^4 L^2 q \log^2 q)}$, where $m = O\left(\left(\frac{d(\log n - \log \varepsilon)}{\varepsilon} \right)^d \right)$, and $L = O(\log n - \log \varepsilon)$. The algorithm can be derandomized efficiently.*

The proposed scheme is QPTAS any time when $q = O(\text{polylog } n)$.

6   Conclusion

In the paper, we extend the famous approximation framework proposed by A. Das and C. Mathieu for the Euclidean Capacitated Vehicle Routing Problem to the case of metric spaces of a fixed doubling dimension. To establish quasi-polynomial time upper bound for our scheme, we replace exhaustive search in the initial algorithm by the internal dynamic program that ensures that the resulting approximation scheme became QPTAS for an arbitrary fixed doubling dimension $d > 1$, at least for $q = \text{polylog } n$.

Nevertheless, the question of whether for any metric space of any fixed doubling dimension there exists a QPTAS without any restriction on the capacity growth, still remains open. We believe that we will manage to bridge this gap in future work.

References

1. Abraham, I., Bartal, Y., Neiman, O.: Advances in metric embedding theory. Adv. Math. **228**(6), 3026–3126 (2011). https://doi.org/10.1016/j.aim.2011.08.003
2. Adamaszek, A., Czumaj, A., Lingas, A.: PTAS for k-tour cover problem on the plane rof moderately large values of k. Int. J. Found. Comput. Sci. **21**(6), 893–904 (2010). https://doi.org/10.1142/S0129054110007623
3. Arora, S.: Polynomial time approximation schemes for Euclidean traveling salesman and other geometric problems. J. ACM **45**, 753–782 (1998)

4. Arora, S., Safra, S.: Probabilistic checking of proofs: a new characterization of NP. J. ACM **45**, 70–122 (1998). https://doi.org/10.1145/273865.273901
5. Asano, T., Katoh, N., Tamaki, H., Tokuyama, T.: Covering points in the plane by k-tours: towards a polynomial time approximation scheme for general k. In: Proceedings of the Twenty-Ninth Annual ACM Symposium on Theory of Computing, STOC 1997, pp. 275–283. ACM, New York (1997). https://doi.org/10.1145/258533.258602
6. Avdoshin, S., Beresneva, E.: Local search metaheuristics for capacitated vehicle routing problem: a comparative study. Proc. Inst. Syst. Program. RAS **31**, 121–138 (2019). https://doi.org/10.15514/ISPRAS-2019-31(4)-8
7. Bartal, Y., Gottlieb, L.A., Krauthgamer, R.: The traveling salesman problem: low-dimensionality implies a polynomial time approximation scheme. SIAM J. Comput. **45**(4), 1563–1581 (2016). https://doi.org/10.1137/130913328
8. Becker, A., Klein, P.N., Schild, A.: A PTAS for bounded-capacity vehicle routing in planar graphs. In: Friggstad, Z., Sack, J.-R., Salavatipour, M.R. (eds.) WADS 2019. LNCS, vol. 11646, pp. 99–111. Springer, Cham (2019). https://doi.org/10.1007/978-3-030-24766-9_8
9. Chen, J., Gui, P., Ding, T., Zhou, Y.: Optimization of transportation routing problem for fresh food by improved ant colony algorithm based on Tabu search. Sustainability **11** (2019). https://doi.org/10.3390/su11236584
10. Dantzig, G.B., Ramser, J.H.: The truck dispatching problem. Manage. Sci. **6**(1), 80–91 (1959)
11. Das, A., Mathieu, C.: A quasipolynomial time approximation scheme for Euclidean capacitated vehicle routing. Algorithmica **73**(1), 115–142 (2014). https://doi.org/10.1007/s00453-014-9906-4
12. Demir, E., Huckle, K., Syntetos, A., Lahy, A., Wilson, M.: Vehicle routing problem: past and future. In: Wells, P. (ed.) Contemporary Operations and Logistics, pp. 97–117. Springer, Cham (2019). https://doi.org/10.1007/978-3-030-14493-7_7
13. Frifita, S., Masmoudi, M.: VNS methods for home care routing and scheduling problem with temporal dependencies, and multiple structures and specialties. Int. Trans. Oper. Res. **27**(1), 291–313 (2020). https://doi.org/10.1111/itor.12604
14. Gupta, A., Krauthgamer, R., Lee, J.R.: Bounded geometries, fractals, and low-distortion embeddings. In: 44th Annual IEEE Symposium on Foundations of Computer Science 2003, Proceedings, pp. 534–543 (2003). https://doi.org/10.1109/SFCS.2003.1238226
15. Haimovich, M., Rinnooy Kan, A.H.G.: Bounds and heuristics for capacitated routing problems. Math. Oper. Res. **10**(4), 527–542 (1985). https://doi.org/10.1287/moor.10.4.527
16. Hokama, P., Miyazawa, F.K., Xavier, E.C.: A branch-and-cut approach for the vehicle routing problem with loading constraints. Expert Syst. Appl. **47**, 1–13 (2016). https://doi.org/10.1016/j.eswa.2015.10.013
17. Khachai, M.Y., Dubinin, R.D.: Approximability of the Vehicle Routing Problem in finite-dimensional Euclidean spaces. Proc. Steklov Inst. Math. **297**(1), 117–128 (2017). https://doi.org/10.1134/S0081543817050133
18. Khachai, M., Ogorodnikov, Y.: Haimovich–Rinnooy Kan polynomial-time approximation scheme for the CVRP in metric spaces of a fixed doubling dimension. Trudy instituta matematiki i mekhaniki UrO RAN **25**(4), 235–248 (2019). https://doi.org/10.21538/0134-4889-2019-25-4-235-248
19. Khachai, M.Y., Ogorodnikov, Y.Y.: Polynomial-time approximation scheme for the capacitated vehicle routing problem with time windows. Proc. Steklov Inst. Math. **307**(1), 51–63 (2019). https://doi.org/10.1134/S0081543819070058

20. Khachay, M., Ogorodnikov, Y.: Efficient PTAS for the Euclidean CVRP with time windows. In: van der Aalst, W.M.P., et al. (eds.) AIST 2018. LNCS, vol. 11179, pp. 318–328. Springer, Cham (2018). https://doi.org/10.1007/978-3-030-11027-7_30

21. Khachay, M., Ogorodnikov, Y.: Approximation scheme for the capacitated vehicle routing problem with time windows and non-uniform demand. In: Khachay, M., Kochetov, Y., Pardalos, P. (eds.) MOTOR 2019. LNCS, vol. 11548, pp. 309–327. Springer, Cham (2019). https://doi.org/10.1007/978-3-030-22629-9_22

22. Khachay, M., Dubinin, R.: PTAS for the Euclidean capacitated vehicle routing problem in R^d. In: Kochetov, Y., Khachay, M., Beresnev, V., Nurminski, E., Pardalos, P. (eds.) DOOR 2016. LNCS, vol. 9869, pp. 193–205. Springer, Cham (2016). https://doi.org/10.1007/978-3-319-44914-2_16

23. Khachay, M., Zaytseva, H.: Polynomial time approximation scheme for single-depot Euclidean capacitated vehicle routing problem. In: Lu, Z., Kim, D., Wu, W., Li, W., Du, D.-Z. (eds.) COCOA 2015. LNCS, vol. 9486, pp. 178–190. Springer, Cham (2015). https://doi.org/10.1007/978-3-319-26626-8_14

24. Nazari, M., Oroojlooy, A., Takáč, M., Snyder, L.V.: Reinforcement learning for solving the vehicle routing problem. In: Proceedings of the 32nd International Conference on Neural Information Processing Systems, NIPS 2018, pp. 9861–9871. Curran Associates Inc., Red Hook (2018)

25. Necula, R., Breaban, M., Raschip, M.: Tackling dynamic vehicle routing problem with time windows by means of ant colony system. In: 2017 IEEE Congress on Evolutionary Computation (CEC), pp. 2480–2487 (2017). https://doi.org/10.1109/CEC.2017.7969606

26. Papadimitriou, C.: Euclidean TSP is NP-complete. Theoret. Comput. Sci. **4**, 237–244 (1977)

27. Pessoa, A.A., Sadykov, R., Uchoa, E.: Enhanced branch-cut-and-price algorithm for heterogeneous fleet vehicle routing problems. Eur. J. Oper. Res. **270**(2), 530–543 (2018). https://doi.org/10.1016/j.ejor.2018.04.009

28. Qiu, M., Fu, Z., Eglese, R., Tang, Q.: A tabu search algorithm for the vehicle routing problem with discrete split deliveries and pickups. Comput. Oper. Res. **100**, 102–116 (2018). https://doi.org/10.1016/j.cor.2018.07.021

29. Talwar, K.: Bypassing the embedding: algorithms for low dimensional metrics. In: Proceedings of the Thirty-Sixth Annual ACM Symposium on Theory of Computing, STOC 2004, pp. 281–290. Association for Computing Machinery, New York (2004). https://doi.org/10.1145/1007352.1007399

30. Toth, P., Vigo, D.: Vehicle Routing: Problems, Methods, and Applications. MOS-SIAM Series on Optimization, 2nd edn. SIAM, Philadelphia (2014)

31. Vidal, T., Crainic, T.G., Gendreau, M., Prins, C.: A hybrid genetic algorithm with adaptive diversity management for a large class of vehicle routing problems with time windows. Comput. Oper. Res. **40**(1), 475–489 (2013). https://doi.org/10.1016/j.cor.2012.07.018

Using Integer Programming to Search for Counterexamples: A Case Study

Giuseppe Lancia[1(✉)], Eleonora Pippia[1,2], and Franca Rinaldi[1]

[1] Dipartimento di Scienze Matematiche, Informatiche e Fisiche, University of Udine,
Via delle Scienze 206, 33100 Udine, Italy
giuseppe.lancia@uniud.it
[2] The Research Hub by Electrolux Professional,
Viale Treviso 15, 33170 Pordenone, PN, Italy

Abstract. It is known that there exist 4-regular, 1-tough graphs which are non-hamiltonian. The smallest such graph known has $n = 18$ nodes and was found by Bauer et al., who conjectured that all 4-regular, 1-tough graphs with $n \leq 17$ are hamiltonian. They in fact proved that this is true for $n \leq 15$, but left open the possibility of non-hamiltonian graphs of 16 or 17 nodes. By using ILP for modeling a counterexample, and then finding out that the model has no solutions, we give an algorithmic proof that their conjecture was indeed correct.

1 Introduction

Graph theory, as well as many other areas of mathematics, is rich of challenging open problems. Many times the study of these problems seems to suggest that their answer leans towards a particular side, so that one ventures to conjecture that this is indeed the case. The conjecture then remains open until someone either proves it (usually by means of some clever mathematical argument) or disproves it, which is in general a simpler task. Indeed, to disprove a conjecture it is sufficient to exhibit a counterexample, but finding one must be difficult or otherwise the conjecture would not still be open. The search of a counterexample becomes then an interesting problem on its own, which can be solved by an ingenious construction or, in lack thereof, by some (apparently) "brute-force" approaches which, however, require a certain degree of sophistication to succeed.

In this paper we adopt the latter strategy to prove a relatively minor, but still challenging, conjecture on graphs by Bauer, Broesma and Veldman (1990) stating that *"Every 4-regular, 1-tough graph with at most 17 nodes is hamiltonian."* Prior to our work it was known that the statement holds for graphs of at most 15 nodes, and also that it is false for 18 nodes. Our work has confirmed that the conjecture was correct.

In order to settle the above conjecture we have formulated the search of a counterexample as an Integer Linear Program (ILP), to be solved by standard branch-and-bound techniques. The ILP models a feasibility problem, whose solution would be an n-vertex graph (for $n = 16, 17$) which is 4-regular, 1-tough but

© Springer Nature Switzerland AG 2020
A. Kononov et al. (Eds.): MOTOR 2020, LNCS 12095, pp. 69–84, 2020.
https://doi.org/10.1007/978-3-030-49988-4_5

non-hamiltonian. Since at the end of the computation the model turned out to be infeasible, we have obtained a proof that the conjecture is indeed true.

Exploiting computers for settling conjectures and/or for theorem proving is not new and has been increasingly adopted in the recent years. The most noteworthy example of this type of approach is the proof that all planar graphs are 4-colorable, also known as the 4-colors theorem, by Appel and Haken in 1976 [1]. Our approach promotes the use of ILP for building combinatorial structures with some given properties (or proving that such structures do not exist) which seems, all considered, a potentially viable line of attack. Clearly, we are well aware that this line of approach has its limitations (mostly due to the high running times required by "large" instances), but we believe that there are still some "small" conjectures in fields such as combinatorics or graph theory for which this strategy could be worth trying.

1.1   Notation and Problem Statement

Let $G = (V, E)$ be an undirected graph. The graph is called k-$regular$ if every node has degree k. For each $S \subseteq V$ we denote by $\delta(S)$ the set of edges of G having an endpoint in S and the other in $V \backslash \{S\}$. Moreover, we denote by $G[S]$ the subgraph of G induced by S, i.e., the graph with vertex set S and edge set $E(S)$ containing all the edges of E with both endpoints in S. The graph is called $connected$ if for each pair of vertices i and j there is a path between i and j in the graph. The graph is called 2-$connected$ if for each $i \in V$ the graph $G[V \setminus \{i\}]$ is connected. A hamiltonian circuit of G is a circuit that visits each node of V exactly once. A graph is called $hamiltonian$ if contains a hamiltonian circuit. The complete graph K_n is the simple graph with n vertices and edges $\{i, j\}$ for each pair $i, j \in V$. We denote the sets of vertices and edges of K_n by, respectively, V^n and E^n. Without loss of generality, we assume $V^n := \{1, \ldots, n\}$.

Let $\omega(G)$ denote the number of components of the graph G. The graph is called t-tough if $|S| \geq t \omega(G[V \setminus S])$ for every subset $S \subseteq V$ with $\omega(G[V \setminus S]) > 1$. The $toughness$ of G, denoted $\tau(G)$, is the maximum value of t for which G is t-tough. Computing the toughness of a graph is an NP-hard problem [2].

The concept of toughness was introduced over 40 years ago by Chvátal [7]. A lot of research has since then been done, mostly investigating the relation between toughness conditions and the existence of cycle structures (see [3,6] for a survey). The original paper by Chvátal contained a number of conjectures, the most challenging of which, still open, states that there exists a finite constant t_0 such that every t_0-tough graph is hamiltonian. Originally it was also believed that in fact $t_0 = 2$ (the "2-tough conjecture") but in 2000 this conjecture was shown to be wrong by Bauer, Broesma and Veldman [5] who provided a 42-nodes counterexample.

Although a small value of toughness is not enough to imply that a graph is hamiltonian, there might be simple additional conditions to force the existence of a hamiltonian cycle in a t-tough graph for a small t, such as $t = 1$.

The property of 1-toughness can be stated in a simple way, namely, a connected graph is 1-tough if one cannot create c components by removing less than c vertices. The following are easy to prove immediate consequences of the definition:

Proposition 1. *If a graph G is 1-tough, then G is 2-connected.*

Proposition 2. *If a graph G is hamiltonian, then G is 1-tough.*

Although hamiltonian implies 1-tough, the converse is in general not true (see, e.g., the Petersen graph). However, a simple condition such as being k-regular might be enough to force a small 1-tough graph to be hamiltonian. In [4] the authors considered the non-hamiltonian k-regular 1-tough graphs for $k \geq 3$ (called (n, k)-graphs where n is the number of the vertices) and studied the problem of finding the minimum order $n = f(k)$ for which there exists an (n, k)-graph. For the case $k = 4$, they provided the $(18, 4)$-graph represented in Fig. 1, proving that $f(4) \leq 18$.

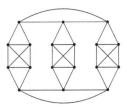

Fig. 1. The $(18, 4)$-graph by Bauer, Broersma and Veldman [4]

On the other hand, the following result of Hilbig implies $f(4) \geq 16$.

Theorem 1. *([11]) Let G be a 2-connected k-regular graph on at most $3k + 3$ vertices. Then G is hamiltonian or G is the Petersen graph P or G is the 3-regular graph obtained by P by replacing one vertex by a triangle.*

At the end of the paper [4], Bauer et al. conjectured that $f(4) = 18$. (For a discussion on this conjecture see also [8]). Given that $f(4) \geq 16$, we can restate this statement explicitly as the following theorem:

Theorem 2. *Each 4-regular, 1-tough graph of $n = 16$ or $n = 17$ nodes is hamiltonian.*

In this paper we give a proof of the above theorem showing, by contradiction, that no 4-regular, 1-tough, non-hamiltonian graph of 16 or 17 nodes exists.

1.2 Paper Organization

The remainder of the paper is organized as follows. In Sect. 2 we describe a preliminary analysis about 4-regular 2-connected graphs of 16 and 17 nodes that are not 1-tough. In Sect. 3 we present the ILP model for the search of possible counterexamples to the conjecture. In Sect. 4 we describe the main features of the branch-and-cut procedure used to solve the ILP model. Finally, in Sect. 5 we report the results of the computations and draw some conclusions.

2 Replacing 1-Toughness by 2-Connectedness: A Preliminary Analysis

Our strategy is based on the use of ILP to model a counterexample to the conjecture. The variables of the model will represent the edges of the sought graph, i.e., of the hypothetical counterexample, and there is no objective function (i.e., the objective coefficients are all 0). While the constraints which enforce a graph to be 1-tough are not simple (it is NP-complete to determine if a graph is 1-tough [2]), it is easy to state a set of constraints which imply that a graph is 2-connected. Since by Proposition 1 every 1-tough graph is 2-connected, the search for a counterexample could then be relaxed to a somewhat simpler task, i.e., finding a 4-regular, *2-connected*, non-hamiltonian graph. If we fail, then we can conclude that there is no counterexample at all. However, we might succeed and find a 4-regular, 2-connected, non-hamiltonian graph which is not 1-tough. In this case, we should add constraints to avoid the feasibility of this graph and continue the search.

It becomes then important for our problem to study if it is possible to have 4-regular, 2-connected graphs of 16 or 17 nodes which are not 1-tough and, in that case, characterize their structure (in the following we call a graph with these properties an *R*-graph). In particular, given $n \in \{16, 17\}$ and $k = 2, \ldots, \frac{n}{2}$, we consider the problem of finding the maximum number $v(n, k)$ of connected components that can result by removing k nodes from a 4-regular 2-connected graph of n nodes. Clearly, if $v(n, k) \leq k$ for every k then there do not exist *R*-graphs with n vertices.

Also for this preliminary analysis we have used Integer Linear Programming. Before presenting the ILP model, let us we outline some simple properties.

Given a 4-regular 2-connected graph $G = (V, E)$ and a subset $S \subseteq V$ with $|S| = k$ let $V_1, \ldots, V_t$ be the vertex-sets of the t connected components of the graph $G[V \setminus S]$ and $n_r := |V_r|$, for $r = 1, \ldots, t$.

Proposition 3. *For each $r = 1, \ldots, t$ it is $n_r \geq 5 - k$.*

Proof. Assume $n_r \leq 4 - k$ for some r. Since each node v of V_r has degree 4 and can be adjacent to at most $n_r - 1 \leq 3 - k$ vertices of V_r, v must be adjacent to at least $k + 1$ vertices in S, a contradiction. $\qquad\square$

Proposition 4. *For each $r = 1, \ldots, t$ it is $|\delta(V_r)| \geq m_r := \max\{2, n_r(5 - n_r)\}$. This in particular implies $\sum_{r=1}^{t} m_r \leq 4k$.*

Proof. The 2-connectivity of G implies $|\delta(V_r)| \geq 2$. Moreover, if $n_r \leq 4$ then each vertex of V_r must be adjacent to at least $5 - n_r$ vertices of S, so $|\delta(V_r)| \geq m_r$. Since $\sum_{r=1}^{t} |\delta(V_r)| = |\delta(S)| \leq 4k$ the second statement holds. $\qquad\square$

For each n and k we can compute an upper bound $v'(n, k)$ to the value $v(n, k)$ by solving the following ILP problem. Let x_i be an integer variable representing the number of components of cardinality i in the graph $G[V \setminus S]$ and $m_i := \max\{2, 5i - i^2\}$. By Proposition 3 we can assume that i goes from $s(k) := \max\{1, 5 - k\}$ to $n - k$. Let us consider the model $\mathcal{P}_{n,k}$:

$$v'(n, k) := \max \sum_{i=s(k)}^{n-k} x_i \tag{1}$$

$$\sum_{i=s(k)}^{n-k} i\, x_i = n - k \tag{2}$$

$$\sum_{i=s(k)}^{n-k} m_i x_i \le 4k \tag{3}$$

$$x_i \in \mathbb{N}, \qquad s(k) \le i \le n - k. \tag{4}$$

The objective function counts the number of components of the graph $G[V \setminus S]$, the constraint (2) states that the total number of nodes in these components must be $n - k$ and the constraint (3) requires that the property stated in Proposition 4 is satisfied. If there exists an R-graph with n vertices then it must be $v'(n, k) > k$ for some k.

By solving problem $\mathcal{P}_{n,k}$ for $n = 16$, it turns out that $v'(16, k) > k$ only for $k = 2$, in which case it is $v'(16, 2) = 3$. The optimal solution is $x_4^* = 1$, $x_5^* = 2$, $x_i^* = 0$ for $i \ne 4, 5$. It is easy to verify that there is one R-graph compatible with this solution, namely the graph G_{16} shown in Fig. 2. Indeed let $S = \{v_1, v_2\}$ and $G[V_r]$ be a component of $G[V \setminus S]$ with $|V_r| = 4, 5$. A degree argument implies that when $|V_r| = 4$ then $G[V_r]$ is a complete graph K_4 with $|\delta(V_r)| = 4$, when $|V_r| = 5$ and $|\delta(V_r)| = 2$ then $G[V_r]$ is a graph R_5, where R_5 denotes a graph obtained by removing one edge from K_5. In the latter case, by the 2-connectivity of G, one of the nodes of degree three of $G[V_r]$ is adjacent to v_1, the other one to v_2. If one solves again problem $\mathcal{P}_{16,2}$ by adding either the condition $x_3 \ge 1$ or the condition $\sum_{i \ge 6} x_i \ge 1$, he obtains optimal value 2. This implies that x^* is the unique optimal solution of $\mathcal{P}_{16,2}$.

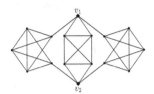

Fig. 2. Graph G_{16}: the unique 4-regular 2-connected graph with 16 vertices that is not 1−tough

By solving problem $\mathcal{P}_{n,k}$ for $n = 17$, it turns out that $v'(17, k) > k$ only for $k = 2, 4$ with optimal values, respectively, $v'(17, 2) = 3$ and $v'(17, 4) = 5$.

For $k = 2$ one optimal solution is $x_5^* = 3$ and $x_i^* = 0$ for $i \neq 5$. This solution determines the four graphs G_{17}^1, G_{17}^2, G_{17}^3 and G_{17}^4 in Fig. 3. Indeed let $G[V_r]$, $r = 1, 2, 3$, be the three components of the graph $G \backslash S$, $S = \{v_1, v_2\}$. Then $|\delta(V_r)| = 2$ for at least two components and thus, by the above argument, we may assume that $G[V_r]$, $r = 2, 3$, is a graph R_5 containing one neighbour of v_1 and one neighbour of v_2. If the nodes v_1 and v_2 are adjacent, the same holds for $G[V_1]$ and G is the graph G_{17}^1. Otherwise $|\delta(V_1)| = 4$ and the number y, w and z of the nodes of $G[V_1]$ of degree 2, 3 and 4, respectively, must satisfy the conditions $2y + 3w + 4z = 16$, $y + w + z = 5$ and $w \geq z - 2$. In particular, w has to be an even number not smaller than two. If $w = 2$ then $y = 1$, $z = 2$ and G is the graph G_{17}^4, otherwise $w = 4$, $x = 0$, $z = 1$ and $G[V_3]$ contains two nodes adjacent to v_1 and two nodes adjacent to v_2. If the two neighbours of v_1 are adjacent, then G is the graph G_{17}^2, otherwise G is the graph G_{17}^3.

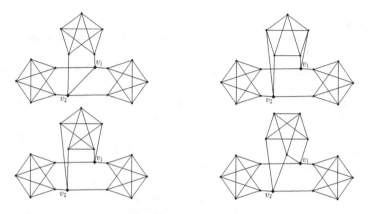

Fig. 3. 4-regular 2-connected graphs with 17 vertices that are not 1−tough: from left to right G_{17}^1, G_{17}^2 in the first line and G_{17}^3, G_{17}^4 in the second

By solving again $\mathcal{P}_{17,2}$ with the additional condition $x_3 + x_4 \geq 1$ one obtains a different solution $\bar{x}$ of value 3 with $\bar{x}_4 = \bar{x}_5 = \bar{x}_6 = 1$ and $\bar{x}_i = 0$ for $i \neq 4, 5, 6$. This solution is compatible only with the two graphs G_{17}^5 and G_{17}^6 in Figs. 4(left) and Fig. 4(center) which differ by the fact that the neighbours of v_1 and v_2 in the component with six nodes are joined by an edge or not. The solutions x^* and $\bar{x}$ are the only solutions of $\mathcal{P}_{17,2}$ of value 3. Indeed, by adding to $\mathcal{P}_{17,2}$ either the condition $x_3 \geq 1$ or the condition $\sum_{i \geq 7} x_i \geq 1$ one obtains 2 as the optimal value. This implies that $x_4 + x_5 + x_6 = 3$ must hold for each solution of value 3.

The optimal solution for $k = 4$ is $\hat{x}_1 = 3$, $\hat{x}_5 = 2$ and $\hat{x}_i = 0$, $i \neq 1, 5$, of value 5 which corresponds to the graph G_{17}^7 in Fig. 4(right). By adding each one of the constraints $x_1 \leq 2$, $x_1 \geq 4$ and $x_5 \leq 1$ to problem $\mathcal{P}_{17,4}$ one obtains an optimal value not grater than 4. So $\hat{x}$ is the unique optimal solution of the problem.

We summarise the results of the section in the following proposition.

Proposition 5. *Let $G = (V, E)$ be a 2-connected 4-regular graph which is not 1-tough. Then, if $n = 16$, G is the graph G_{16} in Fig. 2, while if $n = 17$ G is one of the graphs G_{17}^p, $p = 1, \ldots, 7$, in Figs. 3 and 4.*

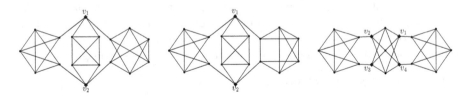

Fig. 4. 4-regular 2-connected graphs with 17 vertices that are not 1−tough: G_{17}^5 (left), G_{17}^6 (center) and G_{17}^7 (right)

3 The ILP Model for Finding a Counterexample

The problem of finding a counterexample to the statement of Theorem 2 can be rephrased as follows. Given the complete graph $K_n = (V^n, E^n)$, where $n \in \{16, 17\}$, find a subset $E \subseteq E^n$ such that the graph $G = (V^n, E)$ is 4-regular, 1-tough and non-hamiltonian. This problem can in turn be modeled by an Integer Linear Program as follows. Let us introduce a binary variable x_e for each edge $e \in E^n$ where $x_e = 1$ if $e \in E$, 0 otherwise. The following three families of constraints guarantee that a solution x defines a graph $G(x) = (V^n, E(x))$ with the required properties.

- **4-degree constraints**:

$$\sum_{e \in \delta(i)} x_e = 4 \qquad \forall\, i = 1, \ldots, n. \qquad (5)$$

These conditions force every node of $G(x)$ to have degree four.

- **non-hamiltonian constraints**:

$$\sum_{e \in H} x_e \leq n - 1 \quad \forall \text{ hamiltonian circuit } H \in \mathcal{H}^n \qquad (6)$$

where $\mathcal{H}^n$ denotes the set of the hamiltonian circuits of K_n. These conditions guarantee that the graph $G(x)$ is non-hamiltonian.

- **1-toughness constraints**:

$$\sum_{1 \leq a < b \leq t} \sum_{e \in \delta(V_a) \cap \delta(V_b)} x_e \geq 1 \quad \forall \text{ partition } S, V_1, \ldots, V_t \text{ of } V^n \text{ with } t > |S|.$$

$$(7)$$

These conditions require that for each partition $S, V_1, \ldots, V_t$ of V^n with $|S| < t$ there exists at least one edge in $E(x)$ joining two nodes in two distinct subsets V_a and V_b. This property is clearly equivalent to the 1-tough property.

The search for a counterexample to the conjecture is therefore equivalent to the search of a binary vector $x \in \{0, 1\}^{|E^n|}$ satisfying the constraints (5), (6) and (7). We call this feasibility ILP problem $\mathcal{M}_n$, for $n = 16$ and $n = 17$. As shown in Sect. 2, apart from a few cases, 4-regular graphs with 16 and 17 nodes that are 2-connected are also 1-tough. For this reason we consider an additional family of constraints that impose the 2-connectivity of the solutions.

- **2-connectivity constraints:**

$$\sum_{e \in \delta(S) \setminus \delta(v)} x_e \geq 1 \quad \forall \, v \in V, \; \forall \, S \subset V \setminus \{v\}, \; 0 < |S| < n - 1 \qquad (8)$$

These conditions require that for each node $v \in V$ and each subset $S \subseteq V \setminus \{v\}$, the cut $(S, V \setminus (S \cup \{v\}))$ in $G[V \setminus \{v\}]$ is not empty, i.e., that $G[V \setminus \{v\}]$ is connected.

The number of constraints (6), (7) and (8) grows exponentially with the number of nodes. Actually, even in the cases $n = 16$ and $n = 17$ this number is very high and makes it impossible to create all the constraints in order to input them to an ILP solver. To solve the problem $\mathcal{M}_n$, we will have to adopt a cutting plane approach (a *branch-and-cut* procedure [9, 16]) in which we do not to introduce all these constraints initially in the model but we generate them dynamically only when needed. The problem of determining if a missing inequality has in fact to be added to the model at any stage of the solving process is called the *separation* problem, and is described in the next section.

3.1   The Separation Problems

In this section we consider the separation problems associated to the constraints of the model $\mathcal{M}_n$. In general, given a family $\mathcal{L}$ of inequalities, the separation problem with respect to $\mathcal{L}$ is the following: Given a vector $\bar{x}$ (not necessarily integer), find an inequality of $\mathcal{L}$ violated by $\bar{x}$ or determine that no such inequalities do exist. An algorithm which solves this problem is called a separation algorithm for $\mathcal{L}$.

- **The separation problem for the non-hamiltonian constraints.** A vector $\bar{x}$ does not satisfy the non-hamiltonian constraints (6) if and only if there exists a hamiltonian circuit $H \in \mathcal{H}^n$ such that

$$\sum_{e \in H} \bar{x}_e > n - 1 \Leftrightarrow \sum_{e \in H} (\bar{x}_e - 1) > -1 \Leftrightarrow \sum_{e \in H} (1 - \bar{x}_e) < 1.$$

Thus the separation problem for the non-hamiltonian constraints can be solved by finding the shortest hamiltonian circuit in K_n with respect to the lengths

$c_e := 1 - \bar{x}_e$ for each $e \in E^n$. This is a Traveling Salesman Problem. If the optimal TSP solution H^* has value smaller than 1 then the constraint

$$\sum_{e \in H^*} x_e \leq n - 1$$

is violated by $\bar{x}$ (which, in our branch-and-cut approach, implies that it must be added to the model). Otherwise, no (6) constraint is violated by $\bar{x}$.

Being the TSP, the separation problem for the non-hamiltonian constraints is NP-hard for a general n. In our case, however, n is fixed (and rather small). Solving the TSP problem on a ≤ 17-node graph is quite simple (see the result on computational experiments) and there are several effective algorithms to this end. In particular, we have used a simple branch-and-bound procedure.

– The Separation for the 1-Tough Constraints. Since the problem of recognizing if a graph is 1-tough is NP-complete, the separation problem with respect to the 1-toughness constraints (7) is NP-hard. We will separate these constraints only for binary vectors $\bar{x}$, corresponding to graphs $G(\bar{x})$ that are 4-regular and 2-connected. As shown in the previous section there exist only one graph on 16 nodes and seven graphs on 17 nodes satisfying these conditions that are not 1-tough. A first way to separate the 1-toughness constraints for an integer $\bar{x}$ is that of testing if the graph $G(\bar{x})$ is isomorphic to G_{16} when $n = 16$ or is isomorphic to one of the graphs G_{17}^p, $p = 1, \ldots, 7$, when $n = 17$. Alternatively, one can model as an ILP the problem of finding a partition $S, V_1, \ldots, V_t$ of V with $|S| = k < t$ and such that $G(\bar{x})[V \backslash S]$ has t connected components. Let us introduce a binary variable z_i, $i \in V$, where $z_i = 1$ if $i \in S$ and 0 otherwise and a binary variable y_{ir}, $i \in V$ and $r = 1, \ldots, t$, where $y_{i,r} = 1$ if $i \in V_r$ and 0 otherwise. Any partition with the required properties defines a feasible solution of the ILP model

$$\mathcal{T}_{k,t}: \qquad \sum_{r=1}^{t} y_{ir} + z_i = 1 \qquad\qquad i \in V \qquad\qquad (9)$$

$$y_{ir} \leq y_{jr} + z_j \qquad \{i, j\} \in E(\bar{x}), \quad r = 1, \ldots, t \quad (10)$$

$$y_{jr} \leq y_{ir} + z_i \qquad \{i, j\} \in E(\bar{x}), \quad r = 1, \ldots, t \quad (11)$$

$$\sum_{i \in V} y_{ir} \geq 1 \qquad\qquad r = 1, \ldots, t \qquad\qquad (12)$$

$$\sum_{i \in V} z_i = k \qquad\qquad\qquad\qquad (13)$$

$$z_i \in \{0, 1\} \quad i \in V \qquad\qquad\qquad (14)$$

$$y_{ir} \in \{0, 1\} \quad i \in V, r = 1., \ldots, t. \qquad (15)$$

The conditions (9) impose that each node belongs to exactly one set of the partition. The conditions (10) and (11) guarantee that each edge $e \in E(\bar{x})$ has either both endpoints in a same set V_r or at least one endpoint in S. Finally, all the sets V_r are not empty by the constraints (12) and $|S| = k$ by the contraint (13). We observe that, on the base of the results in Sect. 2, it will be sufficient to solve the problem $T_{2,3}$ for $n = 16$ and the problems $T_{2,3}$ and $T_{4,5}$ for $n = 17$. Any feasible solution of these problems defines a 1-toughness constraint violated by the vector $\bar{x}$.

- **The Separation for the 2-Connectivity Constraints.** A vector $\bar{x}$ does not satisfy the 2-connectivity constraints if and only if there exists a node $\bar{v} \in V$ and a subset $\bar{S} \subseteq V \setminus \{\bar{v}\}$ such that the sum of the components $\bar{x}_e$ over the edges of the cut $\delta(\bar{S})$ in $K_n[V \setminus \{v\}]$ is strictly smaller than 1. Thus the separation problem for the 2-connectivity constraints can be solved in $O(n^4)$ by solving for each $v \in V_n$ a minimum-cut problem on the graph $K_n[V \setminus \{v\}]$ with respect to the weights $w_e := \bar{x}_e$. If for some $\bar{v}$ the optimal value is smaller than 1, the 2-connectivity constraint defined by $\bar{v}$ and the optimal solution $\bar{S}$ is violated by $\bar{x}$, while if the optimal value is ≥ 1 there are no violated 2-connectivity constraints. If $\bar{x}$ is integer, the separation problem can be alternatively solved in time $O(n)$ (see [19]) by searching for the articulation points of the graph $G(\bar{x})$, i.e., the nodes whose removal disconnects the graph.

4   The Branch-and-Cut Procedure for Solving $\mathcal{M}_n$

As already remarked, the model $\mathcal{M}_n$ has exponential size with respect to n and must therefore be solved with a constraint-generation approach. The standard way to do this is called *branch-and-cut*. Branch-and-Cut is a branch and bound in which the constraint matrix at each node of the search tree contains only a (small) subset of the constraints of the original model. Let us denote by N a node of the search tree and by $\mathcal{M}(N)$ the set of constraints of the subproblem corresponding to this node. These are the constraints that were input at the root node, plus all the constraints which were added in the nodes on the path from the root to N, plus all the branching constraints (fixing variables to 0 or 1) along the path to N.

Whenever the LP-relaxation of $\mathcal{M}(N)$ is solved, yielding a solution $\bar{x}$, the feasibility of $\bar{x}$ with respect to $\mathcal{M}_n$ must be checked (if $\bar{x}$ is feasible for $\mathcal{M}_n$ we could stop the search, since the counterexample would have been found). The solution $\bar{x}$ could be infeasible because it is fractional, but also an integer solution could be infeasible since many constraints of $\mathcal{M}_n$ are missing at N. Before branching from N and creating more subproblems we must then check if there are any constraints of $\mathcal{M}_n$ which are violated by $\bar{x}$. If that is the case, we add one or more of these constraints to $\mathcal{M}(N)$ and solve $\mathcal{M}(N)$ again. In this phase we use the separation procedures mentioned in Sect. 3.1 to find such inequalities. This phase is called constraint- (or cut-) generation. The processing of the node would terminate only when $\bar{x}$ is integer and feasible for $\mathcal{M}_n$ (which, however, never occurred), or when $\bar{x}$ is fractional but does not violate any constraint of $\mathcal{M}_n$. In this case, a branching is performed from N, by picking a fractional component $\bar{x}_j$ and creating two new subproblems, N' in which we fix $x_j = 0$, and N'' in which we fix $x_j = 1$.

Our branch and cut procedure was implemented within the framework SCIP [10] for solving ILPs. The overall computation was rather long, but it could have been *much* longer had we not adopted some steps in order to make the search as effective as possible. We now briefly describe some of the implementation decisions that we had to take.

4.1   Symmetries and Orbital Branching

Consider a solution $x \in \{0,1\}^{|E_n|}$ and the associated graph $G(x)$. For every permutation π of V_n we can define a new solution $\pi(x)$ by setting $\pi(x)_{ij} = x_{(\pi(i)\pi(j))}$ for each $\{i,j\} \in E_n$. The graph $G(\pi(x))$ is clearly isomorphic to $G(x)$. Since the 4-regularity, the 1-toughness and the property to be not hamiltonian are preserved by graph isomorphisms, x is feasible for $\mathcal{M}_n$ if and only if $\pi(x)$ is feasible. This shows that every permutation π of V_n induces a symmetry of the model $\mathcal{M}_n$. It is well known that even relatively small instances of ILP problems with large groups of symmetries can be extremely difficult to solve via branch and cut. For this reason several techniques have been proposed in literature to reduce the impact of symmetries (see for instance the surveys of Margot [13] and Pfetsch and Rehn [18]). Among these techniques a very effective one is Orbital Branching by Ostrovski and al. [17]. This method requires to compute at each node N of the branch and bound tree 1) the group S^N of the symmetries that stabilize the sets of the indices of the variables fixed at 0 and 1 at N and 2) the orbits induced by S^N, where the orbit of an edge $\bar{e}$ is the set $O(\bar{e}) = \{e \in E^n :$ there exists $\pi \in S^N : e = \pi(\bar{e})\}$. Then, given a free variable $x_{\bar{e}}$, two new nodes are created according to the disjunction $(x_{\bar{e}} = 1) \vee (\sum_{e \in O(\bar{e})} x_e = 0)$. Clearly, the additional computational effort required by the method is worthwhile as long as it returns orbits of rather large size, in which case the orbital branching rule significantly limits the visit of isomorphic solutions. Since the branching constraints tends to reduce the symmetries of the problem, orbital branching is usually performed only at the first levels of the branch and bound tree.

The strong impact of the orbital branching method in solving our problem is highlighted by the computational results in Table 1 obtained by solving problem $\mathcal{M}_n$ for $8 \le n \le 13$ using the default branching rule vs using orbital branching in the first eight levels of the search tree. For these values of n the 1-toughness constraints are implied by the 2-connectivity constraints and are not included in the model. The extremely high times required for $n = 12, 13$ with the default branching rule make it clear that it is unlikely that the model can be solved in a reasonable time for larger values of n without resorting to a suitable method to handle its symmetries.

4.2   A Decomposition to Overcome the 1-Toughness Constraints

A main issue in solving the problem $\mathcal{M}_n$ is the time required by the separation routine for the 1-toughness constraints (7). This is true either if we choose to separate them by solving the ILP problems $\mathcal{T}_{k,t}$ introduced in Subsect. 3.1 or by checking the isomorphism of $G(\bar{x})$ with one of the graphs described in Sect. 2.

Table 1. Times and nodes required by solving model $\mathcal{M}_n$ using the default branching rule of SCIP and the orbital branching (up to level 8).

	SCIP		SCIP+OB	
n	Time (sec)	# Nodes	Time	# Nodes
8	0,03	31	0,01 s	13
9	0,12	209	0,06 s	104
10	1,18	2253	0,46 s	809
11	83,20	78441	4,15 s	5550
12	18827,38	2257286	12,34	22896
13	$>5 \times 10^5$	$>15 \times 10^6$	212.33	187572

To overcome the complexity of dealing with the 1-toughness constraints, we have chosen to adopt a decomposition scheme, based on the analysis of Sect. 2. This decomposition has a twofold positive effect: (1) it allows us to restrict the cases when we have to actually separate the constraints (7) to a limited set of possibilities, in which the value of many variables can be fixed before the computation even starts; (2) it allows us to strengthen the formulation for the remaining cases by the introduction of a new family of constraints. We now briefly describe this decomposition scheme for $n = 16$ and $n = 17$.

Case $n = 16$. We know from the results of Sect. 2 that the only 4-regular 2-connected graph which is not 1-tough is the graph G_{16} of Fig. 2. This graph contains three disjoint induced subgraphs: a complete graph K_4 and two graphs R_5, where, as in Sect. 2, R_5 denotes a graph obtained by removing one edge from K_5. We decided to partition the solution set $\mathcal{F}_{16}$ of $\mathcal{M}_{16}$ into three sets $\mathcal{F}_{16}(2R5)$, $\mathcal{F}_{16}(1R5)$ and $\mathcal{F}_{16}(NOR5)$ which are, respectively, the solutions corresponding to graphs containing at least two (disjoint) copies, a single copy or no copy of R_5.

This partitioning allows us to fix many variables in the model. In particular, since the solutions in $\mathcal{F}_{16}(2R5)$ must contain two disjoint copies of R_5 we can fix to 1 the variables corresponding to the subsets of edges shown in Fig. 5(left). Similarly, since the solutions in $\mathcal{F}_{16}(1R5)$ contain one copy of R_5, and considering the 2-connectivity property, in this case we can fix to 1 the variables corresponding to the edges in Fig. 5(right).

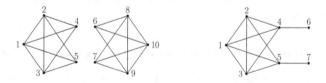

Fig. 5. Variables fixed to 1 for the solutions in $\mathcal{F}_n(2R5)$ (left) and $\mathcal{F}_n(1R5)$ (right).

Furthermore, we can add to the model a set of inequalities which forbid the existence of any further R_5 other than the one possibly fixed as above. We call these inequalities *noR5-constraints (NOR5)* and they are

$$\sum_{e \in E(V')} x_e \leq 8 \qquad \forall\, V' \subseteq W,\ |V'| = 5, \tag{17}$$

where $W = \{6, \ldots, 16\}$ in the case of $\mathcal{F}_{16}(1R5)$, and and $W = V$ in the case of $\mathcal{F}_{16}(NOR5)$. Notice that, for a general n, the number of these constraints is $O(n^5)$ but this number is not small, not even when $n = 16$. Therefore, we have decided not to add all these constraints to the models, but to separate them only when needed.

Based on the above partitioning, we have solved three problems, but only in the one for $\mathcal{F}_{16}(2R5)$ we have used the separation of 1-toughness constraints, since in the remaining two cases 2-connectivity constraints were sufficient.

Case $n = 17$. We have adopted the same partitioning of the solutions also for the case $n = 17$, namely we have decomposed the feasible set $\mathcal{F}_{17}$ into three sets $\mathcal{F}_{17}(2R5)$, $\mathcal{F}_{17}(1R5)$ and $\mathcal{F}_{17}(NOR5)$. We observe that the case $\mathcal{F}_{17}(2R5)$ includes the five problematic graphs G_{17}^p, $p = 1, 2, 3, 4, 7$, and hence separation of 1-toughness constraints is needed, done by solving the problems $T_{2,3}$ and $T_{4,5}$. The case $\mathcal{F}_{17}(1R5)$ includes the two problematic graphs G_{17}^5 and G_{17}^6 and hence separation of 1-toughness constraints is needed, done by solving the problem $T_{2,3}$. The case $\mathcal{F}_{17}(NOR5)$ does not require separation of 1-toughness inequalities.

Also the fixing of the initial variables is the same as the one used for $n = 16$. Furthermore, we know from the Hoffman-Singleton Theorem [12] that there do not exist 4-regular graphs with diameter 2 on 17 nodes. This fact allows us to fix to 1 in $\mathcal{F}_{17}(NOR5)$ the variables for two disjoint sets of edges, namely a star with center 1 and neighbors $\{2, 3, 4, 5\}$, and another with center 6 and neighbors $\{7, 8, 9, 10\}$.

4.3   Solution of Each Subproblem

For each one of the sets defined in the previous subsection, we initialized an integer linear programming problem including only the 4-degree constraints and those defining the initial fixing of some variables. The remaining constraints were added only when generated by the separation routines. In particular we called the separation procedures for the non-hamiltonian constraints, the 2-connectivity constraints, possibly the noR_5-constraints and the 1-toughness constraints in this order. Indeed an (integer) solution that does violates a non-hamiltonian constraint corresponds to a hamiltonian graph and so it cannot violate any 2-connectivity or 1-toughness constraint. Moreover, our separation routine for the 1-toughness constraints is based on the analysis of Sect. 2 and thus it works only if applied to integer vectors that satisfy the 2-connectivity constraints. As described in the previous section, we actually considered (and separated) the 1-toughness constraints only for the instances defined by $\mathcal{F}_{16}(2R5)$, $\mathcal{F}_{17}(1R5)$ and $\mathcal{F}_{17}(2R5)$.

For the separation of the non-hamiltonian constraints we have used a combinatorial branch and bound algorithm for the TSP. Although not the best possible for large TSP instances, this simple algorithm turned out to be very fast for $n = 16, 17$. We have separated the 2-connectivity constraints only for integer solutions using a combinatorial procedure that searches for the articulation points of a graph. Finally, we separated the noR_5-constraints by an exhaustive search for subsets of nodes that could violate the conditions stated in (17).

5   Computational Experiences and Conclusions

In this paper we have modeled the set of counterexamples to Bauer et al.'s conjecture as the integer points of a polytope contained in $[0, 1]^{n(n-1)/2}$, with variables x_e associated to the edges of a complete graph of $n = 16$ and 17 nodes. We easily modeled the 4-regular degree constraints, while the non-existence of a hamiltonian cycle or the property of being 1-tough were guaranteed by a set of exponentially many inequalities, dealt with implicitly through a branch-and-cut procedure.

The algorithm was implemented within SCIP, a framework for constraint integer programming and branch-cut-and-price developed at ZIB (Zuse Institute Berlin). We solved all the instances corresponding to the decomposition described in Subsect. 4.2 for $n = 16$ and $n = 17$, with an overall computation which took about a week (see Table 2 for a detailed account of the time needed to solve each of the subproblems in the decomposition). This might appear as a long time, but it is in fact very small if compared to a brute force approach that should have considered about 30,000 billions of billions of 4-regular graphs of 17 nodes. Even assuming that checking if a graph is 1-tough and hamiltonian takes 1/1000 s, the task would have taken roughly 1 billion centuries. In order to solve the problem in a reasonable amount of time some aspects proved to be fundamental, especially the preliminary analysis and the use of orbital branching.

Since all the instances were found infeasible, we conclude that the conjecture by Bauer, Broesma and Veldman is true.

Table 2. Times and number of nodes required to solve problem $\mathcal{M}_n$ for the sets defined in Subsect. 4.2.

	Time (seconds)	# Search tree nodes
$\mathcal{F}_{16}(2R5)$	2,17	28
$\mathcal{F}_{16}(1R5)$	2171,22	624623
$\mathcal{F}_{16}(NOR5)$	250904	16167197
$\mathcal{F}_{17}(2R5)$	171,03	367
$\mathcal{F}_{17}(1R5)$	1946,81	243540
$\mathcal{F}_{17}(NOR5)$	422303	37750576

While testing our implementation, we also decided to use it for solving $\mathcal{M}_{18}$ for which it is known that there exists a counterexample. Our program did in fact find a feasible solution, i.e., the same graph of Fig. 1 described by Bauer et al. who, on the other hand, had to find it in a (more clever) "old-fashioned" way.

We conclude by remarking how the use of ILP modeling seems promising for building small combinatorial structures (such as graphs) with some given properties, or proving that none exists. We expect to see more applications of this type in the future.

Funding Information. This research has been carried out in the framework of the departmental research project ICON: Innovative Combinatorial Optimization in Networks, Department of Mathematics, Computer Science and Physics (PRID 2017–2018), University of Udine, Italy.

References

1. Appel, K., Haken, W.: Every planar map is four colorable. I. Discharging. Illinois J. Math. **21**(3), 429–490 (1977)
2. Bauer, D., Hakimi, S.L., Schmeichel, E.: Recognizing tough graphs is NP-hard. Discrete Appl. Math **28**, 191–195 (1990)
3. Bauer, D., Broersma, H.J., Schmeichel, E.: Toughness in graphs - a survey. Graphs Comb. **22**, 1–35 (2006)
4. Bauer, D., Broersma, H.J., Veldman, H.J.: On smallest nonhamiltonian regular tough graphs. Congressus Numerantium **70**, 95–98 (1990)
5. Bauer, D., Broersma, H.J., Veldman, H.J.: Not every 2-tough graph is hamiltonian. Discrete Appl. Math **99**, 317–321 (2000)
6. Broersma, H.J.: How tough is toughness? Bull. Eur. Assoc. Theoret. Comput. Sci. **117**, 28–52 (2015)
7. Chvátal, V.: Tough graphs and hamiltonian circuits. Discrete Math. **5**, 215–228 (1973)
8. DeLeon, M.: A study of sufficient conditions for hamiltonian cycles. Rose-Hulman Und. Math. J. **1**(1), 1–19 (2000)
9. Desrosiers, J., Lübbecke, M.E.: Branch-Price-and-Cut Algorithms, in the Wiley Encyclopedia of Operations Research and Management Science. Wiley, Chichester (2010)
10. Gamrath, G., et al.: The SCIP Optimization Suite 3.2.1, ZIB-Report, pp. 15–60 (2016)
11. Hilbig, F.: Kantenstrukturen in nichthamiltonschen Graphen. Ph.D. thesis, Technische Universït at Berlin (1986)
12. Hoffman, A.J., Singleton, R.R.: On Moore graphs of diameter two and three. IBM J. Res. Dev. **4**, 497–504 (1960)
13. Margot, F.: Symmetry in integer linear programming. In: Jünger, M., et al. (eds.) 50 Years of Integer Programming 1958-2008, pp. 647–686. Springer, Heidelberg (2010). https://doi.org/10.1007/978-3-540-68279-0_17
14. McKay, B.D., Piperno, A.: Practical graph isomorphism, II. J. Symbolic Comput. **60**, 94–112 (2014)
15. McKay, B.D., Piperno, A.: nauty and Traces User's Guide (Version 2.6) (2016)

16. Mitchell, J.E.: Branch-and-cut algorithms for combinatorial optimization problems. In: Pardalos, P.M., Resende, M.G.C. (eds.) Handbook of Applied Optimization, pp. 65–77. Oxford University Press, Oxford (2002)
17. Ostrowski, J., Linderoth, J., Rossi, F., Smriglio, S.: Orbital branching. Math. Program. **126**(1), 147–178 (2011)
18. Pfetsch, M.E., Rehn, T.: A computational comparison of symmetry handling methods for mixed integer programs (2015). http://www.optimization-online.org
19. Skiena, S.S.: The Algorithm Design Manual. Springer, London (2008). https://doi.org/10.1007/978-1-84800-070-4

On Asymptotically Optimal Solvability of Max m-k-Cycles Cover Problem in a Normed Space

Edward Kh. Gimadi[1,2(✉)] [iD] and Ivan A. Rykov[1,2]

[1] Sobolev Institute of Mathematics, SB RAS, Novosibirsk, Russia
gimadi@math.nsc.ru, rykovweb@gmail.com
[2] Novosibirsk State University, Novosibirsk, Russia

Abstract. We consider the intractable problem of finding m edge-disjoint vertex covers in d-dimensional normed space with maximum total weight, such that each of them has exactly k cycles. We construct a polynomial-time approximation algorithm for solving this problem and derive conditions of its asymptotical optimality.

Keywords: Cycles cover · m-PSP · Asymptotically optimal · Normed space · Polynomial-time algorithm

1 Introduction

We consider the following problem: given a complete undirected weighted graph $G = (V, E)$, where the set V consists of n vertices represented by points in d-dimensional space R^d.

We assume that the vertexes of the graph G belong to a normed space R^d and the weight of an edge (x, y) is equal to $\|x - y\|$, where $\| \cdot \|$ is a given norm on R^d. Following [19] we define a concept of an angle in an arbitrary normed space, defining the an angle α between the vectors x and y as the distance between the vector $x/\|x\|$ and the closest of two vectors $\pm y/\|y\|$, that is, $\alpha(x, y) = \min\{\|x/\|x\| - y/\|y\|\|, \|x/\|x\| + y/\|y\|\|\}$. For $x = \lambda y$ or if the norm of one of the vectors equals zero, the angle between x and y is assumed to be zero.

A cycle cover of a graph is a spanning subgraph which consists of one or several cycles.

A problem is to find a union $\mathcal{C} = \{C_1 \cup C_2 \cup \cdots \cup C_m\}$ of m edge-disjoint cycle covers, with maximum total weight of all the edges in the union $\mathcal{C}$, such that each cycle cover C_i consists of exactly k cycles.

The work is supported by the Russian Ministry of Science and Education under the 5–100 Excellence Programme and by the program of fundamental scientific researches of the SB RAS I.5.1).

2   Related Works

The problem is closely related to the well-known Traveling Salesman Problem (TSP), being a direct generalization of k-Cycles Cover Problem (k-CCP). In the latter problem, given a fixed natural number k and a complete weighted graph $G = (V, E)$, it is required to find an extremal (minimum, or maximum)-weight vertex cover of G with k vertex-disjoint cycles. k-CCP is in turn a generalization of TSP (which is k-CCP for $k = 1$).

The following results are known for minimization and maximization versions of TSP and its generalizations.

2.1   Minimization Problem

It is known [16] that the TSP is NP-hard even in the Euclidean case, i.e., its optimal solution can not be found in polynomial time, unless $P = NP$. Although TSP is hard to approximate [17] in the general case, polynomial-time approximation algorithms are developed for some special cases. For instance, the Metric TSP [4] can be approximated in polynomial time with a ratio $3/2$, and, for Euclidean TSP, a polynomial-time approximations scheme [1] and an asymptotically correct algorithm [21] and [11] are developed.

Other well-known generalizations of the TSP are the m-Peripatetic Salesman Problem (m-PSP) and the Min-L-Cycle Cover Problem (Min-L-CCP).

m-PSP is a problem of finding m edge-disjoint Hamiltonian cycles of minimum or maximum total weight in a complete edge-weighted n-vertex graph. The problem was first introduced by Krarup in [13] and is known to be NP-hard [5]. Being generalization of TSP, the problem remains intractable in metric and Euclidean special cases.

In the Min-L-Cycle CP it is required to find the cycle cover, such that the length of every cycle belongs to the set $L \subseteq \{3, 4, 5, ...\}$. Here the length of a cycle is the number of its edges. The problem is NP-hard and APX-hard for almost all sets L [15].

In [9] and [10] it is shown that the Min k-cycles CP is NP-hard in the strong sense, both in general and in particular cases, Metric and Euclidean. For the case $k = 2$ efficient 2-approximate algorithm is proposed, and for the Euclidean problem on the plane a polynomial-time approximation scheme is derived.

2.2   Maximization Problem

An asymptotically optimal algorithm was introduced for Euclidean Max m-PSP in [2]. In [8] a geometric variant of the problem in normed space is considered and a polynomial time approximation algorithm for the m-PSP in a normed space with fixed dimension is suggested. It is established that the algorithm is asymptotically optimal for $m = o(n)$.

In [12] the approximation polynomial-time algorithms for the Euclidean Max k-cycles CP with a given lengths of cycles in a multidimensional Euclidean space and the Random Max k-cycles CP with random instances UNI$(0, 1)$ are considered. It is shown that both algorithms have time complexity $\mathcal{O}(n^3)$ and are asymptotically optimal for the number of covering cycles $k = o(n)$ and $k \leq n^{1/3} \ln n$, respectively.

In [18] author considers a problem of finding cycle cover with maximum total weight which satisfies an given upper limit on the number of cycles and a lower limit on the number of edges in each cycle. He suggests a polynomial-time algorithm for solving this problem in the geometric case when the vertexes of the graph are points in a multidimensional real space and the distances between them are induced by a positive homogeneous function, such that its unit ball is an arbitrary convex polyhedron with a fixed number of facets.

3 Preliminary

The input for the normed max-m-k-cycles cover problem is defined by integer numbers d, n, m, k and the points $v_1 = (v_{11}, \ldots, v_{1d}), \ldots, v_n = (v_{n1}, \ldots, v_{nd})$ constituting the set V. The points define complete weighted graph $G = (V, E)$, where the weight of edge is equal to the distance (i.e., the norm of difference) between corresponding points.

This problem is a generalization of two problems mentioned above, both of them being generalizations of Max TSP:

- problem of finding m edge-disjoint Hamiltonian cycles in complete graph (Max m-PSP): Max TSP is a particular case with $m = 1$;
- problem of finding a maximum-weight single covering of complete graph with k cycles (the Max k-cycles CP): Max TSP is a particular case with $k = 1$.

First of these problems was considered in [2], where an asymptotically optimal algorithm for the problem it was introduced.

Definition 1. *An approximation algorithm A for some maximization problem has a guaranteed relative error ε, if the algorithm is a $(1 - \varepsilon)$-approximation. Namely,*

$$\frac{OPT(X) - F_A(X)}{OPT(X)} \leq \varepsilon,$$

for any input X, where $OPT(X)$ is the optimal value of the objective function for the input X and $F_A(X)$ is the value of the objective function obtained by the algorithm A.

Notation $\varepsilon_A(n)$ is used for the worst relative error on all cases of size n:

$$\frac{OPT(X) - F_A(X)}{OPT(X)} \leq \varepsilon_A(n) \quad \forall X : |X| = n$$

(i.e., for any input X consisting of n vertexes). A following theorem was proved:

Theorem 1. *Euclidean Max m-PSP problem is solved with relative error*

$$\varepsilon_A(n) \leq \frac{1}{n} + \alpha_d \left(\frac{m}{n}\right)^{\frac{2}{d+1}},$$

α_d *being a constant depending only on the dimension d.*

In [12] an algorithm for the latter problem (of finding k-cycles coverage of maximal length) was introduced, with the following estimation for the relative error:

Theorem 2. *The Euclidean Max k-cycles CP is solved with relative error*

$$\varepsilon_{\mathcal{A}}(n) \le \frac{2k}{n} + \beta_d \left(\frac{1}{n}\right)^{\frac{2}{d+1}},$$

β_d *being a constant depending only on the dimension d.*

In current work we present algorithm for the general geometric problem in normed space with both m and k being given as a part of input. We establish the sufficient conditions of asymptotical optimality of the suggested algorithm.

4   Algorithm Prerequisites

The ideas of both algorithms mentioned above, as well as the new algorithm are relied on techniques used to solve Euclidean Max TSP problem presented in [21] (later simplified in [11]), and Max TSP in geometric normed space presented in [19].

First, a maximal weighted matching $\mathcal{M}^*$ is found and used as a basis for building either a Hamiltonian cycle, several Hamiltonian cycles or a k-cycles coverage. In fact, the resulting Hamiltonian cycle is edge-disjoint with $\mathcal{M}^*$ (all of them are substituted with different edges in the construction process), but the procedure of substitution allows to estimate the loss in the weight of edges w.r.t. the matching. In particular, the following lemma is being used:

Lemma 1 ([19]). *Among any t vectors ($t < n/2$) in a normed space R^d there exist two vectors such that the angle between them doesn't exceed the value*

$$\alpha(d,t) = \frac{2d}{\lfloor (2t-1)^{1/d} \rfloor}. \tag{1}$$

In short, algorithm for solving Euclidean Max TSP takes the most parallel edges of the matching and substitute them with the best of two pairs of edges joining their end vertices. It is easy to see, that $\cos \alpha(d,t)$ is an estimation for the fraction of weight which remains after such operation. This is done until we have less than t unpaired edges left (t is a parameter of the algorithm), so t "lightest" edges are inserted between t paired chains separately.

In the case of $m = 1$ the maximum m-PSP is the well-known maximum Travelling Salesman Problem (maximum TSP) [3,14], which is NP-hard even in Euclidean d-dimensional space, $d \ge 3$ [20]. In [19] a polynomial asymptotically exact algorithm is introduced for the max-TSP in a normed space. This algorithm uses ideas of known algorithms [21] and [7] for the TSP in a Euclidean space. The algorithm is based on the following geometric facts:

Lemma 2 ([19]). *Let AB and CD be two intervals in R^d and α be the angle between them. Then*

$$1 \geq \frac{\max(\|AC\| + \|BD\|, \|AD\| + \|BC\|)}{\|AB\| + \|CD\|} \geq 1 - \alpha/2.$$

Definition 2. *Algorithm A is* asymptotically optimal *if $\varepsilon_A(n) \to 0$ as the size n of the problem indefinitely increases.*

In order to describe the algorithm for the problem under consideration, we will use the spatial graph structure called the pseudo-prism P (see Fig. 1).

Definition 3. *A subgraph of a graph G forms a* pseudo-prism *if it consists of two edge-disjoint cycles $(u_1, \ldots, u_\mu, u_1)$ and $(v_1, \ldots, v_\mu, v_1)$, such that each pair of vertexes with corresponding indexes is connected with an edge $(u_j.v_j)$, $j = 1, \ldots, \mu$, where $\mu = \lfloor \frac{n}{2} \rfloor$ ("connector"). Thus, the prism consists of 2μ vertices and 3μ edges.*

Note that unlike a regular spatial prism, in the case of a pseudo-prism the opposite edges in side quadrangles can be non-parallel, and the vertexes of the cycles do not have to belong to the same plane.

Further, the edges of the maximum matching of the original graph G will be used as connectors in the constructed pseudo-prism.

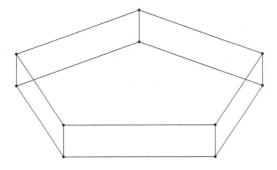

Fig. 1. The example of the pseudo-prism P

5   Description of the Algorithm $\mathcal{A}$ for the Maximum m-k-CsCP in a Normed Space

Let $w(u, v)$ be the weight of edge $e = (u, v) \in E$ and $W(G') = \sum_{e \in E'} w(e)$ be the total weight of a subgraph $G' = (V; E')$ of the initial graph $G = (V; E)$ with the set of edges $E' \subseteq E$. The goal of the algorithm $\mathcal{A}$ for the maximum m-k-CCP is to find a subset of edges $\widetilde{C} \subset E$, consisting of m edge-disjoint k-Cycles Covers $C_1, \ldots, C_m$. Initially, $\widetilde{C}$ is set empty.

Let $\mathcal{M}^* = \{I_1, \ldots, I_\mu\}$ be the set of edges (intervals in R^d) of a maximum weight matching in G; $\mu = \lfloor n/2 \rfloor$.

Definition 4. *Two edges $e_1, e_2 \in E$ are linked (with respect to set $\widetilde{C}$), if there exists an edge $e \in E$ ($e \in \widetilde{C}$), that connects the end vertices of e_1 and e_2.*

Definition 5. *An I-chain is a sequence of edges, where each two neighboring edges are linked.*

Definition 6. *Two I-chains are linked (with respect to set $\widetilde{C}$) if their end edges are linked.*

We refer to one of the end edges of an I-chain as a master edge and to the other one as the inferior edge.

Definition 7. *An α-chain is an I-chain, where the angle between any two neighboring edges of the chain is less or equal α.*

Now let's describe the approximation algorithm $\mathcal{A}$.

5.1   Algorithm $\mathcal{A}$

Preliminary Steps
Find a matching $\mathcal{M}^* = \{I_1, \ldots, I_\mu\}$ of maximum weight in graph G, where $\mu = \lfloor n/2 \rfloor$ is the number of its edges (intervals).

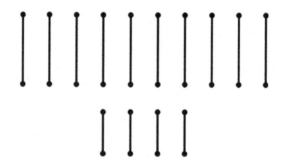

Fig. 2. $\mu - t$ heavy and t light edges of the maximal matching.

Put $\widetilde{C} = \emptyset$ and fix a parameter $t \le \mu/2$. Sort the edges of $\mathcal{M}^*$ in the non-increasing order. We will refer to the first (i.e., heaviest) $(\mu - t)$ edges of $\mathcal{M}^*$ as heavy edges, and to the last t edges as the light edges (Fig. 2).

Phase $i = 1, \ldots, m$
Phase i, consisting of Stages 1–5, constructs a k-Cycles Covering C_i.

Stage 1. Constructing a sequence $\mathcal{S} = \{S_1, \ldots, S_t\}$ of t α-chains.

Define angle $\alpha = \alpha(k, t)$ according to the relation (1):

$$
\alpha(d, t) = \begin{cases} \frac{2d}{\lfloor (2t-1)^{1/d} \rfloor}, & \text{if } i = 1; \\ \frac{2d}{\lfloor (2t/(2i-2)-1)^{1/d} \rfloor}, & \text{if } 1 < i \le m, \end{cases} \tag{2}
$$

where t is the number of available edges.

We are going to build a set $\mathcal{I}$ of α-chains, $|\mathcal{I}| = t$. Recall that we denoted the $(\mu - t)$ heaviest edges of $\mathcal{M}^*$ as the heavy edges. Each α-chain will consist only of the heavy edges. Note that an edge is a one-element α-chain. We will start with $\mathcal{I}_t$ consisting of the first t heaviest edges of $\widetilde{\mathcal{M}^*}$: $\mathcal{I}_t = \{I_1, I_2, \ldots, I_t\}$.

Put $j = t$.

In the current t-chain $\mathcal{I}_j$, find a pair of non-linked (with respect to the set $\widetilde{C}$) I-chains such that the angle between their master edges is at most α.

Join these chains into one α-chain by setting their master edges to be neighbors and assign one of the end edges of the joined chain (one of the former inferior edges) to be the new master edge.

Put $j := j + 1$. If $j < \mu - t$. Append one more heavy edge I_j to the current set $\mathcal{I}$ and repeat Stage 1.

Now we have obtained a sequence $\mathcal{S} = \{S_1, \ldots, S_t\}$ of t α-chains such that each of them consists of a sequence of heavy edges with the angle between any consecutive (neighboring) pair of edges at most $\alpha = \alpha(d, \tau)$ (Fig. 3).

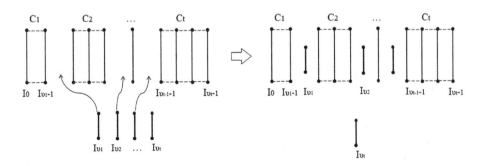

Fig. 3. The bold lines correspond to the edges of $\mathcal{M}^*$, while the dashed lines indicates the α-chains. The t light edges of $\mathcal{M}^*$ were placed to the positions between the α-chains. The last light edge I_{ν_t} is placed between the first and the t-th α-chains.

Stage 2. Constructing a pseudo-prism P_i.

Now consider the sequence $\mathcal{S}$ as a cycle, i.e. the α-chain S_t is followed by the α-chain S_1. Let the edges of the α-chains $S_1, \ldots, S_t$ be enumerated so that $S_r = \{I_{\nu_{r-1}+1}, \ldots, I_{\nu_r-1}\}$, $1 \le r \le t$, where $\nu_1 < \nu_2 < \ldots < \nu_t$ are the numbers reserved for the remaining light edges of the maximum matching $\mathcal{M}^*$ ($\nu_0 = 0$, $\nu_t = \mu$).

Place t light edges of the maximum matching $\mathcal{M}^*$ to the positions $\nu_1, \nu_2, \ldots, \nu_t = \mu$, such that no light edge is linked to the neighboring end edges of the α-chains.

Construct a pseudo-prism P_i such that $P_i \setminus \mathcal{M}^*$ is edge-disjoint with $P_1 \setminus \mathcal{M}^*, P_2 \setminus \mathcal{M}^*, \ldots, P_{i-1} \setminus \mathcal{M}^*$ in the following way.

We assume that the sequence of intervals $\{I_1, I_2, \ldots, I_\mu\}$ of edges of the maximum matching $\mathcal{M}^*$ is given according to their order in the sequence $\mathcal{S}$, $I_j = (x_j, y_j)$, $j = 1, \ldots, \mu$.

Now we are going to construct a pseudo-prism P_i.

For $j = 1, \ldots, \mu$ execute the following operator:

> if $w(u_{j-1}, x_j, u_{j+1}) + w(v_{j-1}, y_j, v_{j+1}) \geq w(u_{j-1}, y_j, u_{j+1}) + w(v_{j-1}, x_j, v_{j+1})$,
> then set $u_j = x_j$; $v_j = y_j$; otherwise, set $u_j = y_j$ and $v_j = x_j$.

As a result of the Stage 2 we have obtained a pseudo-prism P_i consisting of the two non-intersecting circuits $(u_1, u_2, \ldots, u_\mu.u_1)$ and $(v_1, v_2, \ldots, v_\mu.v_1)$, and of the maximum matching $\mathcal{M}^* = \{I_1, I_2, \ldots, I_\mu\}$ where $I_j = (u_j, v_j)$, $j = 1, \ldots, \mu$ (See Figs. 4 and 5).

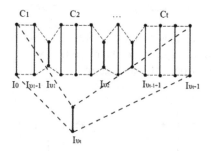

Fig. 4. The t light edges of $\mathcal{M}^*$ are placed to the positions between the α-chains. The last light edge I_{ν_t} is placed between the first and the t-th α-chains.

Stage 3. Constructing k-Cycles Covering C_i.

Randomly choose exactly k pairs of adjacent edges in the corresponding ordering of matching $\mathcal{M}^*$ and perform "reverse operation", i.e. return edges of matching into the solution C_i and remove pair of edges that connected endpoints of this pair of adjacent edges. (see Fig. 6).

On further iterations these $2ik$ edges of matching $\mathcal{M}^*$ added to $C_1, \ldots C_i$ are marked as forbidden.

Stage 4. Cases of evenness and oddness of n.

In the case of even n we have k-Cycles Covering C_i of graph G and go to Stage 5.

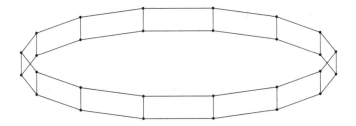

Fig. 5. The example of the pseudo-prism P_i with $n = 28$ after Stage 2.

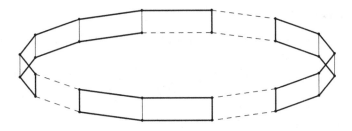

Fig. 6. The example of a solution on the pseudo-prism P_i with $n = 28$. Thick lines highlightes one of 3-Cycles Coverings.

If n is odd, there exists a vertex x_0 that is not in $\mathcal{M}^*$. In this case replace one of the matching edge (u, v) of the constructed covering by the pair of edges $(u, x0)$ and $(v, x0)$ so that none of these edges intersects the set $\widetilde{C}$. The triangle inequality guarantees that the weight of the cycle will not decrease.

Stage 5. Append the edges of the obtained k-covering C_i to the set $\widetilde{C}$.

The description of the algorithm $\mathcal{A}$ is complete.

6   Analysis of Algorithm $\mathcal{A}$

The algorithm $\mathcal{A}$ produces m k-coverings $C_1, C_2, \ldots, C_m$. On each phase we arrange the edges in $\mathcal{S}$ so that they are not linked with respect to the edges of $\widetilde{E}$, hence the obtained coverings are edge-disjoint. The running-time of the algorithm is determined by the time one needs to construct a maximum weight matching, which is $O(n^3)$.

6.1   Correctness of Stage 3

We search for another k pairs of adjacent edges. In the worst case $2ik$ edges are forbidden and $2ik - 1$ edges are located between them, one between each two forbidden and thus can't be used. So, in order to be able to finish this operation on m-th coverage, we need that

$$2k + 4(m - 1)k - 1 \leq n/2.$$

This implies the following condition on the parameters of the problem:

$$k \leq \frac{n}{8m}.$$

If such condition holds, we are guaranteed to be able to build all m covers.

6.2   Error Bound

Note that the total weight $W(\widetilde{\mathcal{M}^*})$ of the first $(\mu - t)$ heaviest edges of $\mathcal{M}^*$ satisfies the inequality [2]:

$$W(\widetilde{\mathcal{M}^*}) \geq W(\mathcal{M}^*)\left(1 - \frac{t}{\mu}\right). \tag{3}$$

Therefore, the weight of single k-CCP satisfies

$$W(C_i) \geq 2W(\mathcal{M}^*)\left(1 - \frac{t}{\mu}\right)(1 - \alpha_i/2),$$

where the angle α_i is defined by (2).

On the other hand, the upper bound of optimum for the coverage problem is estimated as

$$2W(\mathcal{M}^*) \geq \left(1 - \frac{k}{n}\right)W(C_i^*), \tag{4}$$

(the worst case of all odd cycles), used in [12].

Thus,

$$\frac{W(C_i)}{W(C^*)} \geq 1 - \frac{2(k + t) + 1}{n} - d\left(2\tau_i - 1\right)^{-1/d},$$

where C^* is the solution of the maximum k-CCP in the given graph and

$$\tau_i = \begin{cases} t, & \text{if } i = 1; \\ t/(2i - 2) & \text{for } 1 < i \leq m. \end{cases} \tag{5}$$

Using (3) and (5), for the approximation ratio $\varepsilon_{\mathcal{A}}(n)$ of algorithm $\mathcal{A}$ we have

$$\varepsilon_{\mathcal{A}}(n) = 1 - \frac{W_{\mathcal{A}}}{OPT} \leq 1 - \frac{W_{\mathcal{A}}}{mW(C^*)}$$
$$= 1 - \frac{W(C_1) + ... + W(C_m)}{mW(C^*)} \leq \frac{2(k+t)+1}{n} + \frac{d}{m}\sum_{i=1}^{m}(2\tau_i - 1)^{-1/d}. \tag{6}$$

Lemma 3. For $m \leq t$, $d \geq 2$ the following inequality holds:

$$\sum_{i=1}^{m}(2\tau_i - 1)^{-1/d} \leq 2m\left(\frac{m}{t}\right)^{\frac{1}{d}}.$$

Proof. Using the definition (5) of τ_i, we have:

$$\sum_{i=1}^{m}(2\tau_i - 1)^{-1/d} \leq m\left(\frac{t}{m} - 1\right)^{-1/d} \leq m\frac{(m/t)^{1/d}}{(1 - m/t)^{1/d}} \leq m\frac{(m/t)^{1/d}}{1 - m/(dt)} \leq 2m\left(\frac{m}{t}\right)^{1/d},$$

since under lemma assumptions $m/(dt) \leq 1/2$, hence $1/(1 - m/(dt)) \leq 2$.

Theorem 3. *The maximum m-k-CCP is solved by the approximation algorithm $\mathcal{A}$ with relative error satisfying*

$$\varepsilon_{\mathcal{A}}(n) \leq \frac{2(k+t)+1}{n} + 2d\left(\frac{m}{t}\right)^{1/d}. \tag{7}$$

Theorem 4. *The algorithm $\mathcal{A}$ with parameters $t^* = m\left(\frac{n}{m}\right)^{1/d}$ and $k \leq t^*$ gives asymptotically optimal solutions for the maximum m-k-CCP in the considered metric space with a fixed dimension d and $m = o(n)$.*

Proof. Setting $t = t^*$ in (7), we obtain

$$\varepsilon_{\mathcal{A}}(n) \leq \frac{1}{n} + 2\left(\frac{m}{n}\right)^{1-1/d} + 2(d+1)\left(\frac{m}{n}\right)^{1/(d+1)} \to 0$$

as $n \to \infty$.

7 Conclusion

Using angle estimations obtained in [19], we construct an algorithm for the maximum m-k-CCP, which gives asymptotically optimal solution for the problem in an arbitrary normed space of fixed dimension, given that $m = o(n)$ and $k \leq m\left(\frac{n}{m}\right)^{1/d}$.

As a topic for further research it is interesting to extend this approach to different modifications of the considered problem. For example, it is of natural interest to construct an asymptotically optimal algorithm for the m-k-CCP that would essentially rely on the specifics of problem statement and would have a better relative error or less tight condition for the numbers m and k of coverings.

References

1. Arora, S.: Polynomial-programming: methods and applications. In: Reeves, C.R. (ed.) NATO Advanced Study Institutes Series, Series C: Mathematical and Physical Sciences, vol. 19, pp. 1730–178. Reidel, Dordrecht (1975)
2. Baburin, A.E., Gimadi, E.Kh.: On the asymptotic optimality of an algorithm for solving the maximum m-PSP in a multidimensional Euclidean space. Proc. Steklov Inst. Math. **272**(Suppl. 1), 1–13 (2011)
3. Barvinok, A.I., Gimadi, E.Kh., Serdyukov, A.I.: The maximum TSP. In: Punnen, A., Gutin, G. (eds.) The Traveling Salesman Problem and Its Variations, pp. 585–608. Kluwer Acad. Publ., Dordrecht (2002)

4. Christofides, N.: Worst-case analysis of a new heuristic for the traveling salesman problem. In: Symposium on New Directions and Recent Results in Algorithms and Complexity, p. 441. Academic Press, New York (1976)

5. De Kort, J.B.J.M.: Upper bounds and lower bounds for the symmetric K peripatetic salesman problem. Optimization **23**(4), 357–367 (1992)

6. Duhenne, E., Laporte, G., Semet, F.: The undirected m-capacitated peripatetic salesman problem. Eur. J. Oper. Res. **223**(3), 637–643 (2012)

7. Gimadi, E.Kh.: A new version of the asymptotically optimal algorithm for solving the Euclidean maximum traveling salesman problem. In: Proceedings of the 12th Baykal International Conference 2001, Irkutsk, vol. 1, pp. 117–123 (2001). (in Russian)

8. Gimadi, E.Kh, Tsidulko, O.Yu.: Asymptotically optimal algorithm for the maximum m-peripatetic salesman problem in a normed space. In: Battiti, R., Brunato, M., Kotsireas, I., Pardalos, P.M. (eds.) LION 12 2018. LNCS, vol. 11353, pp. 402–410. Springer, Cham (2019). https://doi.org/10.1007/978-3-030-05348-2_33

9. Khachay, M., Neznakhina, E.: Approximation of Euclidean k-size cycle cover problem. Croatian Oper. Res. Rev. (CRORR) **5**, 177–188 (2014)

10. Khachay, M., Neznakhina, E.: Polynomial-time approximations scheme for a Euclidean problem on a cycle covering of agraph. Tr. Inst. Matematiki i mehaniki UrORAN **20**(4), 297–311 (2014). (in Russian)

11. Gimadi, E.Kh.: A new version of the asymptotically optimal algorithm for solving the Euclidean maximum traveling salesman problem. In: Proceedings of the 12th Baykal International Conference, Irkutsk, vol. 1, pp. 117–123 (2001). (in Russian)

12. Gimadi, E.Kh., Rykov, I.A.: Asymptotically optimal approach to the approximate solution of several problems of covering a graph by nonadjacent cycles. Proc. Steklov Inst. Math. **295**, 57–67 (2016)

13. Krarup, J.: The peripatetic salesman and some related unsolved problems. In: Roy, B. (ed.) Combinatorial Programming: Methods and Applications. ASIC, vol. 19, pp. 173–178. Springer, Dordrecht (1995). https://doi.org/10.1007/978-94-011-7557-9_8

14. Lawler, E.L., Lenstra, J.K., Rinnooy Kan, A.H.G., Shmoys, D.: The Traveling Salesman Problem: A Guided Tour of Combinatorial Optimization. Wiley, Chichester (1985)

15. Manthey, B.: Minimum-weight cycle covers and their approximability. In: Brandstädt, A., Kratsch, D., Müller, H. (eds.) WG 2007. LNCS, vol. 4769, pp. 178–189. Springer, Heidelberg (2007). https://doi.org/10.1007/978-3-540-74839-7_18

16. Papadimitriou, C.: Euclidean TSP is NP-complete. Theor. Comput. Sci. **4**, 237–244 (1977)

17. Sahni, S., Gonzales, T.: P-complete approximation problems. J. ACM **23**(3), 555–565 (1976)

18. Shenmaier, V.V.: An algorithm for the polyhedral cyclic covering problem with restrictions on the number and length of cycles. Tr. IMM UrORAN **24**(3), 272–280 (2018)

19. Shenmaier, V.V.: Asymptotically optimal algorithms for geometric Max TSP and Max m-PSP. Discrete Appl. Math. **163**(2), 214–219 (2014)

20. Barvinok, A.I., Fekete, S.P., Johnson, D.S., Tamir, A., Woeginger, G., Woodroofe, R.: The geometric maximum traveling salesman problem. J. ACM **50**(5), 641–664 (2003)
21. Serdyukov, A.I.: An asymptotically optimal algorithm for the maximum traveling salesman problem in Euclidean space. Upravlyaemye sistemy, Novosibirsk **27**, 79–87 (1987). (in Russian)

Mathematical Programming

D.C. Constrained Optimization Approach for Solving Metal Recovery Processing Problem

Rentsen Enkhbat[1] , Tatiana V. Gruzdeva[2(✉)] , and Jamsranjav Enkhbayr[3]

[1] Institute of Mathematics and Digital Technology,
Mongolian Academy of Sciences, Ulaanbaatar, Mongolia
renkhbat46@yahoo.com
[2] Matrosov Institute for System Dynamics and Control Theory of SB of RAS,
Irkutsk, Russia
gruzdeva@icc.ru
[3] National University of Mongolia, Ulaanbaatar, Mongolia

Abstract. This paper was motivated by an industrial optimization problem arisen at the Erdenet Mining Corporation (Mongolia). The problem involved real industrial data turned out to be a quadratically constrained quadratic programming problem, which we solve by applying the global search theory for general DC programming. According to the theory, first, we obtain an explicit DC representation of the nonconvex functions involved in the problem. Second, we perform a local search that takes into account the structure of the problem in question. Further, we construct procedures for escaping critical points provided by the local search method. In particular, we propose a new way of constructing an approximation of the level set based on conjugated vectors. The computational simulation demonstrates that the proposed method is a quite flexible tool which can fast provide operations staff with good solutions to achieve the best performance according to specific requirements.

Keywords: DC programming · Quadratic programming · Inequality constraints · Manufacturing processes · Linearized problem · Local search · Global search

1 Introduction

The general quadratic programming problems of the following form

$$f(x) = \langle x, Cx \rangle + \langle d, x \rangle + q \;\rightarrow\; \max(\min), \;\; x \in D, \tag{1}$$

where C is an $n \times n$ matrix, $d, x \in R^n$, and D is a nonempty polyhedral subset of R^n, play an important role in mathematical optimization. For example, quadratic programs appear as auxiliary problems for nonlinear programming (linearized problems) or in optimization problems that are approximated by

© Springer Nature Switzerland AG 2020
A. Kononov et al. (Eds.): MOTOR 2020, LNCS 12095, pp. 101–114, 2020.
https://doi.org/10.1007/978-3-030-49988-4_7

quadratic functions. Also, they have many applications in science, technology, statistics, economics, and industry. For instance, a problem of maximizing the rougher concentrate grade in the mining industry was considered as an optimization problem with quadratic objective function over a box constraint and solved numerically in [5].

Furthermore, many combinatorial optimization problems may be formulated as quadratic program, e.g., integer programs, quadratic assignment problems, linear complementary problems and network flow problems [13].

Moreover, in theory of optimal experiments and response surface problems $f(\cdot)$ is regarded as a criterion and D is interpreted as a feasible experimental region which might be even nonconvex set. Let us here remind briefly the main idea of the response surface. It is assumed that the experimenter is concerned with a technological process involving some response $f(\cdot)$ which depends on the input variables x_1, $x_2, \ldots, x_n$ from a given experimental region. The standard assumptions on function $f(\cdot)$ are that $f(\cdot)$ is a twice differentiable function on the experimental region and the independent variables x_1, $x_2, \ldots, x_n$ are controlled in the experimental process and measured with a negligible error. As a rule, the researcher has the second-order regression or a quadratic model expressed by a quadratic function that adequately represents the experimental data. It is important for the experimenter to find global solutions in the extremal problems. If one applies gradient methods or other local search methods, one might fail in finding a global solution, because problem (1), in some cases, becomes nonconvex. Unless he checks the nonconvexity of the problem (1), it is difficult to choose an appropriate algorithm in advance for solving it.

If C is indefinite then problem (1) is a nonconvex quadratic program. It is well known that in this case problem (1) is also NP-hard [14] and it can be reduced to a DC programming problem [11,18,21,22]. Classical optimality conditions for this problem yield only stationary points not guaranteed to be global solutions. When D is a polyhedral set, global search methods based on the gradient projection, generalized Bender's cut methods, linear relaxation methods, co-positivity procedure, and branch and bound algorithms, etc., have been developed for solving the problem (1) [1,3,11,13,14,23].

On the other hand, noncovexity of the feasible set D makes the problem very hard. One way to handle the problem is based on DC decomposition of the objective function and constraints in order to apply DC programming approach.

The aim of the paper is to develop a DC algorithm for solving a real-world problem arisen in the mining industry. The paper is organized as follows. In Sect. 2 we consider a real-world optimization problem from the mining industry which is a quadratic program with a quadratic constraint. In Sect. 3 we describe a local search algorithm for the DC programming problem. Section 4 is devoted to global optimality conditions for DC programming. The implementation issues are discussed in Sect. 5. In the last section, we present computational experiments on real industrial data.

2   Model Formulation

First, we consider the process of metal recovery in bulk and copper flotations under specified technological requirements. We examine the best-operating conditions (e.g., reagent dosages) of the flotation process at the Erdenet Mining Corporation Mineral Processing Plant (Erdenet, Mongolia).

Erdenet Mining Corporation is a joint Russian-Mongolian enterprise founded (together with the city of Erdenet) in 1974 and aimed at the commercial exploitation of Asia's largest porphyry copper-molybdenum deposit: Erdenetiyn-Ovoo ("the treasure mountain"). Now the corporation is one of the biggest mining and mineral processing companies in Asia that produces over 530 thousand tons of copper concentrate and about 4.5 thousand tons of molybdenum concentrate by processing 26 million tons of raw ore per year.

In order to model the flotation process of the copper-molybdenum ore, an orthogonal central composite design (for more details see [5]) was conducted. The response (dependent) variables were the metal recovery in copper and bulk flotations. Based on multi-variable linear regression model, we choose the following 11 factors as the independent variables:

x_1 — addition of collector agent AeroMix (in grams per ton);

x_2 — addition of collector agent VK-901 (in grams per ton);

x_3 — consumption of foaming agent MIBK (in grams per ton);

x_4 — content of -74 micrometer grain class in the hydrocyclone overflow (in % of mass);

x_5 — total copper grade in the feed (in % of mass);

x_6 — total content of primary copper in the feed (in % of mass);

x_7 — total content of oxidized copper in the feed (in % of mass);

x_8 — addition of Lime, Lime milk (in grams per ton);

x_9 — content of iron in bulk concentrate (in % of mass);

x_{10} — density of feed-in Cu-Mo flotation (in % of mass);

x_{11} — addition of frother (in grams per ton).

According to theory of design experiment [2,6], in order to construct the objective function it is important to normalize the original variables to the unitary box:

$$-1 \leq x_i \leq 1, \quad i = 1,\ldots,11.$$

Exploring in total 5000 normalized tests providing the responses, the effects of the independent variables were analysed with multiple regression, fitting the two following empirical equations

$$\begin{aligned} f_1(x) &= \langle x, A^1 x \rangle + \langle b^1, x \rangle + 59.2073; \\ f_2(x) &= \langle x, A^2 x \rangle + \langle b^2, x \rangle + 15.3208. \end{aligned} \tag{2}$$

They represent the metal recoveries in copper and bulk flotations (measured in % of mass) as the functions $f_1(\cdot)$ and $f_2(\cdot)$, respectively, f_1, $f_2 : \mathbb{R}^{11} \to \mathbb{R}$.

The data in functions (2) are as follows

$A_1 =$

$$\begin{pmatrix}
7.4844 & 2.8621 & -1.3563 & 2.8452 & 0.8702 & 2.7041 & -1.8030 & 0.6038 & 0.8183 & -1.0549 & 1.5769 \\
2.8621 & 5.2969 & 0.5824 & 1.1256 & -1.4358 & 0.2105 & -0.9234 & -0.3883 & 5.3931 & -3.9475 & 2.1795 \\
-1.3563 & 0.5824 & 6.2678 & 0.1079 & 2.0562 & 2.5337 & -4.6174 & 2.6590 & 1.6805 & -1.5660 & 1.6390 \\
2.8452 & 1.1256 & 0.1080 & 7.0252 & -0.1282 & 1.2684 & 5.3151 & 3.7357 & 2.2280 & 2.2736 & 2.3403 \\
0.8702 & -1.4358 & 2.0562 & -0.1282 & 10.4069 & 0.9900 & 2.8919 & -2.8438 & -0.3838 & -1.6025 & -0.7952 \\
2.7041 & 0.2105 & 2.5337 & 1.2684 & 0.9900 & 6.5465 & -1.1000 & 3.2553 & 0.7692 & -0.9368 & -0.5096 \\
-1.8030 & -0.9234 & -4.6174 & 5.3151 & 2.8919 & -1.1000 & 10.5364 & -0.2095 & -0.7646 & 5.4391 & -1.0138 \\
0.6038 & -0.3883 & 2.6590 & 3.7357 & -2.8438 & 3.2552 & -0.2095 & 7.8553 & 2.0523 & 2.7186 & 0.2416 \\
0.8183 & 5.3930 & 1.6805 & 2.2280 & -0.3839 & 0.7692 & -0.7646 & 2.0523 & 3.5162 & 2.5147 & 4.6300 \\
-1.0550 & -3.9475 & -1.5660 & 2.2736 & -1.6025 & -0.9368 & 5.4391 & 2.7186 & 2.5147 & 6.2143 & -1.8107 \\
1.5769 & 2.1795 & 1.6390 & 2.3403 & -0.7952 & -0.5096 & -1.0138 & 0.2417 & 4.6300 & -1.8107 & 1.3092
\end{pmatrix}$$

$b^1 = (-8.1709, 0.4504, -9.1872, -4.0239, -2.9332, 3.4769, -2.7074, 3.9024, 0.2118, 12.9762, 2.9217),$

$A_2 =$

$$\begin{pmatrix}
2.5153 & 1.3321 & 0.3640 & 1.4082 & 0.9246 & 1.5638 & 0.3606 & 0.8108 & 1.1713 & 0.3147 & 1.0303 \\
1.3321 & 1.4373 & 0.5775 & 0.3157 & -0.0496 & 0.3530 & 0.3119 & 0.4475 & 1.9302 & -1.0077 & 0.6264 \\
0.3640 & 0.5775 & 2.2432 & 0.3594 & 1.0691 & 1.2804 & -0.4989 & 1.0219 & 0.9656 & -0.0746 & 0.8145 \\
1.4082 & 0.3157 & 0.3594 & 2.5899 & 0.1974 & 0.4206 & 1.8583 & 1.2385 & 1.3147 & 0.9326 & 0.5067 \\
0.9246 & -0.0496 & 1.0691 & 0.1974 & 2.9530 & 0.7173 & 1.3171 & -0.1865 & 0.5984 & -0.3380 & -0.0875 \\
1.5638 & 0.3530 & 1.2804 & 0.4206 & 0.7173 & 2.1809 & 0.5357 & 1.4077 & 0.8527 & -0.0540 & 0.1202 \\
0.3606 & 0.3119 & -0.4989 & 1.8583 & 1.3171 & 0.5357 & 3.1876 & 0.5896 & 0.7835 & 1.9548 & 0.1820 \\
0.8108 & 0.4475 & 1.0219 & 1.2385 & -0.1865 & 1.4077 & 0.5896 & 2.6578 & 0.9982 & 1.1706 & 0.6078 \\
1.1713 & 1.9302 & 0.9656 & 1.3147 & 0.5985 & 0.8527 & 0.7835 & 0.9982 & 1.5865 & 1.0152 & 1.6965 \\
0.3147 & -1.0077 & -0.0746 & 0.9326 & -0.3380 & -0.0540 & 1.9548 & 1.1706 & 1.0152 & 1.6928 & -0.5454 \\
1.0303 & 0.6264 & 0.8145 & 0.5068 & -0.0875 & 0.1201 & 0.1820 & 0.6078 & 1.6965 & -0.5454 & 0.3747
\end{pmatrix}$$

$b^2 = (-1.5059, 1.1312, -1.4275, 0.0527, 0.3412, 1.7277, 0.4336, 2.1780, 0.4884, 4.3844, 1.9798).$

Note that the matrices A_1 and A_2 are indefinite (see their eigenvalues in Table 1). Hence, the corresponding quadratic functions $f_1(\cdot)$ and $f_2(\cdot)$ turn out to be nonconvex; however, they happen to be DC functions. So, we can find an explicit DC representation of the non-convex functions $f_l(\cdot) = h_l(\cdot) - g_l(\cdot)$, $l = 1, 2$, using a simple method of identifying convex functions $h(\cdot)$ and $g(\cdot)$ for a general quadratic function: $f(\cdot) = h(\cdot) - g(\cdot)$ [10, 18].

Table 1. The eigenvalues of the matrices A_1 and A_2

	λ_1	λ_2	λ_3	λ_4	λ_5	λ_6	λ_7	λ_8	λ_9	λ_{10}	λ_{11}
A^1	19.834	18.240	−5.350	13.086	11.499	8.236	−1.393	−0.171	3.825	3.256	1.398
A^2	9.632	5.158	−2.165	3.644	2.965	1.909	−0.233	−0.004	1.188	0.788	0.538

It is known that any symmetric quadratic matrix Q may be represented as the difference of two symmetric positive definite matrices, i.e. $Q = Q_1 - Q_2$. Thus, we can get the following DC representation of the quadratic function $f(x) = \langle Qx, x \rangle$:

$$f(x) = \langle Q_1 x, x \rangle - \langle Q_2 x, x \rangle \overset{\triangle}{=} h(x) - g(x), \qquad (3)$$

where $h(\cdot)$ and $g(\cdot)$ are strongly convex functions (since Q_1, Q_2 are positive definite matrices) [5].

Therefore, we consider the following DC minimization problem

$$-f_1(x) = g_1(x) - h_1(x) \downarrow \min, \quad x \in \Pi, \\ -f_2(x) = g_2(x) - h_2(x) \leq -\gamma, \quad \biggr\} \tag{$\mathcal{P}$}$$

where

$$g_1(x) = \langle x, A_2^1 x \rangle - \langle b^1, x \rangle - 59.2073, \quad h_1(x) = \langle x, A_1^1 x \rangle;$$
$$g_2(x) = \langle x, A_2^2 x \rangle - \langle b^2, x \rangle - 15.3208, \quad h_2(x) = \langle x, A_1^2 x \rangle$$

$(A^l = A_1^l - A_2^l, \; l = 1, 2, \text{ see } (2))$. The parameter $\gamma = 22.5$ denotes a given fixed level of metal recovery in the bulk flotation.

Thus, the above problem is maximization of metal recovery in flotation under the constraint of a certain level for metal recovery in a bulk flotation.

3   Local Search

In order to find a local solution to the problem $(\mathcal{P})$, we develop a version of the special local search method for the general DC optimization problem [7,8,16,21]. Its main idea consists in the linearization of the function $h_1(\cdot)$, which defines "the basic non-convexity" of the problem $(\mathcal{P})$, at a current point with the subsequent minimization of the convex approximation of the objective function over the convex set obtained by replacing nonconvex constraints with their linearizations. Observe that the algorithm designed in that way provides critical points by employing only tools and methods of convex programming.

Assume that a feasible starting point $x^0 \in \Pi$ is given and, furthermore, that after several successive iterations we find a current point $x^s \in \Pi$, $s \in \{1, 2, \ldots\}$, so the linearized problem at the point x^s can be written as follows:

$$\Phi_{1s}(x) = g_1(x) - \langle \nabla h_1(x^s), x \rangle \downarrow \min_x, x \in \Pi, \\ \Phi_{2s}(x) = g_2(x) - \langle \nabla h_2(x^s), x - x^s \rangle - h_2(x^s) + \gamma \leq 0. \quad \biggr\} \tag{$\mathcal{PL}_s$}$$

Note that the problem $(\mathcal{PL}_s)$ is convex, since both its objective function and feasible set

$$D_s = \{x \in \Pi \mid g_2(x) - \langle \nabla h_2(x^s), x - x^s \rangle - h_2(x^s) + \gamma \leq 0\} \tag{4}$$

are convex, meanwhile the problem $(\mathcal{P})$ was a nonconvex one. Hence, the problem $(\mathcal{PL}_s)$ can be solved with a suitable convex optimization method [12] at any given precision. Let us compute a new iteration x^{s+1} as an approximate solution to the linearized problem $(\mathcal{PL}_s)$, so that x^{s+1} is feasible, i.e. $x^{s+1} \in D_s$, and satisfies the following inequality:

$$\Phi_{1s}(x^{s+1}) = g_1(x^{s+1}) - \langle \nabla h_1(x^s), x^{s+1} \rangle \leq \mathcal{V}_s + \Delta_s, \tag{5}$$

where $\mathcal{V}_s := \mathcal{V}(\mathcal{PL}_s)$ is the optimal value of the problem $(\mathcal{PL}_s)$, i.e.

$$\mathcal{V}_s = \inf_x \{\Phi_{1s}(x) \mid x \in \Pi, \; \Phi_{2s}(x) \leq 0\},$$

while a given sequence $\{\Delta_s\}$ is such that

$$\Delta_s \geq 0, \ s = 0, 1, 2, \ldots; \ \sum_{s=0}^{\infty} \Delta_s < \infty.$$

One can see that $D_s \subset D$. Hence x^{s+1} is feasible not only for the linearized problem $(\mathcal{PL}_s)$, but also for the original problem $(\mathcal{P})$, since, due to convexity of $h_2(\cdot)$, we have

$$0 \geq g_2(x^{s+1}) - \langle \nabla h_2(x^s), x^{s+1} - x^s \rangle - h_2(x^s) + \gamma$$
$$= \Phi_{2s}(x^{s+1}) \geq g_2(x^{s+1}) - h_2(x^{s+1}) + \gamma = -f_2(x^{s+1}) + \gamma.$$

As was proposed in [16], we can use one of the following inequalities as the stopping criterion of the local search:

$$f_1(x^{s+1}) - f_1(x^s) \leq \frac{\tau}{2}, \quad \Delta_s \leq \frac{\tau}{2}, \tag{6}$$

or

$$\Phi_{1s}(x^s) - \Phi_{1s}(x^{s+1}) \leq g_1(x^s) - g_1(x^{s+1}) - \langle \nabla h_1(x^s), x^s - x^{s+1} \rangle \leq \frac{\tau}{2},$$
$$\Delta_s \leq \frac{\tau}{2}, \tag{7}$$

Thus, if one of the inequalities (6)–(7) holds, the point x^s turns out to be a critical point for the problem $(\mathcal{P})$ with the accuracy τ under the assumption that $\Delta_s \leq \frac{\tau}{2}$. Indeed, (6)–(7) and the inequality (5) imply that

$$g_1(x^s) - \langle \nabla h_1(x^s), x^s \rangle \leq \frac{\tau}{2} + g_1(x^{s+1}) - \langle \nabla h_1(x^s), x^{s+1} \rangle \leq \mathcal{V}_s + \frac{\tau}{2} + \Delta_s.$$

Therefore, if $\Delta_s \leq \frac{\tau}{2}$, the point x^s is a τ-solution to the problem $(\mathcal{PL}_s)$.

In the next paragraph we show how to escape from critical points provided by the local search method and how to develop a global search algorithm based on the global optimality conditions.

4   Optimality Conditions and the Global Search Scheme

First, in order to accumulate all of the "basic non-convexities" of the problem, let us introduce the l_∞-penalty function [4,12] for the problem $(\mathcal{P})$:

$$W(x) := \max\{0, g_2(x) - h_2(x) + \gamma\}, \tag{8}$$

and consider the following penalized problem

$$\theta_\sigma(x) = g_1(x) - h_1(x) + \sigma W(x) \downarrow \min, \quad x \in \Pi. \tag{$\mathcal{P}_\sigma$}$$

As well-known [4,9], if $z \in Sol(\mathcal{P}_\sigma)$, and $z \in D := \{x \in \Pi \mid g_2(x) - h_2(x) + \gamma \leq 0\}$, then $z \in Sol(\mathcal{P})$. On the other hand, if $z \in Sol(\mathcal{P})$, then, under supplementary conditions [4,9,12], for some $\sigma_* \geq \|\lambda_z\|_1$ (where $\lambda_z \in I\!\!R^m$ is the KKT-multiplier corresponding to z), we have $z \in Sol(\mathcal{P}_\sigma)$ provided that $\sigma \geq \sigma_*$.

Furthermore, according to [9, Lemma 1.2.1, Chapt. VII], $Sol(\mathcal{P}) = Sol(\mathcal{P}_\sigma)$, the sets of optimal solutions coincide, and the problems $(\mathcal{P})$ and $(\mathcal{P}_\sigma)$ happen to be equivalent $\forall \sigma \geq \sigma_*$. Hence, the use of the exact penalty approach allows us to solve a single unconstrained problem instead of a sequence of unconstrained problems at $\sigma_k \to \infty$.

It can be seen that the objective function $\theta_\sigma(\cdot)$ of the penalized problem is a DC function, since the functions $f_i(\cdot)$, $i = 1, 2$, are the same. Moreover, since $\sigma > 0$ and

$$\max\{0; g_2(x) - h_2(x) + \gamma\} = \max\{g_2(x) + \gamma; h_2(x)\} - h_2(x),$$

the function $\theta_\sigma(x)$ can be represented as follows

$$\theta_\sigma(x) = G_\sigma(x) - H_\sigma(x), \tag{9}$$

where the functions

$$H_\sigma(x) := h_1(x) + \sigma h_2(x), \tag{10}$$

$$G_\sigma(x) := \theta_\sigma(x) + H_\sigma(x) = g_1(x) + \sigma \max\{h_2(x); g_2(x) + \gamma\}, \tag{11}$$

are obviously convex.

Observe that for any point $z \in \Pi$ and feasible for $(\mathcal{P})$, we have

$$W(z) \stackrel{\triangle}{=} \max\{0, g_2(z) - h_2(z) + \gamma\} = 0.$$

Let us also denote $\zeta := -f_1(z) = g_1(z) - h_1(z)$, so that

$$\theta_\sigma(z) = g_1(z) - h_1(z) + \sigma W(z) = \zeta.$$

Now, similarly to the global optimality conditions for DC minimization problems [15,19], we can state the following theorem.

Theorem 1. *If $z \in Sol(\mathcal{P}_\sigma)$, then*

$$\forall(y, \beta) : H_\sigma(y) = \beta - \zeta \tag{12}$$

the following inequality holds

$$G_\sigma(x) - \beta \geq \langle \nabla h_1(y) + \sigma \nabla h_2(y), x - y \rangle \quad \forall x \in \Pi. \tag{13}$$

Note that Theorem 1 reduces the process of searching for global optimal solutions to the non-convex problem $(\mathcal{P}_\sigma)$ to solving a family of convex linearized problems of the form

$$G_\sigma(x) - \langle \nabla H_\sigma(y), x \rangle \downarrow \min_x, \quad x \in \Pi, \tag{$\mathcal{P}_\sigma\mathcal{L}(y)$}$$

by one of the well known convex optimization methods [12]. The problem $(\mathcal{P}_\sigma\mathcal{L}(y))$ depends on the 'perturbation' parameters (y, β) satisfying (12).

It is worth pointing out that the linearization is performed with respect to the 'unified' nonconvexity of Problem $(\mathcal{P})$ expressed by the function $H_\sigma(\cdot)$ that accumulates the functions $h_1(\cdot)$, $h_2(\cdot)$ responsible for generating all nonconvexities in $(\mathcal{P})$.

If for a pair (y, β) and some $u \in \Pi$ (u may be a solution to $(\mathcal{P}_\sigma \mathcal{L}(y)))$, the inequality (13) is violated, i.e.

$$G_\sigma(u) < \beta + \langle \nabla H_\sigma(y), u - y \rangle, \tag{14}$$

then, using (12) and the fact of convexity of $H_\sigma(\cdot)$, we have

$$G_\sigma(u) < \beta + H_\sigma(u) - H_\sigma(y) = H_\sigma(u) + \zeta.$$

The latter implies that $\theta_\sigma(u) = G_\sigma(u) - H_\sigma(u) < \zeta := \theta_\sigma(z)$, so that a feasible point $u \in \Pi$ is better than z. Therefore, we conclude that $z \in \Pi$ is not optimal: $z \notin Sol(\mathcal{P}_\sigma)$.

Hence, the global optimality conditions (12), (13) described in Theorem 1 are of algorithmic interest: once the conditions are violated, one can find a feasible point which is better than the current one. Thus, using this useful algorithmic feature, one can develop local and global search methods for the problem $(\mathcal{P}_\sigma)$ [16, 17] allowing one to escape from a local "pit" in order to find a global optimal solution to $(\mathcal{P}_\sigma)$.

The issue is to find out whether such a set of parameters (y, β, u) exists. Now the following theorem holds as well.

Theorem 2. *[15] Let for a point $z \in \Pi$ there exists $w \in \mathbb{R}^n$ such that*

$$\theta_\sigma(w) > \theta_\sigma(z).$$

If z is not an optimal solution to the problem $(\mathcal{P}_\sigma)$, then one can find a pair $(y, \beta) \in \mathbb{R}^{n+1}$, satisfying (12), and a point $u \in \Pi$ such that the inequality (14) holds.

Moreover, on each level $\zeta_k = \theta_\sigma(z^k)$, there is no need to investigate all pairs of (y, β) satisfying (12), but only to identify one pair $(\tilde{y}, \tilde{\beta})$ and a point $u \in \Pi$ violating the inequality (13).

The global search strategy consists of two main components:

1. Local search, which provides an approximately critical point (a local solution);
2. Procedure of escaping from critical points, which is based on the global optimality conditions.

Let λ be the Lagrange multiplier for the DC constraint in the problem $(\mathcal{P})$: λ corresponds to the point z^k, $k \in \{1, 2, \ldots\}$.

Global Search Scheme (GSS)

Step 1. Using the local search method from Sect. 3, find a critical point z^k for the problem $(\mathcal{P})$.

Step 2. Set $\sigma_k := \lambda$. Choose a number $\beta : \inf(G_\sigma, \Pi) \le \beta \le \sup(G_\sigma, \Pi)$.
Step 3. Construct a finite approximation

$$R_k(\beta) = \{v^1, \ldots, v^{N_k} \mid H_\sigma(v^i) = \beta + \zeta_k, \ i = 1, \ldots, N_k, \ N_k = N_k(\beta)\}$$

of the level set $\{H_\sigma(x) = \beta + \zeta_k\}$ of the function $H_\sigma(\cdot)$.
Step 4. Find a δ_k-solution $\bar{u}^i$ of the following Linearized Problem:

$$G_\sigma(x) - \langle \nabla H_\sigma(v^i), x \rangle \downarrow \min_x, \ x \in \Pi, \qquad (P_\sigma L_i)$$

so that $G_\sigma(\bar{u}^i) - \langle \nabla H_\sigma(v^i), \bar{u}^i \rangle - \delta_k \le \inf_x\{G_\sigma(x) - \langle \nabla H_\sigma(v^i), x \rangle\}$.
Step 5. Starting from the point $\bar{u}^i$, find a critical point u^i by means of the local search method from Sect. 3.
Step 6. Choose the point u^j : $f_1(u^j) \ge \max_{i=1,\ldots,N} f_1(u^i)$.
Step 7. If $f_1(u^j) > f_1(z^k)$, then set $z_{k+1} := u^j$, $k := k+1$ and go to Step 2.
Step 8. Otherwise, choose a new value of β (for instance, $\beta + \Delta\beta$) and go to Step 3.

5   Implementation Issues

One of the principal features of the Global Search Scheme (GSS) is an approximation of the level set $\{H_\sigma(x) = \beta + \zeta_k\}$ of the convex function $H_\sigma(\cdot)$ which accumulates the nonconvexities of Problem (P) (Step 3). In particular, an approximation $R1_k(\beta)$ of the level set for each pair (β, ζ_k), $\zeta_k = f_1(z^k)$ can be constructed by the following rule [5,18,20]

$$v^i = \mu_i e^i, \ i = 1, \ldots, n, \qquad (15)$$

where e^i is the unit vector from the Euclidean basis of $\mathbb{R}^n$.

The search of μ_i turns out to be rather simple and, moreover, analytical (i.e. it reduces itself to the solution of a quadratic equation of one variable) for the quadratic function. When $H_\sigma = h_1(x) + \sigma h_2(x) = \langle A_1^1 x, x \rangle + \sigma \langle A_1^2 x, x \rangle$ the number μ_i for each $i = 1, \ldots, n$, is computed by the following formula

$$\mu_i = \pm\sqrt{\frac{\beta + \zeta_k}{H_\sigma(e^i)}}.$$

The set (approximation) (15) has proven to be rather competitive [5,18,20] during the computational simulations.

A new way of constructing an approximation of the level set $\{H_\sigma(x) = \beta + \zeta_k\}$ on Step 3 of GSS is to use vectors conjugate with respect to the matrix $M = A_1^1 + \sigma A_1^2$ which defines the function $H_\sigma(\cdot)$, i.e. an approximation $R2_k(\beta)$ of the level set can be constructed by the following rule

$$v^i = z^k + \nu_i p^i, \ i = 1, \ldots, n \qquad (16)$$

where z^k is the current critical point and p^i, $i = 1, \ldots, n$ are conjugate vectors. A search for ν_i turns out to be rather simple and moreover analytical for the quadratic function $H_\sigma(\cdot)$, because it reduces itself to the solution of a quadratic equation in one variable $H_\sigma(z^k + \nu_i p^i) = \beta + \zeta_k$, $i = 1, \ldots, n$.

Therefore we constructed the set $P = \{p^1, \ldots, p^n\}$ of vectors verifying the following condition

$$\langle p^i, Mp^j \rangle = 0 \ \forall i \neq j. \tag{17}$$

In other words, the vectors p^i are conjugated one to another with respect to the matrix M. Such vectors can be constructed by the algorithm as follows (see [12]).

Step 0. Set $i = 1$, $p^i := (1, 1, \ldots, 1)^\top$, $r^i := -p^i$.

Step 1. Compute the number $\alpha_i^1 := \dfrac{\langle r^i, r^i \rangle}{\langle p^i, Mp^i \rangle}$.

Step 2. Set $r^{i+1} := r^i + \alpha_i^1 \cdot Mp^i$.

Step 3. Compute the number $\alpha_{i+1}^2 = \dfrac{\langle r^{i+1}, r^{i+1} \rangle}{\langle r^i, r^i \rangle}$.

Step 4. Set $p^{i+1} := -r^{i+1} + \alpha_{i+1}^2 p^i$.

Step 5. Set $i := i + 1$. If $i < n$, then loop to Step 1, else STOP.

According to results in [12] the set $P = \{p^1, \ldots, p^n\}$ constructed by the algorithm just described has the properties as follows.

1) The condition (17) holds.

2) For every $i \in \{1, \ldots, n\}$ the following identity takes place
$$\langle r^i, p^j \rangle = 0 \quad \forall j = 1, 2, \ldots, (i-1).$$

3) For each $i \in \{1, \ldots, n\}$ the equality as follows holds:

$$\mathrm{Lin}\{p^1, p^2, \ldots, p^i\} = \mathrm{Lin}\{r^1, r^2, \ldots, r^i\}.$$

We use these two methods for constructing approximations during the computational experiment.

Besides, to implement Step 4 of the global search scheme, we have to solve the following linearized (at the point v^i) problem:

$$\left.\begin{aligned} g_1(x) + \sigma \max\{g_2(x) + \gamma; h_2(x)\} - \langle \nabla h_1(v^i) + \sigma \nabla h_2(v^i), x \rangle \downarrow \min_x, \\ x \in \Pi. \end{aligned}\right\} \tag{18}$$

Note that its objective function is nonsmooth, hence we use a new extra variable t and rewrite the linearized problem (18) in the following equivalent formulation:

$$\left.\begin{aligned} g_1(x) - \langle \nabla h_1(v) + \sigma \nabla h_2(v), x \rangle + \sigma t \downarrow \min_{(x,t)}, \quad x \in \Pi, \ t \in \mathbb{R}, \\ g_2(x) + \gamma \leq t, \quad h_2(x) \leq t. \end{aligned}\right\} \tag{19}$$

Since the problem (19) is convex, it can be solved with any suitable commercial solver.

6   Computational Experiments

Problem (P) has been solved by the algorithm developed on the basis of the Global Search Scheme (GSS). All computational experiments were performed on the Intel Core i7-4790K CPU 4.0 GHz. All convex auxiliary (linearized) problems arising when carrying out the GSS (Steps 1, 4, 5) were solved by the optimization software package IBM ILOG CPLEX 12.6.2. (Concert library) freely available for non-commercial research[1].

Table 2 represents the results of computational testing of two variants of the algorithm: GSS-R1 uses approximation $R1$ constructed by the rule (15) and GSS-R2 takes approximation $R2$ with the rule (17). The following denotations are employed in the table: # is the number of starting point; $f_1(x_0)$ is the value of the goal function to Problem (P) at the starting point; $f_l(x^*)$ stands for the value of the goal function at the solution provided by both GSS-R1 and GSS-R2. Furthermore for each algorithm, we use the following denotations: ST is the number of the local solution passed by algorithm; PL stands for the number of Linearized Problems solved; $Time$ is the CPU time of computing solutions (seconds).

Table 2. The testing GSS-R1 and GSS-R2 algorithms

#	$f_1(x_0)$	$f_1(x^*)$	GSS-R1			GSS-R2		
			St	PL	Time	St	PL	Time
1	15.3208	**275.6464**	4	320	1.47	4	136	0.64
2	18.1831	**275.6464**	4	306	1.38	2	92	0.43
3	26.8545	**275.6464**	4	300	1.38	3	100	0.47
4	36.5119	**275.6464**	4	292	1.34	3	95	0.45
5	45.8025	**275.6464**	2	257	1.18	4	141	0.67
6	49.2538	**275.6464**	3	292	1.36	3	96	0.46
7	49.8730	**275.6464**	4	298	1.37	3	114	0.60
8	104.0111	**275.6464**	4	297	1.39	3	94	0.45
9	123.5783	**275.6464**	4	296	1.40	3	112	0.59

Of all the starting points, the same value $f_1(x^*) = 275.6464$ was reached. In this case $f_2(x^*) = 78.1001$,

$$x_i^* = -1, \quad i \in \{1, 2, 3, 4, 6, 8, 11\}, \quad x_i^* = 1, \quad i \in \{5, 7, 10\}.$$

The GSS-R2 algorithm's run-time turned out to be much less than the run-time of GSS-R1 algorithm. We can conclude that approximation $R2$ constructed on the basis of conjugate vectors is more efficient in terms of CPU time.

---

[1] http://www-01.ibm.com/software/commerce/optimization/cplex-optimizer/.

Further, we compared the goal function value and CPU time obtained using GSS-R1, GSS-R2 and the others solvers (see Table 3) starting from the point $x_0 = (0, \ldots, 0)^\top$.

Entries that correspond to the global solution are boldfaced.

So, as to the attainability of global solution, GSS shows itself as rather competitive with respect to the well-known solvers.

Table 3. The comparison with solvers

	Goal function value	CPU time
CONOPT	250.2522	0.01
COUENNE	**275.6464**	0.47
KNITRO	220.4866	0.02
LINDOGLOBAL	250.2522	0.01
MINOS	228.5796	0.01
SCIP	**275.6464**	1.00
GSS-R1	**275.6464**	1.39
GSS-R2	**275.6464**	0.45

7   Conclusions

In this paper, we formulated a real-world optimization problem in Mongolian mining industry as an indefinite quadratic programming problem with a nonconvex quadratic constraint. We examine this problem from a viewpoint of theory for DC constrained DC programming problems. We apply the global optimality conditions developed by A.S. Strekalovsky and propose the specialized local and global search algorithms.

The numerical results are provided using real industrial data. The global (best-known) solution obtained by the algorithm meets the technological requirements given by the Erdenet Mining Corporation.

Acknowledgements. This work supported by the project " P2019-3751" of National University of Mongolia.

References

1. Barrients, O., Correa, R.: An algorithm for global minimizatoion of linearly constrained quadratic functions. J. Global Optim. **16**, 77–93 (2000). https://doi.org/10.1023/A:1008306625093
2. Boer, E.P.J., Hendrih, E.M.T.: Global optimization problems in optimal design of experiments in regression models. J. Global Optim. **18**, 385–398 (2000). https://doi.org/10.1023/A:1026552318150

3. Bomze, I., Danninger, G.: A finite algorithm for solving general quadratic problem. J. Global Optim. **4**(1), 1–16 (1994). https://doi.org/10.1007/BF01096531

4. Bonnans, J.-F., Gilbert, J.C., Lemaréchal, C., Sagastizábal, C.A.: Numerical Optimization: Theoretical and Practical Aspects, 2nd edn. Springer-Verlag, Heidelberg (2006). https://doi.org/10.1007/978-3-540-35447-5

5. Enkhbat, R., Gruzdeva, T.V., Barkova, M.V.: D.C. programming approach for solving an applied ore-processing problem. J. Ind. Manag. Optim. **14**(2), 613–623 (2018)

6. Fedorov, V.V.: Theory of Optimal Experiments. Academic Press, New-York (1972)

7. Gruzdeva, T.V., Strekalovsky, A.S.: Local Search in Problems with Nonconvex Constraints. Comput. Math. Math. Phys. **47**, 381–396 (2007). https://doi.org/10.1134/S0965542507030049

8. Gruzdeva, T., Strekalovsky, A.: An approach to fractional programming via D.C. constraints problem: local search. In: Kochetov, Y., Khachay, M., Beresnev, V., Nurminski, E., Pardalos, P. (eds.) DOOR 2016. LNCS, vol. 9869, pp. 404–417. Springer, Cham (2016). https://doi.org/10.1007/978-3-319-44914-2_32

9. Hiriart-Urruty, J.-B., Lemaréchal, C.: Convex Analysis and Minimization Algorithms. Springer-Verlag, Heidelberg (1993). https://doi.org/10.1007/978-3-662-02796-7

10. Horst, R., Pardalos, P., Thoai, N.V.: Introduction to Global Optimization. Kluwer Academic Publishers, Dordrecht (1995)

11. Horst, R., Pardalos, P.M.: Handbook of Global Optimization. Kluwer Academic, Dordrecht (1995)

12. Nocedal, J., Wright, S.J.: Numerical Optimization. Springer, New York (2006). https://doi.org/10.1007/978-0-387-40065-5

13. Pardalos, P.M., Rosen, J.B. (eds.): Constrained Global Optimization: Algorithms and Applications. LNCS, vol. 268. Springer, Heidelberg (1987). https://doi.org/10.1007/BFb0000035

14. Pardalos, P.M., Schnitger, J.: Checking local optimality in constrained quadratic programming is NP-hard. Oper. Res. Lett. **7**, 33–35 (1988)

15. Strekalovsky, A.S.: On the merit and penalty functions for the D.C. Optimization. In: Kochetov, Y., Khachay, M., Beresnev, V., Nurminski, E., Pardalos, P. (eds.) DOOR 2016. LNCS, vol. 9869, pp. 452–466. Springer, Cham (2016). https://doi.org/10.1007/978-3-319-44914-2_36

16. Strekalovsky, A.S.: On local search in d.c. optimization problems. Appl. Math. Comput. **255**, 73–83 (2015)

17. Strekalovsky, A.S.: On solving optimization problems with hidden nonconvex structures. In: Rassias, T.M., Floudas, C.A., Butenko, S. (eds.) Optimization in Science and Engineering, pp. 465–502. Springer, New York (2014). https://doi.org/10.1007/978-1-4939-0808-0_23

18. Strekalovsky, A.S.: Elements of Nonconvex Optimization. Nauka, Novosibirsk (2003). (in Russian)

19. Strekalovsky, A.S.: Global optimality conditions and exact penalization. Optim. Lett. **13**(3), 597–615 (2019). https://doi.org/10.1007/s11590-017-1214-x

20. Strekalovsky, A.S., Yakovleva, T.V.: On a local and global search involved in nonconvex optimization problems. Autom. Remote Control. **65**, 375–387 (2004). https://doi.org/10.1023/B:AURC.0000019368.45522.7a

21. Tao, P.D., Le Thi, H.A.: The DC (difference of convex functions) programming and DCA revisited with DC models of real world nonconvex optimization problems. Ann. Oper. Res. **133**, 23–46 (2005). https://doi.org/10.1007/s10479-004-5022-1

22. Toland, J.F.: Duality in nonconvex optimization. J. Math. Anal. Appl. **66**(2), 399–415 (1978)
23. Yajima, Y., Fujie, T.A.: Polyhedral approach for nonconvex quadratic programming problems with box constraints. J. Global Optim. **13**, 151–170 (1998). https://doi.org/10.1023/A:1008293029350

On Solving the Quadratic Sum-of-Ratios Problems

Tatiana V. Gruzdeva$^{(\boxtimes)}$ and Alexander S. Strekalovsky

Matrosov Institute for System Dynamics and Control Theory of SB of RAS,
Irkutsk, Russia
{gruzdeva,strekal}@icc.ru

Abstract. This paper addresses the numerical solution of fractional programs with quadratic functions in the ratios. Instead of considering a sum-of-ratios problem directly, we developed an efficient global search algorithm, which is based on two approaches to the problem. The first one adopts a reduction of the fractional minimization problem to the solution of an equation with an optimal value of a parametric d.c. minimization problem. The second approach reduces the original problem to the optimization problem with nonconvex (d.c.) constraints. Hence, the fractional programs can be solved by applying the Global Search Theory of d.c. optimization.

The global search algorithm developed for sum-of-ratios problems was tested on the examples with quadratic functions in the numerators and denominators of the ratios. The numerical experiments demonstrated that the algorithm performs well when solving rather complicated quadratic sum-of-ratios problems with up to 100 variables or 1000 terms in the sum.

Keywords: Fractional optimization · Nonconvex problem · Difference of two convex functions · Quadratic functions · Global search algorithm · Computational testing

1 Introduction

The global minimization of the sum of fractional functions has attracted the interest of researchers and practitioners for a number of years, because these problems have a large number of important real-life applications. From the theoretical viewpoint, the solution of these problems implies facing significant challenges, because, in general, fractional programs are nonconvex problems, i.e., they generally have several (often a huge number of) local optimal solutions that are not globally optimal [18]. It was proven that the sum-of-ratios program is NP-complete [6]. Various specialized methods and algorithms have been proposed for solving these problems globally (see, for example, surveys in [3,18]), but the development of new efficient methods for the following sum-of-ratios problem [3,7,18]

$$(\mathcal{FP}): \qquad f(x) := \sum_{i=1}^{m} \frac{\psi_i(x)}{\varphi_i(x)} \downarrow \min_x, \quad x \in S,$$

© Springer Nature Switzerland AG 2020
A. Kononov et al. (Eds.): MOTOR 2020, LNCS 12095, pp. 115–127, 2020.
https://doi.org/10.1007/978-3-030-49988-4_8

where $S \subset I\!\!R^n$ is a closed convex set and $\psi_i, \varphi_i : I\!\!R^n \to I\!\!R$, are continuous function such that

$$(\mathcal{H}_0) : \qquad \psi_i(x) > 0, \ \varphi_i(x) > 0 \ \forall x \in S, \ i = 1, \ldots, m;$$

still remains an important field of research in the mathematical optimization. If functions $\psi_i(\cdot)$, $\varphi_i(\cdot)$, $i = 1, \ldots, m$ are quadratic functions, then we classify Problem $(\mathcal{F}\mathcal{P})$ as a quadratic fractional program.

On the other hand, the following general problem of d.c. optimization

$$(\mathcal{P}) : \qquad \begin{cases} f_0(x) := g_0(x) - h_0(x) \downarrow \min_x, & x \in S, \\ f_j(x) := g_j(x) - h_j(x) \leq 0, & j = 1, \ldots, m, \end{cases}$$

where the functions $g_j(\cdot)$, $h_j(\cdot)$, $j = 1, \ldots, m$, are convex on $I\!\!R^n$, S is a closed and convex set, and $S \subset I\!\!R^n$, remained over recent years one of the attractive objects in nonconvex optimization [13–15,19,27,28].

It is worth noting that any continuous optimization problem can be approximated by a d.c. problem with any prescribed accuracy [13,14,28]. In addition, the space of d.c. functions, being a linear space, is then closed under most of operations usually considered in optimization (see, e.g., [19,28]). Moreover, any twice differentiable function belong to the space of d.c. functions [13–15,19,27,28]. The convexity of the two convex components g and h of the d.c. function f is widely used to develop appropriate theoretical and algorithmic tools.

We develop a new efficient global search method for the fractional optimization problems, which is based on the two following ideas [9,11,12]. First, generalizing the Dinkelbach's approach [4], we propose to reduce the sum-of-ratios problem with d.c. functions to solving an equation with the optimal value function of an auxiliary parametric problem with the vector parameter that satisfies the nonnegativity assumption. Secondly, we also use the reduction of the fractional program to a problem of type $(\mathcal{P})$, where $f_0(x)$ is a linear function, i.e. an optimization problem over nonconvex feasible set given by d.c. inequality constraints.

Furthermore, based on the Global Search Theory for d.c. optimization [19,22] and on the solution of these two particular cases of the general d.c. optimization problem $(\mathcal{P})$, we develop a two-method technology for solving a sum-of-ratios problem and verify it on test problems with nonconvex quadratic functions in the numerators and denominators of the ratios.

Most of the approaches and techniques for solving fractional programs are designed for problems with affine functions in the numerators and denominators of the ratios [1,5,16,17]. Problems with nonconvex quadratic (d.c.) functions are more complex problems since it concerns a finding a global solution.

The outlines of the paper are as follows. In Sect. 2 and 3, we recall two approaches to solving Problem $(\mathcal{F}\mathcal{P})$ using auxiliary d.c. minimization problems and problems with d.c. inequality constraints, respectively. In Sect. 4, we show how to represent explicitly the nonconvex functions, describing the goal function and the constraints of the auxiliary d.c. problems as differences of two convex functions (the d.c. representation). Further, in Sect. 5, we propose the algorithm which combines the two approaches to solving the fractional program via d.c. optimization. The final section offers computational testing of the developed algorithm on fractional program instances with up to 100 variables or 1000 terms in the sum generated by the approach from [2].

2   Reduction to the D.C. Minimization Problem

Consider the following auxiliary problem

$$(\mathcal{P}_\alpha): \qquad \Phi_\alpha(x) := \sum_{i=1}^{m} [\psi_i(x) - \alpha_i \varphi_i(x)] \downarrow \min_x, \quad x \in S,$$

where $\alpha = (\alpha_1, \ldots, \alpha_m)^\top \in \mathbb{R}_+^m$ is the vector parameter and the set $S \subset \mathbb{R}^n$ is, as above closed and convex.

First, let us recall some results from [9] about the relations between Problems $(\mathcal{FP})$ and $(\mathcal{P}_\alpha)$. Further introduce the function $\mathcal{V}(\alpha)$ of the optimal value to Problem $(\mathcal{P}_\alpha)$:

$$\mathcal{V}(\alpha) := \inf_x \{\Phi_\alpha(x) \mid x \in S\} = \inf_x \left\{ \sum_{i=1}^{m} [\psi_i(x) - \alpha_i \varphi_i(x)] : x \in S \right\}. \quad (1)$$

In addition, suppose that the following assumptions are fulfilled:

$$(\mathcal{H}_1): \quad \begin{cases} (a) \ \mathcal{V}(\alpha) > -\infty \ \forall \alpha \in \mathcal{K}, \text{where } \mathcal{K} \text{ is a convex set from } \mathbb{R}^m; \\ (b) \ \forall \alpha \in \mathcal{K} \subset \mathbb{R}^m \text{ there exists a solution } z = z(\alpha) \text{ to Problem } (\mathcal{P}_\alpha). \end{cases}$$

In what follows, we say that a given parameter vector $\alpha = (\alpha_1, \ldots, \alpha_m)^\top \in \mathbb{R}^m$ satisfies "the nonnegativity condition" in Problem $(\mathcal{FP})$, if the following inequalities hold

$$(\mathcal{H}(\alpha)): \qquad \psi_i(x) - \alpha_i \varphi_i(x) \geq 0 \ \forall x \in S, \ i = 1, \ldots, m.$$

Theorem 1. *[9] Suppose that in Problem $(\mathcal{FP})$ the assumptions $(\mathcal{H}_0), (\mathcal{H}_1)$ are satisfied. In addition, let there exist a vector $\alpha_0 = (\alpha_{01}, \ldots, \alpha_{0m})^\top \in \mathcal{K} \subset \mathbb{R}^m$ at which "the nonnegativity condition" $(\mathcal{H}(\alpha_0))$ holds.*

Finally, suppose that in Problem $(\mathcal{P}_{\alpha_0})$ the following equality takes place:

$$\mathcal{V}(\alpha_0) \overset{\triangle}{=} \min_x \left\{ \sum_{i=1}^{m} [\psi_i(x) - \alpha_{0i} \varphi_i(x)] : x \in S \right\} = 0. \quad (2)$$

Then, any solution $z = z(\alpha_0)$ to Problem $(\mathcal{P}_{\alpha_0})$ is a solution to Problem $(\mathcal{FP})$, so that $z \in Sol(\mathcal{P}_{\alpha_0}) \subset Sol(\mathcal{FP})$.

Hence, in order to check up the equality (2), we should be able to solve globally Problem $(\mathcal{P}_\alpha)$ at a current $\alpha \in \mathbb{R}_+^m$. Since $\psi_i(\cdot), \varphi_i(\cdot), i = 1, \ldots, m$, are convex or d.c. functions it can be readily seen that Problem $(\mathcal{P}_\alpha)$ turns out to be a parametric d.c. minimization problem. As a consequence, in order to solve the auxiliary Problem $(\mathcal{P}_\alpha)$, we can apply the global search strategy for d.c. minimization problems [19,22]. Its theoretical foundation can be provided by the following Global Optimality Condition written here in the terms of Problem $(\mathcal{P}_\alpha)$ under the assumption that $\psi_i(\cdot)$, $\varphi_i(\cdot)$, $i = 1, \ldots, m$, are convex functions and therefore $g(x) = \sum\limits_{i=1}^{m} \psi_i(x)$, $h_\alpha(x) = \sum\limits_{i=1}^{m} \alpha_i \varphi_i(x)$,

$$\Phi_\alpha(x) = \sum_{i=1}^{m} (\psi_i(x) - \alpha_i \varphi_i(x)) = g(x) - h_\alpha(x).$$

If $\psi_i(\cdot)$, $\varphi_i(\cdot)$, $i = 1, \ldots, m$, are d.c. functions, it is easy to see that the goal function of $(\mathcal{P}_\alpha)$ is d.c., as well, but having another d.c. decomposition.

Theorem 2. *[19,22] Suppose, $z(\alpha)$ is a global solution to $(\mathcal{P}_\alpha)$, $\zeta := \Phi_\alpha(z(\alpha))$. Then,*

$$(\mathcal{E}): \quad \begin{cases} (a) \quad \forall (y, \beta) \in \mathbb{R}^n \times \mathbb{R}: \quad h_\alpha(y) = \beta - \zeta, \\ (b) \quad g(x) - \beta \geq \langle \nabla h_\alpha(y), x - y \rangle \ \forall x \in S. \end{cases} \tag{3}$$

The meaning of Theorem 2 lies in the fact that by selecting the "perturbation parameters" (y, β) (satisfying (3a)) and solving the linearized problem

$$(\mathcal{P}_\alpha \mathcal{L}): \qquad g(x) - \langle \nabla h_\alpha(y), x \rangle \downarrow \min_x, \quad x \in S,$$

we try to violate the principal inequality (3b) (where $y \in \mathbb{R}^n$ is not obligatory feasible).

Furthermore, according to the Theorem 1, we are able to avoid the direct solution of Problem $(\mathcal{FP})$ and address the parametrized problem $(\mathcal{P}_\alpha)$ with $\alpha \in \mathbb{R}_+^m$. Hence, we propose to combine a solution of Problem $(\mathcal{P}_\alpha)$ with a search of the parameter $\alpha \in \mathbb{R}_+^m$ in order to find $\alpha_0 \in \mathbb{R}_+^m$ such that $\mathcal{V}(\alpha_0) = 0$. This idea can be implemented by the following algorithm. Let $[\alpha_-^0, \alpha_+^0]$ be an initial segment for varying α, and $x_0 \in S$ stands for the starting point.

Algorithm 1 for solving the fractional problem via d.c. minimization

Stage 0. (Initialization) $k := 0$. Set $x^k := x_0$, $\alpha_-^k := 0$, $\alpha_+^k := \alpha_+^0$, $\alpha^k := \frac{\alpha_+^0}{2} \in [\alpha_-^0, \alpha_+^0]$.

Stage I. (Local search) Starting at x^k find a critical point to Problem $(\mathcal{P}_k) := (\mathcal{P}_{\alpha^k})$ using the special local search method for d.c. minimization [19].

Stage II. (Global search) Find a global solution $z(\alpha^k)$ to Problem $(\mathcal{P}_k)$ using the global search scheme for the parametric d.c. minimization [19,22] Problem $(\mathcal{P}_k)$.

Stage III. (Stopping criterion) If $\mathcal{V}(\alpha^k) = 0$ and $\min\limits_i \{\psi_i(z(\alpha^k)) - \alpha_i^k \varphi_i (z(\alpha^k))\} \geq 0$, then STOP: $z(\alpha^k) \in Sol(\mathcal{FP})$.

Stage IV. (Parameter variation) Find new parameters α^{k+1}, α_-^{k+1} and α_+^{k+1}; $k := k + 1$ and go to Stage I.

Remark 1. The algorithm for solving the fractional program $(\mathcal{FP})$ consists of three basic stages: (a) a local and (b) a global searches in Problem $(\mathcal{P}_\alpha)$ with a fixed vector parameter α (Stages I, II) and (c) the method for finding the vector α (Stage IV).

Let $[\alpha_-, \alpha_+]$ be a segment for varying α. In addition, assume that we are able to compute the value $\mathcal{V} := \mathcal{V}(\alpha)$ and let a solution $z(\alpha)$ to Problem $(\mathcal{P}_\alpha)$ be given.

In order to calculate a new parameter α^{new} and new boundaries α_-^{new} and α_+^{new} at the current iteration k for a segment for varying α on Stage IV of Algorithm 1 one can use the following procedure, where $\alpha := \alpha^k$ from Stage III of Algorithm 1.

Stage IV. Parameter variation algorithm

Step 1. If $\mathcal{V} > 0$, then set $\alpha_-^{new} := \alpha$, $\alpha^{new} := \frac{1}{2}(\alpha_+ + \alpha)$, $\alpha_+^{new} := \alpha_+$.

Step 2. If $\mathcal{V} < 0$, then set $\alpha_+^{new} := \alpha$, $\alpha^{new} := \frac{1}{2}(\alpha_- + \alpha)$, $\alpha_-^{new} := \alpha_-$.

Step 3. If $\mathcal{V} = 0$ and $\min_i \Phi_i(z(\alpha), \alpha) < 0$, then set

$$\alpha_i^{new} := \frac{\psi_i(z(\alpha))}{\varphi_i(z(\alpha))} \ \forall i : \psi_i(z(\alpha)) - \alpha_i\varphi_i(z(\alpha)) < 0;$$
$$\alpha_i^{new} := \alpha_i \ \forall i : \psi_i(z(\alpha)) - \alpha_i\varphi_i(z(\alpha)) \geq 0.$$

In addition, set $\alpha_-^{new} := 0$, $\alpha_+^{new} := t\alpha^{new}$, where $t = \dfrac{\max\{\alpha_{+1}^0; \ldots; \alpha_{+m}^0\}}{\max\{\alpha_1; \ldots; \alpha_m\}}$.

Step 4. $\alpha^{k+1} := \alpha^{new}$, $\alpha_-^{k+1} := \alpha_-^{new}$ and $\alpha_+^{k+1} := \alpha_+^{new}$; $k := k + 1$ and go to Stage I of Algorithm 1.

Remark 2. To choose an initial segment $[\alpha_-^0, \alpha_+^0]$ for varying α, we should take into account the following considerations. According to "the nonnegativity condition" $(\mathcal{H}(\alpha))$ and the assumption $(\mathcal{H}_0)$, we have

$$\alpha_i \leq \frac{\psi_i(x)}{\varphi_i(x)} \leq \sum_{i=1}^{m} \frac{\psi_i(x)}{\varphi_i(x)} \ \forall x \in S, \ \forall i = 1, \ldots, m,$$

therefore one can choose $\alpha_{+\ i}^0 := \dfrac{\psi_i(x^0)}{\varphi_i(x^0)}$, $\alpha_{-\ i}^0 = 0$, $i = 1, \ldots, m$.

The performed computational experiment [8,9,11] showed that solving fractional problem via d.c. minimization takes a large number of iterations, generated by an ineffective work of Stage IV. Therefore, it is very important to choose a suitable parameter α in order to reduce the total number of iterations and, therefore, the corresponding run-time of Algorithm 1.

Using the reduction of the sum-of-ratios problem to the optimization problem with nonconvex constraints, we will look not only for a starting value of the parameter α for Problem $(\mathcal{P}_\alpha)$, but also for a better vector for Problem $(\mathcal{FP})$.

3   Reduction to the Problem with D.C. Constraints

In this section we consider the following optimization problem

$$(\mathcal{DCC}) : \quad \begin{cases} \sum_{i=1}^{m} \alpha_i \downarrow \min_{(x,\alpha)}, \ \ x \in S, \\ \psi_i(x) - \alpha_i \varphi_i(x) \leq 0, \ \ i \in I = \{1, \dots, m\}. \end{cases}$$

The relations between Problems $(\mathcal{FP})$ and $(\mathcal{DCC})$ are as follows.

Proposition 1. *[10] For any global solution $(x_*, \alpha_*) \in I\!\!R^n \times I\!\!R^m$ to Problem $(\mathcal{DCC})$, the point x_* will be a global solution to Problem $(\mathcal{FP})$ and*
$$\alpha_{*i} = \frac{\psi_i(x_*)}{\varphi_i(x_*)}, \ i \in I.$$

Remark 3. It is clear that, in contrast to Theorem 1 from Sect. 2, the vector $\alpha_* = (\alpha_{*1}, \dots, \alpha_{*m})^\top \in I\!\!R^m$ must be found simultaneously with the solution vector x_*.

It is easy to see that Problem $(\mathcal{DCC})$ is a nonconvex optimization problem with the linear goal function and the nonconvex feasible set (see, e.g., [14,25]). So, we can solve Problem $(\mathcal{DCC})$ using the exact penalization approach for d.c. optimization developed in [20,23,24]

However, the computational experiments showed [11,12] that solving fractional program via problem with d.c. constraints $(\mathcal{DCC})$ took more run-time than using the parametric d.c. minimization, i.e. Problem $(\mathcal{P}_\alpha)$.

Notwithstanding, in low-dimensional test problems of fractional programming the known global solutions were found just by the local search method (LSM) [21] for Problem $(\mathcal{DCC})$ (see [10]). Therefore, we apply here only the LSM, based on the classical idea of linearization with respect to the basic nonconvexity of the problem [19,21,26].

The LSM for Problem $(\mathcal{DCC})$ is based on the consecutive solutions of the following partially linearized (at the point (x^s, α^s)) problem [10,21,26]:

$$(\mathcal{DCCL}_s) : \quad \begin{cases} \sum_{i=1}^{m} \alpha_i \downarrow \min_{(x,\alpha)}, \ \ x \in S, \\ g_i(x, \alpha_i) - \langle \nabla h_i(x^s, \alpha_i^s), (x, \alpha_i) - (x^s, \alpha_i^s) \rangle - h_i(x^s, \alpha_i^s) \leq 0, \ \ i \in I, \end{cases}$$

where the functions $g_i(\cdot)$ and $h_i(\cdot)$, $i \in I$, are the convex functions obtained by the d.c. representation of the constraint functions $\psi_i(x) - \alpha_i \varphi_i(x)$, $i \in I$ (see, for example, Sect. 4 for quadratic $\psi_i(\cdot)$ and $\varphi_i(\cdot)$) and where $\nabla h_i(x, \alpha_i) = (\nabla_{x_1} h_i(x, \alpha_i), \dots, \nabla_{x_n} h_i(x, \alpha_i), \nabla_{\alpha_i} h_i(x, \alpha_i))^\top \in I\!\!R^{n+1}$.

Due to the consecutive solutions of linearized convex problems $(\mathcal{DCCL}_s)$ starting at the point (x^0, α^0) we generate the sequence $\{(x^s, \alpha^s)\}$: $(x^{s+1}, \alpha^{s+1}) \in Sol(\mathcal{DCCL}_s)$. As it was proven in [21], the cluster point $(x_*, \alpha_*) \in \{x \in S \mid g_i(x, \alpha_i) - \langle \nabla h_i(x_*, \alpha_{*i}), (x, \alpha_i) - (x_*, \alpha_{*i}) \rangle - h_i(x_*, \alpha_{*i}) \leq 0, \ i \in I\}$ of the sequence $\{(x^s, \alpha^s)\}$ generated by the LSM, is a solution to the linearized Problem $(\mathcal{DCCL}_*)$ (which is Problem $(\mathcal{DCCL}_s)$ with (x_*, α_*) instead of (x_s, α_s)), and a critical point of Problem $(\mathcal{DCC})$: $(x_*, \alpha_*) \in Sol(\mathcal{DCCL}_*)$.

In order to implement the LSM [21, 26] and the global search scheme for d.c. minimization [19, 22] (Stage II of Algorithm 1), we need an explicit d.c. representation of functions $\psi_i(x) - \alpha_i \varphi_i(x)$, $i \in I$. As well-know the d.c. decomposition of function is not unique. The next section presents several possible d.c. representations for the goal function of Problem $(\mathcal{P}_\alpha)$ and constraint's functions of Problem $(\mathcal{DCC})$ (in the case of quadratic $\psi_i(\cdot)$ and $\varphi_i(\cdot)$).

4 D.C. Representations of the Goal Function and Constraint's Functions

Consider the following quadratic functions $\psi_i(x) > 0$, $\varphi_i(x) > 0$ $\forall x \in S$):

$$\psi_i(x) = \langle x, A^i x \rangle + \langle a^i, x \rangle + \xi_i, \quad \varphi_i(x) = \langle x, B^i x \rangle + \langle b^i, x \rangle + \gamma_i,$$

where the matrices A^i and B^i are $(n \times n)$ positive definite, $a^i, b^i \in \mathbb{R}^n$, $\xi_i, \gamma_i \in \mathbb{R}$, $i \in I$. Therefore, the functions $\psi_i(\cdot)$ and $\varphi_i(\cdot)$ are convex functions, $i \in I$.

In this case, the d.c. representation of the goal function $\Phi_\alpha(x)$ of Problem $(\mathcal{P}_\alpha)$ (where $\alpha = (\alpha_1, \ldots, \alpha_m) \in \mathbb{R}_+^m$ is the vector parameter) can be rather simple: $\Phi_\alpha(x) = g(x) - h_\alpha(x)$, where $g(x) = \sum\limits_{i=1}^{m} \psi_i(\cdot)$, $h_\alpha(x) := \sum\limits_{i=1}^{m} \alpha_i \varphi_i(\cdot)$, or, in another way, $g_\alpha^1(x) = \sum\limits_{i=1}^{m} \left[\langle x, A^i x \rangle + \langle (a^i - \alpha_i b^i), x \rangle \right] + \sum\limits_{i=1}^{m} (\xi_i - \alpha_i \gamma_i)$,

$h_\alpha^1(x) = \sum\limits_{i=1}^{m} \alpha_i \langle x, B^i x \rangle.$

Remark 4. If the symmetric matrices A^i or/and B^i are indefinite, then they can be represented as the difference of two symmetric positive definite matrices $A^i = A_1^i - A_2^i$, $A_1^i, A_2^i > 0$, $B^i = B_1^i - B_2^i$, $B_1^i, B_2^i > 0$, using, for example, the method from [19]. After this it is possible to construct functions $g(\cdot)$ and $h(\cdot)$ by adding for all $i \in I$ a convex part with the matrix A_1^i or/and B_1^i into $g(\cdot)$ and a nonconvex part with the matrix A_2^i or/and B_2^i into $h(\cdot)$, i.e.

$$g_\alpha^2(x) = \sum_{i=1}^{m} \left[\langle x, A_1^i x \rangle + \alpha_i \langle x, B_2^i x \rangle + \langle (a^i - \alpha_i b^i), x \rangle \right] + \sum_{i=1}^{m} (\xi_i - \alpha_i \gamma_i),$$

$$h_\alpha^2(x) = \sum_{i=1}^{m} \left[\langle x, A_2^i x \rangle + \alpha_i \langle x, B_1^i x \rangle \right].$$

A more complicated d.c. representation of functions appears in Problem $(\mathcal{DCC})$, because $\alpha = (\alpha_1, \ldots, \alpha_m) \in \mathbb{R}_+^m$ is a variable and the problem has the following nonconvex term

$$\alpha_i \varphi_i(x) = \alpha_i \langle x, B^i x \rangle + \alpha_i \langle b^i, x \rangle + \alpha_i \gamma_i, \tag{4}$$

which generates the bilinearity, and as a consequence, the nonconvexity in every constraint $(i \in I)$.

The term $\alpha_i \langle b^i, x \rangle$ in (4) can be represented in the d.c. form as follows

$$\langle \alpha_i b^i, x \rangle = \frac{1}{4} \parallel \alpha_i b^i + x \parallel^2 - \frac{1}{4} \parallel \alpha_i b^i - x \parallel^2, \quad i \in I. \tag{5}$$

Further, the product $\alpha_i \langle x, B^i x \rangle$ can be expressed by formula (5):

$$\alpha_i \langle x, B^i x \rangle = \frac{1}{4} \left(\alpha_i + \langle x, B^i x \rangle \right)^2 - \frac{1}{4} \left(\alpha_i - \langle x, B^i x \rangle \right)^2, \quad i \in I,$$

if B^i, $i \in I$, are positive definite matrices and the following conditions hold

$$\alpha_i + \langle x, B^i x \rangle \geq 0, \quad \alpha_i - \langle x, B^i x \rangle \geq 0 \ \forall x \in S, \ i \in I. \tag{6}$$

Then
$$g_i(x, \alpha_i) = \psi_i(x) + \frac{1}{4} \left(\alpha_i - \langle x, B^i x \rangle \right)^2 + \frac{1}{4} \parallel \alpha_i b^i - x \parallel^2 - \alpha_i \gamma_i,$$
$$h_i(x, \alpha_i) = \frac{1}{4} \left(\alpha_i + \langle x, B^i x \rangle \right)^2 + \frac{1}{4} \parallel \alpha_i c^i + x \parallel^2$$

are convex functions and their difference present the constraints in Problem $(\mathcal{DCC})$ in the following d.c. form:

$$\psi_i(x) - \alpha_i \varphi_i(x) = g_i(x, \alpha_i) - h_i(x, \alpha_i) \leq 0, \quad i \in I. \tag{7}$$

5   Global Search Scheme for Solving the Sum-of-Ratios Problems

The previous computational experiments [9,11,12] have demonstrated that Algorithm 1 developed for solving fractional programs via d.c. minimization (see Sect. 2) is quite efficient when applied to problems with affine functions in the ratios. The algorithm has also shown its effectiveness for problems with quadratic numerators and linear denominators [9,11,12]. However, the algorithm wastes a lot of run-time on finding the vector parameter α at which the optimal value of Problem $(\mathcal{P}_\alpha)$ is equal to zero. The shortcoming of the approach of the reduction of the sum-of-ratios problems $(\mathcal{FP})$ to problems with d.c. constraints $(\mathcal{DCC})$ is a lot of run-times spent on solving Problem $(\mathcal{DCC})$.

The results of computational experiments suggest a combination of the two approaches for solving the fractional programs. So, we propose to use the local search for Problem $(\mathcal{DCC})$ to find a starting value α^0 of the parameter α, which takes less time to reduce the optimal value function of Problem $(\mathcal{P}_\alpha)$ to zero. This idea could be implemented by the method which consists of 3 basic parts: the local search in Problem $(\mathcal{DCC})$ with d.c. inequality constraints, the global search in d.c. minimization Problem $(\mathcal{P}_\alpha)$ with a fixed vector parameter α (found by the LSM for Problem $(\mathcal{DCC})$) and the method for finding the vector α at which the optimal value of Problem $(\mathcal{P}_\alpha)$ is equal to zero.

We denote $g^k(\cdot) = g_{\alpha^k}(\cdot)$, $h^k(\cdot) = h_{\alpha^k}(\cdot)$, $\Phi_k(\cdot) = \Phi_{\alpha^k}(\cdot)$.

Let an initial point $x_0 \in S$, a vector $\alpha_0 = (\alpha_{01}, \ldots, \alpha_{0m}) \in \mathbb{R}_+^m$ and an initial segment $[\alpha_-^0, \alpha_+^0]$ for varying α be given. In addition, let there be given the number sequences $\{\tau_s\}$, $\{\delta_s\}$, $\{\varepsilon_s\}$, such that τ_s, δ_s, $\varepsilon_s > 0$, $s = 0, 1, 2, \ldots, \tau_s \downarrow 0$, $\delta_s \downarrow 0 \, \varepsilon_s \downarrow 0 \, (s \to \infty)$.

Global search scheme for fractional program (F-GSS)

Stage 0. (**Initialization**) $k := 0$, $\vartheta^k := (x_0, \alpha_0)$, $\alpha_-^k := 0$, $\alpha_+^k := \alpha_+^0$.

Stage I. (**Local search**) Starting from the point ϑ^k, find by the LSM from [21] a critical point $(x(\alpha^k), \alpha^k)$ in the d.c. constrained Problem $(\mathcal{DCC})$.

Stage II. (**Global search**) Starting from the point $x(\alpha^k)$ find a solution $z(\alpha^k)$ to Problem $(\mathcal{P}_k)$ with the help of the global search scheme for d.c. minimization [19, 22], which consists of the following steps:

Step 0. $s := 0$, $z^s := x(\alpha^k)$, $\zeta_s := \Phi_k(z^s)$.

Step 1. Choose a number $\beta \in [\beta_-, \beta_+]$, where the numbers $\beta_- = \inf(g^k, S)$, $\beta_+ = \sup(g^k, S)$ can be approximated by rather rough estimates. Set $\beta_0 := g^k(z^0)$.

Step 2. Construct an approximation $\mathcal{A}(\beta) = \{y^1, \ldots, y^N \mid h^k(y^j) = \beta - \zeta_s, \ j = 1, \ldots, N = N(\beta)\}$ of the level surface $\{y \in \mathbb{R}^n \mid h^k(y) = \beta - \zeta_s\}$ of the function $h^k(\cdot)$ and, according to Theorem 2, form a collection of indices J_s defined as follows $J_s = J_s(\beta) = \{j \in \{1, \ldots, N_s\} \mid g^k(y^j) \le \beta\}$.

Step 3. If $J_s = \emptyset$, then set $\beta := \beta + \Delta\beta \in [\beta_-, \beta_+]$, and loop to Step 2.

Step 4. For every $j \in J_s$ find a global $2\delta_s$-solution $\overline{u}^j \in S$ to the following linearized convex problem
$$g^k(x) - \langle \nabla h^k(y^j), x \rangle \downarrow \min_x, \quad x \in S.$$
and after that, starting at $\overline{u}^j \in S$, apply the LSM from [19] to produce a $2\tau_s$-critical vector $u^j \in S$, so that $g^k(u^j) - \langle \nabla h^k(u^j), u^j \rangle - 2\tau_s \le \inf_x\{g^k(x) - \langle \nabla h^k(u^j), x \rangle \mid x \in S\}$.

Step 5. For every $j \in J_s$ find a global $2\delta_s$-solution $v^j \colon h^k(v^j) = \beta - \zeta_s$, to the following level problem (see Global Optimality Condition $(\mathcal{E})$)
$$\langle \nabla h^k(v), u^j - v \rangle \uparrow \max_v, \quad h^k(v) = \beta - \zeta_s.$$
Note that for a quadratic function $h^k(\cdot)$, this problem can be solved manually.

Step 6. Compute the number $\eta_s(\beta) := \eta_s^0(\beta) - \beta$, where $\eta_s^0(\beta) := g^k(u^p) - \langle \nabla h^k(v^p), u^p - v^p \rangle := \min_{j \in J_s}\{g^k(u^j) - \langle \nabla h^k(v^j), u^j - v^j \rangle\}$.

Step 7. If $\eta_s(\beta) < 0$, then set $s := s + 1$, $z^s := u^p$ and loop to Step 2.

Step 8. (Else) set $\beta := \beta + \Delta\beta \in [\beta_-, \beta_+]$ and go to Step 2. If $\eta_s(\beta) \ge 0 \, \forall \beta \in [\beta_-, \beta_+]$ (i.e. the one-dimensional search on β is terminated) and $\delta_s \le \delta_*, \tau_s \le \tau_*, \varepsilon_s \le \varepsilon_*$, where $\delta_* > 0, \tau_* > 0$, $\varepsilon_* > 0$ are the fixed accuracies of corresponding computations, then the global search (Stage II) has been terminated: $z(\alpha^k) := z^s$.

Stage III. (**Stopping criterion**) If $\mathcal{V}(\alpha^k) = 0$ and $\min_i \{\psi_i(z(\alpha^k)) - \alpha_i^k \varphi_i (z(\alpha^k))\} \ge 0$, then STOP: $z(\alpha^k) \in Sol(\mathcal{FP})$.

Stage IV. (**Parameter variation**) Implement the parameter variation algorithm to find new parameters α^{k+1}, α_-^{k+1} and α_+^{k+1}. Set $\vartheta^{k+1} = (z(\alpha^k), \alpha^{k+1})$, $k = k + 1$ and go to Stage I.

6   Computational Simulations

The algorithm developed on the basis of Global search scheme F-GSS from Sect. 5 combining two approaches for solving the fractional programs $(\mathcal{FP})$ via d.c. optimization problems was coded in C++ language and applied to solve sum-of-ratios problems with quadratic functions in the ratios.

The set of test examples was generated by the technique from [2]. The method of generation was based on the Calamai's and Vicente's idea [29] to construct a nonconvex quadratic problem with known local and global solutions and on the reduction Theorem 1 to obtain fractional problem with quadratic functions in the numerators and denominators of the ratios.

All computational experiments were performed on the Intel Core i7-4790K CPU 4.0 GHz. All auxiliary convex (linearized) problems arising not only during the implementation of the LSM for a d.c. constrained problem (Stage I of the F-GSS) but also during the global search procedures for a d.c. minimization problem (Stage II of the F-GSS) were solved by the software package IBM ILOG CPLEX 12.6.2.

Table 1 shows the results of computational testing of the F-GSS and employs the following denotations: n is the number of variables (problem's dimension); m is the number of terms in the sum; $f(x_0)$ is the value of the goal function of Problem $(\mathcal{FP})$ at the starting point x_0; f_{glob} stands for the value of the function at the solution provided by the F-GSS; St is the number of critical points passed by the algorithm; it-α is the number of variation of the parameter α in the F-GSS; PL is the number of solved auxiliary (linearized) problems; T stands for the CPU time; f_M is the value of the goal function of Problem $(\mathcal{FP})$ provided by the fmincon solver of MATLAB.

The test problems (with known global solution) constructed with up to 100 variables and 1000 terms in the sum were successfully solved (see Table 1). We got the global solution in all tests. At the same time, the fmincon solver of MATLAB fails to find the global solution in all test problems (the starting point was the same for both algorithms) and the gap was from 1% ($n = 100$, $m = 70$) to 4.3% ($n = 10$, $m = 10$), and 1.8% on average.

Thus, we conclude that new computational results on solving of the sum-of-ratios problems with quadratic functions in the ratios are rather promising. The computational experiment showed that the algorithm developed for solving fractional programs is quite efficient.

Table 1. Results of computational testing of the F-GSS

n	m	$f(x_0)$	f_{glob}	St	it -α	PL	T(hh:mm:ss)	f_M
10	10	12.412	10	3	62	742	00:00:00.96	10.444
10	20	24.184	20	4	87	984	00:00:01.27	20.747
10	30	36.600	30	3	135	1241	00:00:01.68	31.185
10	40	45.905	40	6	160	1967	00:00:02.68	41.055
10	50	59.338	50	5	163	1961	00:00:02.38	51.579
10	100	117.097	100	5	125	2171	00:00:03.00	102.925
20	10	10.985	10	7	73	2850	00:00:07.11	10.163
20	50	53.140	50	8	175	6853	00:00:17.13	50.923
20	100	109.900	100	7	216	6070	00:00:14.86	101.779
20	200	217.532	200	8	231	12648	00:00:33.45	203.212
20	300	331.148	300	8	437	18183	00:00:52.05	305.493
30	300	317.672	300	6	256	41390	00:02:58.92	303.219
30	400	429.763	400	8	501	36858	00:02:22.43	405.392
30	500	540.361	500	3	405	27851	00:02:48.37	507.485
30	600	648.150	600	7	477	134532	00:12:01.93	609.279
30	700	760.107	700	5	307	444514	00:59:51.79	711.176
30	800	843.491	800	4	498	674305	01:39:07.62	808.296
30	900	912.228	900	6	495	3691396	09:21:47.13	912.228
30	1000	1089.367	1000	3	500	2165762	06:25:22.73	1016.486
50	100	106.017	100	5	181	71973	00:11:37.92	101.043
50	200	213.638	200	7	276	56980	00:12:06.29	203.447
50	300	320.066	300	8	324	133878	00:25:20.29	303.882
50	400	429.090	400	6	298	85923	00:17:37.12	405.265
50	500	532.842	500	9	308	189331	00:42:18.27	506.385
50	1000	1059.655	1000	53	104	369255	01:38:07.78	1013.276
70	100	105.464	100	8	214	65055	00:22:52.85	102.133
70	200	210.726	200	8	226	188620	01:08:06.39	203.113
70	300	316.092	300	6	214	233466	01:32:50.30	303.042
70	400	425.458	400	6	214	115923	00:51:24.14	404.868
100	70	73.042	70	110	150	281305	03:26:19.46	70.693
100	90	95.764	90	6	181	139464	01:42:55.57	92.530
100	100	105.456	100	8	253	248367	04:47:31.88	101.156

7   Conclusions

In this paper, we showed how fractional programs with nonconvex quadratic functions can be solved by applying the Global Search Theory for d.c.

optimization problems. Instead of considering a sum-of-ratios problem directly, we developed an efficient global search algorithm, which is based on two approaches. The first one adopts a reduction of the fractional minimization problem to the solution of an equation with an optimal value of the d.c. minimization problem with a vector parameter. The second method is based on the reduction of the sum-of-ratios problem to the optimization problem with nonconvex constraints.

The global search algorithm developed for fractional program was tested on examples with nonconvex quadratic functions in the numerators and denominators of the ratios. The numerical experiments demonstrated that the algorithm performs well when solving rather complicated sum-of-ratios problems with up to 100 variables or 1000 terms in the sum.

References

1. Ashtiani, A.M., Ferreira, P.A.V.: A branch-and-cut algorithm for a class of sum-of-ratios problems. Appl. Math. Comput. **268**, 596–608 (2015)
2. Barkova, M.V.: On generating nonconvex optimization test problems. In: Khachay, M., Kochetov, Y., Pardalos, P. (eds.) MOTOR 2019. LNCS, vol. 11548, pp. 21–33. Springer, Cham (2019). https://doi.org/10.1007/978-3-030-22629-9_2
3. Bugarin, F., Henrion, D., Lasserre, J.B.: Minimizing the sum of many rational functions. Math. Programm. Comput. **8**(1), 83–111 (2015). https://doi.org/10.1007/s12532-015-0089-z
4. Dinkelbach, W.: On nonlinear fractional programming. Manag. Sci. **13**, 492–498 (1967)
5. Dur, M., Horst, R., Thoai, N.V.: Solving sum-of-ratios fractional programs using efficient points. Optimization **49**(5–6), 447–466 (2001)
6. Freund, R.W., Jarre, F.: Solving the sum-of-ratios problem by an interior-point method. J. Glob. Optim. **19**(1), 83–102 (2001). https://doi.org/10.1023/A:1008316327038
7. Frenk, J.B.G., Schaible, S.: Fractional programming. In: Hadjisavvas, S.S.N., Komlosi, S. (eds.) Handbook of Generalized Convexity and Generalized Monotonicity. Series Nonconvex Optimization and Its Applications, vol. 76, pp. 335–386. Springer, Heidelberg (2002). https://doi.org/10.1007/b101428
8. Gruzdeva, T.V., Enkhbat, R., Tungalag, N.: Fractional programming approach to a cost minimization problem in electricity market. Yugoslav J. Oper. Res. **29**(1), 43–50 (2019)
9. Gruzdeva, T.V., Strekalovskiy, A.S.: On solving the sum-of-ratios problem. Appl. Math. Comput. **318**, 260–269 (2018)
10. Gruzdeva, T., Strekalovsky, A.: An approach to fractional programming via D.C. constraints problem: local search. In: Kochetov, Y., Khachay, M., Beresnev, V., Nurminski, E., Pardalos, P. (eds.) DOOR 2016. LNCS, vol. 9869, pp. 404–417. Springer, Cham (2016). https://doi.org/10.1007/978-3-319-44914-2_32
11. Gruzdeva, T., Strekalovsky, A.: A D.C. programming approach to fractional problems. In: Battiti, R., Kvasov, D.E., Sergeyev, Y.D. (eds.) LION 2017. LNCS, vol. 10556, pp. 331–337. Springer, Cham (2017). https://doi.org/10.1007/978-3-319-69404-7_27

12. Gruzdeva, T.V., Strekalovsky, A.S.: On a Solution of Fractional Programs via D.C. Optimization Theory. In: CEUR Workshop Proceedings, OPTIMA-2017, vol. 1987, pp. 246–252 (2017)
13. Hiriart-Urruty, J.B.: Generalized differentiability, duality and optimization for problems dealing with difference of convex finctions. In: Ponstein, J. (ed.) Convexity and Duality in Optimization. Lecture Notes in Economics and Mathematical Systems, vol. 256, pp. 37–69. Springer, Berlin (1985). https://doi.org/10.1007/978-3-642-45610-7_3
14. Horst, R., Tuy, H.: Global Optimization: Deterministic Approaches. Springer, Heidelberg (1996). https://doi.org/10.1007/978-3-662-03199-5
15. Horst, R., Pardalos, P.M. (eds.): Handbook of Global Optimization. Nonconvex Optimization and its Applications. Kluwer Academic Publishers, Dordrecht (1995)
16. Kuno, T.: A branch-and-bound algorithm for maximizing the sum of several linear ratios. J. Glob. Optim. **22**, 155–174 (2002). https://doi.org/10.1023/A:1013807129844
17. Ma, B., Geng, L., Yin, J., Fan, L.: An effective algorithm for globally solving a class of linear fractional programming problem. J. Softw. **8**(1), 118–125 (2013)
18. Schaible, S., Shi, J.: Fractional programming: the sum-of-ratios case. Optim. Meth. Softw. **18**, 219–229 (2003)
19. Strekalovsky, A.S.: Elements of Nonconvex Optimization. Nauka, Novosibirsk (2003). (in Russian)
20. Strekalovsky, A.S.: Global optimality conditions and exact penalization. Optim. Lett. **13**(3), 597–615 (2019). https://doi.org/10.1007/s11590-017-1214-x
21. Strekalovsky, A.S.: On local search in d.c. optimization problems. Appl. Math. Comput. **255**, 73–83 (2015)
22. Strekalovsky, A.S.: On solving optimization problems with hidden nonconvex structures. In: Rassias, T.M., Floudas, C.A., Butenko, S. (eds.) Optimization in Science and Engineering, pp. 465–502. Springer, New York (2014). https://doi.org/10.1007/978-1-4939-0808-0_23
23. Strekalovsky, A.S.: On the merit and penalty functions for the D.C. Optimization. In: Kochetov, Y., Khachay, M., Beresnev, V., Nurminski, E., Pardalos, P. (eds.) DOOR 2016. LNCS, vol. 9869, pp. 452–466. Springer, Cham (2016). https://doi.org/10.1007/978-3-319-44914-2_36
24. Strekalovsky, A.S.: Global optimality conditions in nonconvex optimization. J. Optim. Theory Appl. **173**(3), 770–792 (2017). https://doi.org/10.1007/s10957-016-0998-7
25. Strekalovsky, A.S.: Minimizing sequences in problems with D.C. constraints. Comput. Math. Math. Phys. **45**(3), 418–429 (2005)
26. Strekalovsky, A.S., Gruzdeva, T.V.: Local Search in Problems with Nonconvex Constraints. Comput. Math. Math. Phys. **47**, 381–396 (2007)
27. Strongin, R.G., Sergeyev, Y.D.: Global Optimization with Non-Convex Constraints. Sequential and Parallel Algorithms. Springer, New York (2000)
28. Tuy, H.: Parametric decomposition. Convex Analysis and Global Optimization. SOIA, vol. 110, pp. 283–336. Springer, Cham (2016). https://doi.org/10.1007/978-3-319-31484-6_9
29. Vicente, L.N., Calamai, P.H., Judice, J.J.: Generation of disjointly constrained bilinear programming test problems. Comput. Optim. Appl. **1**(3), 299–306 (1992). https://doi.org/10.1007/BF00249639

Adaptive Descent Splitting Method for Decomposable Optimization Problems

Igor Konnov[1] and Olga Pinyagina[2]

[1] Department of System Analysis and Information Technologies,
Institute of Computational Mathematics and Information Technologies,
Kazan Federal University, Kazan, Russia
konn-igor@ya.ru

[2] Department of Data Mining and Operations Research, Institute of Computational
Mathematics and Information Technologies, Kazan Federal University, Kazan, Russia
Olga.Piniaguina@kpfu.ru

Abstract. We suggest a modified descent splitting method for optimization problems having a special decomposable structure. The proposed modification maintains the basic convergence properties but enables one to reduce computational efforts per iteration and to provide computations in a distributed manner. On the one hand, it consists in component-wise choice of descent directions together with a special threshold control. On the other hand, it involves a simple adaptive step-size choice, which takes into account the problem behavior along the iteration sequence. Preliminary computational tests confirm the efficiency of the proposed modification.

Keywords: Descent splitting method · Adaptive step-size choice · Decomposable optimization problem · Threshold control · Coordinate-wise step

1 Introduction

Over the past decades, many methods have been developed for solving optimization problems. Currently, the most interesting are not universal methods, but those that use the peculiarities of optimization problems. Taking into account the specific properties of the objective function and admissible set allows us to create more efficient processes, which is especially important for solving large dimensional problems. The splitting method is a good example of such methods.

The forward-backward splitting method was proposed first in [1] and further developed, for example, in [2–9]. This method is well-suited for decomposable optimization problems whose objective function can be split into two parts: the

The results of the first author in this work were obtained within the state assignment of the Ministry of Science and Education of Russia, project No. 1.460.2016/1.4. In this work, the authors were also supported by the RFBR grant, project No. 19-01-00431.

A. Kononov et al. (Eds.): MOTOR 2020, LNCS 12095, pp. 128–140, 2020.
https://doi.org/10.1007/978-3-030-49988-4_9

one part is differentiable but can be non-convex, and the other is convex but non-differentiable in general.

Usually, the iterative process of the splitting method requires a certain step-size line-search procedure, exact or inexact. Recently, in paper [13], using the approach of [11,12], we proposed a modification of splitting method without line-search iterative procedure. This modification takes into account the behavior of the problem along the iterative sequence. In this approach, a majorant step-size sequence converging to zero is given and the next decreased value of the step-size is taken only when the current iterate does not give a sufficient descent.

On the other hand, in [10], a modification of the splitting method was proposed for decomposable composite optimization problems, whose structure allows one to construct a component-wise descent method. In the present paper, for such problems we combine these two approaches and describe the adaptive splitting method with component-wise choice of descent directions, a special threshold control, and a simple adaptive step-size choice.

The paper is organized as follows. In Sect. 2 we recall the general scheme and main properties of the splitting method. Section 3 contains a description of a class of decomposable optimization problems. Section 4 describes and substantiates the coordinate splitting method with the adaptive step-size choice for decomposable optimization problems. In Sect. 5 we present some results of numerical tests.

2   The General Splitting Method and Its Properties

Let $f : \mathbf{R}^N \to \mathbf{R}$ be a smooth but not necessarily convex function, $h : \mathbf{R}^N \to \mathbf{R}$ be a non necessarily smooth but rather simple and convex function, $X \in \mathbf{R}^N$ be a nonempty convex closed set. Many applications can be presented in the form

$$\min_{\mathbf{x} \in X} \longrightarrow f(\mathbf{x}) + h(\mathbf{x}). \tag{1}$$

We denote by X^0 the set of stationary points of this problem, they are also solutions to the following variational inequality

$$\langle f'(\mathbf{x}), \mathbf{y} - \mathbf{x} \rangle + h(\mathbf{y}) - h(\mathbf{x}) \geq 0 \quad \forall \mathbf{y} \in X. \tag{2}$$

At first we recall the general scheme of the forward-backward splitting method for problem (1). Let a current iterate $\mathbf{x}^k \in X$ be given. Then the next iterate $\mathbf{x}^{k+1} \in X$ is defined as a solution to the variational inequality

$$\langle f'(\mathbf{x}^k) + \alpha^{-1}(\mathbf{x}^{k+1} - \mathbf{x}^k), \mathbf{y} - \mathbf{x}^{k+1} \rangle + h(\mathbf{y}) - h(\mathbf{x}^{k+1}) \geq 0 \quad \forall \mathbf{y} \in X, \tag{3}$$

where $\alpha > 0$ is a given step-size parameter. Scheme (3) generalizes the well-known optimization methods: if the function h is a constant, process (3) reduces to the (explicit) projection method, but if f is a constant, then we obtain the (implicit) proximal method.

Under the given assumptions, there exists a unique solution to the optimization problems

$$\min_{y \in X} \rightarrow \left\{ h(\mathbf{y}) + (2\alpha)^{-1} \|\mathbf{y} - \mathbf{x}\|^2 \right\} \tag{4}$$

and

$$\min_{y \in X} \longrightarrow \left\{ \langle f'(\mathbf{x}), \mathbf{y} \rangle + h(\mathbf{y}) + (2\alpha)^{-1} \|\mathbf{y} - \mathbf{x}\|^2 \right\} \tag{5}$$

for any $\alpha > 0$ and any $\mathbf{x} \in X$. We denote the solutions to problems (4) and (5) by $\bar{\mathbf{y}}(\alpha, \mathbf{x})$ and $\mathbf{y}(\alpha, \mathbf{x})$, respectively. These optimization problems can be equivalently written in the form of variational inequalities:

$$\exists \mathbf{d} \in \partial h(\mathbf{x}), \quad \langle \mathbf{d} + \alpha^{-1}(\mathbf{y}(\alpha, \mathbf{x}) - \mathbf{x}), \mathbf{y} - \mathbf{y}(\alpha, \mathbf{x}) \rangle \geq 0 \quad \forall \mathbf{y} \in X$$

and

$$\exists \mathbf{d} \in \partial h(\mathbf{x}), \quad \langle f'(\mathbf{x} + \mathbf{d} + \alpha^{-1}(\mathbf{y}(\alpha, \mathbf{x}) - \mathbf{x}), \mathbf{y} - \mathbf{y}(\alpha, \mathbf{x}) \rangle \geq 0 \quad \forall \mathbf{y} \in X,$$

respectively. It is clear that

$$\mathbf{y}(\alpha, \mathbf{x}) = \bar{\mathbf{y}}(\alpha, \mathbf{x} - \alpha f'(\mathbf{x})). \tag{6}$$

Hence, the splitting method (3) can equivalently be defined as

$$\mathbf{x}^{k+1} = \bar{\mathbf{y}}(\alpha, \mathbf{x}^k - \alpha f'(\mathbf{x}^k)), \quad \alpha > 0. \tag{7}$$

The main properties of the mapping $\mathbf{x} \rightarrow \mathbf{y}(\alpha, \mathbf{x})$ are composed in the following proposition (see Propositions 4.1 and 4.2 from [3]).

Proposition 1. *The mapping* $\mathbf{x} \rightarrow \mathbf{y}(\alpha, \mathbf{x})$ *has the following properties:*

a) it is continuous;
b) if $\bar{\mathbf{x}} = \mathbf{y}(\alpha, \bar{\mathbf{x}})$ *for some* $\bar{\mathbf{x}} \in X$, *then* $\bar{\mathbf{x}} \in X^0$;
c) $\exists \mathbf{d} \in \partial h(\mathbf{x}), \quad \langle f'(\mathbf{x}) + \mathbf{d}, \mathbf{y}(\alpha, \mathbf{x}) - \mathbf{x} \rangle \leq -\alpha^{-1} \|\mathbf{y}(\alpha, \mathbf{x}) - \mathbf{x}\|^2 \ \forall \mathbf{x} \in X.$

The essential property of the splitting method is that it easily allows one to develop modifications that take into account peculiarities of the problem under consideration. Following [10] and [13], we will apply the splitting method to decomposable optimization problems.

3   A Class of Decomposable Optimization Problems

Let a partition $\mathcal{N} = \sum_{i=1}^{n} \mathcal{N}_i$ be given such that $\mathcal{N} = \{1, \ldots, N\}$, $|\mathcal{N}_i| = N_i$, $N = \sum_{i=1}^{n} N_i$, and $\mathcal{N}_i \cap \mathcal{N}_j = \emptyset$ if $i \neq j$. Also, let the feasible set $X \subset \mathbf{R}^N$ have the form

$$X = \prod_{i=1}^{n} X_i,$$

where X_i is a nonempty convex closed set in $\mathbf{R}^{N_i}$, $i = 1,\ldots,n$. Then each point $\mathbf{x} = (x_1,\ldots,x_N)^T$ can be presented in the form $\mathbf{x} = (\mathbf{x}_1,\ldots,\mathbf{x}_n)^T$, where $\mathbf{x}_i = \{x_j\}_{j\in\mathcal{N}_i}$, $i = 1,\ldots,n$.

In addition, let the function h be given in the decomposable form

$$h(\mathbf{x}) = \sum_{i=1}^{n} h_i(\mathbf{x}_i), \tag{8}$$

we assume that $h_i : X_i \to \mathbf{R}$ are convex functions and have the nonempty subdifferentials $\partial h_i(\mathbf{x}_i)$ at any $\mathbf{x}_i \in X_i$ for all $i = 1,\ldots,n$. Hence each h_i is lower semi-continuous on X_i, the whole function h is lower semi-continuous on X, and

$$\partial h(\mathbf{x}) = \partial h_1(\mathbf{x}_i) \times \cdots \times \partial h_n(\mathbf{x}_n), \quad \forall \mathbf{x} \in X.$$

We will consider the decomposable optimization problem

$$\min_{\mathbf{x} \in X_1 \times \cdots \times X_n} \longrightarrow \varphi(\mathbf{x}) = \left\{ f(\mathbf{x}) + \sum_{i=1}^{n} h_i(\mathbf{x}_i) \right\}. \tag{9}$$

In addition, we assume that the function $f : \mathbf{R}^N \to \mathbf{R}$ is smooth. We set $\mathbf{g}(\mathbf{x}) = f'(\mathbf{x})$, then

$$\mathbf{g}(\mathbf{x}) = (\mathbf{g}_1(\mathbf{x}_1),\ldots,\mathbf{g}_n(\mathbf{x}_n)), \text{where } \mathbf{g}_i(\mathbf{x}_i) = \left(\frac{\partial f(\mathbf{x})}{\partial x_j} \right)_{j\in\mathcal{N}_i} \in \mathbf{R}^{N_i}, i = 1,\ldots,n.$$

We need a general coercivity condition, which will provide the convergence of the method, when the feasible set of the initial problem is unbounded.

(A1) There exists a number $\gamma > \varphi^*$ such that the set

$$X_\gamma = \{\mathbf{x} \in X : \varphi(\mathbf{x}) \leq \gamma\}$$

is bounded.

Now we formulate the optimality condition for problem (9) (see Proposition 2.1 from [10])

Lemma 1. *1) Each solution of (9) is a solution to VI: Find a point $x^* \in X$ such that*

$$\sum_{i=1}^{n}\langle \mathbf{g}_i(\mathbf{x}^*), \mathbf{y} - \mathbf{x}_i^* \rangle + \sum_{i=1}^{n}[h_i(\mathbf{y}) - h_i(\mathbf{x}_i^*)] \geq 0 \ \ \forall \mathbf{y_i} \in X_i, \ \ i = 1,\ldots,n. \tag{10}$$

2) If the function f is convex, each solution to VI (10) is also a solution to (9).

We denote by X^0 the solution set of VI (10) and by X^* and φ^* the solution set of (9) and optimal value of its objective function, respectively.

For problem (9), the mapping $\mathbf{x} \to \mathbf{y}(\alpha, \mathbf{x})$, which was defined in Sect. 2, has the form $\mathbf{y}(\alpha, \mathbf{x}) = (\mathbf{y}_1(\alpha, \mathbf{x}), \ldots, \mathbf{y}_n(\alpha, \mathbf{x}))^T$ and each component $i = 1, \ldots, n$ can be found independently, as a solution to the problem

$$\min_{\mathbf{y}_i \in X_i} \longrightarrow \left\{ \langle \mathbf{g}_i(\mathbf{x}), \mathbf{y}_i \rangle + h_i(\mathbf{y}_i) + (2\alpha)^{-1} \|\mathbf{y}_i - \mathbf{x}_i\|^2 \right\}.$$

Let us formulate the descent component-wise property (see Lemma 2.1 from [10]). For brevity, we set $M = \{1, \ldots, n\}$.

Lemma 2. *Set*

$$\mathbf{d}_s = \begin{cases} \mathbf{y}_s(\alpha, \mathbf{x}) - \mathbf{x}_s & \text{if } s = i, \\ 0 & \text{if } s \neq i. \end{cases}$$

for any $\mathbf{x} \in X$ *and any* $i \in M$. *Then*

$$\varphi'(\mathbf{x}; \mathbf{d}) \leq -\alpha^{-1} \|\mathbf{y}_i(\alpha, \mathbf{x}) - \mathbf{x}_i\|^2. \tag{11}$$

For brevity, we set $\Delta(\mathbf{x}) = \|\mathbf{x} - \mathbf{y}(\alpha, \mathbf{x})\|$ and $\Delta_i(\mathbf{x}_i) = \|\mathbf{x}_i - \mathbf{y}_i(\alpha, \mathbf{x})\|$, $i = 1, \ldots, n$.

Now we are ready to formulate the splitting method with component-wise choice of descent directions and adaptive step-size.

4   Adaptive Component-Wise Splitting Method for Decomposable Optimization Problems

Adaptive Component-Wise Splitting Method (ACWSM)

Step 0. Choose a point $\mathbf{z}^0 \in X_\gamma$, numbers $\alpha > 0$, $\beta \in (0, 1)$, and sequences $\{\delta_l\} \searrow 0$, $\{\tau_p\} \to 0$, $\tau_p \in (0, 1)$. Set $l = 1$, $p = 0$, $\mathbf{u}^0 = \mathbf{z}^0$, choose a number $\lambda_0 \in (0, \tau_0]$.

Step 1. Set $k = 1$, $\mathbf{x}^k = \mathbf{z}^{l-1}$, $\mathbf{u}^k = \mathbf{z}^{l-1}$.

Step 2. Choose $i \in M$ such that $\Delta_i(\mathbf{x}^k) \geq \delta_l$, set $i_k = i$,

$$\mathbf{d}_s^k = \begin{cases} \mathbf{y}_s(\alpha, \mathbf{x}^k) - \mathbf{x}_s^k & \text{if } s = i_k, \\ 0 & \text{if } s \neq i_k. \end{cases} \tag{12}$$

If this is not the case, set $\mathbf{z}^l = \mathbf{x}^k$, $l = l + 1$, and go to Step 1.

Step 3. Set $\mathbf{v}^{k+1} = \mathbf{x}^k + \lambda_k \mathbf{d}^k$. If

$$\varphi(\mathbf{v}^{k+1}) - \varphi(\mathbf{x}^k) \leq -\beta \alpha^{-1} \lambda_k \Delta^2(\mathbf{x}^k), \tag{13}$$

then take $\lambda_{k+1} \in [\lambda_k, \tau_p]$, set $\mathbf{x}^{k+1} = \mathbf{v}^{k+1}$, and go to Step 5.

Step 4. Set $\lambda'_{k+1} = \min\{\lambda_k, \tau_{p+1}\}$, $p = p + 1$, and take $\lambda_{k+1} \in (0, \lambda'_{k+1}]$. If $\varphi(\mathbf{v}^{k+1}) \leq \gamma$, then set $\mathbf{x}^{k+1} = \mathbf{v}^{k+1}$ and go to Step 5. Otherwise set $\mathbf{x}^{k+1} = \mathbf{u}^k$, $\mathbf{u}^{k+1} = \mathbf{u}^k$, $k = k + 1$, and go to Step 2.

Step 5. If $\varphi(\mathbf{x}^{k+1}) < \varphi(\mathbf{u}^k)$, then set $\mathbf{u}^{k+1} = \mathbf{x}^{k+1}$. Set $k = k + 1$ and go to Step 2.

This method does not contain any iterative linesearch procedure, but expression (13) similar to the condition of the Armijo inexact linesearch is used: with the help of this condition we make a decision to decrease the step-size at the next iteration. In addition, the auxiliary sequence $\{\mathbf{u}^k\}$ contains the "best" points of the inner process (Steps 2–5), which is necessary only if the feasible set X is unbounded.

Let us show the finiteness of the inner iterative process (Steps 2–5) of ACWSM.

Theorem 1. *Under the above assumptions, the inner iterative process (Steps 2–5) of ACWSM is finite.*

Proof. Assume the contrary that the sequence $\{\mathbf{x}^k\}$ is infinite. By construction, $\{\mathbf{x}^k\}$ belongs to the bounded set X_γ, therefore it has limit points. Since the set M is finite, there exists an index $i_k = i$, which is chosen infinitely in the iterative process. We take the corresponding subsequence $\{\mathbf{x}^{k_s}\}$. In the rest of the proof, we will take subsequences of this subsequence.

We take a subsequence of points with indices $\{j_s\}$ such that

$$\varphi(\mathbf{v}^{j_s+1}) > \varphi(\mathbf{x}^{j_s}) - \beta\alpha^{-1}\lambda_{j_s}\|\mathbf{d}^{j_s}\|^2, \tag{14}$$

$$\varphi(\mathbf{v}^{j_s+1}) > \gamma, \quad \varphi(\mathbf{x}^{j_s}) \leq \gamma. \tag{15}$$

In other words, j_s are indices of such iterations, which do not give a sufficient descent and the step value will decrease at the next iterate. However, we do not take $\mathbf{v}^{j_s+1}$ as the next iterative point, because the objective function value is too large at this point.

Then several cases are possible.

Case 1. The subsequence $\{j_s\}$ is infinite.
Take an arbitrary limit point $\mathbf{x}'$ of subsequence $\{\mathbf{x}^{j_s}\}$. Without loss of generality we can assume that

$$\mathbf{x}' = \lim_{s\to\infty} \mathbf{x}^{j_s}, \quad \mathbf{d}' = \lim_{s\to\infty} \mathbf{d}^{j_s},$$

by construction of $\{\mathbf{d}^{j_s}\}$ and because the mapping $\mathbf{x} \to \mathbf{y}(\alpha, \mathbf{x})$ is continuous due to assertion a) of Proposition 1. We also note that we take the next value from the sequence $\{\tau_p\}$ for the next iterate at each i_s, i.e.,

$$\lambda_{j_s} \in (0, \tau_{p_s}], \quad \lambda_{j_s+1} \in (0, \tau_{p_s+1}],$$

for some infinite subsequence of indices p_s, where $\lim_{p\to\infty} \tau_{p_s} = 0$. Therefore $\lim_{s\to\infty} \lambda_{j_s} = 0$. The sequence $\{\mathbf{d}^{j_s}\}$ is bounded, then by the construction of $\{\mathbf{v}^{j_s+1}\}$ the limit points of subsequences $\{\mathbf{v}^{j_s+1}\}$ and $\{\mathbf{x}^{j_s}\}$ coincide. Since the function h is convex by definition, from assumption (14) we obtain

$$f(\mathbf{x}^{j_s} + \lambda_{j_s}\mathbf{d}^{j_s}) - f(\mathbf{x}^{j_s}) + \lambda_{j_s}\langle\mathbf{q}(\mathbf{v}^{j_s+1}), \mathbf{d}^{j_s}\rangle > -\beta\alpha^{-1}\lambda_{j_s}\|\mathbf{d}^{j_s}\|^2$$

for some subgradient $\mathbf{q}(\mathbf{v}^{j_s+1}) \in \partial h(\mathbf{v}^{j_s+1})$. Taking the limit $s \to \infty$ in the previous inequality yields

$$\langle f'(\mathbf{x}') + \mathbf{q}(\mathbf{x}'), \mathbf{d}' \rangle \geq -\beta \alpha^{-1} \|\mathbf{d}'\|^2$$

for some subgradient $\mathbf{q}(\mathbf{x}') \in \partial h(\mathbf{x}')$. Using Lemma 2, we obtain

$$\beta \|\mathbf{d}'\|^2 \geq \|\mathbf{d}'\|^2.$$

Hence it follows that $(1 - \beta)\|\mathbf{d}'\|^2 \leq 0$. On the other hand, $\|\mathbf{d}^k\| \geq \delta_l > 0$ for all k. We obtain a contradiction, the subsequence $\{j_s\}$ cannot be infinite.

Case 2: The subsequence $\{j_s\}$ is finite.
We above assumed that the sequence $\{\mathbf{x}^k\}$ is infinite. Then $\mathbf{v}^k = \mathbf{x}^k$ for sufficiently large k. The further proof depends on the properties of the sequence $\{\lambda_k\}$.

Case 2a: The number of changes of the index p is finite.
Then we have $\lambda_{k_s} \geq \bar{\lambda} > 0$ for number k_s large enough, therefore we obtain from condition (13) that

$$\varphi(\mathbf{x}^{k_s+1}) - \varphi(\mathbf{x}^{k_s}) \leq -\beta\alpha^{-1}\lambda_{k_s}\|\mathbf{d}^{k_s}\|^2 \leq -\beta\alpha^{-1}\bar{\lambda}\|\mathbf{d}^{k_s}\|^2$$

for k_s large enough. Since $\varphi(x_s^k) \geq \varphi^* > -\infty$, we obtain

$$\lim_{k_s \to \infty} \|\mathbf{d}^{k_s}\| = 0. \tag{16}$$

Then (16) contradicts the condition $\|\mathbf{d}^k\| \geq \delta_l > 0$, which holds for all k. Hence, the number of changes of the index p cannot be finite.

Case 2b: The number of changes of the index p is infinite.
In this case, in the subsequence $\{\mathbf{x}^{k_s}\}$ there exists an infinite subsequence of points with indices $\{l_s\}$ such that $\mathbf{x}^{l_s+1} = \mathbf{x}^{l_s} + \lambda_{l_s}\mathbf{d}^{l_s}$ and condition (13) is violated:

$$\varphi(\mathbf{x}^{l_s} + \lambda_{l_s}\mathbf{d}^{l_s}) - \varphi(\mathbf{x}^{l_s}) = \varphi(\mathbf{x}^{l_s+1}) - \varphi(\mathbf{x}^{l_s}) > -\beta\alpha^{-1}\lambda_{l_s}\|\mathbf{d}^{l_s}\|^2. \tag{17}$$

In addition,

$$\lambda_{l_s} \in (0, \tau_{p_s}], \quad \lambda_{l_s+1} \in (0, \tau_{p_s+1}],$$

and $\lim_{p_s \to \infty} \tau_{p_s} = 0$. Therefore, $\lim_{l \to \infty} \lambda_{l_s} = 0$. Note that since the subsequence $\{\mathbf{d}^{l_s}\}$ is bounded, the limit points of the subsequences $\{\mathbf{x}^{l_s+1}\}$ and $\{\mathbf{x}^{l_s}\}$ coincide. Let us take an arbitrary limit point $\mathbf{x}'$ of this subsequene $\{\mathbf{x}^{l_s}\}$. Without loss of generality we can assume that

$$\mathbf{x}' = \lim_{s \to \infty} \mathbf{x}^{l_s}, \quad \mathbf{d}' = \lim_{s \to \infty} \mathbf{d}^{l_s}.$$

by construction of $\{\mathbf{d}^{j_s}\}$ and because the mapping $\mathbf{x} \to \mathbf{y}(\alpha, \mathbf{x})$ is continuous due to assertion a) of Proposition 1. Since the function h is convex by definition, we obtain from assumption (17) that

$$f(\mathbf{x}^{l_s} + \lambda_{l_s}\mathbf{d}^{l_s}) - f(\mathbf{x}^{l_s}) + \lambda_{l_s}\langle \mathbf{q}(\mathbf{x}^{l_s+1}), \mathbf{d}^{l_s} \rangle > -\beta\alpha^{-1}\lambda_{l_s}\|\mathbf{d}^{l_s}\|^2$$

for some subgradient $\mathbf{q}(\mathbf{x}^{l_s+1}) \in \partial h(\mathbf{x}^{l_s+1})$. Taking the limit $s \to \infty$ in the previous inequality yields

$$\langle f'(\mathbf{x}') + \mathbf{q}(\mathbf{x}'), \mathbf{d}' \rangle \geq -\beta\alpha^{-1}\|\mathbf{d}'\|^2$$

for some subgradient $\mathbf{q}(\mathbf{x}') \in \partial h(\mathbf{x}')$. Using Lemma 2, we obtain

$$\beta\|\mathbf{d}'\|^2 \geq \|\mathbf{d}'\|^2.$$

Hence it follows that $(1 - \beta)\|\mathbf{d}'\|^2 \leq 0$, then $\|\mathbf{d}'\| = 0$. On the other hand, $\|\mathbf{d}^k\| \geq \delta_l > 0$ for all k. We obtain a contradiction, and the number of changes of the index p cannot be infinite. This contradiction completes the proof.

Theorem 2. *(See also Theorem 3.1 from [10].) Let assumption (A1) be fulfilled. Then the sequence $\{\mathbf{z}^l\}$ generated by ACWSM has limit point, all of them belong to X^0. If, in addition, f is convex, then all the limit points of $\{\mathbf{z}^l\}$ belong to X^*.*

Proof. First, we note that the sequence $\{\mathbf{z}^l\}$ is bounded and has limit points. Take an arbitrary point $\bar{\mathbf{z}}$ of $\{\mathbf{z}^l\}$, then

$$\lim_{s\to\infty} \mathbf{z}^{l_s} = \bar{\mathbf{z}}.$$

For all $l > 0$ and for all $i \in M$ we have

$$\Delta_i(\mathbf{z}^l) \leq \delta_l,$$

hence $\Delta(\mathbf{z}^l) \leq \delta_l\sqrt{n}$. Taking the limit $l = l_s \to \infty$, we obtain $\Delta(\bar{\mathbf{z}}) = 0$ and $\mathbf{y}(\alpha, \bar{\mathbf{z}}) = \bar{\mathbf{z}}$. Due to Proposition 1, b) it follows that $\bar{\mathbf{z}}$ belongs to X^0. If, in addition, f is convex, then $\bar{\mathbf{z}}$ also belongs to X^*.

5   Test Calculations

We compared the proposed modification of the splitting method with adaptive step-size and component-wise choice of descent direction (ACWSM) with the component-wise version of this method (CWSM) proposed in [10], which uses the inexact line-search procedure.

The component-wise splitting method with inexact step-size line-search (CWSM)
    Step 0. Choose an initial point $\mathbf{z}^0 \in X_\gamma$, coefficients $\alpha > 0$, $\beta \in (0,1)$, $\theta \in (0,1)$, and sequences $\{\delta_l\} \searrow 0$. Set $l = 1$.
    Step 1. Set $k = 1$, $\mathbf{x}^k = \mathbf{z}^{l-1}$.
    Step 2. Choose $i \in M$ such that $\Delta_i(\mathbf{x}^k) \geq \delta_l$, set $i_k = i$,

$$\mathbf{d}_s^k = \begin{cases} \mathbf{y}_s(\mathbf{x}^k) - \mathbf{x}_s^k & \text{if } s = i_k, \\ 0 & \text{if } s \neq i_k. \end{cases}$$

If it is impossible, set $\mathbf{z}^l = \mathbf{x}^k$, $l = l + 1$, and go to Step 1.

Step 3. Choose the smallest non-negative integer m such that

$$\varphi(\mathbf{x}^k + \theta^m \mathbf{d}^k) - \varphi(\mathbf{x}^k) \le -\beta \alpha^{-1} \theta^m \Delta_i^2(\mathbf{x}^k),$$

set $\lambda_k = \theta^m$, $\mathbf{x}^{k+1} = \mathbf{x}^k + \lambda_k \mathbf{d}^k$, $k = k + 1$ and go to Step 2.

The computational results are presented in the tables, which have the following structure. The first column contains the dimensions of problems. Each row presents the aggregate results of 100 problem instances: it contains mean values (mean val.) and standard deviations (st.dev.) of calculation time and iterations numbers. All the randomly generated data were uniformly distributed.

The smooth part of the objective function had the form

$$f(x) = 1/2\langle Ax, x\rangle - \langle b, x\rangle, \tag{18}$$

where $A = B^T B$. Coefficients b_{ij} and b_i were randomly generated numbers from the segment $[-1, 1]$, $i, j = 1, \ldots, n$. The coefficients of methods were $\alpha = 1$, $\beta = 0.5$, $\theta = 0.5$, $\tau_0 = 1$, $\tau_{k+1} = 0.5\tau_k$. The stopping criterion was $\delta_l < 0.01$.

We remind that if the current iterate gives a sufficient descent, we can even take an increasing step-size value at the next iterate. At each 20-th iterate, we increased the step-size value $\lambda_{k+1} = \lambda_k/0.5$.

Example 1. The first series of experiments contained simple nonsmooth functions, which were defined as follows

$$h(x) = \sum_{i=1,\ldots,n} |x_i|. \tag{19}$$

For the sake of simplicity we considered the unconditional optimization problem, i.e., $X = \mathbf{R}^N$ (Table 1).

Table 1. Results for Example 1.

n	CWSM				ACWSM			
	Time (s)		Iterations		Time (s)		Iterations	
	Mean val.	St.dev.	Mean val.	St.dev.	Mean val.	St.dev.	Mean val.	St.dev.
50	0,012	0,004	6 124	552	0,006	0,005	3 031	519
100	0,092	0,015	23 042	1 663	0,047	0,009	11 936	1 590
150	0,193	0,029	32 362	3 341	0,126	0,019	20 983	1 968
200	0,680	0,066	85 512	5 418	0,379	0,054	46 404	5 088
250	1,074	0,095	101 420	5 630	0,695	0,088	63 215	5 083
300	1,518	0,126	117 630	5 676	1,040	0,108	78 389	5 499
350	4,214	0,245	280 706	11 808	1,568	0,148	97 092	6 003

Example 2. In the next tests we took conditional optimization problems, where the functions f and h were defined in (18) and (19), and the feasible set was a parallelepiped

$$X = \{x \in \mathbf{R}^N : d_i \leq x_i \leq e_i, \quad i = 1, \ldots, n\}, \tag{20}$$

where the coefficients d_i, e_i were randomly generated numbers from the segment $[-10, 10]$, $i = 1, \ldots, n$, taking into account that $d_i \leq e_i$ for all $i = 1, \ldots, n$ (Table 2).

Table 2. Results for Example 2.

n	CWSM				ACWSM			
	Time (s)		Iterations		Time (s)		Iterations	
	Mean val.	St.dev.	Mean val.	St.dev.	Mean val.	St.dev.	Mean val.	St.dev.
50	0,010	0,002	5 584	281	0,006	0,005	2 949	371
100	0,071	0,005	21 740	914	0,032	0,005	9 399	891
150	0,153	0,018	32 004	3 538	0,072	0,006	14 950	911
200	0,551	0,030	84 585	2 502	0,189	0,017	28 279	1 966
250	0,864	0,038	104 403	2 752	0,310	0,024	36 827	2 181
300	1,223	0,037	123 923	3 132	0,446	0,023	43 820	2 056
350	3,193	0,360	273 583	30 951	0,620	0,027	51 511	2 041

Example 3. This series of tests contained conditional optimization problems with nonsmooth functions

$$h(x) = \sum_{i=1,\ldots,n} \alpha_i |x_i|,$$

where α_i were randomly generated numbers from the segment $[0, 20]$, $i = 1, \ldots, n$. The functions f were defined in (18) and X was given in (20) (Table 3).

Example 4. This example used problems with nonsmooth functions

$$h(x) = \sum_{i=1,\ldots,n} |x_i - c_i|,$$

where c_i were randomly generated numbers from the segment $[-10, 10]$, $i = 1, \ldots, n$. The functions f were defined in (18) and X was given in (20) (Table 4).

Example 5. The next series of tests contained problems with nonsmooth functions

$$h(x) = \sum_{i=1,\ldots,n} \alpha_i |x_i - c_i|,$$

where α_i were randomly generated numbers from the segment $[0, 20]$, $i = 1, \ldots, n$, c_i were randomly generated numbers from the segment $[-10, 10]$, $i = 1, \ldots, n$. The functions f were defined in (18) and X was given in (20) (Table 5).

Table 3. Results for Example 3.

n	CWSM				ACWSM			
	Time (s)		Iterations		Time (s)		Iterations	
	Mean val.	St.dev.	Mean val.	St.dev.	Mean val.	St.dev.	Mean val.	St.dev.
50	0,010	0,002	5 613	324	0,005	0,003	2 699	334
100	0,074	0,005	21 843	868	0,031	0,005	8 779	897
150	0,157	0,010	31 874	1 082	0,073	0,008	14 493	1 022
200	0,561	0,020	85 300	2 313	0,182	0,013	26 961	1 789
250	0,895	0,050	105 094	2 685	0,314	0,031	35 761	2 205
300	1,263	0,040	125 479	3 339	0,434	0,029	41 784	2 264
350	3,258	0,345	276 851	28 978	0,602	0,033	49 923	2 607

Table 4. Results for Example 4.

n	CWSM				ACWSM			
	Time (s)		Iterations		Time (s)		Iterations	
	Mean val.	St.dev.	Mean val.	St.dev.	Mean val.	St.dev.	Mean val.	St.dev.
50	0,010	0,003	5 610	339	0,005	0,005	3 064	382
100	0,076	0,007	21 845	909	0,033	0,006	9 606	963
150	0,157	0,009	31 525	1 112	0,077	0,008	15 316	993
200	0,569	0,019	84 843	2 344	0,193	0,013	28 610	1 709
250	0,876	0,028	103 991	2 701	0,321	0,025	37 524	2 369
300	1,255	0,037	124 067	3 414	0,446	0,026	43 614	2 170
350	3,257	0,367	272 637	30 604	0,628	0,035	51 803	2 593

Table 5. Results for Example 5.

n	CWSM				ACWSM			
	Time (s)		Iterations		Time (s)		Iterations	
	Mean val.	St.dev.	Mean val.	St.dev.	Mean val.	St.dev.	Mean val.	St.dev.
50	0,010	0,001	5 636	370	0,005	0,001	2 929	335
100	0,072	0,005	21 741	848	0,031	0,004	9 334	934
150	0,154	0,017	32 201	3 398	0,073	0,006	14 980	1 028
200	0,559	0,022	85 173	2 446	0,188	0,014	28 056	1 825
250	0,865	0,027	104 094	2 474	0,313	0,023	36 939	2 056
300	1,279	0,079	124 740	3 273	0,458	0,043	43 495	2 296
350	3,299	0,287	277 123	22 813	0,624	0,038	51 245	2 348

The program was written in Visual C# with double precision, tested on an Intel i3-4170 CPU at 3.7 GHz, 4 Gb, running under Windows 7.

These preliminary results of computational tests show the efficiency of the proposed modification. It is more flexible in the choice of parameters in comparison with the version including the inexact line-search. The proposed approach allows one to construct efficient implementations and deserves further study.

6 Conclusion

In the present work, we propose the modified splitting method for decomposable optimization problems and prove its convergence under rather mild assumptions. The proposed modification maintains the basic convergence properties but enables one to reduce computational efforts per iteration and provides computations in a distributed manner. On the one hand, it consists in choosing coordinate-wise descent directions together with a special threshold control. On the other hand, it involves a simple adaptive step-size choice, which takes into account the problem behavior along the iterative sequence and needs no iterative line-search sequence. Preliminary numerical tests confirm the efficiency of the proposed modification.

References

1. Lions, P.L., Mercier, B.: Splitting algorithms for the sum of two monotone operators. SIAM. J. Num. Anal. **16**(6), 964–979 (1979)
2. Gabay, D.: Application of the method of multipliers to variational inequalities. In: Fortin, M., Glowinski, R. (eds.) Augmented Lagrangian Methods: Applications to the Numerical Solution of Boundary-value Problems, pp. 299–331. North-Holland, Amsterdam (1983)
3. Fukushima, M., Mine, H.: A generalized proximal point algorithm for certain nonconvex minimization problems. Int. J. Syst. Sci. **12**, 989–1000 (1981)
4. Patriksson, M.: Cost approximations: a unified framework of descent algorithms for nonlinear programs. SIAM J. Optim. **8**(2), 561–582 (1998)
5. Patriksson, M.: Nonlinear Programming and Variational Inequality Problems: A Unified Approach. Kluwer, Dordrecht (1999)
6. Konnov, I.V., Kum, S.: Descent methods for mixed variational inequalities in a Hilbert space. Nonlinear Anal. Theory Methods Appl. **47**(1), 561–572 (2001)
7. Konnov, I.V.: Iterative solution methods for mixed equilibrium problems and variational inequalities with non-smooth functions. In: Haugen, I.N., Nilsen, A.S. (eds.) Game Theory: Strategies, Equilibria, and Theorems, pp. 117–160. NOVA, Hauppauge (2008)
8. Konnov, I.V.: Descent methods for mixed variational inequalities with non-smooth mappings. In: Reich, S., Zaslavski, A.J. (eds.) Optimization Theory and Related Topics. Contemporary Mathematics, vol. 568, pp. 121–138 (2012). Amer. Math. Soc., Providence
9. Konnov, I.V.: Salahuddin: two-level iterative method for non-stationary mixed variational inequalities. Russ. Math. (Iz. VUZ) **61**(10), 44–53 (2017)

10. Konnov, I.V.: Sequential threshold control in descent splitting methods for decomposable optimization problems. Optim. Methods Softw. **30**(6), 1238–1254 (2015)
11. Konnov, I.: Conditional gradient method without line-search. Russ. Math. **62**(1), 82–85 (2018)
12. Konnov, I.: A simple adaptive step-size choice for iterative optimization methods. Adv. Model. Optim. **20**(2), 353–369 (2018)
13. Konnov, I., Pinyagina, O.: Splitting method with adaptive step-size. In: Khachay, M., Kochetov, Y., Pardalos, P. (eds.) MOTOR 2019. LNCS, vol. 11548, pp. 46–58. Springer, Cham (2019). https://doi.org/10.1007/978-3-030-22629-9_4

Convergence Analysis of Penalty Decomposition Algorithm for Cardinality Constrained Convex Optimization in Hilbert Spaces

Michael Pleshakov, Sergei Sidorov$^{(\boxtimes)}$ ⓘD, and Kirill Spiridonov

Saratov State University, Saratov, Russian Federation
sidorovsp@sgu.ru
http://www.sgu.ru

Abstract. The paper examines an algorithm for finding approximate sparse solutions of convex cardinality constrained optimization problem in Hilbert spaces. The proposed algorithm uses the penalty decomposition (PD) approach and solves sub-problems on each iteration approximately. We examine the convergence of the algorithm to a stationary point satisfying necessary optimality conditions. Unlike other similar works, this paper discusses the properties of PD algorithms in infinite-dimensional (Hilbert) space. The results showed that the convergence property obtained in previous works for cardinality constrained optimization in Euclidean space also holds for infinite-dimensional (Hilbert) space. Moreover, in this paper we established a similar result for convex optimization problems with cardinality constraint with respect to a dictionary (not necessarily the basis).

Keywords: Nonlinear optimization · Convex optimization · Sparsity · Cardinality constraint · Penalty decomposition

1 Introduction

Let H be a Hilbert space with norm $\| \cdot \|_H$. Let E be a convex function defined on H. The problem of convex optimization is to find an approximate solution to the problem

$$E(x) \to \min_{x \in H} . \tag{1}$$

Let $\mathcal{B}$ the orthonormal basis in H. In many applied problems it is necessary to find the solutions of problem (1) that are sparse with respect to $\mathcal{B}$, i.e. it is necessary to solve the following problem:

$$E(x) \to \inf_{x \in \Sigma_m} , \tag{2}$$

This work was supported by the Ministry of science and education of the Russian Federation in the framework of the basic part of the scientific research state task, project FSRR-2020-0006.

A. Kononov et al. (Eds.): MOTOR 2020, LNCS 12095, pp. 141–153, 2020.
https://doi.org/10.1007/978-3-030-49988-4_10

where Σ_m is the set of all m-term polynomials with respect to $\mathcal{B} = \{e_i\}_{i \in \mathbb{N}}$:

$$\Sigma_m = \Sigma_m(\mathcal{B}) = \Big\{ x \in H \ : \ x = \sum_{i \in I} x_i e_i, \ \mathrm{card}(I) = m \Big\}. \tag{3}$$

Denote for any $x \in H$, $x = \sum_{i \in I} x_i e_i$,

$$\|x\|_0 := \mathrm{card}(I),$$

i.e. $\|x\|_0$ denotes a number of non-zero elements of x with respect to the basis $\mathcal{B}$.

Thus, problem (2)–(3) can be rewritten as follows:

$$E(x) \to \inf_{\|x\|_0 \leq m} . \tag{4}$$

Obtaining sparse solutions is of interest in many real applications including (among many others)

- machine learning where one has to find a sparse solutions for a loss function minimization problem [2,18,24];
- compressed sensing in which one need to decode a sparse signal by means of a small number of linear measurements [3,5,8];
- logistic regression models where one is interested in feature selection in classification problems in which a sparse solution is needed to minimize the average logistic loss [12,15];
- sparse inverse covariance selection in which one has to obtain the conditional independence in graphical models [1,4,25];
- portfolio selection and economic applications [10,20,21].

One of the approaches to finding a sparse solution to problem (1) is to solve a problem with imposed an additional l_1-regularization on the original problem [6,11]. The paper [17] employs another popular approach to deal with cardinality constrained optimization problems which is based on the penalty decomposition (PD) method. The paper [17] obtains the first-order optimality conditions for these problems. Moreover, the authors of [17] proposed to use the penalty decomposition method with the idea of solving a penalty subproblems via a block coordinate descent method. They proved that under some suitable assumptions, any limit point of the sequence generated by the penalty decomposition algorithm fulfils the first-order optimality conditions.

It should be noted that the interest in the penalty decomposition method has increased recently since the method had proved to be an effective and skillful algorithm for solving the sparse optimization problems in various applications. For example, the extension of the method (which embeds the accelerated iteration hard thresholding) was proposed in [7] combining the advantages of the PD method and the AIHT method while averting their disadvantages. In the paper [23] the authors proposed a penalty proximal alternating linearized minimization method for the large-scale sparse portfolio problems in which a sequence of penalty subproblems is solved by utilizing the proximal alternating linearized minimization framework and sparse projection techniques. Other efficient developments of the penalty decomposition method as well as various technics for solving the l_0-norm minimization problems have been proposed recently in [9,16,19,26].

Following ideas of paper [17], this study examines an algorithm for finding approximate sparse solutions of convex cardinality constrained optimization problem (4) in Hilbert space H both with respect to the orthonormal basis of H (Sect. 3) and with respect to a dictionary in H (Sect. 5). The proposed algorithm uses the penalty decomposition approach proposed in [17] and solves sub-problems on each iteration approximately. We examine the convergence of the algorithm to a stationary point satisfying necessary optimality conditions.

2 Optimality Condition

For a functional $F \in H^*$ and an element $f \in H$ in this paper we will use an appropriate bracket notation $F(f) = \langle F, f \rangle$.

Let $\Omega(E) := \{x \in H : E(x) \leq E(0)\}$. We will suppose that function E is Fréchet differentiable on Ω. We note that it follows from convexity of E that for any $x, y \in \Omega$

$$E(y) \geq E(x) + \langle E'(x), y - x \rangle,$$

where $E'(x)$ denotes Fréchet differential of E at x.

Necessary optimality conditions for cardinality constrained problems with additional nonlinear constraints have been studied in [17]. Such conditions have been used to study the convergence of the PD method proposed in the same work.

Definition 1. *We say that a point*

$$x^* = \sum_{i \in I^*} x_i^* e_i \in H,$$

satisfies first order optimality conditions of Lu–Zhang type [17] for problem (4) *if*

- $\mathrm{card}(I^*) = m$,
- $x_i^* = 0$ *for all* $i \in \mathbb{N} \setminus I^*$,
- $\langle E'(x^*), e_i \rangle = 0$ *for all* $i \in I^*$.

Necessary optimality conditions for cardinality constrained problem (4) are given in the following proposition.

Lemma 1. *Let a point* x^* *be the solution to the cardinality constrained problem* (4). *Then* x^* *satisfies first order optimality conditions of Lu–Zhang type defined in Definition 1.*

Proof. Let $x^* = \sum_{i \in \mathbb{N}} x_i^* e_i$ be the solution to the cardinality constrained problem (4), i.e.

$$E(x^*) = \inf_{\|x\|_0 \leq m} E(x).$$

There exists $I^* \subset \mathbb{N}$, such that $\mathrm{card}(I^*) \leq m$ and

$$x^* = \sum_{i \in I^*} x_i^* e_i.$$

Thus, we have $x_i^* = 0$ for all $i \in \mathbb{N} \setminus I^*$.

It follows from necessary conditions for an extremum, that for all $i \in I^*$

$$\langle E'(x^*), e_i \rangle = 0.$$

Thus, first order optimality conditions of Lu–Zhang type are fulfilled.   □

3 Penalty Decomposition Method for the Cardinality Constrained Problem in Hilbert Space

Applying the classical variable splitting technique [14], Problem (4) can be equivalently expressed as

$$E(x) \rightarrow \inf_{x \in H, y \in H}, \tag{5}$$

where x, y such that

$$\|y\|_0 \leq m, \tag{6}$$

$$x = y. \tag{7}$$

Let us define, for any $\delta > 0$, the function

$$F(x, y, \delta) = E(x) + \frac{\delta}{2} \|x - y\|_H^2, \tag{8}$$

associated with problem (5)–(7). F represents a penalized objective function which gets lower values when the unconstrained variable x is a good solution w.r.t. E and at the same time is close to a point y which is feasible for the original constrained problem. The penalty parameter δ balances the importance of the two components.

Lemma 2. *Let the function $F(x, y, \delta)$ be defined in (8) and suppose that E is convex and for any $x \in H$ we have $E(\alpha x) \rightarrow \infty$ if $\alpha \rightarrow \infty$. Then the set $\Omega(F) := \{x, y \in H : F(x, y, \delta) \leq F(0, 0, \delta)\}$ is compact in H for any δ.*

Proof. First we will show that under the assumption that $E(x)$ is coercive, it also holds that $F(x, y, \delta)$ defined in (8) is coercive as well.

We have two possible cases:

1. If $\|x\|_H \rightarrow \infty$, then $E(x) \rightarrow \infty$ and therefore $F(x, y, \delta) \rightarrow \infty$;
2. If $\|y\|_H \rightarrow \infty$, then $\|x - y\|_H^2 \rightarrow \infty$ and therefore $F(x, y, \delta) \rightarrow \infty$.

Since $F(x, y, \delta)$ is coercive on $H \times H$ and convex in $H \times H$, the set $\Omega(F)$ is compact. □

In [17], an algorithm was proposed to solve the cardinality constrained problem in R^n. In this section, we extend the algorithm to solve the problem in Hilbert space with respect to basis $\mathcal{B}$, i.e. problem (5)–(7).

The combination of reformulation (5)–(7) and Algorithm 1 is what can be referred to as the penalty decomposition method (PDM) for the cardinality constrained problem in Hilbert space.

The output of Algorithm 1 is two sequences $\{x^k\}, \{y^k\}$. At each iteration (for each $k = 1, 2, \ldots$), the algorithm finds x_k, which is an approximate solution to the optimization problem consisting in minimizing the penalty function F (with no restriction on cardinality). Block coordinate descent (BCD) is used until a stationary solution is obtained, i.e. the point in which the gradient of F is bounded by ϵ_{k-1}. Then the point y_k is found as an approximation to x_k with the given cardinality m. The algorithm starts from an arbitrary feasible point (x^0, y^0). After each iteration, the penalty parameter δ_k increases, and the quantity of the approximation ϵ_k decreases. Before each iteration, it is necessary to verify that the points obtained earlier are at the appropriate level. If this test succeeds, the BCD starts from the point (x^k, y^k) obtained at the previous iterations; otherwise, the pair (x^k, y^{k-1}) is used as the starting point for BCD step. This is necessary in order to ensure the existence of an accumulation point of the generated sequence.

Algorithm 1: PDM IN HILBERT SPACE

---

begin
  · Input $x^0 = y^0 \in \Omega(F)$, s.t. $\|x^0\|_0 \le m$, $\{\delta_k\}$ s.t. $\delta_k \to \infty$ and $\delta_k > 0$, $\{\epsilon_k\}$
  s.t. $\epsilon_k \to 0$ and $\epsilon_k > 0$, $A \ge \max\{E(x^0), \inf_x F(x, y^0, \delta_0)\}$;
  for each $k \ge 0$ **do**
    · $l = 0$;
    · $u^0 = x^k$;
    · **if** $\inf_x F(x, y^k, \delta_k) \le A$ **then**
      $\quad \llcorner\ v^0 = y^k$
    $v^0 = y^{k-1}$
    · **while** $\|F'_u(u^l, v^l, \delta_k)\|_H > \epsilon_k$ **do**
      · $u^{l+1} = \arg\min_u F(u, v^l, \delta_k)$;
      · $v^{l+1} = \arg\min_{v \in \Sigma_m} F(u^{l+1}, v, \delta_k)$;
      · $l := l + 1$;
    · $x^{k+1} = u^l$, $y^{k+1} = v^l$;
  · Output $\{x^k\}, \{y^k\}$;
end

The algorithm presented in the paper has no significant differences with the original method proposed in [17]. However, the main emphasis in the work is on the study of the convergence of the PD algorithm in the case of infinite-dimensional (Hilbert) spaces.

Various numerical optimization methods can be used to solve the auxiliary minimization problem $\arg\min_u F(u, v^l, \delta_k)$. The algorithm implementation and the demonstration of the numerical results of its application for solving applied problems are important. However, we leave these questions outside the scope of this paper and restrict ourselves to a theoretical study of the convergence of the algorithm.

It should be noted that the problem $\arg\min_{v\in\Sigma_m} F(u^{l+1}, v, \delta_k)$ has a closed-form solution in Hilbert space and is a vector v^{l+1} obtained as follows: m components with largest absolute values are selected among the components of the vector u^{l+1}, and then the solution v^{l+1} to the problem is a vector, all components of which are set to be 0, except for the m selected components that are set to be equal to the corresponding components of the vector u^{l+1}.

4   Convergence Analysis

Note that the while loop of the algorithm does not loop infinitely for any k. Indeed, let us suppose the opposite, i.e. there is an iteration k for which the sequence $\{u^l, v^l\}$ is infinite. From the definition of the algorithm we get

$$F(u^l, v^l, \delta_k) \le F(u^0, v^0, \delta_k).$$

Consequently, we have $\{u^l, v^l\} \in \Omega(F)$, which is compact from Lemma 2. Thus, there is a subsequence $K \subset \{0, 1, \ldots\}$ such that $(u^l, v^l) \to_K (u^*, v^*)$. From the definition of Algorithm 1 we get

$$F(u^{l+1}, v^{l+1}, \delta_k) \le F(u^{l+1}, v^l, \delta_k) \le F(u^l, v^l, \delta_k).$$

Since F is continuous and is convex on the compact set, the function reaches its infimum on the set. Note that F is Fréchet differentiable, and therefore there exists $k_0 \in K$ such that

$$\left\| F'_u(u^l, v^l, \delta_k) \right\|_H < \epsilon_k$$

for all $k \ge k_0$, $k \in K$. We obtain that

$$\lim_{l\to\infty,\, l\in K} \left\| F'_u(u^l, v^l, \delta_k) \right\|_H = 0,$$

which contradicts the initial suggestion.

Lemma 3. *Suppose that the sequence $\{x^k, y^k\}$ is generated by Algorithm 1. Then $\{x^k, y^k\}$ has at least one limit (accumulation) point.*

Proof. Let us consider an arbitrary iteration k of Algorithm 1. The choice of values u^{l+1}, v^{l+1} described in the algorithm yields the non-increasing of F, therefore we have

$$F(x^{k+1}, y^{k+1}, \delta_k) \le F(u^1, v^0, \delta_k) \le F(u^0, v^0, \delta_k).$$

From the definition of (u^0, v^0) we have only two possibilities: $(u^0, v^0) = (x^k, y^k)$, or $(u^0, v^0) = (x^k, y^{k-1})$, where

$$F(u^0, v^0, \delta_k) \leq A \tag{9}$$

in both cases. Note that by the definition of F we have $E(x^{k+1}) \leq F(x^{k+1}, y^{k+1}, \delta_k)$, i.e. $E(x^{k+1}) \leq A$.

It holds for every k, since $\{x^k\} \subset \Omega(F)$ is compact due to Lemma 2. Thus, the sequence $\{x^k\}$ is bounded. From (9) we have

$$F(x^{k+1}, y^{k+1}, \delta_k)) = E(x^{k+1}) + \frac{\delta_k}{2} \|x^{k+1} - y^{k+1}\|_H^2 \leq A.$$

Then

$$\|x^{k+1} - y^{k+1}\|_H^2 \leq 2 \frac{A - E(x^{k+1})}{\delta_k}.$$

Taking $k \to \infty$, using the boundness of $\{x^k\}$, and $\delta_k \to \infty$, we get that $\{y^k\}$ is bounded as well. Thus, the sequence $\{x^k, y^k\}$ is bounded and has a accumulation point. $\qquad \square$

There may be several limit (accumulation) points for the sequence $\{x^k, y^k\}$ generated by Algorithm 1.

Theorem 1. *Let $\{x^k, y^k\}$ be the sequence generated by Algorithm 1. Suppose that (x^*, y^*) is a limit point of $\{x^k, y^k\}$, i.e. there is $K \subset \{1, 2, \ldots\}$, such that $(x^k, y^k) \to_K (x^*, y^*)$. Then*

1. *(x^*, y^*) is a feasible point for problem (5)–(7), and x^* is feasible for problem (4).*
2. *x^* satisfies Lu–Zhang necessary conditions for problem (4).*

Proof. 1) First we will show the feasibility of the limit point (x^*, y^*). From the definition of Algorithm 1, for each k we have

$$\|F_u'(x^k, y^k, \delta_{k-1})\|_H \leq \epsilon_{k-1},$$

or

$$\left\| \frac{E'(x^k)}{\delta_{k-1}} + (x^k - y^k) \right\|_H \leq \frac{\epsilon_{k-1}}{\delta_{k-1}}.$$

Taking $k \to \infty$, $k \in K$, using $\delta_k \to \infty$, $\epsilon_k \to 0$, and the boundness of $\{E'(x^k)\}$ (since $E'(x)$ is bounded), we get

$$\|x^* - y^*\|_H = \lim_{k \in K, k \to \infty} \|x^k - y^k\|_H = 0. \tag{10}$$

Thus, the first proposition of Theorem holds since $\|y^*\|_0 \leq m$ and norm $\|.\|_0$ is semi-continuous.

2) Since the sequence $\{x^k\}$, with $x^k = \sum_{i=1}^{\infty} x_i^{(k)} e_i$, $e_i \in \mathcal{B}$, has a limit point, for any real ϵ, there are $k^* = k^*(\epsilon) \in \mathbb{N}$ and an integer $n^* = n^*(\epsilon)$ such that for all $k > k^*$

$$\left\| x^k - \sum_{i=1}^{n^*} x_i^{(k)} e_i \right\|_H \leq \epsilon.$$

Let $J(x^k) \subset \{1, 2, \dots, n^*\}$ be the index set defined by the following rules:

- $\text{card}(J(x^k)) \leq m$;
- if $i \in J(x^k)$ then $x_i^{(k)} \neq 0$;
- if $i \in J(x^k)$ then $|x_i^{(k)}| > |x_j^{(k)}|$ for all $j \notin J(x^k)$.

It follows from (10) that $x^* = y^*$ and therefore $\|x^*\|_0 \leq m$.

We have

$$\left\| E'(x^k) + \delta_{k-1}(x^k - y^k) \right\|_H \leq \epsilon_{k-1} \tag{11}$$

where $y_i^{(k)} = x_i^{(k)}$ for all $i \in J(x^k)$, while for all $\{1, 2, \dots, n^*\} \setminus J(x^k)$ we have $y_i^{(k)} = 0$.

Consider the sequence $\{J(x^k)\}_{k > k^*}$. Note that n^* is fixed and therefore the number of different index sets with cardinality m in the sequence is finite (in fact it is equal to $\binom{n^*}{m}$). Denote J^* the index set that is included in this sequence infinitely many times and let $K_1 \subset K$ (K_1 is infinite) be such that $J(x^k) = J^*$ for all indices $k \in K_1$.

From convergence of (x^k, y^k) to (x^*, y^*) for $k \in K_1$ we have

$$J(x^*) \subset J^*. \tag{12}$$

It follows from (11) that for any $i \in J^*$

$$\frac{dE(x^k)}{dx_i} + \delta_{k-1}(x_i^{(k)} - y_i^{(k)}) \to_{K_1} 0.$$

By definition the equality $x_i^{(k)} = y_i^{(k)}$ holds for every $i \in J^*$, and therefore

$$\frac{dE(x^k)}{dx_i} \to_{K_1} 0, \tag{13}$$

for all $i \in J^*$. It follows from (12) that (13) holds for all $i \in J(x^*)$. Then we have

$$\frac{dE(x^*)}{dx_i} = 0,$$

for all $i \in J(x^*)$, i.e. Lu–Zhang type optimality condition (see Lemma 1) holds.

□

5   Sparse Convex Optimization with Respect to a Dictionary in Hilbert Space

A set of elements $\mathcal{D}$ from the space H is called a *dictionary* (see, e.g. [22]) if each element $g \in \mathcal{D}$ has norm bounded by one, $\|g\|_H \leq 1$, and the closure of span $\mathcal{D}$ is H, i.e. $\overline{\text{span}\mathcal{D}} = H$. A dictionary $\mathcal{D}$ is called symmetric if $-g \in \mathcal{D}$ for every $g \in \mathcal{D}$. In this paper we assume that the dictionary $\mathcal{D}$ is symmetric.

Many problems in machine learning can be reduced to problem (1) with E as a loss function [2]. In many real applications it is required that the optimal solution x^* of (1) should have a simple structure, e.g. be a *finite* linear combination of elements from a dictionary $\mathcal{D}$ in H. In other words, x^* should be a sparse element with respect to the dictionary $\mathcal{D}$ in H. Of course, one can substitute the requirement of sparsity by a constraint on cardinality (i.e. the limit on the number of elements used in linear combinations of elements from the dictionary $\mathcal{D}$ to construct a solution of problem (1)). However, in many cases the optimization problems with cardinality-type constraint are NP-complete.

Note that the orthonormal basis $\mathcal{B}$ of the inner product space H is a special case of the dictionary $\mathcal{D}$.

We are interested in finding the solutions of problem (1) that are sparse with respect to $\mathcal{D}$, i.e. we are looking for solving the following problem:

$$E(x) \to \inf_{x \in \Sigma_m(\mathcal{D})}, \tag{14}$$

where $\Sigma_m(\mathcal{D})$ is the set of all m-term polynomials with respect to $\mathcal{D}$:

$$\Sigma_m(\mathcal{D}) = \Big\{ x \in H \ : \ x = \sum_{g \in \Lambda} c_g g, \ \text{card}(\Lambda) = m, \ \Lambda \subset \mathcal{D} \Big\}. \tag{15}$$

Denote for any $x \in H$

$$\|x\|_{0,\mathcal{D}} := \min \Big\{ \text{card}(\Lambda) \ : \ x = \sum_{g \in \Lambda \subset \mathcal{D}} c_g g \Big\},$$

i.e. $\|x\|_{0,\mathcal{D}}$ denotes the number of non-zero elements of x with respect to the dictionary $\mathcal{D}$.

Definition 2. *We say that a point*

$$\overline{x} = \sum_{g \in \overline{\Lambda}} \overline{c}_g g \in H, \ \overline{\Lambda} \subset \mathcal{D},$$

satisfies first order optimality conditions of Lu-Zhang type for problem (14) *if*

- *card*$(\overline{\Lambda}) = m$,
- $\overline{c}_g = 0$ *for all* $g \in \mathcal{D} \setminus \overline{\Lambda}$,
- $\langle E'(\overline{x}), g \rangle = 0$ *for all* $g \in \overline{\Lambda}$.

Necessary optimality conditions for cardinality constrained problem (14) are given in the following proposition.

Proposition 1. *Let a point $\bar{x}$ be the solution to the cardinality constrained problem (14). Then $\bar{x}$ satisfies first order optimality conditions of Lu–Zhang type defined in Definition 2.*

Applying the classical variable splitting technique [14], problem (4) can be equivalently expressed as

$$E(x) \rightarrow \inf_{x \in H, y \in H}, \tag{16}$$

where x, y such that

$$\|y\|_{0,\mathcal{D}} \leq m, \tag{17}$$

$$x = y. \tag{18}$$

Let

$$\|x\|_{2,\mathcal{D}} := \inf \left\{ \left(\sum_{g \in \mathcal{D}} c_g^2 \right)^{\frac{1}{2}} : x = \sum_{g \in \mathcal{D}} c_g g \right\}$$

be the l_2-type norm with respect to the set $\mathcal{D}$. It should be noted that dictionary expansion of an element from H over dictionary $\mathcal{D}$ may be not unique, and therefore in the definition of $\|\cdot\|_{2,\mathcal{D}}$ we choose the expansion that minimizes the sum of squares of the expansion coefficients.

If the orthonormal basis $\mathcal{B}$ of H is the part of the dictionary, i.e. $\mathcal{B} \subset \mathcal{D}$, $\|x\|_{2,\mathcal{D}} \leq \|x\|_H$.

Let us define, for any $\delta > 0$, the function

$$Q(x, y, \delta) = E(x) + \frac{\delta}{2} \|x - y\|_{2,\mathcal{D}}^2, \tag{19}$$

associated with problem (16)–(18). Here Q is a penalized objective function, which gets lower values when the unconstrained variable x is a good solution w.r.t. E and at the same time is close to a point y which is feasible for the original constrained problem. The penalty parameter δ reflects the balance between the two components.

Then the penalty decomposition method (PDM) for the cardinality constrained problem in Hilbert space with respect to dictionary $\mathcal{D}$ is presented by the reformulation (5)–(7) and Algorithm 2.

Penalty decomposition method (PDM) for the cardinality constrained problem in Hilbert space with respect to dictionary $\mathcal{D}$ have some differences (with compare with PDM w.r.t. $\mathcal{B}$). We redefine $\|\cdot\|_0$ norm with respect to a dictionary $\mathcal{D}$. Dictionary expansion of an element from H over dictionary $\mathcal{D}$ may be not unique, and therefore in our new definition we choose the expansion that have minimum number of non-zero coefficients. At the same reason we introduce $\|\cdot\|_{2,\mathcal{D}}$ which is l_2-type norm with respect to a dictionary $\mathcal{D}$ instead classic Hilbert space norm $\|\cdot\|_H$.

The following proposition can be proved in the same way as Theorem 1.

Algorithm 2: PDM IN HILBERT SPACE W.R.T. THE DICTIONARY $\mathcal{D}$

---

begin
  · Input $x^0 = y^0 \in \Omega(Q)$, s.t. $\|x^0\|_{0,\mathcal{D}} \leq m$, $\{\delta_k\}$ s.t. $\delta_k \to \infty$ and $\delta_k > 0$,
   $\{\epsilon_k\}$ s.t. $\epsilon_k \to 0$, $\epsilon_k > 0$, $A \geq \max\{E(x^0), \inf_x Q(x, y^0, \delta_0)\}$;
  for each $k \geq 0$ **do**
     · $l = 0$;
     · $u^0 = x^k$;
     · **if** $\inf_x Q(x, y^k, \delta_k) \leq A$ **then**
       $\lfloor\ v^0 = y^k$
     $v^0 = y^{k-1}$
     · **while** $\|Q_u'(u^l, v^l, \delta_k)\|_{2,\mathcal{D}} > \epsilon_k$ **do**
        · $u^{l+1} = \arg\min_u Q(u, v^l, \delta_k)$;
        · $v^{l+1} = \arg\min_{v \in \Sigma_m(\mathcal{D})} Q(u^{l+1}, v, \delta_k)$;
        · $l := l + 1$;
     · $x^{k+1} = u^l$, $y^{k+1} = v^l$;
  · Output $\{x^k\}$, $\{y^k\}$;
end

Theorem 2. *Let $\{x^k, y^k\}$ be the sequence generated by Algorithm 2. Suppose that (x^*, y^*) is a limit point of $\{x^k, y^k\}$, i.e. there is $K \subset \{1, 2, \ldots\}$, such that $(x^k, y^k) \to_K (x^*, y^*)$. Then*

1. *(x^*, y^*) is a feasible point for problem (16)–(18), and x^* is feasible for problem (14).*
2. *x^* satisfies Lu–Zhang necessary conditions for problem (14).*

6 Conclusion

This paper presents the results on the convergence of a sequence of points generated by PD algorithms to a point satisfying the Lu–Zhang type necessary optimality conditions. Unlike other similar works, this paper discusses the properties of PD algorithms in infinite-dimensional (Hilbert) space. The results showed that the convergence property obtained in the pioneering work [17] for cardinality constrained optimization in Euclidean space also holds for infinite-dimensional (Hilbert) space. Moreover, in this paper we established a similar result for convex optimization problems with cardinality constraint with respect to a dictionary (not necessarily the basis in H). Note that consideration of the problem in infinite-dimensional spaces is justified for two reasons. Firstly, many applied problems are distinguished by the fact that their dimension is extremely large and can practically be considered infinite. Secondly, the results and estimates of the rate of convergence obtained for problems in infinite-dimensional spaces are independent of dimension.

   Future research may include obtaining estimates on the convergence rate for the PD algorithms under some restrictions on the moduli of smoothness for the objective function, as well as the study of issues related to sufficient optimality conditions for the accumulation points of sequences generated by the proposed algorithms. It would also be interesting to develop software based on the PD algorithms and to perform it in real-world applications including sparse k-monotone regression [13].

References

1. Bollhofer, M., Eftekhari, A., Scheidegger, S., Schenk, O.: Large-scale sparse inverse covariance matrix estimation. SIAM J. Sci. Comput. **41**(1), A380–A401 (2019). https://doi.org/10.1137/17M1147615

2. Bubeck, S.: Convex optimization: algorithms and complexity. Found. Trends Mach. Learn. **8**(3–4), 231–358 (2015)

3. Chen, Z., Huang, C., Lin, S.: A new sparse representation framework for compressed sensing MRI. Knowl.-Based Syst. **188**, 104969 (2020). https://doi.org/10.1016/j.knosys.2019.104969. http://www.sciencedirect.com/science/article/pii/S0950705119303983

4. Dempster, A.P.: Covariance selection. Biometrics **28**(1), 157–175 (1972). https://doi.org/10.2307/2528966

5. Deng, Q., et al.: Compressed sensing for image reconstruction via back-off and rectification of greedy algorithm. Sig. Process. **157**, 280–287 (2019). https://doi.org/10.1016/j.sigpro.2018.12.007. http://www.sciencedirect.com/science/article/pii/S0165168418303980

6. Dereventsov, A., Temlyakov, V.N.: Biorthogonal greedy algorithms in convex optimization. CoRR abs/2001.05530 (2020). https://arxiv.org/abs/2001.05530

7. Dong, Z., Zhu, W.: An improvement of the penalty decomposition method for sparse approximation. Sig. Process. **113**, 52–60 (2015). https://doi.org/10.1016/j.sigpro.2015.01.012. http://www.sciencedirect.com/science/article/pii/S0165168415000353

8. Donoho, D.L.: Compressed sensing. IEEE Trans. Inf. Theor. **52**(4), 1289–1306 (2006). https://doi.org/10.1109/TIT.2006.871582

9. Dou, H.X., Huang, T.Z., Deng, L.J., Zhao, X.L., Huang, J.: Directional l_0 sparse modeling for image stripe noise removal. Remote Sens. **10**(3) (2018). https://doi.org/10.3390/rs10030361. https://www.mdpi.com/2072-4292/10/3/361

10. Fan, J., Lv, J., Qi, L.: Sparse high-dimensional models in economics. Ann. Rev. Econ. **3**(1), 291–317 (2011). https://doi.org/10.1146/annurev-economics-061109-080451

11. Figueiredo, M.A.T., Nowak, R.D., Wright, S.J.: Gradient projection for sparse reconstruction: application to compressed sensing and other inverse problems. IEEE J. Sel. Top. Sig. Process. **1**(4), 586–597 (2007). https://doi.org/10.1109/JSTSP.2007.910281

12. Gajare, S., Sonawani, S.: Improved logistic regression approach in feature selection for EHR. In: Abraham, A., Cherukuri, A.K., Melin, P., Gandhi, N. (eds.) ISDA 2018 2018. AISC, vol. 940, pp. 325–334. Springer, Cham (2020). https://doi.org/10.1007/978-3-030-16657-1_30

13. Gudkov, A.A., Mironov, S.V., Sidorov, S.P., Tyshkevich, S.V.: A dual active set algorithm for optimal sparse convex regression. Vestn. Samar. Gos. Tekhn. Univ. Ser. Fiz.-Mat. Nauki (J. Samara State Tech. Univ. Ser. Phys. Math. Sci.) **23**(1), 113–130 (2019). https://doi.org/10.14498/vsgtu1673

14. Holmberg, K.: Creative modeling: variable and constraint duplicationin primal - dual decomposition methods. Ann. Oper. Res. **82**, 355–390 (1998). https://doi. org/10.1023/A:1018927123151

15. Kampa, K., Mehta, S., Chou, C.A., Chaovalitwongse, W.A., Grabowski, T.J.: Sparse optimization in feature selection: application in neuroimaging. J. Global Optim. **59**(2), 439–457 (2014). https://doi.org/10.1007/s10898-013-0134-2

16. Lu, Z., Li, X.: Sparse recovery via partial regularization: models, theory, and algorithms. Math. Oper. Res. **43**(4), 1290–1316 (2018). https://doi.org/10.1287/moor. 2017.0905

17. Lu, Z., Zhang, Y.: Sparse approximation viaSparse approximation via penalty decomposition methods. SIAM J. Optim. **23**(4), 2448–2478 (2013). https://doi. org/10.1137/100808071

18. Luo, X., Chang, X., Ban, X.: Regression and classification using extreme learning machine based on l1-norm and l2-norm. Neurocomputing **174**, 179–186 (2016). https://doi.org/10.1016/j.neucom.2015.03.112. http://www.sciencedirect.com/ science/article/pii/S092523121501139X

19. Pan, L.L., Xiu, N.H., Fan, J.: Optimality conditions for sparse nonlinear programming. Sci. China Math. **60**(5), 759–776 (2017). https://doi.org/10.1007/s11425-016-9010-x

20. Pun, C.S., Wong, H.Y.: A linear programming model for selection of sparse high-dimensional multiperiod portfolios. Eur. J. Oper. Res. **273**(2), 754–771 (2019). https://doi.org/10.1016/j.ejor.2018.08.025. http://www.sciencedirect.com/ science/article/pii/S0377221718307203

21. Sidorov, S.P., Faizliev, A.R., Khomchenko, A.A.: Algorithms for l_1-norm minimisation of index tracking error and their performance. Int. J. Math. Oper. Res. **11**(4), 497–519 (2017). https://ideas.repec.org/a/ids/ijmore/v11y2017i4p497-519.html

22. Temlyakov, V.N.: Greedy approximation in convex optimization. Constr. Approx. **41**(2), 269–296 (2015). https://doi.org/10.1007/s00365-014-9272-0

23. Teng, Y., Yang, L., Yu, B., Song, X.: A penalty palm method for sparse portfolio selection problems. Optim. Methods Softw. **32**(1), 126–147 (2017). https://doi. org/10.1080/10556788.2016.1204299

24. Wipf, D.P., Rao, B.D.: Sparse Bayesian learning for basis selection. IEEE Trans. Signal Process. **52**(8), 2153–2164 (2004). https://doi.org/10.1109/TSP. 2004.831016

25. Xu, F., Deng, R.: Fast algorithms for sparse inverse covariance estimation. Int. J. Comput. Math. **96**(8), 1668–1686 (2019). https://doi.org/10.1080/00207160.2018. 1506108

26. Zhu, W., Dong, Z., Yu, Y., Chen, J.: Lagrange dual method for sparsity constrained optimization. IEEE Access **6**, 28404–28416 (2018). https://doi.org/10. 1109/ACCESS.2018.2836925

Game Theory

Dixit-Stiglitz-Krugman Model
with Nonlinear Costs

Ivan Belyaev[1] and Igor Bykadorov[1,2,3(✉)]

[1] Novosibirsk State University, Novosibirsk, Russia
ivanbelyaev1708@gmail.com
[2] Sobolev Institute of Mathematics SB RAS, Novosibirsk, Russia
bykadorov.igor@mail.ru
[3] Novosibirsk State University of Economics and Management, Novosibirsk, Russia

Abstract. We study the market equilibrium in international trade monopolistic competition model a'la Dixit-Stiglitz-Krugman with homogeneous firms. The utility of consumers are additive separable. Transport costs are of "iceberg type." The only production factor is labor. The concrete functional form of sub-utility function is assumed unknown. Thus, it is not possible to get the equilibrium in closed form. We examine the local symmetric comparative statics of consumption, prices, firms masses and firms sizes with respect to transport costs. For linear production costs, the results about equilibria near free trade and autarky are known. We show that many of these results are true for the case of non-linear production costs.

Keywords: Dixit-Stiglitz-Krugman Model · Market equilibrium · Free trade · Autarky · Comparative statics

1 Introduction

Differences in firms productivity are important and have various explanations. Recent empirical researches examining industry markets[1]. The modern theoretical interpretation of empirical phenomenons is based on some variations of the international trade model under the monopolistic competition.

The concept of monopolistic competition, introduced by Chamberlin [12,13], widely develops now, starting with the paper by Dixit and Stiglitz [14] for the case of a closed economy, by Krugman [16,17] for the international trade, as well as its modifications with heterogeneous firms, considered by Melitz [18]. The Dixit-Stiglitz-Krugman model describes the impact of economies of scale on monopolistic competition and on the trade of two absolutely identical countries.

As it is usual in monopolistic competition (see, e.g., [4,14,24]), we assume that consumers are identical, firms are identical, labor is the only production factor; moreover, free entry, labor and trade balances hold[2].

---

[1] Some discussion of the stylized facts (see, e.g., [11,15,21,22]) can be found in [8].

[2] See the details, e.g., in [1,6–8].

© Springer Nature Switzerland AG 2020
A. Kononov et al. (Eds.): MOTOR 2020, LNCS 12095, pp. 157–169, 2020.
https://doi.org/10.1007/978-3-030-49988-4_11

In this paper we consider trade between two groups of countries: a group of big countries and a group of small countries. Countries differ only in the number of consumers. The number of countries in the group of big countries is $(K+1)$, the number of countries in the group of small countries is $(k+1)$. It is proposed to consider the case of free trade and the case of autarky. Previously, results were obtained for trade between two countries with a nonlinear production costs [5] and results for two groups of countries with linear production costs [8]. We generalize these results for the case of trade between two groups of countries with non-linear costs.

The paper is organized as follows. In Sect. 2 we set the model and the main notations, describe the problem of consumers (Sect. 2.1) and the problem of producers (Sect. 2.2), labor and trade balances (Sect. 2.3), equilibrium (Sect. 2.4). Moreover, here we define the concept of free trade (Sect. 2.5) and the concept of total autarky (Sect. 2.6). In Sect. 3 we study the local comparative statics of equilibrium on transport costs: in free trade (Sect. 3.1) and in total autarky (Sect. 3.2). Section 4 concludes.

2   Problem

Let us introduce the basic concepts and notation:

- L the number of consumers (the quantity of labor) in each big country,
- l the number of consumers (the quantity of labor) in each small country, $l \leq L$, therefore, $\Gamma = (K+1)L + (k+1)l$ is the total number of consumers,
- N the number (mass) of firms in each big country,
- n the number (mass) of firms in each small country,
- X_i the individual domestic consumption produced by i-th firm, $i \in [0, N]$,
- Y_i the individual import consumption in a big country produced in another big country by i-th firm, $i \in [0, N]$,
- Z_i the individual import consumption in a small country produced in a big country by i-th firm, $i \in [0, N]$,
- x_i the individual domestic consumption produced by i-th firm, $i \in [0, n]$,
- y_i the individual import consumption in a small country produced in another small country by i-th firm, $i \in [0, n]$,
- z_i the individual import consumption in a big country produced in a small country by i-th firm, $i \in [0, n]$,
- $P_i^X, P_i^Y, P_i^Z, p_i^x, p_i^y, p_i^z$ the corresponding prices,
- $w^B = w$ the wage in each big country,
- $w^s \equiv 1$ the wage in each small country.

Figure 1 illustrates the notations in the case of three big and three small countries.

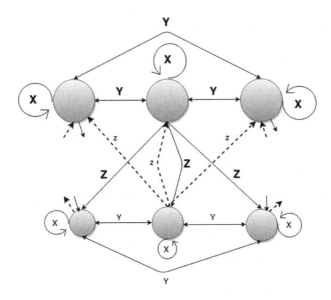

Fig. 1. Graphical representation of the case of three big and three small countries

2.1   Consumers

The problem of a representative consumer in a big country is

$$\int_{i\in[0,N]} u\left(X_i\right) di + K \cdot \int_{i\in[0,N]} u\left(Y_i\right) di + (k+1) \cdot \int_{i\in[0,n]} u\left(z_i\right) di \to max,$$

$$\int_{i\in[0,N]} P_i^X X_i di + K \cdot \int_{i\in[0,N]} P_i^Y Y_i di + (k+1) \cdot \int_{i\in[0,n]} p_i^z z_i di \leq w,$$

while the problem of a representative consumer in a small country is

$$\int_{i\in[0,n]} u\left(x_i\right) di + k \cdot \int_{i\in[0,n]} u\left(y_i\right) di + (K+1) \cdot \int_{i\in[0,N]} u\left(Z_i\right) di \to max,$$

$$\int_{i\in[0,n]} p_i^x x_i di + k \cdot \int_{i\in[0,n]} p_i^y y_i di + (K+1) \cdot \int_{i\in[0,N]} P_i^Z Z_i di \leq 1,$$

where the elementary utility function $u\left(\cdot\right)$ is three times differentiable and satisfies (cf. [24])

$$u(0) = 0,\ u'(\xi) > 0,\ u''(\xi) < 0,$$

i.e., it is everywhere strictly increasing and strictly concave.

From the consumer's first order condition (FOC), one has the inverse demand functions

$$P_i^X = \frac{u'(X_i)}{\Lambda}, \quad P_i^Y = \frac{u'(Y_i)}{\Lambda}, \quad p_i^z = \frac{u'(Z_i)}{\lambda}, \quad i \in [0, N],$$

$$p_i^x = \frac{u'(x_i)}{\lambda}, \quad p_i^y = \frac{u'(y_i)}{\lambda}, \quad P_i^z = \frac{u'(z_i)}{\Lambda}, \quad i \in [0, n],$$

where λ, Λ are Lagrange multipliers.

2.2 Producers

To sell in foreign country, the firm spend the transport costs $\tau \geq 1$ of iceberg type[3]. Then

$$Q_i = L \cdot X_i + K \cdot \tau \cdot L \cdot Y_i + (k+1) \cdot \tau \cdot l \cdot Z_i, \quad i \in [0, N],$$

is the output (the size) of the firm $i \in [0, N]$ in a big country, while

$$q_i = l \cdot x_i + k \cdot \tau \cdot l \cdot y_i + (K+1) \cdot \tau \cdot L \cdot z_i, \quad i \in [0, n],$$

is the output (the size) of the firm $i \in [0, n]$ in a small country.

Each firm has a twice differentiable strictly increasing production cost function $V(\cdot)$, $V'(\cdot) > 0$.

In symmetric case, we omit the index i in individual consumptions

$$X_i = X, \quad Y_i = Y, \quad Z_i = Z, \quad i \in [0, N],$$

$$x_i = x, \quad y_i = y, \quad z_i = z, \quad i \in [0, n],$$

sizes of the firms

$$Q_i = Q = L \cdot X + K \cdot \tau \cdot L \cdot Y + (k+1) \cdot \tau \cdot l \cdot Z, \quad i \in [0, N],$$

$$q_i = q = l \cdot x + k \cdot \tau \cdot l \cdot y + (K+1) \cdot \tau \cdot L \cdot z, \quad i \in [0, n],$$

and prices (inverse demand functions)

$$P^X = \frac{u'(X)}{\Lambda}, \quad P^Y = \frac{u'(Y)}{\Lambda}, \quad p^z = \frac{u'(z)}{\Lambda}, \tag{1}$$

$$p^x = \frac{u'(x)}{\lambda}, \quad p^y = \frac{u'(y)}{\lambda}, \quad P^Z = \frac{u'(Z)}{\lambda}. \tag{2}$$

Using the "normalized" revenue

$$R(\xi) = u'(\xi) \cdot \xi,$$

we write the profit of a firm in a big country (Π) and in a small country (π) as

$$\Pi = L \cdot \frac{R(X)}{\Lambda} + K \cdot L \cdot \frac{R(Y)}{\Lambda} + (k+1) \cdot l \cdot \frac{R(Z)}{\lambda} - w \cdot V(Q),$$

$$\pi = l \cdot \frac{R(x)}{\lambda} + k \cdot l \cdot \frac{R(y)}{\lambda} + (K+1) \cdot L \cdot \frac{R(z)}{\Lambda} - V(q).$$

---

[3] To sell a unit, the firm produce the $\tau \cdot 1$.

2.3 Labor and Trade Balances

In symmetric case, labor balances in the two kind of countries can be written as

$$N \cdot V(Q) = L, \qquad n \cdot V(q) = l. \tag{3}$$

Trade balances in big and small countries are, respectively

$$N \cdot \left(K \cdot L \cdot \frac{R(Y)}{\Lambda} + (k+1) \cdot l \cdot \frac{R(Z)}{\lambda} \right) = N \cdot K \cdot L \cdot \frac{R(Y)}{\Lambda} + n \cdot (k+1) \cdot L \cdot \frac{R(z)}{\Lambda}$$

and

$$n \cdot \left(k \cdot l \cdot \frac{R(y)}{\lambda} + (K+1) \cdot L \cdot \frac{R(z)}{\Lambda} \right) = n \cdot k \cdot l \cdot \frac{R(y)}{\lambda} + N \cdot (K+1) \cdot l \cdot \frac{R(Z)}{\Lambda}.$$

Note that these two trade balances can both be rewritten as (one!) equation

$$N \cdot l \cdot \frac{R(Z)}{\lambda} = n \cdot L \cdot \frac{R(z)}{\Lambda},$$

i.e., by substituting N and n from the labor balances (3),

$$TB \equiv \frac{R(Z)}{\lambda \cdot V(Q)} - \frac{R(z)}{\Lambda \cdot V(q)} = 0. \tag{4}$$

2.4 Equilibrium

Standardly in monopolistic competition framework, we assume that firms freely enter the market while their profit remains positive, which implies a zero-profit (free-entry) conditions. In symmetric case we get

$$\Pi = 0, \qquad \pi = 0. \tag{5}$$

Producer's First order conditions (FOC) in symmetric case are

$$\frac{\partial \Pi}{\partial X} = 0, \quad \frac{\partial \Pi}{\partial Y} = 0, \quad \frac{\partial \Pi}{\partial Z} = 0, \quad \frac{\partial \pi}{\partial x} = 0, \quad \frac{\partial \pi}{\partial y} = 0, \quad \frac{\partial \pi}{\partial z} = 0. \tag{6}$$

Second order conditions (SOC) are

$$\frac{\partial^2 \Pi}{\partial X^2} < 0, \qquad \frac{\partial^2 \Pi}{\partial X^2} \cdot \frac{\partial^2 \Pi}{\partial Y^2} - \left(\frac{\partial^2 \Pi}{\partial X \partial Y} \right)^2 > 0, \qquad \det \Pi'' < 0, \tag{7}$$

where

$$\Pi'' = \begin{pmatrix} \dfrac{\partial^2 \Pi}{\partial X^2} & \dfrac{\partial^2 \Pi}{\partial X \partial Y} & \dfrac{\partial^2 \Pi}{\partial X \partial Z} \\[2mm] \dfrac{\partial^2 \Pi}{\partial X \partial Y} & \dfrac{\partial^2 \Pi}{\partial Y^2} & \dfrac{\partial^2 \Pi}{\partial Y \partial Z} \\[2mm] \dfrac{\partial^2 \Pi}{\partial X \partial Z} & \dfrac{\partial^2 \Pi}{\partial Y \partial Z} & \dfrac{\partial^2 \Pi}{\partial Z^2} \end{pmatrix},$$

and

$$\frac{\partial^2 \pi}{\partial x^2} < 0, \qquad \frac{\partial^2 \pi}{\partial x^2} \cdot \frac{\partial^2 \pi}{\partial y^2} - \left(\frac{\partial^2 \pi}{\partial x \partial y}\right)^2 > 0, \qquad \det \pi'' < 0, \qquad (8)$$

where

$$\pi'' = \begin{pmatrix} \dfrac{\partial^2 \pi}{\partial x^2} & \dfrac{\partial^2 \pi}{\partial x \partial y} & \dfrac{\partial^2 \pi}{\partial x \partial z} \\[3mm] \dfrac{\partial^2 \pi}{\partial x \partial y} & \dfrac{\partial^2 \pi}{\partial y^2} & \dfrac{\partial^2 \pi}{\partial y \partial z} \\[3mm] \dfrac{\partial^2 \pi}{\partial x \partial z} & \dfrac{\partial^2 \pi}{\partial y \partial z} & \dfrac{\partial^2 \pi}{\partial z^2} \end{pmatrix}.$$

Symmetric equilibrium is a bundle

$$\left(X^*, Y^*, Z^*, x^*, y^*, z^*, \Lambda^*, \lambda^*, w^*, N^*, n^*, P^{X^*}, P^{Y^*}, P^{Z^*}, p^{x^*}, p^{y^*}, p^{z^*}\right)$$

satisfying the following:

- utility maximization (1) and (2);
- profit maximization (6) and (7), (8), and free entry conditions (5);
- labor balances (3) and trade balance (4).

Note that the equilibrium prices $P^{X^*}, P^{Y^*}, P^{Z^*}, p^{x^*}, p^{y^*}, p^{z^*}$ and masses of firms N^*, n^* can be obtained from (1)–(3). So equilibrium individual consumptions $X^*, Y^*, Z^*, x^*, y^*, z^*$, Lagrange multipliers Λ^*, λ^* and wage in each big country w^* can be defined from the reduced system of equilibrium (4)–(6), i.e.,

$$\frac{\partial \Pi}{\partial X} = 0, \quad \frac{\partial \Pi}{\partial Y} = 0, \quad \frac{\partial \Pi}{\partial Z} = 0, \quad \frac{\partial \pi}{\partial x} = 0, \quad \frac{\partial \pi}{\partial y} = 0, \quad \frac{\partial \pi}{\partial z} = 0,$$

$$\Pi = 0, \qquad \pi = 0, \qquad TB = 0.$$

In the following we will use for function $f(\xi)$, elasticity

$$\mathcal{E}_f(\xi) = \frac{f'(\xi) \cdot \xi}{f(\xi)}$$

and Arrow-Pratt measure

$$r_f(\xi) = -\frac{f''(\xi) \cdot \xi}{f'(\xi)} = -\mathcal{E}_{f'}(\xi).$$

Note that we study mainly such functions u that their Arrow-Pratt measure $r_u(\cdot)$ increases, i.e., consider "pro-competitive" case[4].

---

[4] The meaning of "pro-competitiveness" have been explained, e.g. in [24], where closed economy (only one country) was considered. It turns out that, in equilibrium, mass of firms increases w.r.t. market size; price decreases if $r_u(\cdot)$ increases, and price increases if $r_u(\cdot)$ decreases (of course, price is constant if $r_u(\cdot)$ is constant, i.e., in CES-case). Increasing of mass of firms means increasing of competition. Therefore, if $r'_u(\cdot) > 0$ then price decreases ("pro-competitive" case) while if $r'_u(\cdot) < 0$ then price increases ("anti-competitive" case). In our opinion, "pro-competitive" case is more natural.

2.5 The Case of Free Trade

Consider the situation of Free Trade, i.e., let $\tau = 1$. Then

$$X = Y = Z = x = y = z,$$

$$w = 1, \qquad \Lambda = \lambda,$$

$$Q = q = ((K+1) \cdot L + (k+1) \cdot l) \cdot X = \Gamma \cdot X.$$

As usual, let us denote

$$R = R(X) = R(Y), \; R' = R'(X), \; V = V(Q), \; \mathcal{E}_R = \mathcal{E}_R(X), \; \mathcal{E}_V = \mathcal{E}_V(Q),$$

etc.

2.6 The Case of Total Autarky

Autarky means that trade stops. Let us assume that there are three types of (*"partial"*) autarky:

a) Trade among big countries stops;
b) Trade among big and small countries stops;
c) Trade among countries s stops.

Let us say the autarky is "total" if $a), b), c)$ hold together, i.e., all (international) trades stop. Depending on the order (w.r.t. τ) in which $a), b), c)$ happen, we can get six cases[5]:

- Case $a - b - c$,
- Case $b - a - c$,
- Case $a - c - b$,
- Case $c - a - b$,
- Case $b - c - a$,
- Case $c - b - a$.

For the study of total autarky, cases $a - b - c$ and $b - a - c$, $a - c - b$ and $c - a - b$, $b - c - a$ and $c - b - a$ coincide, i.e., we can consider three different cases:

- Case I) the last trade that stops, is among small countries.
- Case II) the last trade that stops, is among big and small countries.
- Case III) the last trade that stops, is among big countries.

Obviously, in Case I), X, P^X, Q, N, w, W^B are constant w.r.t. τ on $[\tau_2, \tau_3]$. Analogously, in Case III), x, p^x, q, n, w, W^s are constant w.r.t. τ on $[\tau_2, \tau_3]$.

---

[5] See details, e.g., in [8].

3   Local Comparative Statics of Equilibrium on Transport Costs

Comparative statics of equilibrium is an analysis of the reaction of equilibrium variables to changes in parameters. Local comparative statics is a reaction to a small change in parameters. Thus, to obtain local comparative statics on transport costs, it is necessary to determine the signs of the derived equilibrium quantities on the parameter τ. This section examines the behavior of the equilibrium variables (individual consumptions X, Y, Z, x, y, z, Lagrange multipliers Λ, λ and wage w) with a small change on transportation costs τ. To do this, we can consider a system of the following nine equations:

- FOCs (6);
- free entry conditions (5);
- trade balance (4).

This way the equilibrium system can be written as

$$A\left(\Xi\right) = \Theta_9, \tag{9}$$

where

$$A = \begin{pmatrix} \dfrac{\partial \Pi}{\partial X} \\ \dfrac{\partial \Pi}{\partial Y} \\ \dfrac{\partial \Pi}{\partial Z} \\ \dfrac{\partial \pi}{\partial x} \\ \dfrac{\partial \pi}{\partial y} \\ \dfrac{\partial \pi}{\partial z} \\ \Pi \\ \pi \\ TB \end{pmatrix}, \quad \Xi = \begin{pmatrix} X \\ Y \\ Z \\ x \\ y \\ z \\ \Lambda \\ \lambda \\ w \end{pmatrix}, \quad \Theta_9 = \begin{pmatrix} 0 \\ 0 \\ 0 \\ 0 \\ 0 \\ 0 \\ 0 \\ 0 \\ 0 \end{pmatrix} \in \mathbb{R}^9.$$

We assume that system (9) has a solution[6]. By the total differentiation of system (9) w.r.t. τ, we get the equations of the local comparative statics w.r.t. τ as

$$\frac{d}{d\tau}\left(A\left(\Xi\right)\right) = \Theta_9,$$

i.e.,

$$\frac{\partial A}{\partial \Xi} \cdot \frac{d\Xi}{d\tau} = -\frac{\partial A}{\partial \tau},$$

---

[6] The question of the existence of equilibrium is a separate problem (often not quite simple), which is not the subject of this study.

where

$$
\frac{d\Xi}{d\tau} =
\begin{pmatrix}
\dfrac{dX}{d\tau} \\[2mm]
\dfrac{dY}{d\tau} \\[2mm]
\dfrac{dZ}{d\tau} \\[2mm]
\dfrac{dx}{d\tau} \\[2mm]
\dfrac{dy}{d\tau} \\[2mm]
\dfrac{dz}{d\tau} \\[2mm]
\dfrac{d\Lambda}{d\tau} \\[2mm]
\dfrac{d\lambda}{d\tau} \\[2mm]
\dfrac{dw}{d\tau}
\end{pmatrix},
\qquad
\frac{\partial A}{\partial \tau} =
\begin{pmatrix}
\dfrac{\partial^2 \Pi}{\partial X \partial \tau} \\[2mm]
\dfrac{\partial^2 \Pi}{\partial Z \partial \tau} \\[2mm]
\dfrac{\partial^2 \pi}{\partial x \partial \tau} \\[2mm]
\dfrac{\partial^2 \pi}{\partial Z \partial \tau} \\[2mm]
\dfrac{\partial \Pi}{\partial \tau} \\[2mm]
\dfrac{\partial \pi}{\partial \tau} \\[2mm]
\dfrac{\partial T B}{\partial \tau}
\end{pmatrix}.
$$

Note that, due to the form of profits and free entry conditions, matrix $\dfrac{\partial A}{\partial \Xi}$ is rather sparse because at least 42 elements (from 81) are zero. More precisely,

$$
\frac{\partial A}{\partial \Xi} =
\begin{pmatrix}
* & * & * & 0 & 0 & 0 & * & 0 & * \\
* & * & * & 0 & 0 & 0 & * & 0 & * \\
* & * & * & 0 & 0 & 0 & 0 & * & * \\
0 & 0 & 0 & * & * & * & 0 & * & 0 \\
0 & 0 & 0 & * & * & * & 0 & * & 0 \\
0 & 0 & 0 & * & * & * & * & 0 & 0 \\
0 & 0 & 0 & 0 & 0 & 0 & * & 0 & * \\
0 & 0 & 0 & 0 & 0 & 0 & 0 & * & 0 \\
* & * & * & * & * & * & * & * & 0
\end{pmatrix}.
$$

3.1 Comparative Statics in Free Trade

To formulate the results for free trade, let us define the elasticity of a variable ξ with respect to transport costs τ

$$
E_\xi = E_{\xi/\tau} = \frac{d\xi}{d\tau} \cdot \frac{\tau}{\xi}.
$$

It turned out that the results of comparative statistics for free trade, obtained in [5,8], are naturally transferred to the case of trade of two groups of countries at non-linear production costs. The signs of the elasticities of firm's sizes and the masses of firms are determined uniquely by the monotony of r_u. More precisely, the following proposition holds.

Proposition 1. *In free trade, the elasticities of equilibrium firm sizes in small countries (q*) and big (Q*) countries and the elasticities of equilibrium masses of firms in big countries (N*) and small (n*) countries can be ordered depending on the monotony of the Arrow-Pratt measure of sub-utility u as follows:*

1. *in pro-competitive case, when $r'_u > 0$,*

$$r'_u > 0 \Longrightarrow E_q > E_Q > 0 > E_N > E_n,$$

   i.e., the size of firms increases, while the mass of firms decreases;
2. *in anti-competitive case, when $r'_u < 0$,*

$$r'_u < 0 \Longrightarrow E_q < E_Q < 0 < E_N < E_n.$$

   i.e., the size of firms decreases, while the mass of firms increases.

Moreover, if $r'_u \geq 0$ then $E_\Lambda < E_\Lambda + E_w < 0$.

3.2   Comparative Statics in Autarky

In total autarky we have three cases, see Sect. 2.6:

- Case I): the last trade that stops, is among small countries, $y = 0$;
- Case II): the last trade that stops, is among big and small countries, $Z = 0, z = 0$;
- Case III): the last trade that stops, is among big countries, $Y = 0$.

Proposition 2. *Under convex production costs (i.e., $V''(\cdot) > 0$), in total autarky, the behavior of the equilibrium variable is as follows.*

1. *In Case I), in each small country, domestic consumption, the masses of firms, prices for imported goods increase:*

$$\frac{dx}{d\tau} > 0, \quad \frac{dn}{d\tau} > 0, \quad \frac{dp^y}{d\tau} > 0,$$

   while consumption of imports within a group of small countries, outputs and prices of domestic goods decrease:

$$\frac{dy}{d\tau} < 0, \quad \frac{dq}{d\tau} < 0, \quad \frac{dp^x}{d\tau} < 0.$$

2. *In Case II), in each (both big and small) country, domestic consumption, the masses of firms and the prices for imported goods produced in a large country and consumed in a small one increase:*

$$\frac{dX}{d\tau} > 0, \quad \frac{dx}{d\tau} > 0, \quad \frac{dN}{d\tau} > 0, \quad \frac{dn}{d\tau} > 0, \quad \frac{dp^Z}{d\tau} > 0,$$

   while import consumption, outputs and the prices of domestic goods in a small country decrease:

$$\frac{dZ}{d\tau} < 0, \quad \frac{dz}{d\tau} < 0, \quad \frac{dQ}{d\tau} < 0, \quad \frac{dq}{d\tau} < 0 \quad \frac{dp^x}{d\tau} < 0.$$

3. *In Case III), in each big country, there are growing: domestic consumption, the masses of firms, prices for imported goods (produced and consumed within a group of big countries) increase:*

$$\frac{dX}{d\tau} > 0, \quad \frac{dN}{d\tau} > 0, \quad \frac{dp^Y}{d\tau} > 0,$$

while import consumption within a group of big countries, outputs and prices of domestic goods decrease:

$$\frac{dY}{d\tau} < 0, \quad \frac{dQ}{d\tau} < 0, \quad \frac{dp^X}{d\tau} < 0.$$

4 Conclusion

In this paper, we study the international trade between two groups of countries: a group of big (identical) countries and a group of small (identical) countries under monopolistic competition of producers. The transport costs τ is of "iceberg type." The utility is additive separable. The firms are homogeneous, the production costs are assumed not necessary linear.

The consideration of two groups of countries needs to study new kind of consumption: the import consumption in countries within the same group.

It turn out that the most results for linear production costs [8] can be generalized to the case of convex costs.

Besides, we generalize the results for two countries [5] to the case of two groups of countries.

The obtained results can be useful for comparative statics of social welfare, cf. [2,3,8,19,20].

Moreover, it seems interesting to consider consider the international trade of of a larger number of different countries (or groups of countries). In this case, we need to consider more then one trade balance (cf. (4)). More precisely, for M different countries (or M groups of countries), we need to consider $M - 1$ trade balances.

Besides, we can consider the impact of investments in R&D [10].

Finally, it is interesting to consider retailing questions [9,23].

Acknowledgments. The authors are very grateful to the anonymous reviewers for very useful suggestions for improving the text. The study was carried out within the framework of the state contract of the Sobolev Institute of Mathematics (project no. 0314-2019-0018). The work was supported in part by the Russian Foundation for Basic Research, projects 18-010-00728 and 19-010-00910 and by the Russian Ministry of Science and Education under the 5-100 Excellence Programme.

References

1. Antoshchenkova, I.V., Bykadorov, I.A.: Monopolistic competition model: the impact of technological innovation on equilibrium and social optimality. Autom. Remote Control **78**(3), 537–556 (2017). https://doi.org/10.1134/S0005117917030134
2. Arkolakis, C., Costinot, A., Donaldson, D., Rodríguez-Clare, A.: The elusive pro-competitive effects of trade. Rev. Econ. Stud. **86**(1), 46–80 (2019)
3. Arkolakis, C., Costinot, A., Rodríguez-Clare, A.: New trade models, same old gains? Am. Econ. Rev. **102**(1), 94–130 (2012)
4. Behrens, K., Murata, Y.: General equilibrium models of monopolistic competition: a new approach. J. Econ. Theory **136**, 776–787 (2007)
5. Belyaev, I., Bykadorov, I.: International trade models in monopolistic competition: the case of non-linear costs. IEEE Xplore, pp. 12–16 (2019)
6. Bykadorov, I.: Monopolistic competition model with different technological innovation and consumer utility levels. In: CEUR Workshop Proceeding, vol. 1987, pp. 108–114 (2017)
7. Bykadorov, I.: Monopolistic competition with investments in productivity. Optim. Lett. **13**(8), 1803–1817 (2018). https://doi.org/10.1007/s11590-018-1336-9
8. Bykadorov, I., Ellero, A., Funari, S., Kokovin, S., Molchanov, P.: Painful birth of trade under classical monopolistic competition. National Research University Higher School of Economics, Basic Research Program Working Papers, Series: Economics, WP BRP 132/EC/2016. https://www.hse.ru/data/2016/04/06/1127112143/132EC2016.pdf
9. Bykadorov, I., Ellero, A., Funari, S., Kokovin, S., Pudova, M.: Chain store against manufacturers: regulation can mitigate market distortion. In: Kochetov, Y., Khachay, M., Beresnev, V., Nurminski, E., Pardalos, P. (eds.) DOOR 2016. LNCS, vol. 9869, pp. 480–493. Springer, Cham (2016). https://doi.org/10.1007/978-3-319-44914-2_38
10. Bykadorov, I., Kokovin, S.: Can a larger market foster R&D under monopolistic competition with variable mark-ups? Res. Econ. **71**(4), 663–674 (2017)
11. Campbell, J.R., Hopenhayn, H.A.: Market size matters. J. Ind. Econ. **53**(1), 1–25 (2005)
12. Chamberlin, E.H.: The Theory of Monopolistic Competition: A Re-Orientation of the Theory of Value. Harvard University Press, Cambridge (1933)
13. Chamberlin, E.H.: The Theory of Monopolistic Competition. Harvard University Press, Cambridge (1962)
14. Dixit, A., Stiglitz, J.: Monopolistic competition and optimum product diversity. Am. Econ. Rev. **67**(3), 297–308 (1977)
15. Hummels, D., Klenow, P.T.: The variety and quality of a nation's exports. Am. Econ. Rev. **95**(3), 704–723 (2005)
16. Krugman, P.: Increasing returns, monopolistic competition and international trade. J. Int. Econ. **9**(4), 469–479 (1979)
17. Krugman, P.: Scale economies, product differentiation, and the pattern of trade. Am. Econ. Rev. **70**(5), 950–959 (1980)
18. Melitz, M.J.: The impact of trade on intra-industry reallocations and aggregate industry productivity. Econometrica **71**(6), 1695–1725 (2003)
19. Melitz, M.J., Redding, S.J.: Missing gains from trade? Am. Econ. Rev. **104**(5), 317–321 (2014)

20. Melitz, M.J., Redding, S.J.: New trade models, new welfare implications. Am. Econ. Rev. **105**(3), 1105–1146 (2015)

21. Redding, S.: Theories of heterogeneous firms and trade. Ann. Rev. Econ. **3**, 77–105 (2011)

22. Syverson, C.: Prices, spatial competition, and heterogeneous producers: an empirical test. J. Ind. Econ. **55**(2), 197–222 (2007)

23. Tilzo, O., Bykadorov, I.: Retailing under monopolistic competition: a comparative analysis. IEEE Xplore, pp. 156–161 (2019)

24. Zhelobodko, E., Kokovin, S., Parenti, M., Thisse, J.-F.: Monopolistic competition in general equilibrium: beyond the constant elasticity of substitution. Econometrica **80**(6), 2765–2784 (2012)

Investments in R&D in Monopolistic Competitive Trade Model

Igor Bykadorov[1,2,3](✉) (iD)

[1] Sobolev Institute of Mathematics SB RAS, Novosibirsk, Russia
bykadorov.igor@mail.ru
[2] Novosibirsk State University, Novosibirsk, Russia
[3] Novosibirsk State University of Economics and Management, Novosibirsk, Russia

Abstract. We study a monopolistic competition model in the open economy case. The utility of consumers are additive separable. The producers can choose technology (R&D) endogenously. We examine the local comparative statics of market equilibrium with respect to trade costs (of iceberg type). Our main finding is the following: increasing trade costs has opposite impacts on mass of firms and productivity. Moreover, we study the cases of small trade costs and symmetric (on numbers of consumers) countries.

Keywords: Dixit-Stiglitz-Krugman model · Market equilibrium · Endogenous choice of technology · Comparative statics

1 Introduction

The cross-countries differences in productivity are noticeable and allow for various explanations. The trade literature reports noticeable cross-countries differences in productivity and related indicators[1].

In [11] we get the results for the model of closed economy. To expand this analysis to trade, now we introduce a trade model. The setting of the trade model is as in [12]. As to the main results, they are completely new. The economy consists of two countries, "big" country and "small" country, one production factor (labor) and one differentiated sector including continuum of varieties or brands[2].

The paper is organized as follows. In Sect. 2 we set the model, the main notations, describe the equilibrium and discuss the main equilibrium equations. In Sect. 3 we present the preliminary results for comparative statics. In particular, we get one of the main result of the paper: *increasing trade cost has opposite impacts on mass of firms and productivity*. In Sects. 4 and 5 we study the comparative statics with respect to trade costs in two situations: when these costs

---

[1] Some stylized facts that we believe to be a challenge for theory are, e.g., in [2,14, 18,22,23]. The details see, e.g., in [12,13].

[2] There can be a different approach, see, e.g., [15,20,21,26]. The details see, e.g., in [12].

© Springer Nature Switzerland AG 2020
A. Kononov et al. (Eds.): MOTOR 2020, LNCS 12095, pp. 170–183, 2020.
https://doi.org/10.1007/978-3-030-49988-4_12

are "'infinitesimally" small and when the countries are symmetric in population. Section 6 concludes.

2   Model

The setting of the model is as in [12].

The model of monopolistic competition [3,4,10,16,17,19,21,27] is based on the following assumptions:

- the manufacturers produce goods of the same nature, but not completely interchangeable (product diversity);
- each firm produces one type of product diversity and sets its price;
- the number (mass) of firms is large enough;
- the firms enter the market as long as their profits are positive.

In this paper, we consider the model of monopolistic competition in relation to the trade of two countries, country B ("big") and country s ("small"), different in population. There are one industry and one production factor, interpreted as labor. We introduce the basic concepts and notation. Let

- L be the number of consumers in big country,
- l be the number of consumers in small country, $L \geq l$;
- N be the mass of firms in big country;
- n be the mass of firms in small country;
- $X_i = X(i)$ be the individual domestic consumption in big country of the goods produced by firm $i \in [0, N]$;
- Z_i be the individual foreign consumption in small country of the goods produced by firm $i \in [0, N]$ in big country;
- x_i be the individual domestic consumption in small country of the goods produced by firm $i \in [0, n]$;
- z_i be the individual foreign consumption in big country of the goods produced by firm $i \in [0, n]$ in small country;
- Q_i be the output (size)of firm $i \in [0, N]$ in big country;
- q_i be the output (size) of firm $i \in [0, n]$ in small country;
- w be the wage in a big country, while the wage in small country be normalized to 1.

2.1   Consumers

We assume that each consumer share the same twice differentiable sub-utility function, such that
$$u(0) = 0,\ u'(\xi_i) > 0,\ u''(\xi_i) < 0.$$
Thus, function u is increasing and strictly concave.

The Problem of a Representative Consumer in the Big Country is

$$\int_0^N u\left(X_i\right) di + \int_0^n u\left(z_i\right) di \rightarrow \max_{X_i \geq 0, i \in [0,N], z_i \geq 0, i \in [0,n]}$$

subject to

$$\int_0^N p_i^X\left(X_i\right) di + \int_0^n p_i^z z_i di \leq w.$$

The Problem of a Representative Consumer in the Small Country is

$$\int_0^n u\left(x_i\right) di + \int_0^N u\left(Z_i\right) di \rightarrow \max_{x_i \geq 0, i \in [0,n], Z_i \geq 0, i \in [0,N]}$$

subject to

$$\int_0^n p_i^x x_i di + \int_0^N p_i^Z Z_i di \leq 1.$$

From the consumer's First Order Conditions (FOC), we get the inverse demand functions

$$p_i^X\left(X_i, \Lambda\right) = \frac{u'\left(X_i\right)}{\Lambda}, \qquad p_i^Z\left(Z_i, \lambda\right) = \frac{u'\left(Z_i\right)}{\lambda}, \qquad i \in [0, N], \qquad (1)$$

$$p_i^x\left(x_i, \lambda\right) = \frac{u'\left(x_i\right)}{\lambda}, \qquad p_i^z\left(z_i, \Lambda\right) = \frac{u'\left(z_i\right)}{\Lambda}, \qquad i \in [0, n], \qquad (2)$$

where Λ and λ are the Lagrange multipliers of the two above defined problems.

2.2   Producers

Let F and f be fixed costs in countries B and s (chosen endogenously); $c\left(F\right)$ and $c\left(f\right)$ be the corresponding marginal costs. We assume that $c'\left(\cdot\right) < 0$. Besides, let us assume, standardly, that the trade incurs some trade costs of "iceberg type"[3]. Then the sizes of the firms in countries B and s are

$$Q_i = LX_i + \tau l Z_i, \quad i \in [0, N], \qquad (3)$$

$$q_i = lx_i + \tau L z_i, \quad i \in [0, n], \qquad (4)$$

while the production costs are

$$V\left(Q_i, F\right) = c(F)Q_i + F, \quad i \in [0, N], \qquad (5)$$

$$V\left(q_i, f\right) = c(f)Q_i + f, \quad i \in [0, n]. \qquad (6)$$

Let

$$R\left(\xi\right) = u'\left(\xi\right) \cdot \xi \qquad (7)$$

---

[3] To sell a unit, the firm produce the $\tau \cdot 1$.

be the "normalized" revenue. Note that, due to (1) and (2), the normalized revenue equals to the costs that one consumer spends on the purchase of products of one company (divided by the corresponding Lagrange multiplier). Moreover, $R'(\xi) = u'(\xi)(1 - r_u(\xi))$, where

$$r_g(\xi) = -\frac{g''(\xi)\xi}{g'(\xi)} = -\varepsilon_{g'}(\xi) \tag{8}$$

is Arrow-Pratt measure of function g while ε_h is the elasticity of function h :

$$\varepsilon_h(\xi) = \frac{h'(\xi)\xi}{h(\xi)}. \tag{9}$$

Using the inverse demand functions (1), (2) and "normalized" revenue (7), the profit of firm i in the big country can be written as

$$\Pi_i = L \cdot \frac{R(X_i)}{\Lambda} + l \cdot \frac{R(Z_i)}{\lambda} - wV(Q_i, F), \quad i \in [0, N], \tag{10}$$

while the profit of firm i in the small country can be written as

$$\pi_i = L \cdot \frac{R(z_i)}{\Lambda} + l \cdot \frac{R(x_i)}{\lambda} - V(q_i, f), \quad i \in [0, n]. \tag{11}$$

Labor and Trade Balances. In countries B and s, the labor balances ("total production costs equal total labor") are

$$\int_0^N V(Q_i, F)\, di = L, \tag{12}$$

$$\int_0^n V(q_i, f)\, di = l. \tag{13}$$

Let us assume that trade balance ("export equals import") holds:

$$l \cdot \int_0^N p_i^Z(Z_i, \lambda) Z_i di = L \cdot \int_0^n p_i^z(z_i, \Lambda) z_i di, \tag{14}$$

where inverse demand functions $p_i^Z(Z_i, \lambda)$ and $p_i^z(z_i, \Lambda)$ are defined in (1), (2).

2.3   Symmetric Case

In each country, all consumers are assumed identical. So we will consider the symmetric case, omitting index i. This way, the individual consumptions are

$$X_i = X, \quad Z_i = Z, \quad i \in [0, N],$$

$$x_i = x, \quad z_i = z, \quad i \in [0, n].$$

Therefore, we rewrite the inverse demand functions (1) and (2) as

$$p(X, \Lambda) = \frac{u'(X)}{\Lambda}, \qquad p(Z, \lambda) = \frac{u'(Z)}{\lambda}, \tag{15}$$

$$p(x, \lambda) = \frac{u'(x)}{\lambda}, \qquad p(z, \Lambda) = \frac{u'(z)}{\Lambda}, \tag{16}$$

sizes of the firms (3) and (4) as

$$Q = LX + \tau l Z, \tag{17}$$

$$q = lx + \tau L z, \tag{18}$$

production costs (5) and (6) as

$$V(Q, F) = c(F)Q + F, \tag{19}$$

$$V(q, f) = c(f)q + f, \tag{20}$$

profits (10) and (11) as

$$\Pi = L \cdot \frac{R(X)}{\Lambda} + l \cdot \frac{R(Z)}{\lambda} - wV(Q, F), \tag{21}$$

$$\pi = L \cdot \frac{R(z)}{\Lambda} + l \cdot \frac{R(x)}{\lambda} - V(q, f), \tag{22}$$

labor balances (12) and (13) as

$$N \cdot V(Q, F) = L, \tag{23}$$

$$n \cdot V(q, f) = l. \tag{24}$$

As to trade balance (14), it is

$$l \cdot N \cdot p(Z, \lambda) Z = L \cdot n \cdot p(z, \Lambda) z,$$

i.e., due to (15), (16), (23), (24) and (7),

$$\frac{R(Z)}{V(Q, F)\lambda} = \frac{R(z)}{V(q, f)\Lambda}. \tag{25}$$

2.4   Symmetric Equilibrium

Here, sizes of the firms are (17) and (18), production costs are (19) and (20), profits are (21), (22).

As it is usual in monopolistic competition, we assume that firms enter the market while their profit remains positive, which implies a zero-profit (free-entry) conditions. In symmetric case we get

$$\Pi = 0, \qquad \pi = 0. \tag{26}$$

Each firm maximizes its profits:

$$\Pi \to \max_{X,Z,F \geq 0}, \qquad \pi \to \max_{x,z,f \geq 0}.$$

Thus, the Producer's First order conditions (FOC) in symmetric case are

$$\frac{\partial \Pi}{\partial X} = 0, \quad \frac{\partial \Pi}{\partial Z} = 0, \quad \frac{\partial \Pi}{\partial F} = 0, \quad \frac{\partial \pi}{\partial x} = 0, \quad \frac{\partial \pi}{\partial z} = 0, \quad \frac{\partial \pi}{\partial f} = 0. \qquad (27)$$

Second order conditions (SOC) are

$$\frac{\partial^2 \Pi}{\partial X^2} < 0, \qquad \frac{\partial^2 \Pi}{\partial X^2} \cdot \frac{\partial^2 \Pi}{\partial Z^2} - \left(\frac{\partial^2 \Pi}{\partial X \partial Z} \right)^2 > 0, \qquad \det \Pi'' < 0, \qquad (28)$$

where

$$\Pi'' = \begin{vmatrix} \dfrac{\partial^2 \Pi}{\partial X^2} & \dfrac{\partial^2 \Pi}{\partial X \partial Z} & \dfrac{\partial^2 \Pi}{\partial X \partial F} \\[2mm] \dfrac{\partial^2 \Pi}{\partial X \partial Z} & \dfrac{\partial^2 \Pi}{\partial Z^2} & \dfrac{\partial^2 \Pi}{\partial Z \partial F} \\[2mm] \dfrac{\partial^2 \Pi}{\partial X \partial F} & \dfrac{\partial^2 \Pi}{\partial Z \partial F} & \dfrac{\partial^2 \Pi}{\partial F^2} \end{vmatrix},$$

and

$$\frac{\partial^2 \pi}{\partial x^2} < 0, \qquad \frac{\partial^2 \pi}{\partial x^2} \cdot \frac{\partial^2 \pi}{\partial z^2} - \left(\frac{\partial^2 \pi}{\partial x \partial z} \right)^2 > 0, \qquad \det \pi'' < 0, \qquad (29)$$

where

$$\pi'' = \begin{vmatrix} \dfrac{\partial^2 \pi}{\partial x^2} & \dfrac{\partial^2 \pi}{\partial x \partial z} & \dfrac{\partial^2 \pi}{\partial x \partial f} \\[2mm] \dfrac{\partial^2 \pi}{\partial x \partial z} & \dfrac{\partial^2 \pi}{\partial z^2} & \dfrac{\partial^2 \pi}{\partial z \partial f} \\[2mm] \dfrac{\partial^2 \pi}{\partial x \partial f} & \dfrac{\partial^2 \pi}{\partial z \partial f} & \dfrac{\partial^2 \pi}{\partial f^2} \end{vmatrix}.$$

Symmetric equilibrium is a bundle

$$(X^*, Z^*, F^*, x^*, z^*, f^*, w^*, N^*, n^*, \Lambda^*, \lambda^*)$$

satisfying the following:

- profit maximization (27) and (28), (29);
- free entry conditions (26);
- labor balances (23) and (24);
- trade balance (25).

Note that the equilibrium prices (inverse demand functions)

$$p(X^*,\Lambda^*), \quad p(Z^*,\lambda^*), \quad p(x^*,\lambda^*), \quad p(z^*,\Lambda^*)$$

can be obtained from (15) and (16). Moreover, First Order Conditions (27) are

$$\frac{R'(X)}{\Lambda} = wc(F), \qquad \frac{R'(Z)}{\lambda} = w\tau c(F), \qquad c'\left(f^H\right)q^H = -1, \tag{30}$$

$$\frac{R'(x)}{\lambda} = c(f), \qquad \frac{R'(z)}{\Lambda} = \tau c(f), \qquad c'\left(f^F\right)q^F = -1. \tag{31}$$

Following tradition and using (30), (31), we can express Λ^*, λ^* via another equilibrium variables. Therefore, after some calculations we lead to

Proposition 1. *Trade equilibrium nine-elements bundle*

$$(X^*, Z^*, F^*, x^*, z^*, f^*, w^*, N^*, n^*)$$

satisfy the following nine equations[4]:

$$\frac{R'(X)}{R'(z)} \cdot \frac{c(f)}{c(F)} \cdot \frac{\tau}{w} = 1, \tag{32}$$

$$\frac{R'(x)}{R'(Z)} \cdot \frac{c(F)}{c(f)} \cdot \tau w = 1, \tag{33}$$

$$\left(\frac{LX}{\varepsilon_R(X)} + \frac{l\tau Z}{\varepsilon_R(Z)}\right) \cdot \frac{\varepsilon_V(Q)}{Q} = 1, \tag{34}$$

$$\left(\frac{lx}{\varepsilon_R(x)} + \frac{L\tau z}{\varepsilon_R(z)}\right) \cdot \frac{\varepsilon_V(q)}{q} = 1, \tag{35}$$

$$-c'(F)Q = 1, \tag{36}$$

$$-c'(f)q = 1, \tag{37}$$

$$\frac{N}{L} \cdot V(Q) = 1, \tag{38}$$

$$\frac{n}{l} \cdot V(q) = 1, \tag{39}$$

$$\frac{Z}{\varepsilon_R(Z)} \cdot \frac{\varepsilon_R(z)}{z} \cdot \frac{q}{\varepsilon_V(q)} \cdot \frac{\varepsilon_V(Q)}{Q} \cdot w = 1. \tag{40}$$

Note that in Proposition 1,

---

[4] Here $\varepsilon_V(\cdot)$ is the "partial" elasticity of production costs with respect to firm size:

$$\varepsilon_V(\xi) = \frac{\partial V(\xi,\eta)}{\partial \xi} \cdot \frac{\xi}{V(\xi,\eta)}.$$

- (32) and (33) mean the optimality in consumption in each country: the ratio of marginal utilities equals the ratio of the prices (inverse demands);
- (34) and (35) are the generalizations of well-known condition of monopolistic competition in closed economy: "elasticity of revenue equals elasticity of costs;"
- (36) and (37) determine optimal choice of technology in each country;
- (38) and (39) mean labor balances (23) and (24);
- (40) means trade balance in terms of elasticities of revenue and production costs.

Note that the question of the existence of equilibrium is a separate problem (often not quite simple), which is not the subject of this study[5].

Actually, the aim is to study the impact of trade costs (τ) on investments per firm and total investments, and on other equilibrium variables. Is it true that when country B is bigger ($L > l$) then this country invests more per firm and has less costs per unit? What is the ratio of total investments in countries B and s? What we can say about the relation between the size of the firm, the mass of the firms and prices in these two countries? Is it true that the wage is bigger in larger country?

To answer all these questions, we plan to study comparative statics of the above equations. At the moment it seems incredible to get complete global comparative statics for these equations. Therefore, let us concentrate on local comparative statics[6].

2.5   Second Order Conditions

The conditions (28),(29) seem heavy. But in terms of revenue and Arrow-Pratt measure, the Second Order Conditions (SOC) have a rather elegant form, as we can see in the following

Proposition 2. *1. For country B, Second Order Conditions are*

$$R''(X) < 0, \qquad R''(Z) < 0, \qquad \frac{LX}{r_R(X)} + \frac{l\tau Z}{r_R(Z)} < \frac{r_c(F)}{\varepsilon_c(F)\,c'(F)}. \tag{41}$$

2. For country s, Second Order Conditions are

$$R''(x) < 0, \qquad R''(z) < 0, \qquad \frac{lx}{r_R(x)} + \frac{L\tau z}{r_R(z)} < \frac{r_c(f)}{\varepsilon_c(f)\,c'(f)}. \tag{42}$$

---

[5] The discuss this system of equations can be found, e.g., in [12].

[6] Namely, the solution of these equations can be considered as implicit function of two parameters, $\dfrac{L}{L+l}$ (the share of the population of big country in total population) and $\dfrac{1}{\tau}$ (the parameter of trade liberalization). Without loss of generality (cf. [12]) we can assume that couple $\left(\dfrac{L}{L+l}, \dfrac{1}{\tau}\right)$ belongs to the rectangle $\left[\dfrac{1}{2}, 1\right] \times [0, 1]$. The idea is to get the answers to the questions above for the boundary of this rectangle, see details in [12].

Note that, due to (36) and (37), the last inequalities in (41) and (41) can be written as

$$\frac{LXc'(F)}{r_R(X)} - \frac{1 + LXc'(F)}{r_R(Z)} > \frac{r_c(F)}{\varepsilon_c(F)}, \tag{43}$$

$$\frac{lxc'(f)}{r_R(x)} - \frac{1 + lxc'(f)}{r_R(z)} > \frac{r_c(f)}{\varepsilon_c(f)}. \tag{44}$$

Therefore, *SOC depend explicitly only on individual consumptions and on investments in R&D, do not depend on wage w and trade costs τ.*

3   Preliminary Results for Comparative Statics

Let us substitute (36) in (38), and (37) in (39):

$$\left(F - \frac{c(F)}{c'(F)}\right) N = L, \tag{45}$$

$$\left(f - \frac{c(f)}{c'(f)}\right) n = l. \tag{46}$$

Note that (45) and (46) do not depend explicitly from τ. These help us to obtain some preliminary results on comparative statics for investments F and f, masses of firms N and n, and also for the outputs of a firm (size of the firm) in countries B and s.

In what follows, we will consider not only elasticity of function with respect to a variable (see (8) and (9)), but also elasticity of a variable with respect to a parameter, i.e.,

$$E_{\xi/\varsigma} = \frac{d\xi}{d\varsigma} \cdot \frac{\varsigma}{\xi}.$$

Proposition 3. *In symmetric equilibrium, masses of firms N^*, n^*, investments F^*, f^* and sizes of firms Q^*, q^* are such that*

1. elasticities w.r.t. τ:
   (a) in country B:

$$E_{N^*/\tau} \cdot E_{F^*/\tau} < 0, \qquad E_{Q^*/\tau} \cdot E_{F^*/\tau} > 0,$$

   (b) in country s:

$$E_{n^*/\tau} \cdot E_{f^*/\tau} < 0, \qquad E_{q^*/\tau} \cdot E_{f^*/\tau} > 0.$$

2. elasticities w.r.t. market sizes:
   (a) in country B:

$$(E_{N^*/L} - 1) \cdot E_{F^*/L} < 0, \qquad E_{Q^*/L} \cdot E_{F^*/L} > 0,$$

(b) in country s:

$$\left(E_{n^*/l} - 1\right) \cdot E_{f^*/l} < 0, \qquad E_{q^*/l} \cdot E_{f^*/l} > 0.$$

Proposition 3 shows that comparative statics in trade cost coefficient τ makes mass of firms negatively correlated with investment in *R&D* which is always positively connected with the size of the firms. As to the market size impact, when we call elasticity relations $E_{N^*/L} > 1$ and $E_{n^*/l} > 1$ as the "Home market effect," it turns out *negatively* correlated with increasing individual investment, i.e., entails $E_{F^*/L} < 0$ and $E_{f^*/l} < 0$.

4   Comparative Statics w.r.t. τ: The Case $\tau \approx 1$

Let us study the comparative statics w.r.t. τ in the point $\tau = 1$, i.e., consider the case of "infinitesimally" small trade costs. Then in equilibrium one has $w^* = 1$ and moreover

$$X^* = Z^* = x^* = z^*, \qquad F^* = f^*.$$

Further, let us denote

$$D = \frac{X^*}{((2 - r_{u'}(X^*)) r_c(F^*) - 1) r_u(X^*)} > 0. \tag{47}$$

Remark that $D > 0$ due to *SOC*.

The following Proposition presents the comparative statics w.r.t. τ in the neighborhood of $\tau = 1$ for wage w^*, investments F^* and f^*, masses of firms N^* and n^*, total investments N^*F^* and n^*f^*, "home market effect" (w.r.t. market size)

$$\frac{N^*l}{n^*L},$$

"generalized home market effect" (w.r.t. GDP)

$$\frac{N^*l}{n^*w^*L}$$

and also for production of a firm (size of the firm) Q^* and q^*.

Proposition 4. *In the neighborhood of $\tau = 1$, one has (about D see in (47))*

$$E_{w^*/\tau} = \frac{L-l}{L+l} \cdot (1 - r_u(X^*)) \quad (\Longrightarrow w^* \geq 1)$$

$$E_{F^*/\tau} = \frac{l}{L+l} \cdot D \cdot r_u'(X^*) = \frac{l}{L} \cdot E_{f^*/\tau},$$

$$E_{N^*/\tau} = -\frac{l}{L+l} \cdot D \cdot (1 - r_u(X^*)) \cdot r_u'(X^*) \cdot r_c(F^*) = \frac{l}{L} \cdot E_{n^*/\tau},$$

$$E_{N^*F^*/\tau} = \frac{l}{L+l} \cdot D \cdot (1 - r_u(X^*)) \cdot r_u'(X^*) \cdot (1 - r_{\ln c}(F^*)) = \frac{l}{L} \cdot E_{n^*f^*/\tau},$$

$$E_{\frac{N^*l}{n^*L}/\tau} = E_{\frac{N^*}{n^*}/\tau} = \frac{L-l}{L+l} \cdot D \cdot (1 - r_u(X^*)) \cdot r'_u(X^*) \cdot r_c(F^*),$$

$$E_{\frac{N^*l}{n^*w^*L}/\tau} = E_{\frac{N^*}{n^*w^*}/\tau} = \frac{L-l}{L+l} \cdot D \cdot \frac{(1-r_u(X^*))^2 \cdot r_u(X^*)}{X^*} \cdot (1 - r_{\ln c}(F^*)),$$

$$E_{Q^*/\tau} = \frac{l}{L+l} \cdot D \cdot r'_u(X^*) \cdot r_c(F^*) = \frac{l}{L} \cdot E_{q^*/\tau}$$

and their signs satisfy classification in Table 1:

Table 1. Equilibrium: comparative statics w.r.t. $\tau \approx 1$

	$r'_u < 0$	$r'_u = 0$	$r'_u > 0$		
	$r_{\ln c} > 1$	$r_{\ln c} > 1$	$r_{\ln c} > 1$	$r_{\ln c} = 1$	$r_{\ln c} < 1$
$E_{w^*/\tau}$	>0	>0	>0	>0	>0
$E_{F^*/\tau} = \frac{L}{l} \cdot E_{f^*/\tau}$	<0	$=0$	>0	>0	>0
$E_{N^*/\tau} = \frac{L}{l} \cdot \mathcal{E}_{n^*/\tau}$	>0	$=0$	<0	<0	<0
$E_{N^*f^*/\tau} = \frac{L}{l} \cdot E_{n^*f^*/\tau}$	>0	$=0$	<0	$=0$	>0
$E_{\frac{N^*}{n^*}/\tau}$	>0	$=0$	<0	<0	<0
$E_{\frac{N^*}{n^*w^*}/\tau}$	<0	<0	>0	$=0$	<0
$E_{Q^*/\tau} = \frac{L}{l} \cdot \mathcal{E}_{q^*/\tau}$	<0	$=0$	>0	>0	>0

Due to Proposition 4, in the *neighborhood* of $\tau = 1$ we get

$$r'_u(X^*) > 0 \Longleftrightarrow F^* > f^*, \quad Q^* > q^*, \quad \frac{N^*}{n^*} > \frac{L}{l}.$$

Moreover, since

$$1 - r_{\ln c}(F^*) = \frac{\varepsilon'_c(F^*) F^*}{\varepsilon_c(F^*)} = \varepsilon_{\varepsilon_c}(F^*),$$

we get:

$$\varepsilon'_c(F^*) > 0 \Longleftrightarrow \frac{N^*}{n^*} < \frac{Lw^*}{l}.$$

5 Comparative Statics w.r.t. τ: The Case of Symmetric Countries

Let the countries are symmetric in population, i.e., $L = l$.
Then $X = x, Z = z, F = f, N = n, w = 1$. Therefore,

– (32) and (33) become

$$R'(X) = \tau R'(Z); \tag{48}$$

- (34) and (35) coincide;
- (36) and (37) coincide;
- (38) and (39) coincide;
- trade balance (40) becomes identity,
- SOC (41) and (42) coincide, (43) and (44) coincide.

Thus, equilibrium X^*, Z^*, F^*, N^* satisfies (48), (34), (36), (23), (41). Let us denote

$$T = \left(\left(\frac{1}{r_R(X^*)} + \frac{r_c(F^*)}{\varepsilon_c(F^*)} \right) X^* + \left(\frac{1}{r_R(Z^*)} + \frac{r_c(F^*)}{\varepsilon_c(F^*)} \right) \tau Z^* \right) L < 0 \quad (49)$$

$(T < 0$ due to (43)) and

$$M = \frac{1}{r_R(Z^*)} + \frac{r_c(F^*) c(F^*)}{\varepsilon_c(F^*)} \cdot Q^* N^* - 1. \quad (50)$$

Proposition 5. *If $L = l$ then individual domestic consumption X^*, individual foreign consumption Z^*, investments F^*, mass of firms N^*, total investments $N^* F^*$ and firm size Q^* are such that (definition of T and M see in (49), (50))*

$$E_{X^*/\tau} = \frac{M \tau Z^*}{r_R(X^*) T},$$

$$E_{Z^*/\tau} = \frac{M \tau Z^* - T}{r_R(Z^*) T},$$

$$E_{F^*/\tau} = \frac{(N c(F^*) T - M) \tau Z^*}{\varepsilon_c(F^*) T},$$

$$E_{N^*/\tau} = -\frac{r_c(F^*)}{1 - \varepsilon_c(F^*)} \cdot E_{F^*/\tau},$$

$$E_{N^* F^*/\tau} = \frac{r_{\ln c}(F^*) - 1}{\varepsilon_c(F^*) - 1} \cdot E_{F^*/\tau},$$

$$E_{Q^*/\tau} = r_c(F^*) \cdot E_{F^*/\tau}.$$

Note that, due to (32) and Proposition 5, $E_{N^*/\tau} \cdot E_{F^*/\tau} < 0$, this corresponds to Proposition 3.

6   Conclusion

Endogenous technology choice is a popular topic, but only recently has its theoretical representation been achieved in a rich enough model [25], one demonstrates that both the positive and negative impacts of a big market on investments in productivity.

In [10,12,13] we modify the Vives's model into the monopolistic competition framework, get rid of the quasi-linearity assumption (absent income effect) and strategic oligopolistic considerations. The resulting model becomes simpler and more tractable. Now we extend our study to trade and cross-countries comparisons[7].

Besides, it can be interesting to consider the case of more than two countries. In the case of k countries, we have $k-1$ trade balances instead of (40). Moreover, the case of autarky (cf. [5]) with R&D has not yet been considered. Further, we can consider the impact of technological innovations [1,6,7]. Finally, it is interesting to consider the questions of retailing [9,24] and social optimality [8].

Acknowledgments. The study was carried out within the framework of the state contract of the Sobolev Institute of Mathematics (project no. 0314-2019-0018). The work was supported in part by the Russian Foundation for Basic Research (projects no. 18-010-00728 and no. 19-010-00910) and by the Russian Ministry of Science and Education under the 5-100 Excellence Programme.

References

1. Antoshchenkova, I.V., Bykadorov, I.A.: Monopolistic competition model: the impact of technological innovation on equilibrium and social optimality. Autom. Remote Control **78**(3), 537–556 (2017). https://doi.org/10.1134/S0005117917030134
2. Aw, B.Y., Roberts, M.J., Xu, D.Y.: R&D investments, exporting, and the evolution of firm productivity. Am. Econ. Rev. Pap. Proc. **98**(2), 451–456 (2008)
3. Baldwin, R.E., Forslid, R.: Trade liberalization with heterogeneous firms. Rev. Dev. Econ. **14**(2), 161–176 (2010)
4. Behrens, K., Murata, Y.: General equilibrium models of monopolistic competition: a new approach. J. Econ. Theory **136**(1), 776–787 (2007)
5. Belyaev, I., Bykadorov, I.: International trade models in monopolistic competition: the case of non-linear costs. IEEE Xplore, pp. 12–16 (2019)
6. Bykadorov, I.: Monopolistic competition model with different technological innovation and consumer utility levels. In: CEUR Workshop Proceeding, vol. 1987, pp. 108–114 (2017)
7. Bykadorov, I.: Monopolistic competition with investments in productivity. Optim. Lett. **13**(8), 1803–1817 (2018). https://doi.org/10.1007/s11590-018-1336-9
8. Bykadorov, I.: Social optimality in international trade under monopolistic competition. In: Bykadorov, I., Strusevich, V., Tchemisova, T. (eds.) MOTOR 2019. CCIS, vol. 1090, pp. 163–177. Springer, Cham (2019). https://doi.org/10.1007/978-3-030-33394-2_13
9. Bykadorov, I., Ellero, A., Funari, S., Kokovin, S., Pudova, M.: Chain store against manufacturers: regulation can mitigate market distortion. In: Kochetov, Y., Khachay, M., Beresnev, V., Nurminski, E., Pardalos, P. (eds.) DOOR 2016. LNCS, vol. 9869, pp. 480–493. Springer, Cham (2016). https://doi.org/10.1007/978-3-319-44914-2_38

---

[7]   For policy-making (cf. [13]), our topic may be interesting because of new understanding of gains from trade: technological changes in response to trade liberalization. Furthermore, for modernization and active industrial policy practiced in some countries it can be interesting, which equilibrium outcome in various sectors may follow from some stimulating measures like tax reductions conditional on R&D.

10. Bykadorov, I., Gorn, A., Kokovin, S., Zhelobodko, E.: Why are losses from trade unlikely? Econ. Lett. **129**, 35–38 (2015)
11. Bykadorov, I., Kokovin, S.: Can a larger market foster R&D under monopolistic competition with variable mark-ups? Res. Econ. **71**(4), 663–674 (2017)
12. Bykadorov, I., Kokovin, S., Zhelobodko, E.: Investments in productivity and quality under trade liberalization: monopolistic competition model. Contrib. Game Theory Manage. **5**, 61–72 (2012). Vol. V. Collected papers presented on the Fifth International Conference Game Theory and Management (GTM2011). Graduate School of Management SpbU
13. Bykadorov, I., Kokovin, S., Zhelobodko, E.: Investments in productivity under monopolistic competition: large market advantage. The Economic Education and Research Consortium, Working Paper No 13/08E (2013)
14. Campbell, J.R., Hopenhayn, H.A.: Market size matters. J. Ind. Econ. **53**(1), 1–25 (2005)
15. Davis, D.R.: The home market, trade, and industrial structure. Am. Econ. Rev. **88**(5), 1264–1276 (1998)
16. Dhingra, S., Morrow, J.: Monopolistic competition and optimum product diversity under firm heterogeneity. J. Polit. Econ. **127**(1), 196–232 (2019)
17. Dixit, A., Stiglitz, J.: Monopolistic competition and optimum product diversity. Am. Econ. Rev. **67**(3), 297–308 (1977)
18. Hummels, D., Klenow, P.T.: The variety and quality of a nation's exports. Am. Econ. Rev. **95**(3), 704–723 (2005)
19. Krugman, P.R.: Increasing returns, monopolistic competition, and international trade. J. Int. Econ. **9**(4), 469–479 (1979)
20. Krugman, P.R.: Scale economies, product differentiation, and the pattern of trade. Am. Econ. Rev. **70**(5), 950–959 (1980)
21. Melitz, M.J.: The impact of trade on intra-industry reallocations and aggregate industry productivity. Econometrica **71**(6), 1695–1725 (2003)
22. Redding, S.: Theories of heterogeneous firms and trade. Ann. Rev. Econ. **3**, 1–24 (2011)
23. Syverson, C.: Prices, spatial competition, and heterogeneous producers: an empirical test. J. Ind. Econ. **55**(2), 197–222 (2007)
24. Tilzo, O., Bykadorov, I.: Retailing under monopolistic competition: a comparative analysis. IEEE Xplore, pp. 156–161 (2019)
25. Vives, X.: Innovation and competitive pressure. J. Ind. Econ. **56**(3), 419–469 (2008)
26. Yu, Z.: Trade, market size, and industrial structure: revisiting the home-market effect. Can. J. Econ. **38**(1), 255–272 (2005)
27. Zhelobodko, E., Kokovin, S., Parenti, M., Thisse, J.-F.: Monopolistic competition in general equilibrium: beyond the constant elasticity of substitution. Econometrica **80**(6), 2765–2784 (2012)

On the Cooperative Behavior
in Multistage Multicriteria Game
with Chance Moves

Denis Kuzyutin[1,2] ⓘ, Ekaterina Gromova[1,2(✉)] ⓘ, and Nadezhda Smirnova[1,2] ⓘ

[1] Saint Petersburg State University, Universitetskaya nab. 7/9,
199034 St. Petersburg, Russia
e.v.gromova@spbu.ru
[2] National Research University Higher School of Economics (HSE),
Soyuza Pechatnikov ul. 16, 190008 St. Petersburg, Russia

Abstract. We consider a class of multistage multicriteria games in
extensive form with chance moves where the players cooperate to maximize their expected joint vector payoff. Assuming that the players have
agreed to accept the minimal sum of relative deviations rule in order to
choose a unique Pareto optimal payoffs vector, we prove the time consistency of the optimal cooperative strategy profile and corresponding
optimal bundle of the cooperative trajectories. Then, if the players adopt
a vector analogue of the Shapley value as the solution concept, they need
to design an appropriate imputation distribution procedure to ensure the
sustainability of the achieved cooperative agreement. We provide a generalization of the incremental payment schedule that is applicable for the
games with chance moves and satisfies such advantageous properties as
the efficiency, strict balance condition and the time consistency property
in the whole game. We illustrate our approach with an example of the
extensive-form game tree with chance moves.

Keywords: Multicriteria game · Multistage game · Cooperative
behavior · Time consistency · Shapley value · Chance moves

1 Introduction

Multicriteria games (multiobjective games or games with vector payoffs) are
used to model various real-world interactive decision situations where several objectives have to be taken into account. For example, in multiobjective
environmental games [1,3,17,18,34] a player aims at simultaneously obtaining
large quote for the use of a common resource, increasing production, saving
costs of water purification, saving health care costs, etc. Starting from [36],
much research has been done on non-cooperative multicriteria games (see, e.g.,
[8,12,15,32,37]). A cooperative behavior in games with vector payoffs was examined in [2,11,13,14,16,29,30,34].

The reported study was funded by RFBR under the research project 18-00-00727 (18-
00-00725).

This paper is mainly focused on the dynamic aspects of cooperation in an n-person multistage multicriteria games in extensive form (see, e.g., [7,9,10,12, 20,27]) with chance moves. The methods and the results are based on and closely connected with the previous papers [11,13,15,16]. However, the main difference and challenge here is that due to the chance moves a cooperative pure strategy profile does not generate the unique optimal path in the game tree but rather the whole bundle of optimal cooperative trajectories.

To achieve and implement a long-term cooperative agreement in a multicriteria dynamic game we solve the following problems. First, when players seek to achieve the maximal total vector payoff of the grand coalition, they face the problem of choosing a unique Pareto efficient payoffs vector. In the dynamic setting it is necessary that a specific method the players agreed to accept in order to choose a particular Pareto optimal solution not only takes into account the relative importance of the criteria, but also satisfies time consistency [5–7,10–13,16,21,22,24–28,33], i.e., a fragment of the optimal cooperative bundle of the trajectories in the subgame should remain optimal in this subgame. In this paper, we use the rule of minimal sum of relative deviations (MSRD) from the ideal payoffs vector (see [16] for details) to find a unique optimal cooperative bundle of the trajectories, which is proved to satisfy time consistency.

After choosing the cooperative bundle of the trajectories it is necessary to construct a vector-valued characteristic function. To this end, we use the ζ-characteristic function introduced in [4] as well as the MSRD rule in order to choose a particular Pareto efficient solution for the auxiliary vector optimization problems. To determine the optimal payoff allocation we use the vector analogue of the Shapley value [29,30,35]. Such an approach is based on the assumption that the payoff can be transferred between the players within the same criterion. Note that the main measurable criteria used in multicriteria resource management problems usually satisfy this component-wise transferable utility property.

Lastly, to guarantee the sustainability of the achieved long-term cooperative agreement one needs to design a consistent imputation distribution procedure (IDP) or a payment schedule (see, e.g., [7,11,13,16,21,22,25,27,28,33]) that should satisfy a set of useful p Lastly, to guarantee the sustainability of the achieved long-term cooperative agreement one needs to design a consistent imputation distribution procedure (IDP) or a payment schedule (see, e.g., [7,11,13,16,21,22,25,27,28,33]) that should satisfy a set of useful properties. The detailed review of dynamical properties the IDP may satisfy for multistage multicriteria games is presented in [11,13–16] . In this paper we mainly focus on the efficiency constraint and the strict balance condition as well as time consistency in the whole game (see, e.g., [22,33]) and provide a generalization of the incremental IDP to implement a long-term cooperative agreement in a multistage multicriteria game with chance moves.

The contribution of this paper is twofold:

1. Using the MSRD rule, we provide an approach to choose the optimal bundle of the cooperative trajectories for the class of multicriteria multistage games with chance moves and prove that such Pareto optimal solution satisfies time consistency.

2. We propose an appropriate refinement of the time consistency definition for imputation distribution procedure and design a generalized incremental IDP which is proved to satisfy a number of important properties.

The rest of the paper is organized as follows: Sect. 2 recalls the main ingredients of the class of games of interest, and Sect. 3 deals with the construction of time consistent optimal cooperative strategy profile and corresponding bundle of cooperative trajectories. In Sect. 4, we design the incremental IDP for the Shapley value that satisfies efficiency, strict balance condition and time consistency property in the whole game. Section 5 provides a brief conclusion.

2 Multistage Game with Chance Moves and Vector Payoffs

We consider a finite multistage r-criteria game in extensive form following [7,9–13,16,20,27]. First we define the notation that will be used later on:

- $N = \{1, \ldots, n\}$ is the set of all players.
- K is the game tree with the root x_0 and the set of all nodes P.
- $S(x)$ is the set of all direct successors (descendants) of the node x, and $S^{-1}(y)$ is the unique predecessor (parent) of the node $y \neq x_0$ such that $y \in S(S^{-1}(y))$.
- P_i is the set of all decision nodes of the ith player (at these nodes the player i chooses the following node), $P_i \cap P_j = \emptyset$, for all $i, j \in N$, $i \neq j$.
- $P_{n+1} = \{z^j\}_{j=1}^m$ denotes the set of all terminal nodes (final positions), $S(z^j) = \emptyset \; \forall z^j \in P_{n+1}$.
- P_0 is the set of all nodes at which a chance moves, where $\pi(y|x) > 0$ denotes the probability of transition from node $x \in P_0$ to node $y \in S(x)$. It holds that $\bigcup_{i=0}^{n+1} P_i = P$.
- $\omega = (x_0, \ldots, x_{t-1}, x_t, \ldots, x_T)$ is the trajectory (or the path) in the game tree, $x_{t-1} = S^{-1}(x_t)$, $1 \leq t \leq T$, $x_T = z^j \in P_{n+1}$; where index t in x_t denotes the ordinal number of this node within the trajectory ω and can be interpreted as the "time index".
- $h_i(x) = (h_{i/1}(x), \ldots, h_{i/r}(x))$ is the r-component vector payoff of the ith player at the node $x \in P$. We assume that for all $i \in N$, $k = 1, \ldots, r$, and $x \in P$ the components of vector payoffs are positive, i.e., $h_{i/k}(x) > 0$.

In the following, we will use $MG^{cm}(n, r)$ to denote the class of all finite multistage n-person r-criteria games with chance moves in extensive form defined above, where $\Gamma^{x_0} \in MG^{cm}(n, r)$ denotes a game with root x_0. Note that Γ^{x_0} is an extensive-form game with perfect information (see, e.g., [9,20,27] for details).

Since all the solutions we are interested in throughout the paper are attainable when the players restrict themselves to the class of pure strategies we will focus on this class of strategies. The pure strategy $u_i(\cdot)$ of the ith player is a function with domain P_i that specifies for each node $x \in P_i$ the next node

$u_i(x) \in S(x)$ which the player i has to choose at x. Let U_i denote the (finite) set of all ith player's pure strategies, $U = \prod_{i \in N} U_i$.

Denote by $p(y|x, u)$ the conditional probability that node $y \in S(x)$ is reached if node x has been already reached (the probability of transition from x to y) while the players use the strategies u_i, $i \in N$. Note that for all $x \in P_i$ and for all $y \in S(x)$ $p(y|x, u) = 1$, $u_i(x) = y$ and $p(y|x, u) = 0$ if $u_i(x) \neq y$. For chance moves, i.e., if $x \in P_0$ $p(y|x, u) = \pi(y|x)$ for all $y \in S(x)$ for each $u \in U$.

Then one can calculate the probability $p(\omega, u)$ of realization of the trajectory $\omega = (x_0, \ldots, x_\tau, x_{\tau+1}, \ldots, x_T)$, $x_T \in P_{n+1}$, $x_{\tau+1} \in S(x_\tau)$, $\tau = 0, \ldots, T-1$, when the players use the strategies u_i from the strategy profile $u = (u_1, \ldots, u_n)$.

$$p(\omega, u) = p(x_1|x_0, u) \cdot p(x_2|x_1, u) \cdot \ldots \cdot p(x_T|x_{T-1}, u) = \prod_{\tau=0}^{T-1} p(x_{\tau+1}|x_\tau, u). \quad (1)$$

Denote by $\Omega(u) = \{\omega_k(u)|p(\omega_k, u) > 0\}$ the finite set (or the bundle) of the trajectories ω_k which are generated by strategy profile $u \in U$. Note that for all $\omega_k(u) \in \Omega(u)$, $u_j(x_\tau) = x_{\tau+1}$ for all $x_\tau \in \omega_k(u) \cap P_j$, $j \in N$, $0 \leq \tau \leq T-1$.

Let $\tilde{h}_i(\omega) = \sum_{\tau=0}^{T} h_i(x_\tau)$ denote the ith player's vector payoff corresponding to the trajectory $\omega = (x_0, \ldots, x_t, x_{t+1}, \ldots, x_T)$.

Denote by

$$H_i(u) = \sum_{\omega_k \in \Omega(u)} p(\omega_k, u) \cdot \tilde{h}_i(\omega_k) = \sum_{\omega_k \in \Omega(u)} p(\omega_k, u) \cdot \sum_{\tau=0}^{T(k)} h_i(x_\tau) \quad (2)$$

the (expected) value of the ith player's vector payoff function which corresponds to the strategy profile $u = (u_1, \ldots, u_n)$. Let $\Omega_{n+1}(u) = \{\Omega(u) \cap P_{n+1}\}$ denote the set of all terminal nodes of the trajectories $\omega_k(u) \in \Omega(u)$.

Remark 1. If the pure strategy profiles u and v generate different bundles $\Omega(u)$ and $\Omega(v)$ of the trajectories, i.e., $\Omega(u) \neq \Omega(v)$, then $\Omega_{n+1}(u) \cap \Omega_{n+1}(v) = \emptyset$.

According to [9,20,27] each intermediate node $x_t \in P \backslash P_{n+1}$ generates a subgame Γ^{x_t} with the subgame tree K^{x_t} and the subgame root x_t as well as a factor-game Γ^D with the factor-game tree $K^D = (K \backslash K^{x_t}) \cup \{x_t\}$. Decomposition of the original extensive game Γ^{x_t} at node x_t onto the subgame Γ^{x_t} and the factor-game Γ^D generates the corresponding decomposition of the pure (and mixed) strategies (see [9] for details).

Let $P_i^{x_t}(P_i^D)$, $i \in N$, denote the restriction of P_i on the subgame tree $K^{x_t}(K^D)$, and $u_i^{x_t}(u_i^D)$, $i \in N$, denote the restriction of the ith player's pure strategy $u_i(\cdot)$ in Γ^{x_0} on $P_i^{x_t}(P_i^D)$. The pure strategy profile $u^{x_t} = (u_1^{x_t}, \ldots, u_n^{x_t})$ generates the bundle of the subgame trajectories $\Omega^{x_t}(u^{x_t}) = \{\omega_k^{x_t}(u^{x_t})|p(\omega_k^{x_t}, u^{x_t}) > 0\}$. Similarly to (2), let us denote by

$$H_i^{x_t}(u^{x_t}) = \sum_{\omega_k^{x_t} \in \Omega^{x_t}(u^{x_t})} p(\omega_k^{x_t}, u^{x_t}) \cdot \sum_{\tau=t}^{T(k)} h_i(x_\tau) = \sum_{\omega_k^{x_t} \in \Omega^{x_t}(u^{x_t})} p(\omega_k^{x_t}, u^{x_t}) \cdot \tilde{h}_i(\omega_k^{x_t}) \quad (3)$$

the value of the ith player's vector payoff in Γ^{x_t}, and by $U_i^{x_t}$ the set of all possible ith player's pure strategies in the subgame Γ^{x_t}, $U^{x_t} = \prod_{i \in N} U_i^{x_t}$. Note that for each trajectory $\omega = (x_0, \ldots, x_t, x_{t+1}, \ldots, x_T)$, $1 \leqslant t \leqslant T - 1$, $x_T \in P_{n+1}$,

$$
\begin{aligned}
p(\omega, u) &= \prod_{\tau=0}^{t-1} p(x_{\tau+1}|x_\tau, u) \cdot \prod_{\tau=t}^{T-1} p(x_{\tau+1}|x_\tau, u) \\
&= p(\underline{\omega}^{x_t}, u) \cdot p(\omega^{x_t}, u) = p(\underline{\omega}^{x_t}, u^D) \cdot p(\omega^{x_t}, u^{x_t}),
\end{aligned}
\tag{4}
$$

where $\underline{\omega}^{x_t} = (x_0, x_1, \ldots, x_{t-1}, x_t)$ denotes a fragment of trajectory ω implemented before the subgame Γ^{x_t} starts, and $p(\underline{\omega}^{x_t}, u) = p(x_t, u)$ denotes the probability that node x_t is reached when the players employ the strategies u_i, $i \in N$. It is worth noting that factor-game $\Gamma^D = \Gamma^D(u^{x_t})$ is usually defined for given strategy profile u^{x_t} in the subgame Γ^{x_t} since we assume that

$$
h_i^D(x_0, x_1, \ldots, x_{t-1}, x_t) = \sum_{\tau=0}^{t-1} h_i(x_\tau) + H_i^{x_t}(u^{x_t}) = \tilde{h}_i(\underline{\omega}^{x_t} \setminus \{x_t\}) + H_i^{x_t}(u^{x_t})
\tag{5}
$$

(see, e.g., [9,27] for details). Moreover, given intermediate node x_t, the bundle $\Omega(u) = \{\omega_k(u)|p(\omega_k, u) > 0\}$ can be divided in two subsets, i.e. $\Omega(u) = \{\Psi_m\} \cup \{\chi_l\}$, where $x_t \in \Psi_m$, and $x_t \notin \chi_l$, $\{\Psi_m\} \cap \{\chi_l\} = \emptyset$. Then, taking (1), (3), (4) and (5) into account, we get

$$
\begin{aligned}
H_i(u) &= \sum_m p(\Psi_m, u) \cdot \tilde{h}_i(\Psi_m) + \sum_l p(\chi_l, u) \cdot \tilde{h}_i(\chi_l) \\
&= \sum_m p(x_t, u) \cdot p(\Psi_m^{x_t}, u^{x_t}) \cdot \left[\tilde{h}_i(\underline{\Psi}_m^{x_t} \setminus \{x_t\}) + \tilde{h}_i(\Psi_m^{x_t}) \right] \\
&\quad + \sum_l p(\chi_l, u) \cdot \tilde{h}_i(\chi_l) = p(x_t, u^D) \cdot \tilde{h}_i(x_0, \ldots, x_{t-1}) \cdot \sum_m p(\Psi_m^{x_t}, u^{x_t}) \\
&\quad + p(x_t, u^D) \cdot \sum_m p(\Psi_m^{x_t}, u^{x_t}) \cdot \tilde{h}_i(\Psi_m^{x_t}) + \sum_l p(\chi_l, u) \cdot \tilde{h}_i(\chi_l) \\
&= p(x_t, u^D) \cdot \tilde{h}_i(x_0, \ldots, x_{t-1}) + p(x_t, u^D) \cdot H_i^{x_t}(u^{x_t}) \\
&\quad + \sum_l p(\chi_l, u) \cdot \tilde{h}_i(\chi_l) = p(x_t, u^D) \cdot h_i^D(x_0, \ldots, x_t) + \sum_l p(\chi_l, u) \cdot \tilde{h}_i(\chi_l).
\end{aligned}
\tag{6}
$$

Note that, since $P_i = P_i^{x_t} \cup P_i^D$, one can compose the ith player's pure strategy $W_i = (u_i^D, v_i^{x_t}) \in U_i$ in the original game Γ^{x_0} from her strategies $v_i^{x_t} \in U_i^{x_t}$ in the subgame Γ^{x_t} and $u_i^D \in U_i^D$ in the factor-game Γ^D [9,27].

3   Cooperative Behavior

Let $a, b \in R^m$; we use the following vector preferences: $a \geqq b$ if $a_k \geqslant b_k$, $k = 1, \ldots, m$; $a > b$ if $a_k > b_k$, $k = 1, \ldots, m$; $a \geq b$ if $a \geqq b$, and $a \neq b$. The last vector inequality means that b is Pareto dominated (or Edgeworth-Pareto dominated) by a (and hence b is called "inefficient").

If the players agree to cooperate in the game Γ^{x_0}, first they are expected to maximize w.r.t. the binary relation $\geq$ the total vector payoff $\sum_{i=1}^{n} H_i(u)$ of the grand coalition. Let $PO(\Gamma^{x_0})$ denote the set of all Pareto optimal pure strategy profiles from U, i.e.:

$$u \in PO(\Gamma^{x_0}) \text{ if } \nexists v \in U : \sum_{i \in N} H_i(v) \geq \sum_{i \in N} H_i(u).$$

The set $PO(\Gamma^{x_0})$ is known to be nonempty and in general it contains multiple strategy profiles (see, e.g. [27,31] for details). The players need to agree on a specific rule γ they are going to use to choose a unique optimal cooperative strategy profile $\overline{u} \in PO(\Gamma^{x_0})$ as well as the corresponding optimal bundle of strategies in the game tree. One of such rules that is applicable for a wide class of multicriteria games with positive payoffs – the minimal sum of relative deviations (MSRD) rule, – was formally introduced in [16].

Namely, denote by $H_{N/k}(u) = \sum_{i \in N} H_{i/k}(u)$ the sum of all players' payoffs w.r.t. the criterion k, $h_{N/k}(x_\tau) = \sum_{i \in N} h_{i/k}(x_\tau)$, $x_\tau \in P$. Let $H_k^* = \max_{u \in U} H_{N/k}(u)$. The vector $(H_1^*, \ldots, H_r^*)$ can be interpreted as the vector of ideal payoffs for the grand coalition N (see, e.g., [19,27,31]).

Definition 1. *According to the MSRD rule the players have to choose a Pareto optimal pure strategy profile u which minimizes the sum of relative deviations w.r.t. each criterion from ideal payoffs vector H^*. Namely,*

$$\sum_{k=1}^{r} \frac{H_k^* - H_{N/k}(u)}{H_k^*} = \min_{v \in U} \sum_{k=1}^{r} \frac{H_k^* - H_{N/k}(v)}{H_k^*} = r - \max_{v \in U} \sum_{k=1}^{r} \frac{H_{N/k}(v)}{H_k^*},$$

or

$$u \in \arg \max_{v \in U} \sum_{k=1}^{r} \frac{1}{H_k^*} \cdot H_{N/k}(v) = \arg \max_{v \in U} \sum_{k=1}^{r} \mu_k \cdot H_{N/k}(v), \qquad (7)$$

where $\mu_k = \dfrac{1}{H_k^*} > 0$, $k = 1, \ldots, r$.

Since different criteria may have different scales a simple sum of absolute values of all the deviations is a rather rough estimate of the distance from given Pareto optimal payoffs vector to the vector of ideal payoffs. One approach to obtain more precise estimate is to sum relative deviations, hence, we need to divide the absolute deviation $H_k^* - H_{N/k}(v)$ w.r.t. criterion k by the range H_k^* of this criterion (see [27,31] for details).

Let $\overline{PO}(\Gamma^{x_0})$ denote the nonempty set of pure strategy profiles $u \in U$ which satisfy (7). If all strategy profiles $u \in \overline{PO}(\Gamma^{x_0})$ generate the same bundle of trajectories $\Omega(u)$ (see, e.g., [9,20,27] for discussion on redundancy of the pure strategy definition in extensive game), let the players choose any strategy profile $\overline{u} \in \overline{PO}(\Gamma^{x_0})$ and $\Omega(\overline{u})$ denote the corresponding bundle of the trajectories.

Otherwise, i.e. if the strategy profiles from $\overline{PO}(\Gamma^{x_0})$ generate different (and hence, disjoint) bundles of the trajectories, we assume that the players choose such $\overline{u} \in \overline{PO}(\Gamma^{x_0})$ that $\Omega(\overline{u}) = \{\omega_k(\overline{u}) = (x_0, \ldots, x_{T(k)} = z_l) \mid p(\omega_k, \overline{u}) > 0\}$ contains the trajectory $\omega(\overline{u})$ with minimal number l of the terminal node z_l (see Remark 1).

Henceforth, we will refer to the strategy profile $\overline{u} \in \overline{PO}(\Gamma^{x_0})$ and the bundle of the trajectories $\Omega(\overline{u})$ as the optimal cooperative strategy profile and the optimal bundle of the cooperative trajectories respectively. We will assume in this paper that all the players have agreed to apply the MSRD rule in order to choose the optimal cooperative strategy profile $\overline{u}$ and the corresponding bundle $\Omega(\overline{u})$ of cooperative trajectories. Denote by $\text{Max}_{u \in U}^{\mu} \sum_{i \in N} H_i(u) = \sum_{i \in N} H_i(\overline{u})$ the maximal (namely in the sense of MSRD rule) total vector payoff.

In the dynamic setting it is necessary that a specific method the players agreed to accept in order to choose a particular Pareto optimal solution from $PO(\Gamma^{x_0})$ satisfies time consistency [6,24,27], i.e., a fragment of the optimal bundle of the cooperative trajectories in the subgame should remain optimal in this subgame. Suppose that at every subgame $\Gamma^{\overline{x}_t}$, $\overline{x}_t \in \omega(\overline{u})$, $\omega(\overline{u}) \in \Omega(\overline{u})$, i.e. a subgame along the cooperative trajectories, the players choose the strategy profile $u^{\overline{x}_t} \in U^{\overline{x}_t}$ such that

$$u^{\overline{x}_t} \in \arg \max_{v^{\overline{x}_t} \in U^{\overline{x}_t}} \sum_{k=1}^{r} \mu_k \cdot H_{N/k}^{\overline{x}_t}(v^{\overline{x}_t}), \tag{8}$$

where the coefficients $\mu_k = \dfrac{1}{H_k^*}$ are the same as in (7). Let $\overline{PO}(\Gamma^{\overline{x}_t})$ denote the set of all pure strategy profiles $u^{\overline{x}_t} \in U^{\overline{x}_t}$ that satisfy (8) and the players use the same approach to choose a unique optimal cooperative strategy profile $\overline{u}^{\overline{x}_t} \in \overline{PO}(\Gamma^{\overline{x}_t})$ in the subgame as for the original game Γ^{x_0} (minimal number l of the terminal node z_l on the cooperative trajectory $\omega^{\overline{x}_t} \in \Omega(\overline{u}^{\overline{x}_t})$ generated by $\overline{u}^{\overline{x}_t}$).

Proposition 1. *A Pareto optimal solution for* $\Gamma^{x_0} \in MG^{cm}(n, r)$ *based on the MSRD rule satisfies time consistency. Namely, let* $\overline{u} \in U$ *satisfies* (7), *and* $\Omega(\overline{u})$ *be the optimal bundle of cooperative trajectories. Then for each subgame* $\Gamma^{\overline{x}_t}$, $\overline{x}_t \in \omega(\overline{u}) = (\overline{x}_0, \ldots, \overline{x}_t, \overline{x}_{t+1}, \ldots, \overline{x}_T)$ *with* $\overline{x}_0 = x_0$, $\omega(\overline{u}) \in \Omega(\overline{u})$, *it holds that*

$$\overline{u}^{\overline{x}_t} = (\overline{u}_1^{\overline{x}_t}, \ldots, \overline{u}_n^{\overline{x}_t}) \in \arg \max_{v^{\overline{x}_t} \in U^{\overline{x}_t}} \sum_{k=1}^{r} \mu_k \cdot H_{N/k}^{\overline{x}_t}(v^{\overline{x}_t}), \tag{9}$$

while $\omega^{\overline{x}_t} = (\overline{x}_t, \overline{x}_{t+1}, \ldots, \overline{x}_T) \in \Omega(\overline{u}^{\overline{x}_t})$, *i.e.* $\omega^{\overline{x}_t}$ *belongs to the optimal bundle of cooperative trajectories in the subgame* $\Gamma^{\overline{x}_t}$.

Proof. The optimal bundle of cooperative trajectories $\Omega(\overline{u})$ generated by $\overline{u} \in \overline{PO}(\Gamma^{x_0})$ can be divided onto two subsets $\{\Psi_m\} = \{\omega \in \Omega(\overline{u}) \mid \overline{x}_t \in \omega\}$ and $\{\chi_l\} = \{\omega \in \Omega(\overline{u}) \mid \overline{x}_t \notin \omega\}$ while $\{\Psi_m\} \cap \{\chi_l\} = \emptyset$, $\{\Psi_m\} \cup \{\chi_l\} = \Omega(\overline{u})$. Then, taking (5) and (6) into account we get

$$H_{i/k}(\overline{u}) = \sum_m p(\Psi_m, \overline{u}) \cdot \tilde{h}_{i/k}(\Psi_m) + \sum_l p(\chi_l, \overline{u}) \cdot \tilde{h}_{i/k}(\chi_l)$$

$$= p(\overline{x}_t, \overline{u}) \cdot \left[\tilde{h}_{i/k}(\overline{x}_0, \overline{x}_1, \dots, \overline{x}_{t-1}) + H_{i/k}^{\overline{x}_t}(\overline{u}^{\overline{x}_t})\right] + \sum_l p(\chi_l, \overline{u}^D) \cdot \tilde{h}_{i/k}(\chi_l), \tag{10}$$

and (7) takes the form

$$\sum_{k=1}^r \mu_k \cdot H_{N/k}(\overline{u}) = \sum_{k=1}^r \mu_k \cdot \left[p(\overline{x}_t, \overline{u}) \cdot \left(\tilde{h}_{N/k}(\overline{x}_0, \overline{x}_1, \dots, \overline{x}_{t-1}) + H_{N/k}^{\overline{x}_t}(\overline{u}^{\overline{x}_t})\right)\right.$$

$$\left. + \sum_l p(\chi_l, \overline{u}^D) \cdot \tilde{h}_{N/k}(\chi_l)\right] = \max_{v \in U} \sum_{k=1}^r \mu_k \cdot H_{N/k}(v). \tag{11}$$

Suppose that $\overline{u}^{\overline{x}_t}$ does not satisfy (9), i.e. there exists $v^{\overline{x}_t} \in U^{\overline{x}_t}$ such that

$$\sum_{k=1}^r \mu_k \cdot H_{N/k}^{\overline{x}_t}(\overline{u}^{\overline{x}_t}) < \sum_{k=1}^r \mu_k \cdot H_{N/k}^{\overline{x}_t}(v^{\overline{x}_t}). \tag{12}$$

Denote by $\Omega(v^{\overline{x}_t}) = \{\lambda_m^{\overline{x}_t} = (\overline{x}_t, \dots, \overline{x}_{T(m)}) \mid p(\lambda_m^{\overline{x}_t}, v^{\overline{x}_t}) > 0\}$ the bundle of all trajectories in the subgame $\Gamma^{\overline{x}_t}$ generated by $v^{\overline{x}_t}$. Then (12) takes the form

$$\sum_{k=1}^r \mu_k \cdot \left[\sum_m p(\Psi_m^{\overline{x}_t}, \overline{u}^{\overline{x}_t}) \cdot \tilde{h}_{N/k}^{\overline{x}_t}(\Psi_m^{\overline{x}_t})\right] < \sum_{k=1}^r \mu_k \cdot \left[\sum_m p(\lambda_m^{\overline{x}_t}, v^{\overline{x}_t}) \cdot \tilde{h}_{N/k}^{\overline{x}_t}(\lambda_m^{\overline{x}_t})\right]. \tag{13}$$

Denote by $W_i = (\overline{u}_i^D, v_i^{\overline{x}_t})$, $i \in N$, the ith player's compound pure strategy in Γ^{x_0}. The strategy profile $W = (W_1, \dots, W_n)$ generates the strategy bundle $\Omega(W)$ that can be divided onto two disjoint subsets $\{\lambda_m\} = \{\omega \in \Omega(W) \mid \overline{x}_t \in \omega\}$ and $\{\chi_l\} = \{\omega \in \Omega(W) \mid \overline{x}_t \notin \omega\}$, where the second subset for $\Omega(W)$ coincides with the second subset for $\Omega(\overline{u})$ since $W^D = u^D$, and $\lambda_m = (\overline{x}_0, \dots, \overline{x}_t) \cup (\overline{x}_t, \dots, \overline{x}_{T(m)}) = (\overline{x}_0, \dots, \overline{x}_t) \cup \lambda_m^{\overline{x}_t}$.

Adding $\sum_{k=1}^r \mu_k \cdot \tilde{h}_{N/k}(\overline{x}_0, \overline{x}_1 \dots, \overline{x}_{t-1})$ to both sides of (13) we get

$$\sum_{k=1}^r \mu_k \left(\tilde{h}_{N/k}(\overline{x}_0, \dots, \overline{x}_{t-1}) + H_{N/k}^{\overline{x}_t}(\overline{u}^{\overline{x}_t})\right) < \sum_{k=1}^r \mu_k \left(\tilde{h}_{N/k}(\overline{x}_0, \dots, \overline{x}_{t-1}) + H_{N/k}^{\overline{x}_t}(v^{\overline{x}_t})\right). \tag{14}$$

Then we can multiply both sides of (14) on $p(\overline{x}_t, \overline{u}) = p(\overline{x}_t, \overline{u}^D) = p(\overline{x}_t, W^D) = p(\overline{x}_t, W) > 0$ and rearrange the terms to obtain

$$\sum_{k=1}^r \mu_k \left[p(\overline{x}_t, \overline{u}) \cdot \left(\tilde{h}_{N/k}(\overline{x}_0, \dots, \overline{x}_{t-1}) + H_{N/k}^{\overline{x}_t}(\overline{u}^{\overline{x}_t})\right)\right]$$

$$< \sum_{k=1}^r \mu_k \left[p(\overline{x}_t, W) \cdot \left(\tilde{h}_{N/k}(\overline{x}_0, \dots, \overline{x}_{t-1}) + H_{N/k}^{\overline{x}_t}(v^{\overline{x}_t})\right)\right].$$

Finally, adding $\sum_{k=1}^{r} \mu_k \cdot \sum_l p(\chi_l, \overline{u}^D) \cdot \tilde{h}_{N/k}(\chi_l)$ to both sides of the last inequality and taking into account (4)–(6) and (11) for some $W \in U$ we get

$$\sum_{k=1}^{r} \mu_k \cdot H_{N/k}(\overline{u}) < \sum_{k=1}^{r} \mu_k \cdot H_{N/k}(W)$$

The last inequality contradicts the fact that $\overline{u} \in \overline{PO}(\Gamma^{x_0})$, hence (9) is valid.

Arguing in a similar way (for the case when different strategy profiles from $\overline{PO}(\Gamma^{\overline{x}_t})$ generate different bundles of the trajectories) we can verify that $\omega^{\overline{x}_t} = (\overline{x}_t, \ldots, \overline{x}_T)$—a fragment of the cooperative trajectory $\omega \in \Omega(\overline{u})$, starting at $\overline{x}_t$—remains the cooperative trajectory in the subgame $\Gamma^{\overline{x}_t}$, i.e. $\omega^{\overline{x}_t} \in \Omega(\overline{u}^{\overline{x}_t})$. □

When the players have agreed to choose optimal cooperative strategy profile $\overline{u} = (\overline{u}_1, \ldots, \overline{u}_n)$ that generates the optimal bundle $\Omega(\overline{u})$ of cooperative trajectories in $\Gamma^{x_0} \in MG^{cm}(n, r)$, the next step of cooperation is to define a vector-valued characteristic function $V^{x_0}(S)$. We adopt here a novel approach, proposed in [4], i.e. we assume that when coalition S forms, the players $i \in S$ use cooperative strategies $\overline{u}_i$ while the left-out players $j \in N \setminus S$ seek to minimize (in the sense of MSRD rule) the total payoffs vector of the players from coalition S. Let

$$\operatorname*{Min}_{u_j, j \in N \setminus S}^{\mu} \sum_{i \in S} H_i(\overline{u}_S, u_{N \setminus S}) = \sum_{i \in S} H_i(\overline{u}_S, \underline{u}_{N \setminus S})$$

denote the minimal (in the sense of MSRD rule) total payoffs vector for coalition S. Then, this so-called ζ-characteristic function takes the form:

$$V^{x_0}(S) = \begin{cases} \overline{0} \in R^r, & S = \emptyset, \\ \operatorname*{Min}_{u_j, j \in N \setminus S}^{\mu} \sum_{i \in S} H_i(\overline{u}_S, u_{N \setminus S}), & S \subset N, \\ \operatorname*{Max}_{u \in U}^{\mu} \sum_{i \in N} H_i(u), & S = N. \end{cases} \tag{15}$$

Note that ζ-vector-valued characteristic function is relatively friendly computable and is proved to satisfy the weak superadditivity property for cooperative multistage multicriteria games without chance moves [16]. The characteristic function $V^{\overline{x}_t}$ for the subgame $\Gamma^{\overline{x}_t}$, $\overline{x}_t \in \omega_m(\overline{u}) = (\overline{x}_0, \ldots, \overline{x}_t, \ldots, \overline{x}_{T(m)})$, $\omega_m(\overline{u}) \in \Omega(\overline{u})$ along the optimal bundle of cooperative trajectories can be constructed using the same approach as in (15). Note that

$$V^{\overline{x}_t}(N) = \sum_{\omega_m^{\overline{x}_t} \in \Omega(\overline{u}^{\overline{x}_t})} p(\omega_m^{\overline{x}_t}, \overline{u}^{\overline{x}_t}) \cdot \sum_{\tau=t}^{T(m)} \sum_{i \in N} h_i(\overline{x}_\tau). \tag{16}$$

Let $\Gamma^{x_0}(N, V^{x_0})$ denote multicriteria cooperative game $\Gamma^{x_0} \in MG^{cm}(n, r)$ with vector-valued characteristic function (15), and $\Gamma^{\overline{x}_t}(N, V^{\overline{x}_t})$ denote the corresponding subgame.

Definition 2. *([30,35]) The Shapley value of multicriteria game $\Gamma^{x_0}(N, V^{x_0})$ denoted by φ^{x_0} is defined for every player $i \in N$ as*

$$\varphi_i^{x_0} = \sum_{S \subset N, i \in S} \frac{(n - |S|)!(|S| - 1)!}{n!}(V^{x_0}(S) - V^{x_0}(S \setminus \{i\})). \qquad (17)$$

Remark 2. ([30,35]) The Shapley value in a multicriteria cooperative game was proved to satisfy the efficiency property, i.e.:

$$\sum_{i=1}^{n} \varphi_i^{x_0} = V^{x_0}(N) = \sum_{\omega_m \in \Omega(\overline{u})} p(\omega_m, \overline{u}) \cdot \sum_{\tau=0}^{T(m)} \sum_{i \in N} h_i(\overline{x}_\tau). \qquad (18)$$

Denote by $\left(\varphi_i^{\overline{x}_t}\right)_{i \in N}$ the Shapley value in $\Gamma^{\overline{x}_t}\left(N, V^{\overline{x}_t}\right)$, $\overline{x}_t \in \omega_m(\overline{u})$, $\omega_m(\overline{u}) \in \Omega(\overline{u})$, $t = 0, \dots, T(m)$.

4   Sustainability of the Shapley Value

Let $\beta = \{\beta_{i/k}(\overline{x}_\tau)\}$, $i = 1, \dots, n$; $k = 1, \dots, r$; $\tau = 1, \dots, T(l)$, $\overline{x}(\tau) \in \omega_l(\overline{u})$, $\omega_l(\overline{u}) \in \Omega(\overline{u})$ denote the Imputation Distribution Procedure (IDP) for the Shapley value $\left(\varphi_i^{\overline{x}_0}\right)_{i \in N}$ or the payment schedule (see, i.e., [7,11,13,16,21,22,25,27, 28,33] for details). The IDP approach means that all the players have agreed to allocate the total cooperative vector payoff $V^{x_0}(N)$ between the players along the optimal bundle $\Omega(\overline{u})$ of cooperative trajectories $\omega_l(\overline{u})$ according to some specific rule called IDP. Namely, $\beta_{i/k}(\overline{x}_\tau)$ denotes the actual current payment that the player i receives at node $\overline{x}_\tau$ w.r.t. criterion k (instead of $h_{i/k}(\overline{x}_\tau)$) if the players employ the IDP β. Moreover, one can design such an IDP β that all the players will be interested in cooperation at any intermediate time, i.e. in any subgame $\Gamma^{\overline{x}_\tau}$, $\overline{x}(\tau) \in \omega_l(\overline{u})$, $\omega_l(\overline{u}) \in \Omega(\overline{u})$.

Definition 3. *([13,22,33]) The IDP $\beta = \{\beta_{i/k}(\overline{x}_\tau)\}$ satisfies the efficiency condition at initial node x_0 if*

$$\sum_{\omega_m \in \Omega(\overline{u})} p(\omega_m, \overline{u}) \cdot \sum_{\tau=0}^{T(m)} \beta_i(\overline{x}_\tau) = \varphi_i^{x_0}, i = 1, \dots, n. \qquad (19)$$

Equation (19) means that the expected sum of the payments to player i along the optimal game evolution equals to what she is entitled to in the whole game Γ^{x_0}. Then the IDP for each player can be reasonably interpreted as a rule for step-by-step allocation of the ith player's optimal payoff.

Definition 4. *([13]) The IDP $\beta = \{\beta_{i/k}(\overline{x}_\tau)\}$ satisfies the strict balance condition if for each node $\overline{x}_\tau \in \omega_m(\overline{u})$, $\omega_m(\overline{u}) \in \Omega(\overline{u})$ $\forall t = 0, \dots, T(m); \forall k = 1, \dots, r$*

$$\sum_{i \in N} \beta_{i/k}(\overline{x}_\tau) = \sum_{i \in N} h_{i/k}(\overline{x}_\tau). \qquad (20)$$

Equation (20) ensures the "admissibility" of the IDP, i.e. the sum of payments to the players in any node $\bar{x}_\tau$ is equal to the sum of payoffs that they can collect in this node.

The next advantageous dynamic property of an IDP—the time consistency, introduced in [25]—was extended to multicriteria cooperative games (without chance moves) in [11,13,15,16].

To write down properly the time consistency condition for some intermediate node $\bar{x}_t \in \omega(\bar{u}) = (\bar{x}_0, \bar{x}_1, \ldots, \bar{x}_{t-1}, \bar{x}_t, \bar{x}_{t+1}, \ldots, \bar{x}_T)$, $\omega(\bar{u}) \in \Omega(\bar{u})$, $1 \leqslant t < T$, in multistage game Γ^{x_0} with chance moves we need to pay attention to all chance nodes on the path $(\bar{x}_0, \ldots, \bar{x}_{t-1}) = \underline{\omega}^{x_t} \setminus \{\bar{x}_t\}$.

Namely, let us numerate the chance nodes from $P_0 \cap (\underline{\omega}^{x_t} \setminus \{\bar{x}_t\})$ in order of their occurrence on the path $(\bar{x}_0, \ldots, \bar{x}_{t-1})$, i.e. $y_1 = \bar{x}_{t(1)}$, $y_2 = \bar{x}_{t(2)}, \ldots, y_\theta = \bar{x}_{t(\theta)}$, $0 \leqslant t(1) < t(2) < \ldots < t(\theta) < t$.

Definition 5. *The IDP* $\beta = \{\beta_{i/k}(\bar{x}_\tau)\}$ *for the Shapley value* φ^{x_0} *is called time consistent in the whole game* $\Gamma^{x_0} (N, V^{x_0}) \in MG^{cm}(n,r)$ *if at any intermediate node* $\bar{x}_t \in \omega(\bar{u})$, $\omega(\bar{u}) \in \Omega(\bar{u})$, $1 \leqslant t < T$, *for all* $i \in N$, *it holds that*

case $\theta = 0$ *(no chance nodes on the path* $(\bar{x}_0, \ldots, \bar{x}_{t-1})$*):*

$$\sum_{\tau=0}^{t-1} \beta_i(\bar{x}_\tau) + \varphi_i^{\bar{x}_t} = \varphi_i^{x_0}, \tag{21}$$

case $\theta = 1$ *(only one chance node* $y_1 = \bar{x}_{t(1)}$ *before* $\bar{x}_t$*):*

$$\sum_{\tau=0}^{t(1)} \beta_i(\bar{x}_\tau) + p(\bar{x}_{t(1)+1}, \bar{u}) \cdot \left\{ \sum_{\tau=t(1)+1}^{t-1} \beta_i(\bar{x}_\tau) + \varphi_i^{\bar{x}_t} \right\} + \sum_{x^k \in S(\bar{x}_{t(1)}) \setminus \{\bar{x}_{t(1)+1}\}} p(x^k, \bar{u}) \cdot \varphi_i^{x^k} = \varphi_i^{x_0}, \tag{22}$$

case $\theta = 2$ *(two chance nodes* $y_1 = \bar{x}_{t(1)}$, $y_2 = \bar{x}_{t(2)}$ *before* $\bar{x}_t$*):*

$$\sum_{\tau=0}^{t(1)} \beta_i(\bar{x}_\tau) + p(\bar{x}_{t(1)+1}, \bar{u}) \cdot \left\{ \sum_{\tau=t(1)+1}^{t(2)} \beta_i(\bar{x}_\tau) + p(\bar{x}_{t(2)+1} \mid \bar{x}_{t(2)}, \bar{u}) \right.$$

$$\times \left[\sum_{\tau=t(2)+1}^{t-1} \beta_i(\bar{x}_\tau) + \varphi_i^{\bar{x}_t} \right] + \sum_{x^m \in S(\bar{x}_{t(2)}) \setminus \{\bar{x}_{t(2)+1}\}} p(x^m \mid \bar{x}_{t(2)}, \bar{u}) \cdot \varphi_i^{x^m} \right\} \tag{23}$$

$$+ \sum_{x^k \in S(\bar{x}_{t(1)}) \setminus \{\bar{x}_{t(1)+1}\}} p(x^k, \bar{u}) \cdot \varphi_i^{x^k} = \varphi_i^{x_0},$$

$\ldots$

Note that for partial case when $\bar{x}_t \in S(\bar{x}_{t(1)})$, i.e. if $\bar{x}_t$ follows the chance node $\bar{x}_{t(1)}$ Eq. (22) takes the simpler form

$$\sum_{\tau=0}^{t(1)} \beta_i(\bar{x}_\tau) + \sum_{x^k \in S(\bar{x}_{t(1)})} p(x^k, \bar{u}) \cdot \varphi_i^{x^k} = \varphi_i^{x_0}.$$

Similar note is valid for Eq. (23) etc.

Roughly speaking, Definition 5 implies that the payments collected by the ith player (according to the payment schedule β) before reaching some intermediate node $\overline{x}_t$ plus the expected ith player's component of the Shapley value in the subgame $\Gamma^{\overline{x}_t}$ starting at $\overline{x}_t$ plus this player's expected Shapley value components in other subgames along the cooperative trajectories which do not contain $\overline{x}_t$ corresponds to what the player i is entitled to in the original game Γ^{x_0} (N, V^{x_0}).

Let us use the following example to clarify Definition 5 and then to demonstrate the properties of the incremental IDP (see Definition 6 below).

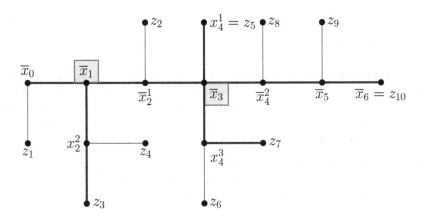

Fig. 1. The game tree.

Example 1. (A 3-player multistage game tree with chance moves).

Let $P_0 = \{\overline{x}_1, \overline{x}_3\}$, $P_1 = \{\overline{x}_0, \overline{x}_4^2\}$, $P_2 = \{\overline{x}_2^1, \overline{x}_5\}$, $P_3 = \{x_2^2, x_4^3\}$, $P_{n+1} = \{z_1, \ldots, z_{10}\}$. Suppose that the optimal bundle $\Omega(\overline{u})$ of cooperative trajectories contains four trajectories (marked in bold in Fig. 1): $\omega_1 = (\overline{x}_0, \overline{x}_1, \overline{x}_2^1, \overline{x}_3, \overline{x}_4^2, \overline{x}_5, \overline{x}_6)$, $\omega_2 = (\overline{x}_0, \overline{x}_1, x_2^2, z_3)$, $\omega_3 = (\overline{x}_0, \overline{x}_1, \overline{x}_2^1, \overline{x}_3, x_4^1)$ and $\omega_4 = (\overline{x}_0, \overline{x}_1, \overline{x}_2^1, \overline{x}_3, x_4^3, z_7)$.

The time consistency conditions for player $i \in N$ at nodes $\overline{x}_1$, $\overline{x}_3$ and $\overline{x}_5$ according to (21), (22) and (23) take the form:

$$\beta_i(\overline{x}_0) + \varphi_i^{\overline{x}_1} = \varphi_i^{x_0},$$

$$\beta_i(\overline{x}_0) + \beta_i(\overline{x}_1) + p(\overline{x}_2^1, \overline{u})\left\{\beta_i(\overline{x}_2^1) + \varphi_i^{\overline{x}_3}\right\} + p(x_2^2, \overline{u}) \cdot \varphi_i^{x_2^2} = \varphi_i^{x_0},$$

$$\beta_i(\overline{x}_0) + \beta_i(\overline{x}_1) + p(\overline{x}_2^1, \overline{u})\Big\{\beta_i(\overline{x}_2^1) + \beta_i(\overline{x}_3) + p(\overline{x}_4^2|\overline{x}_3, \overline{u}) \cdot \big[\beta_i(\overline{x}_4^2) + \varphi_i^{\overline{x}_5}\big]$$

$$+ \ p(x_4^1|\overline{x}_3, \overline{u}) \cdot \varphi_i^{x_4^1} + p(x_4^3|\overline{x}_3, \overline{u}) \cdot \varphi_i^{x_4^3}\Big\} + p(x_2^2, \overline{u}) \cdot \varphi_i^{x_2^2} = \varphi_i^{x_0}.$$

The review of different IDP for multicriteria games (without chance moves) as well as the analysis of their properties can be found in [11,13,15,16]. In this paper we introduce a refinement of so-called incremental IDP (see, e.g.,

[13,15,16,22,25,33]) that can be applied in multicriteria multistage games with chance moves to ensure the sustainability of a cooperative agreement.

Definition 6. *The incremental IDP for the Shapley value φ^{x_0} in multicriteria multistage game with chance moves Γ^{x_0} is defined as follows:*

$$\beta_i(x_t) = \varphi_i^{x_t} - \sum_{x_{t+1}^k \in S(x_t)} p(x_{t+1}^k | x_t, \overline{u}) \cdot \varphi_i^{x_{t+1}^k} \tag{24}$$

for $x_t \in \omega_l(\overline{u}) = (x_0, \ldots, x_t, \ldots, x_{T(l)})$, $\omega_l(\overline{u}) \in \Omega(\overline{u})$, $t = 0, \ldots, T(l) - 1$, and

$$\beta_i(x_{T(l)}) = \varphi_i^{x_{T(l)}} \tag{25}$$

for $x_{T(l)} \in \Omega(\overline{u}) \cap P_{n+1}$.

Remark 3. Formulae (24), (25) are similar to the imputation distribution procedures suggested in [22,23,33] for (single-criterion) stochastic discrete-time dynamic games played over event trees. If $x_t \in P_i$, $i = 1, \ldots, n$ Eq. (24) takes the simpler form $\beta_i(x_t) = \varphi_i^{x_t} - \varphi_i^{x_{t+1}}$, where $\overline{u}_i(x_t) = x_{t+1}$, that coincides with the "classical" incremental IDP.

Proposition 2. *The incremental IDP (24), (25) satisfies strict balance condition (20), the efficiency condition at initial node (19), and the time consistency conditions (21)–(23) in the whole game $\Gamma^{x_0} \in MG^{cm}(n, r)$.*

Proof. To verify that incremental IDP β satisfies strict balance condition (20) at any node $\overline{x}_t \in \omega_m(\overline{u}) = (\overline{x}_1, \ldots, \overline{x}_t, \overline{x}_{t+1}, \ldots, \overline{x}_{T(m)})$, $\omega_m(\overline{u}) \in \Omega(\overline{u})$ first consider the case when $\overline{x}_t \in P_i$, $i = 1, \ldots, n$. Note that $\overline{u}_i(\overline{x}_t) = \overline{x}_{t+1}$, $\omega_m^{\overline{x}_t} = \{\overline{x}_t\} \cup \omega_m^{\overline{x}_{t+1}}$ and $p(\omega_m^{\overline{x}_{t+1}}, \overline{u}^{\overline{x}_{t+1}}) = p(\omega_m^{\overline{x}_t}, \overline{u}^{\overline{x}_t})$ since $p(\overline{x}_{t+1} | \overline{x}_t, \overline{u}) = 1$. Then, taking into account (3) and (16) we get

$$\sum_{i \in N} \beta_i(\overline{x}_t) = \sum_{i \in N} \varphi_i^{\overline{x}_t} - \sum_{i \in N} \varphi_i^{\overline{x}_{t+1}} = V^{\overline{x}_t} - V^{\overline{x}_{t+1}}$$

$$= \sum_{\omega_m^{\overline{x}_t} \in \Omega(\overline{u}^{\overline{x}_t})} p(\omega_m^{\overline{x}_t}, \overline{u}^{\overline{x}_t}) \cdot \left(h_N(\overline{x}_t) + h_N(\overline{x}_{t+1}) + \ldots + h_N(\overline{x}_{T(m)}) \right)$$

$$- \sum_{\omega_m^{\overline{x}_{t+1}} \in \Omega(\overline{u}^{\overline{x}_{t+1}})} p(\omega_m^{\overline{x}_{t+1}}, \overline{u}^{\overline{x}_{t+1}}) \cdot \left(h_N(\overline{x}_{t+1}) + \ldots + h_N(\overline{x}_{T(m)}) \right)$$

$$= h_N(\overline{x}_t) \cdot \sum_{\omega_m^{\overline{x}_t} \in \Omega(\overline{u}^{\overline{x}_t})} p(\omega_m^{\overline{x}_t}, \overline{u}^{\overline{x}_t}) = \sum_{i \in N} h_i(\overline{x}_t).$$

Arguing in a similar way for the case when $\overline{x}_t \in P_0$ one can verify that (20) is satisfied for the chance nodes as well.

The proof that IDP (24), (25) satisfies efficiency (19) and time-consistency conditions (21)–(23) is based on direct calculations but rather cumbersome in general case (i.e., for arbitrary game Γ^{x_0}). Let us demonstrate how it works for the game in Example 1. For instance we verify that the incremental IDP satisfies the time consistency conditions at node $\overline{x}_3$. Note that $\beta_i(\overline{x}_0) = \varphi_i^{\overline{x}_0} - \varphi_i^{\overline{x}_1}$, $\beta_i(\overline{x}_1) = \varphi_i^{\overline{x}_1} - \left(p(\overline{x}_2^1, \overline{u}) \cdot \varphi_i^{\overline{x}_2^1} + p(x_2^2, \overline{u}) \cdot \varphi_i^{x_2^2} \right)$, $\beta_i^{\overline{x}_2^1} = \varphi_i^{\overline{x}_2^1} - \varphi_i^{\overline{x}_3}$.

Then, Eq. (22) takes the form

$$\left(\varphi_i^{\overline{x}_0} - \varphi_i^{\overline{x}_1} \right) + \left\{ \varphi_i^{\overline{x}_1} - \left(p(\overline{x}_2^1, \overline{u}) \cdot \varphi_i^{\overline{x}_2^1} + p(x_2^2, \overline{u}) \cdot \varphi_i^{x_2^2} \right) \right\}$$
$$+ p(\overline{x}_2^1, \overline{u}) \cdot \left\{ \left(\varphi_i^{\overline{x}_2^1} - \varphi_i^{\overline{x}_3} \right) + \varphi_i^{\overline{x}_3} \right\} + p(x_2^2, \overline{u}) \cdot \varphi_i^{x_2^2} = \varphi_i^{\overline{x}_0}.$$

□

According to Proposition 2, the incremental payment schedule (24), (25) can be used to implement a long-term cooperative agreement in a multistage multicriteria game with chance moves.

5 Conclusion

In the paper we adopt a novel ζ-vector-valued characteristic function. It is friendly computable and is proved to satisfy a number of good properties [4,16]. Note that the obtained results do not depend on the particular approach which the players use to construct a vector-valued characteristic function.

We assume that the players use the Shapley value as an allocation mechanism to divide the total cooperative vector payoff at each subgame. However, it is worth noting that Proposition 2 remains valid if the players use another cooperative single-valued solution satisfying the efficiency property (18).

References

1. Climaco, J., Romero, C., Ruiz, F.: Preface to the special issue on multiple criteria decision making: current challenges and future trends. Int. Trans. Oper. Res. **25**, 759–761 (2018). https://doi.org/10.1111/itor.12515
2. Crettez, B., Hayek, N.: A dynamic multi-objective duopoly game with pollution and depollution (2020, submitted to Dynamic Games and Applications)
3. Finus, M.: Game Theory and International Environmental Cooperation. Edward Elgar, Cheltenham (2001)
4. Gromova, E.V., Petrosyan, L.A.: On an approach to constructing a characteristic function in cooperative differential games. Autom. Remote Control **78**(9), 1680–1692 (2017). https://doi.org/10.1134/S0005117917090120
5. Gromova, E.V., Plekhanova, T.M.: On the regularization of a cooperative solution in a multistage game with random time horizon. Discrete Appl. Math. **255**, 40–55 (2019). https://doi.org/10.1016/j.dam.2018.08.008
6. Haurie, A.: A note on nonzero-sum diferential games with bargaining solution. J. Optim. Theory Appl. **18**, 31–39 (1976)
7. Haurie, A., Krawczyk, J.B., Zaccour, G.: Games and Dynamic Games. Scientific World, Singapore (2012)
8. Hayek, N.: Infinite-horizon multiobjective optimal control problems for bounded processes. Discrete Continuous Dyn. Syst. Ser. S **11**(6), 1121–1141 (2018). https://doi.org/10.3934/dcdss.2018064

9. Kuhn, H.: Extensive games and the problem of information. Ann. Math. Stud. **28**, 193–216 (1953)

10. Kuzyutin, D.: On the problem of the stability of solutions in extensive games. Vestnik St. Petersburg Univ. Math. **4**(22), 18–23 (1995). (in Russian)

11. Kuzyutin, D., Gromova, E., Pankratova, Y.: Sustainable cooperation in multicriteria multistage games. Oper. Res. Lett. **46**(6), 557–562 (2018). https://doi.org/10.1016/j.orl.2018.09.004

12. Kuzyutin, D., Nikitina, M., Razgulyaeva, L.: On the A-equilibria properties in multicriteria extensive games. Appl. Math. Sci. **9**(92), 4565–4573 (2015)

13. Kuzyutin, D., Nikitina, M.: Time consistent cooperative solutions for multistage games with vector payoffs. Oper. Res. Lett. **45**(3), 269–274 (2017)

14. Kuzyutin, D., Nikitina, M.: An irrational behavior proof condition for multistage multicriteria games. In: Consrtuctive Nonsmooth Analysis and Related Topics (dedic. to the memory of V.F. Demyanov), CNSA 2017, Proceedings, pp. 178–181. IEEE (2017)

15. Kuzyutin, D., Pankratova, Y., Svetlov, R.: A-subgame concept and the solutions properties for multistage games with vector payoffs. In: Petrosyan, L.A., Mazalov, V.V., Zenkevich, N.A. (eds.) Frontiers of Dynamic Games. SDGTFA, pp. 85–102. Springer, Cham (2019). https://doi.org/10.1007/978-3-030-23699-1_6

16. Kuzyutin, D., Smirnova, N., Gromova, E.: Long-term implementation of the cooperative solution in multistage multicriteria game. Oper. Res. Persp. **6**, 100107 (2019). https://doi.org/10.1016/j.orp.2019.100107

17. Madani, K., Lund, J.R.: A Monte-Carlo game theoretic approach for multi-criteria decision making under uncertainty. Adv. Water Resour. **34**, 607–616 (2011)

18. Mendoza, G.A., Martins, H.: Multi-criteria decision analysis in natural resource management: a critical review of methods and new modelling paradigms. Forest Ecol. Manage. **230**, 1–22 (2006)

19. Moulin, H.: Axioms of Cooperative Decision Making. Cambridge University Press, Cambridge (1988)

20. Myerson, R.: Game Theory. Analysis of Conflict. Harvard University Press, Cambridge (1997)

21. Pankratova, Y., Tarashnina, S., Kuzyutin, D.: Nash equilibria in a group pursuit game. Appl. Math. Sci. **10**(17), 809–821 (2016)

22. Parilina, E., Zaccour, G.: Node-consistent core for games played over event trees. Automatica **55**, 304–311 (2015)

23. Parilina, E., Zaccour, G.: Node-consistent Shapley value for games played over event trees with random terminal time. J. Opt. Theory Appl. **175**(1), 236–254 (2017)

24. Petrosyan, L.: Stable solutions of differential games with many participants. Vestn. Leningrad Univ. **19**, 46–52 (1977). (in Russian)

25. Petrosyan, L., Danilov, N.: Stability of the solutions in nonantagonistic differential games with transferable payoffs. Vestn. Leningrad Univ. **1**, 52–59 (1979). (in Russian)

26. Petrosyan, L.A., Kuzyutin, D.V.: On the stability of E-equilibrium in the class of mixed strategies. Vestnik St. Petersburg Univ. Math. **3**(15), 54–58 (1995). (in Russian)

27. Petrosyan, L., Kuzyutin, D.: Games in Extensive Form: Optimality and Stability. Saint Petersburg University Press, Saint Petersburg (2000). (in Russian)

28. Petrosyan, L., Zaccour, G.: Time-consistent Shapley value allocation of pollution cost reduction. J. Econ. Dyn. Control **27**(3), 381–398 (2003)

29. Pieri, G., Pusillo, L.: Interval values for multicriteria cooperative games. AUCO Czech Econ. Rev. **4**, 144–155 (2010)
30. Pieri, G., Pusillo, L.: Multicriteria partial cooperative games. Appl. Math. **6**, 2125–2131 (2015)
31. Podinovskii, V., Nogin, V.: Pareto-Optimal Solutions of Multicriteria Problems. Nauka, Moscow (1982). (in Russian)
32. Puerto, J., Perea, F.: On minimax and Pareto optimal security payoffs in multicriteria games. J. Math. Anal. Appl. **457**(2), 1634–1648 (2018). https://doi.org/10.1016/j.jmaa.2017.01.002
33. Reddy, P., Shevkoplyas, E., Zaccour, G.: Time-consistent Shapley value for games played over event trees. Automatica **49**(6), 1521–1527 (2013)
34. Rettieva, A.N.: Cooperation in dynamic multicriteria games with random horizons. J. Glob. Optim. **76**(3), 455–470 (2018). https://doi.org/10.1007/s10898-018-0658-6
35. Shapley, L.: A value for n-person games. In: Kuhn, H., Tucker, A.W. (eds.) Contributions to the Theory of Games, II, pp. 307–317. Princeton University Press, Princeton (1953)
36. Shapley, L.: Equilibrium points in games with vector payoffs. Naval Res. Logistics Q. **6**, 57–61 (1959)
37. Voorneveld, M., Vermeulen, D., Borm, P.: Axiomatizations of Pareto equilibria in multicriteria games. Games Econ. Behav. **28**, 146–154 (1999)

On a One-Dimensional Differential Game with a Non-convex Terminal Payoff

Igor' V. Izmest'ev[1,2]($\boxtimes$) (ID) and Viktor I. Ukhobotov[1,2] (ID)

[1] N.N. Krasovskii Institute of Mathematics and Mechanics,
S. Kovalevskaya Street, 16, 620108 Yekaterinburg, Russia
[2] Chelyabinsk State University,
Br. Kashirinykh Street, 129, 454001 Chelyabinsk, Russia
j748e8@gmail.com, ukh@csu.ru

Abstract. A one-dimensional differential game is considered, in which the payoff is determined by the modulus of the deviation of the phase variable at a fixed time from the set value, taking into account the periodicity. The first player seeks to minimize the payoff. The goal of the second player is the opposite. For this problem, the price of the game is calculated and optimal player controls are constructed. As an example, we consider the problem of controlling a rotational mechanical system in which the goal of the first player acquires the meaning of minimizing the modulus of deviation of the angle from the desired state.

Keywords: Control · Differential game · Payoff

1 Introduction

The linear differential game with a given duration, using a linear change of variables [6], can be reduced to the form, in which the dynamics of the new system is determined by the sum of player controls with values that belong to time-dependent sets. In the case when in a linear differential game the quality criterion is defined as the value of the modulus of a linear function at a fixed time moment, using a linear change of variables, we obtain a single-type differential game, in which the sets of player controls are segments that depend on time (see as example [4,5]). In general case, in these problems, the vectograms of player controls are balls with time-dependent radii. Differential games that have this type of dynamics after change of variables, are considered, for example, in [3] and [9]. For differential games of this type, if the target set is a ball of a fixed radius, the alternating integral is constructed in [9]. In [10], optimal controls of players are found. In addition, in [10], a variant of this problem was considered with a terminal payoff determined by the norm of the phase vector at a fixed time moment. For this variant of the problem, the price of the game was found, and the corresponding optimal player controls were constructed.

This work was funded by the Russian Science Foundation (project no. 19-11-00105).

A. Kononov et al. (Eds.): MOTOR 2020, LNCS 12095, pp. 200–211, 2020.
https://doi.org/10.1007/978-3-030-49988-4_14

In [5,11], we consider single-type differential games with non-convex terminal sets. In [11], the terminal set is determined by the condition of belonging the norm of a phase vector to a segment with positive ends. In this paper, a set defined by this condition is called a ring. In [5], the terminal set is the union of an infinite number of disjoint segments of equal length. This terminal set has the meaning of ε-neighbourhood of the target position of the system, taking into account the periodicity.

In present paper, continuing the research begun in [5], we consider a linear differential game with a non-convex payoff, which is determined using the modulus of a linear function of the phase vector. Using a linear change of variables, this linear differential game reduced to a single-type one-dimensional differential game. After change of variables, the game payoff takes the form of the modulus of the deviation of the new one-dimensional variable at a fixed time moment from the given value, taking into account the periodicity. The goal of the first player is to minimize this payoff. The goal of the second player is the opposite. For this problem, the price of the game is calculated, and the corresponding optimal player controls are constructed.

The obtained results can find application in solving problems of controlling rotational mechanical systems (see as example [1,2,4,5,12]) with uncontrolled disturbance, in which the control goal in the original problem acquires the meaning of minimizing the modulus of deviation of the angle from the desired state.

In this paper, as an example of illustrating the theory, we consider a modification of the problem of the turning of a ship [8, p. 103–104] on which uncontrolled external forces act.

2   Problem Statement

Consider antagonistic differential game

$$\dot{x} = A(t)x - \xi + \eta, \quad x(t_0) = x_0; \quad x \in \mathbb{R}^n, \quad t \leq p. \tag{1}$$

Here, control of the first player is $\xi \in W \subset \mathbb{R}^n$, control of the second player is $\eta \in F \subset \mathbb{R}^n$, where W and F are connected compacts; $A(t)$ is a matrix with corresponding dimension whose elements are continuous for $t_0 \leq t \leq p$ functions.

Define payoff as follows

$$\min_{i \in I} |\langle \psi_0, x(p) \rangle - \alpha_* - i\alpha| \to \min_{\xi} \max_{\eta}. \tag{2}$$

Here, $\langle \cdot, \cdot \rangle$ denotes the scalar product in $\mathbb{R}^n$, $I = 0, \pm 1, \pm 2, \pm 3, \ldots$. Vector $\psi_0 \in \mathbb{R}^n$ and numbers $\alpha_*, \alpha \in \mathbb{R}$ are given.

Reduce this problem to one-dimensional single-type differential game.

Denote by $\psi(t)$ the solution of the Cauchy problem

$$\dot{\psi}(t) = -A^*(t)\psi(t), \quad \psi(p) = \psi_0; \quad t \leq p. \tag{3}$$

Here, $A^*(t)$ denotes the transposed of matrix $A(t)$.

Denote

$$a_-(t) = \min_{\xi}\langle\psi(t),\xi\rangle, \quad a_+(t) = \max_{\xi}\langle\psi(t),\xi\rangle, \quad \xi \in W;$$

$$b_-(t) = \min_{\eta}\langle\psi(t),\eta\rangle, \quad b_+(t) = \max_{\eta}\langle\psi(t),\eta\rangle, \quad \eta \in F.$$

Then connectivity of compacts W and F imply [7, p. 333–334, Theorem 4] that

$$\langle\psi(t),\xi\rangle = \frac{a_+(t) + a_-(t)}{2} + a(t)u, \quad |u| \le 1, \quad a(t) = \frac{a_+(t) - a_-(t)}{2} \ge 0;$$

$$\langle\psi(t),\eta\rangle = \frac{b_+(t) + b_-(t)}{2} + b(t)v, \quad |v| \le 1, \quad b(t) = \frac{b_+(t) - b_-(t)}{2} \ge 0.$$

Introduce a new one-dimensional variable

$$z = \langle\psi(t),x\rangle + \frac{1}{2}\int_t^p (b_+(r) + b_-(r) - a_+(r) - a_-(r))dr - \alpha_*.$$

Differentiate z:

$$\dot{z} = \langle -A^*(t)\psi(t),x\rangle + \langle\psi(t),A(t)x - \xi + \eta\rangle + \frac{1}{2}(a_+(t) + a_-(t) - b_+(t) - b_-(t)).$$

Given equality $\langle\psi(t),A(t)x\rangle = \langle A^*(t)\psi(t),x\rangle$, the problem (1), (2) can be written as follows

$$\dot{z} = -a(t)u + b(t)v, \quad |u| \le 1, \quad |v| \le 1, \quad \min_{i\in I}|z(p) - i\alpha| \to \min_u \max_v. \quad (4)$$

Further, for the completeness of the exposition we assume that the functions $a(t) \ge 0$ and $b(t) \ge 0$ are summable on each segment of the semiaxis $(-\infty, p]$.

Admissible controls of players are arbitrary function, which satisfy inequalities

$$|u(t,z)| \le 1, \quad |v(t,z)| \le 1, \quad t \le p, \quad z \in \mathbb{R}. \quad (5)$$

Fix the initial state $t_0 < p$, $z(t_0) \in \mathbb{R}$ and the time moment $t_0 < t_* \le p$. Take partition

$$\omega : t_0 < t_1 < \ldots < t_i < t_{i+1} < \ldots < t_k < t_{k+1} = t_*$$

with diameter $d(\omega) = \max(t_{i+1} - t_i)$, $i = \overline{0,k}$. Construct polygonal line for Eq. (4)

$$z_\omega(t) = z_\omega(t_i) - \left(\int_{t_i}^t a(r)dr\right)u(t_i, z_\omega(t_i)) + \left(\int_{t_i}^t b(r)dr\right)v(t_i, z_\omega(t_i)) \quad (6)$$

for $t_i < t \le t_{i+1}$. Here, $z_\omega(t_0) = z(t_0)$.

The motion of the system $z(t)$ realized with admissible controls (5) from the initial state $z(t_0)$ is defined as any uniform limit of the sequence of the polygonal lines (6), for which diameters of partition tend to zero.

3   Main Result

Denote

$$f(t) = \int_t^p (a(r) - b(r))dr$$

for $t \leq p$.

Define function $G(t, z)$ for $t \leq p$ and $z \in \mathbb{R}$ as follows:

$$G(t, z) = \max \left\{ \min_{i \in I} |z - i\alpha| - f(t); - \min_{t \leq \tau \leq p} f(\tau) \right\}, \tag{7}$$

if

$$\max \left\{ \min_{i \in I} |z - i\alpha| - f(t); - \min_{t \leq \tau \leq p} f(\tau) \right\} < \frac{\alpha}{2} - \max_{t \leq \tau \leq p} f(\tau); \tag{8}$$

$$G(t, z) = \frac{\alpha}{2} - \max_{t \leq \tau \leq p} f(\tau), \tag{9}$$

if

$$- \min_{t \leq \tau \leq p} f(\tau) < \frac{\alpha}{2} - \max_{t \leq \tau \leq p} f(\tau) \text{ and } \frac{\alpha}{2} - \max_{t \leq \tau \leq p} f(\tau) \leq \min_{i \in I} |z - i\alpha| - f(t); \tag{10}$$

$$G(t, z) = \frac{\alpha}{2} - \max_{t_* \leq \tau \leq p} f(\tau) = - \min_{t_* \leq \tau \leq p} f(\tau), \tag{11}$$

where

$$t_* = \min \left\{ r \in [t, p] : - \min_{s \leq \tau \leq p} f(\tau) \leq \frac{\alpha}{2} - \max_{s \leq \tau \leq p} f(\tau) \text{ for all } s \in [r, p] \right\}, \tag{12}$$

if

$$\frac{\alpha}{2} - \max_{t \leq \tau \leq p} f(\tau) \leq - \min_{t \leq \tau \leq p} f(\tau). \tag{13}$$

Lemma 1. *For all $z \in \mathbb{R}$ and $\alpha > 0$ equality*

$$\min_{i \in I} |z - i\alpha| + \min_{i \in I} |z - (i + 0.5)\alpha| = \frac{\alpha}{2} \tag{14}$$

holds.

Corollary 1. *For all $z \in \mathbb{R}$ and $\alpha > 0$ inequality*

$$\min_{i \in I} |z - i\alpha| \leq \frac{\alpha}{2}$$

holds.

Theorem 1. *There exists control of the first player that guarantees the inequality*

$$\min_{i \in I} |z(p) - i\alpha| \leq G(t_0, z(t_0)). \tag{15}$$

Proof. Case 1. Let $G(t_0, z(t_0))$ be defined by formula (7).

Fix $i_* \in I$ such that

$$\min_{i \in I} |z(t_0) - i\alpha| = |z(t_0) - i_*\alpha|. \tag{16}$$

Let's change of the variable

$$z_* = z - i_*\alpha$$

and consider differential game

$$\dot{z}_* = -a(t)u + b(t)v, \quad |u| \leq 1, \quad |v| \leq 1, \quad |z_*(p)| \to \min_u \max_v.$$

In [10], for this game, it was proved that control of the first player

$$u(t, z_*) = \operatorname{sign} z_*$$

guarantees the inequality

$$|z_*(p)| \leq \max \left\{ |z_*(t_0)| - f(t_0); - \min_{t_0 \leq \tau \leq p} f(\tau) \right\}.$$

Making the inverse change of variables, we obtain that control

$$u(t, z) = \operatorname{sign} (z - i_*\alpha)$$

guarantees the inequality

$$|z(p) - i_*\alpha| \leq \max \left\{ |z(t_0) - i_*\alpha| - f(t_0); - \min_{t_0 \leq \tau \leq p} f(\tau) \right\}.$$

Further, using (16) and inequality

$$\min_{i \in I} |z(p) - i\alpha| \leq |z(p) - i_*\alpha|,$$

we obtain inequality (15):

$$\min_{i \in I} |z(p) - i\alpha| \leq \max \left\{ \min_{i \in I} |z(t_0) - i\alpha| - f(t_0); - \min_{t_0 \leq \tau \leq p} f(\tau) \right\} = G(t_0, z(t_0)).$$

Case 2. Let $G(t_0, z(t_0))$ be defined by formula (9).

Take $t^* \in [t_0, p]$ such that

$$\max_{t_0 \leq \tau \leq p} f(\tau) = f(t^*). \tag{17}$$

Further, acting similarly to case 1, we obtain that for any realized $z(t^*)$ there is a control of the first player on the segment $[t^*, p]$, which guarantees the inequality

$$\min_{i \in I} |z(p) - i\alpha| \leq \max \left\{ \min_{i \in I} |z(t^*) - i\alpha| - f(t^*); - \min_{t^* \leq \tau \leq p} f(\tau) \right\}.$$

Let

$$\min_{i \in I} |z(t^*) - i\alpha| - f(t^*) \geq - \min_{t^* \leq \tau \leq p} f(\tau).$$

Using Corollary 1 and (17), we obtain

$$\min_{i \in I} |z(t^*) - i\alpha| - f(t^*) \leq \frac{\alpha}{2} - \max_{t_0 \leq \tau \leq p} f(\tau) = G(t_0, z(t_0)).$$

Let

$$\min_{i \in I} |z(t^*) - i\alpha| - f(t^*) < - \min_{t^* \leq \tau \leq p} f(\tau).$$

Using first inequality from (10), we obtain

$$- \min_{t^* \leq \tau \leq p} f(\tau) \leq - \min_{t_0 \leq \tau \leq p} f(\tau) < \frac{\alpha}{2} - \max_{t_0 \leq \tau \leq p} f(\tau) = G(t_0, z(t_0)).$$

Case 3. Let $G(t_0, z(t_0))$ be defined by formula (11), in which t_* is defined by formula (12).

Take $t^* \in [t_*, p]$ such that

$$\max_{t_* \leq \tau \leq p} f(\tau) = f(t^*). \tag{18}$$

Further, acting similarly to case 1, we obtain that for any realized $z(t^*)$ there is a control of the first player on the segment $[t^*, p]$, which guarantees the inequality

$$\min_{i \in I} |z(p) - i\alpha| \leq \max \left\{ \min_{i \in I} |z(t^*) - i\alpha| - f(t^*); - \min_{t^* \leq \tau \leq p} f(\tau) \right\}.$$

Let

$$\min_{i \in I} |z(t^*) - i\alpha| - f(t^*) \geq - \min_{t^* \leq \tau \leq p} f(\tau).$$

Using Corollary 1 and (18), we obtain

$$\min_{i \in I} |z(t^*) - i\alpha| - f(t^*) \leq \frac{\alpha}{2} - \max_{t_* \leq \tau \leq p} f(\tau) = G(t_0, z(t_0)).$$

Let

$$\min_{i \in I} |z(t^*) - i\alpha| - f(t^*) < - \min_{t^* \leq \tau \leq p} f(\tau).$$

Then

$$- \min_{t^* \leq \tau \leq p} f(\tau) \leq - \min_{t_* \leq \tau \leq p} f(\tau) = G(t_0, z(t_0)).$$

The theorem is proved.

Remark 1. The control constructed in the proof of Theorem 1 is positional control with memory, since for $\hat{t} > t$ its value depends on the value $z(\hat{t})$ of the phase variable realized at time moment $\hat{t}$ ($\hat{t} = t_0$ in the Case 1, $\hat{t} = t_*$ in the Case 2, $\hat{t} = t^*$ in the Case 3).

Theorem 2. *There exists control of the second player that guarantees the inequality*

$$\min_{i \in I} |z(p) - i\alpha| \geq G(t_0, z(t_0)). \tag{19}$$

Proof. Consider auxiliary differential game

$$\dot{z} = -a(t)u + b(t)v, \quad |u| \leq 1, \quad |v| \leq 1, \quad \min_{i \in I} |z(p) - (i + 0.5)\alpha| \to \min_{v} \max_{u}.$$

Acting similarly to proof Theorem 1, we obtain that there is control of the second player that guarantees fulfilment following inequalities:

$$\min_{i \in I} |z(p) - (i + 0.5)\alpha| \leq \max \left\{ \min_{i \in I} |z(t_0) - (i + 0.5)\alpha| + f(t_0); - \min_{t_0 \leq \tau \leq p} (-f(\tau)) \right\}, \tag{20}$$

if

$$\max \left\{ \min_{i \in I} |z(t_0) - (i + 0.5)\alpha| + f(t_0); - \min_{t_0 \leq \tau \leq p} (-f(\tau)) \right\} < \frac{\alpha}{2} - \max_{t_0 \leq \tau \leq p} (-f(\tau)); \tag{21}$$

$$\min_{i \in I} |z(p) - (i + 0.5)\alpha| \leq \frac{\alpha}{2} - \max_{t_0 \leq \tau \leq p} (-f(\tau)), \tag{22}$$

if

$$- \min_{t_0 \leq \tau \leq p} (-f(\tau)) < \frac{\alpha}{2} - \max_{t_0 \leq \tau \leq p} (-f(\tau)) \tag{23}$$

and

$$\frac{\alpha}{2} - \max_{t_0 \leq \tau \leq p} (-f(\tau)) \leq \min_{i \in I} |z(t_0) - (i + 0.5)\alpha| + f(t_0); \tag{24}$$

$$\min_{i \in I} |z(p) - (i + 0.5)\alpha| \leq \frac{\alpha}{2} - \max_{t_* \leq \tau \leq p} (-f(\tau)), \tag{25}$$

where

$$t_* = \min \left\{ r \in [t_0, p] : - \min_{s \leq \tau \leq p} (-f(\tau)) \leq \frac{\alpha}{2} - \max_{s \leq \tau \leq p} (-f(\tau)) \text{ for all } s \in [r, p] \right\}, \tag{26}$$

if

$$\frac{\alpha}{2} - \max_{t_0 \leq \tau \leq p} (-f(\tau)) \leq - \min_{t_0 \leq \tau \leq p} (-f(\tau)). \tag{27}$$

Note also that

$$\min_{t \leq \tau \leq p} (-f(\tau)) = - \max_{t \leq \tau \leq p} f(\tau) \text{ and } \max_{t \leq \tau \leq p} (-f(\tau)) = - \min_{t \leq \tau \leq p} f(\tau). \tag{28}$$

Case 1. Let inequality (21) holds.
Case 1.1. Let

$$\min_{i \in I} |z(t_0) - (i + 0.5)\alpha| + f(t_0) \leq - \min_{t_0 \leq \tau \leq p} (-f(\tau)).$$

Using Lemma 1 and (28), rewrite this inequality, (20) and (21), respectively, as follows:

$$\frac{\alpha}{2} - \max_{t_0 \leq \tau \leq p} f(\tau) \leq \min_{i \in I} |z(t_0) - i\alpha| - f(t_0),$$

$$\frac{\alpha}{2} - \max_{t_0 \leq \tau \leq p} f(\tau) \leq \min_{i \in I} |z(p) - i\alpha|,$$

$$- \min_{t_0 \leq \tau \leq p} f(\tau) < \frac{\alpha}{2} - \max_{t_0 \leq \tau \leq p} f(\tau).$$

From here we obtain (19), where $G(t_0, z(t_0))$ is defined by formulas (9) and (10).

Case 1.2. Let

$$\min_{i \in I} |z(t_0) - (i + 0.5)\alpha| + f(t_0) > - \min_{t_0 \leq \tau \leq p} (-f(\tau)).$$

Using Lemma 1 and (28), rewrite this inequality, (20) and (21), respectively, as follows:

$$\frac{\alpha}{2} - \max_{t_0 \leq \tau \leq p} f(\tau) > \min_{i \in I} |z(t_0) - i\alpha| - f(t_0),$$

$$\min_{i \in I} |z(t_0) - i\alpha| - f(t_0) \leq \min_{i \in I} |z(p) - i\alpha|,$$

$$- \min_{t_0 \leq \tau \leq p} f(\tau) < \min_{i \in I} |z(t_0) - i\alpha| - f(t_0).$$

From here we obtain (19), where $G(t_0, z(t_0))$ is defined by formulas (7) and (8).

Case 2. Let inequalities (23) and (24) hold.

Using Lemma 1 and (28), rewrite (22), (23) and (24), respectively, as follows:

$$- \min_{t_0 \leq \tau \leq p} f(\tau) \leq \min_{i \in I} |z(p) - i\alpha|,$$

$$\frac{\alpha}{2} - \max_{t_0 \leq \tau \leq p} f(\tau) > - \min_{t_0 \leq \tau \leq p} f(\tau),$$

$$- \min_{t_0 \leq \tau \leq p} f(\tau) \geq \min_{i \in I} |z(t_0) - i\alpha| - f(t_0).$$

From here we obtain (19), where $G(t_0, z(t_0))$ is defined by formulas (7) and (8).

Case 3. Let inequality (27) holds. Definition of t_* (26) implies equality

$$- \min_{t_* \leq \tau \leq p} (-f(\tau)) = \frac{\alpha}{2} - \max_{t_* \leq \tau \leq p} (-f(\tau)).$$

Further, using this, Lemma 1 and (28), we obtain the following. Inequality (25) takes the form

$$\min_{i \in I} |z(p) - i\alpha| \geq \frac{\alpha}{2} - \max_{t_* \leq \tau \leq p} f(\tau).$$

Formulas (26) and (27) take the form (12) and (13) for $t = t_0$, respectively. Thus, we obtain (19).

The theorem is proved.

Corollary 2. *Theorem 1 and Theorem 2 imply that $G(t_0, z(t_0))$ is the price of game in problem (4).*

4   Example

Let us consider a modification of the problem of the turning of a ship [8, p. 103–104] on which uncontrolled external forces act.

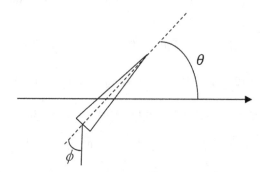

Fig. 1. The problem of the turning of a ship.

Write down the equation of rotation of the ship around a vertical axis, passing through its center of mass perpendicular to the plane of the figure (see Fig. 1). We neglect lateral drift of the ship during its turns. We assume that the value of the ship's speed is constant. Then

$$J\ddot{\theta} = -\beta\dot{\theta} + M(\phi) + N.$$

Here, θ is the deviation angle from the horizontal axis; J is the moment of inertia of the ship relative to its vertical axis; value $\beta\dot{\theta}$ corresponds to the moment of friction viscosity forces, and $\beta > 0$ is the coefficient of viscous friction; $M(\phi)$ is the moment of forces created by the steering wheel; ϕ is the steering angle; N is the moment of uncontrolled external forces. $M(\phi)$ and N are bounded:

$$|M(\phi)| \leq M_*, \quad M_* > 0; \quad |N| \leq N_*, \quad N_* > 0.$$

The goal of control is to minimize value

$$\min_{i \in I} |\theta(p) - \theta_* - 2\pi i|.$$

Here, θ_* is an angle corresponding to the desired direction; p is the end time moment.

Denote

$$x_1 = \theta, \quad x_2 = \dot{\theta}, \quad k = \frac{\beta}{J}, \quad \gamma = \frac{M_*}{J}, \quad \xi = \frac{M(\phi)}{M_*}, \quad \delta = \frac{N_*}{J}, \quad \eta = \frac{N}{N_*}.$$

Write down the considered problem as problem (1)

$$\begin{pmatrix} \dot{x}_1 \\ \dot{x}_2 \end{pmatrix} = \begin{pmatrix} 0 & 1 \\ 0 & -k \end{pmatrix} \begin{pmatrix} x_1 \\ x_2 \end{pmatrix} + \begin{pmatrix} 0 \\ \gamma \end{pmatrix} \xi + \begin{pmatrix} 0 \\ \delta \end{pmatrix} \eta, \quad |\xi| \leq 1, \quad |\eta| \leq 1$$

with quality criterion (2)

$$\min_{i \in I} |x_1(p) - \theta_* - 2\pi i| \to \min_{\xi} \max_{\eta},$$

where

$$\alpha = 2\pi, \quad \psi_0 = \begin{pmatrix} 1 \\ 0 \end{pmatrix}.$$

Further, reduce this problem to one-dimensional single-type problem. We write down Cauchy problem (3)

$$\begin{pmatrix} \dot{\psi}_1 \\ \dot{\psi}_2 \end{pmatrix} = \begin{pmatrix} 0 & 0 \\ -1 & k \end{pmatrix} \begin{pmatrix} \psi_1 \\ \psi_2 \end{pmatrix}, \quad \psi_1(p) = 1, \quad \psi_2(p) = 0$$

and find its solution

$$\psi_1(t) = 1, \quad \psi_2(t) = \frac{1}{k}\left(1 - e^{(t-p)k}\right).$$

Introduce a new variable

$$z = x_1 + \frac{1}{k}\left(1 - e^{(t-p)k}\right)x_2 - \theta_*.$$

Then

$$\dot{z} = -\frac{\gamma}{k}\left(1 - e^{(t-p)k}\right)u + \frac{\delta}{k}\left(1 - e^{(t-p)k}\right)v, \quad u = -\xi, \quad v = \eta.$$

Since $z(p) = x_1(p) - \theta_*$, the quality criterion takes the form

$$\min_{i \in I} |z(p) - 2\pi i| \to \min_{u} \max_{v}.$$

Consider this problem as antagonistic differential game (4), in which u is control of the first player, and v is control of the second player,

$$a(t) = \frac{\gamma}{k}\left(1 - e^{(t-p)k}\right), \quad b(t) = \frac{\delta}{k}\left(1 - e^{(t-p)k}\right).$$

Compute function $f(t)$ in this example:

$$f(t) = \int_t^p (a(r) - b(r))dr = \frac{\gamma - \delta}{k}\left(p - t - \frac{1 - e^{(t-p)k}}{k}\right).$$

Case 1. Let $\gamma = \delta$. Then $f(t) = 0$ for all $t \leq p$. Therefore

$$\min_{t \leq \tau \leq p} f(\tau) = 0, \quad \max_{t \leq \tau \leq p} f(\tau) = 0.$$

Using these equalities and formulas (7)–(13), we obtain in this case that

$$G(t_0, z(t_0)) = \min_{i \in I} |z(t_0) - 2\pi i|.$$

The optimal control ξ can be written as follows:

$$\xi(t, z) = -\text{sign}(z - 2\pi i),$$

where $i \in I$ gives a minimum to the expression $|z(t_0) - 2\pi i|$.

Case 2. Let $\gamma > \delta$. Then $\dot{f}(t) \leq 0$ for all $t \leq p$. Taking into account that $f(p) = 0$, we obtain follows

$$\min_{t \leq \tau \leq p} f(\tau) = 0, \quad \max_{t \leq \tau \leq p} f(\tau) = f(t).$$

Using formulas (7)–(13), we obtain in this case the following:

$$G(t_0, z(t_0)) = \max\left\{\min_{i \in I}|z(t_0) - 2\pi i| - f(t_0); 0\right\}.$$

If $f(t_0) < 0.5\alpha$, then optimal control ξ is defined as in Case 1.
If $f(t_0) \geq 0.5\alpha$, then $\xi(t, z)$ —any with constraint $|\xi(t, z)| \leq 1$ for $t < t_*$ and

$$\xi(t, z) = -\text{sign}(z - i\alpha) \quad \text{for} \quad t_* \leq t,$$

where $i \in I$ gives a minimum to the expression $|z(t_*) - 2\pi i|$. Here, $t_* \in [t_0, p]$ is solution of equation $0.5\alpha = f(t)$.

Case 3. Let $\gamma < \delta$. Then $\dot{f}(t) \geq 0$ for all $t \leq p$. Taking into account that $f(p) = 0$, we obtain follows

$$\min_{t \leq \tau \leq p} f(\tau) = f(t), \quad \max_{t \leq \tau \leq p} f(\tau) = 0.$$

Using formulas (7)–(13), we obtain in this case the following:

$$G(t_0, z(t_0)) = \min\left\{\min_{i \in I}|z(t_0) - 2\pi i| - f(t_0); \pi\right\}.$$

If $0.5\alpha > -f(t_0)$, then optimal control ξ is defined as in Case 1.
If $0.5\alpha \leq -f(t_0)$, then, with any control, the first player cannot guarantee a deviation from the course less than π.

5   Conclusion

In this paper, we consider antagonistic differential game over the state-linear control system and with a non-convex terminal payoff, which is determined using the modulus of a linear function of the phase vector. Using a linear change of variables, this problem is reduced to a single-type one-dimensional differential game.

The main feature of the game is the structure of the payoff. First, it is non-convex, which significantly complicates the game. Second, it allows to taking into account periodicity properties inherent in the problem.

Based on our previous results, we found the price of the game and constructed the corresponding controls of the players.

As an example, we consider a modification of the problem of the turning of a ship on which uncontrolled external forces act. The theoretical results obtained can also find application in solving other problems of controlling rotational mechanical systems.

In the future, this problem can be considered in the case, when dynamics of the first player changes at an unknown time moment (for example, a breakdown occurs).

References

1. Andrievsky, B.R.: Global stabilization of the unstable reaction-wheel pendulum. Control Big Syst. **24**, 258–280 (2009). (in Russian)
2. Beznos, A.V., Grishin, A.A., Lenskiy, A.V., Okhozimskiy, D.E., Formalskiy, A.M.: The control of pendulum using flywheel. In: Workshop on Theoretical and Applied Mechanics, pp. 170–195. Publishing of Moscow State University, Moscow (2009). (in Russian)
3. Isaacs, R.: Differential Games: A Mathematical Theory with Applications to Warfare and Pursuit, Control and Optimization. Wiley, New York (1965)
4. Izmest'ev, I.V., Ukhobotov, V.I.: On a linear control problem under interference with a payoff depending on the modulus of a linear function and an integral. In: 2018 IX International Conference on Optimization and Applications (OPTIMA 2018) (Supplementary Volume), DEStech, pp. 163–173. DEStech Publications, Lancaster (2019). https://doi.org/10.12783/dtcse/optim2018/27930
5. Izmest'ev, I.V., Ukhobotov, V.I.: On a single-type differential game with a non-convex terminal set. In: Khachay, M., Kochetov, Y., Pardalos, P. (eds.) MOTOR 2019. LNCS, vol. 11548, pp. 595–606. Springer, Cham (2019). https://doi.org/10.1007/978-3-030-22629-9_42
6. Krasovskii, N.N., Subbotin, A.I.: Positional Differential Games. Nauka Publishing, Moscow (1974). (in Russian)
7. Kudryavtsev, L.D.: A Course of Mathematical Analysis, vol. 1. Vysshaya Shkola Publishers, Moscow (1981). (in Russian)
8. Leonov, G.A.: Introduction to Control Theory. Publishing St. Petersburg University, St. Petersburg (2004). (in Russian)
9. Pontryagin, L.S.: Linear differential games of pursuit. Math. USSR-Sbornik **40**(3), 285–303 (1981). https://doi.org/10.1070/SM1981v040n03ABEH001815
10. Ukhobotov, V.I.: Synthesis of control in single-type differential games with fixed time. Bull. Chelyabinsk Univ. **1**, 178–184 (1996). (in Russian)
11. Ukhobotov, V.I., Izmest'ev, I.V.: Single-type differential games with a terminal set in the form of a ring. In: Systems dynamics and control process. In: Proceedings of the International Conference, Dedicated to the 90th Anniversary of Acad. N.N. Krasovskiy, Ekaterinburg, 15–20 September 2014, pp. 325–332. Publishing House of the UMC UPI, Ekaterinburg (2015). (in Russian)
12. Ushakov, V.N., Ukhobotov, V.I., Ushakov, A.V., Parshikov, G.V.: On solution of control problems for nonlinear systems on finite time interval. IFAC-PapersOnLine **49**(18), 380–385 (2016). https://doi.org/10.1070/10.1016/j.ifacol.2016.10.195

Open-Loop Based Strategies
for Autonomous Linear Quadratic Game
Models with Continuous Updating

Ildus Kuchkarov[1] and Ovanes Petrosian[1,2(✉)]

[1] St. Petersburg State University, Saint-Petersburg 199034, Russia
kuchkarov_ildus@mail.ru, petrosian.ovanes@yandex.ru
[2] School of Mathematics and Statistics, Qingdao University, 308 Ningxia Road,
Qingdao 266071, People's Republic of China

Abstract. The class of differential games with continuous updating is quite new, there it is assumed that at each time instant, players use information about the game structure (motion equations and payoff functions of players) defined on a closed time interval with a fixed duration. As time goes on, information about the game structure updates. A linear-quadratic case for this class of games is particularly important for practical problems arising in the engineering of human-machine interaction. In this paper, it is particularly interesting that the open-loop strategies are used to construct the optimal ones, but subsequently, we obtain strategies in the feedback form. Using these strategies the notions of Shapley value and Nash equilibrium as optimality principles for cooperative and non-cooperative cases respectively are defined and the optimal strategies for the linear-quadratic case are presented.

Keywords: Differential games with continuous updating · Nash equilibrium · Linear quadratic differential games

1 Introduction

Most conflict-driven processes in real-life evolve continuously in time, and their participants continuously receive updated information and adapt accordingly. The principal models considered in classical differential game theory are associated with problems defined for a fixed time interval (players have all the information within a closed time interval) [7], problems defined for an infinite time interval with discounting (players have all information specified over an infinite time interval) [1], problems defined for a random time interval (players have information over a given time interval, but the duration of this interval is a random variable) [19]. One of the first works in the theory of differential games was devoted to a differential pursuit game (the player's payoff depends on the time

Research was supported by a grant from the Russian Science Foundation (Project No 18-71-00081).

of capture of the opponent) [17]. Another interesting application of dynamic and differential games is the network, [5]. In all the above models and approaches it is assumed that at the onset players process all information about the game dynamics (equations of motion) and players' preferences (cost functions). However, these approaches do not take into account the fact that many real-life conflict-controlled processes are characterized by the fact that players at the initial time instant do not have all the information about the game. Therefore such classical approaches for defining optimal strategies as the Nash equilibrium, the Hamilton-Jacobi-Bellman equation [2], or the Pontryagin maximum principle [18], for example, cannot be directly used to construct a large range of real game-theoretic models.

In this paper, we extend the results of the paper [8], where the class of non-cooperative linear-quadratic autonomous (model's parameters do not depend on time) differential games with continuous updating is considered and the explicit form of the Nash equilibrium derived. It is interesting to construct not only the non-cooperative solution but also to obtain the form of cooperative strategies, characteristic function and cooperative solution for continuous updating case, i.e. to study cooperative differential game model. An important and interesting result that arises in the setting of open-loop strategies for continuous updating is that the open-loop based Nash equilibrium and cooperative open-loop based strategies have a feedback form. This fact is counter-intuitive by nature, but it takes place for a continuous updating approach. The popularity of so-called linear-quadratic differential games [4] on one hand can be explained by practical applications in engineering. To some extent, these kinds of differential games are analytically and numerically solvable.

Most real conflict-driven processes continuously evolve, and their participants constantly adapt. This paper presents the approach of constructing a Nash equilibrium for game models with continuous updating. In the game models with continuous updating, it is assumed that players

- have information about motion equations and payoff functions only on $[t, t + \overline{T}]$, where $\overline{T}$ – information horizon, t – current time instant.
- receive updated information regarding motion equations and payoff functions as time $t \in [t_0, +\infty)$ evolves.

It is difficult to obtain the Nash equilibrium owing to the lack of fundamental approaches to control problems with a moving information horizon. Classical methods such as dynamic programming and Hamilton-Jacobi-Bellman equation do not permit the direct construction of the Nash equilibrium in problems with a moving information horizon.

In the framework of the dynamic updating approach, the following papers were published [6,9,10,12–16,21]. Their authors set the foundations for further study of a class of games with dynamic updating. It is assumed that the information about motion equations and payoff functions is updated in discrete time instants and the interval on which players know the information is defined by the value of the information horizon. Another related paper [20] published in

2019 paper considers repeated games with sliding planning horizons, which is close in nature to dynamic updating approach.

However, the class of games with continuous updating provides new theoretical results. The class of differential games with continuous updating was considered in the papers [8,11], here it is supposed that the updating process evolves continuously in time. In the paper [11], the system of Hamilton-Jacobi-Bellman equations are derived for the Nash equilibrium in a game with continuous updating. In the paper [8] the class of linear-quadratic differential games with continuous updating is considered and the explicit form of the Nash equilibrium is derived. Strategies in the feedback form are considered in subgames in [8] as opposed to strategies in the open-loop form in this paper.

For autonomous linear-quadratic game models with continuous updating, the Nash equilibrium in open-loop form is constructed. The cooperative case is presented as well, cooperative strategies, cooperative trajectory, characteristic function, and Shapley value are constructed. A model example of non-renewable resource extraction with an explicit solution is presented and conclusions are drawn.

The paper is structured as follows. In Sect. 2, a description of the initial differential game model and corresponding game model with continuous updating as well as the concept of a strategy for it are presented. Section 3 is devoted to optimal strategies in both non-cooperative and cooperative cases, i.e. Nash equilibrium and cooperative strategies with continuous updating. In Sect. 4, the explicit form of characteristic function and Shapley value for the linear-quadratic autonomous game model are presented. The illustrative model example and corresponding numerical simulation are presented in Sect. 5. Section 6 presents our conclusions.

2   Game Model

In this section description of the initial linear-quadratic differential autonomous game model and corresponding game model with continuous updating are presented.

2.1   Initial Linear Quadratic Autonomous Game Model

Consider n-player ($|N| = n$) linear quadratic autonomous differential game $\Gamma(x_0, T - t_0)$ defined on the interval $[t_0, T]$:

Motion equations have the form

$$\dot{x}(t) = Ax(t) + B_1 u_1(t, x) + \ldots + B_n u_n(t, x),$$
$$x(t_0) = x_0, \tag{1}$$
$$x \in \mathbb{R}^l, \ u = (u_1, \ldots, u_n), \ u_i = u_i(t, x) \in U_i \subset \mathrm{comp}\mathbb{R}^k, \ t \in [t_0, T].$$

Payoff function of player $i \in N$ is defined as

$$K_i(x_0, t_0, T; u) = \int_{t_0}^{T} \left(x'(t) Q_i x(t) + \sum_{j=1}^{n} u_j'(t, x) R_{ij} u_j(t, x) \right) dt, \ i \in N, \tag{2}$$

where Q_i, R_{ij} are assumed to be symmetric, R_{ii} is positive defined, $(\,\cdot\,)'$ means transpose here and hereafter, A, B_i, Q_i, R_{ij} are constants.

2.2 Linear Quadratic Autonomous Game Model with Continuous Updating

Consider n-player differential game $\Gamma(x, t, \overline{T})$, $t \in [t_0, +\infty)$ defined on the interval $[t, t + \overline{T}]$, where $0 < \overline{T} < +\infty$.

Motion equations of $\Gamma(x, t, \overline{T})$ have the form

$$
\begin{aligned}
&\dot{x}^t(s) = A x^t(s) + B_1 u_1^t(s, x^t) + \ldots + B_n u_n^t(s, x^t), \\
&x^t(t) = x, \\
&x^t \in \mathbb{R}^l, \; u^t = (u_1^t, \ldots, u_n^t), \; u_i^t = u_i^t(s, x^t) \in U_i \subset \mathrm{comp}\mathbb{R}^k, \; t \in [t_0, +\infty).
\end{aligned}
\tag{3}
$$

Payoff function of player $i \in N$ in the game $\Gamma(x, t, \overline{T})$ is defined as

$$
K_i^t(x^t, t, \overline{T}; u^t) = \int\limits_t^{t+\overline{T}} \left((x^t(s))' Q_i x^t(s) + \sum_{j=1}^{n} (u_j^t(s, x^t))' R_{ij} u_j^t(s, x^t) \right) ds, \tag{4}
$$

where $x^t(s)$, $u^t(s, x)$ are trajectory and strategies in the game $\Gamma(x, t, \overline{T})$.

Differential game with continuous updating evolves according to the rule:

Time parameter $t \in [t_0, +\infty)$ evolves continuously, as a result players continuously receive updated information about motion equations and payoff functions under $\Gamma(x, t, \overline{T})$.

Strategies $u(t, x)$ in the game model with continuous updating are defined in the following way:

$$
u(t, x) = u^t(t, x), \; t \in [t_0, +\infty), \tag{5}
$$

where $u^t(s, x)$, $s \in [t, t + \overline{T}]$ are some fixed strategies defined in the subgame $\Gamma(x, t, \overline{T})$.

State $x(t)$ in the model with continuous updating is defined according to

$$
\begin{aligned}
&\dot{x}(t) = A(t)x(t) + B_1(t)u_1(t, x) + \ldots + B_n(t)u_n(t, x), \\
&x(t_0) = x_0, \\
&x \in \mathbb{R}^l
\end{aligned}
\tag{6}
$$

with strategies with continuous updating $u(t, x)$ involved.

The essential difference between the game model with continuous updating and classic differential game $\Gamma(x_0, T - t_0)$ with prescribed duration is that players in the initial game are guided by the payoffs that they will eventually receive on the interval $[t_0, T]$, but in the case of a game with continuous updating, at the time instant t they orient themselves on the expected payoffs (4), which are calculated using information about the game structure defined on the interval $[t, t + \overline{T}]$.

3 Optimal Strategies with Continuous Updating in LQ Differential Games

3.1 General Concept of Optimal Strategies with Continuous Updating

In a game with continuous updating, the player at each moment focuses on the nearest event horizon, trying to maximize profits on it. Comparing the structure of the subgame and the initial game, we see that they are very similar. Thus, the player chooses the optimal strategy for, generally speaking, a new game at each moment in time. And the optimal strategy for the whole game is made up of each particular decision.

For example, consider two time intervals $[t, t+\overline{T}]$ and $[t+\epsilon, t+\overline{T}+\epsilon]$, $\epsilon << \overline{T}$. According to the problem statement, $u^{NE}(t, x)$ at the instant t should coincide with the Nash equilibrium in the game defined on the interval $[t, t + \overline{T}]$ and $u^{NE}(t + \epsilon, x)$ at instant $t + \epsilon$ should coincide with the Nash equilibrium in the game defined on the interval $[t + \epsilon, t + \epsilon + \overline{T}]$. Therefore direct application of classical approaches for determining Nash equilibrium in feedback strategies is not possible.

In order to determine the solution of a game with continuous updating, we introduce the concept of a generalized solution as combination of solutions in subgames. Let $u^{t,*}(s, x)$—some solution of (3), (4) for some t. Define generalized solution as $u^*(t, s, x) \overset{\Delta}{=} u^{t,*}(s, x)$. Now define solution of a game with continuous updating as $u^*(s, x) \overset{\Delta}{=} u^*(t, s, x)|_{s=t}$.

3.2 Concept of Nash Equilibrium for Games with Continuous Updating

For non-cooperative games, we use open-loop Nash equilibrium. Define corresponding concepts for a game with continuous updating.

Definition 1. *Strategy profile* $\widetilde{u}^{NE}(t, s, x) = (\widetilde{u}_1^{NE}(t, s, x), \ldots, \widetilde{u}_n^{NE}(t, s, x))$, $t \in [t_0, +\infty)$, $s \in [t, t + \overline{T}]$ *is a generalized open-loop Nash equilibrium in the game with continuous updating, if for any fixed* $t \in [t_0, +\infty)$ *strategy profile* $\widetilde{u}^{NE}(t, s, x)$ *is Nash equilibrium in open-loop strategies in the game* $\Gamma(x, t, \overline{T})$, $0 < \overline{T} < \infty$.

Using generalized open-loop Nash equilibrium it is possible to define solution concept for a game model with continuous updating.

Definition 2. *Strategy profile* $u^{NE}(t, x)$ *is called the open-loop-based Nash equilibrium with continuous updating, if it is defined in the following way:*

$$u^{NE}(t, x) = \widetilde{u}^{NE}(t, s, x)|_{s=t} = (\widetilde{u}_1^{NE}(t, s, x)|_{s=t}, \ldots, \widetilde{u}_n^{NE}(t, s, x)|_{s=t}), \qquad (7)$$

where $t \in [t_0, +\infty)$, $\widetilde{u}^{NE}(t, s, x)$ *is the generalized open-loop Nash equilibrium defined above.*

Strategy profile $u^{NE}(t, x)$ will be used as a solution concept in the non-cooperative game with continuous updating. Sufficient conditions for the existence of open-loop-based Nash equilibrium with continuous updating are presented below.

Theorem 1. *For an N-person linear-quadratic differential game with $Q_i \geq 0$, $R_{ij} \geq 0$ $(i, j \in N, i \neq j)$, let there exist a solution set $\{M_i^t, i \in N, t \geq t_0\}$ to the matrix Riccati differential equations*

$$\frac{dM_i^t(\tau)}{d\tau} + \overline{T}M_i^t(\tau)A + \overline{T}A'M_i^t(\tau) + Q_i - \overline{T}^2 M_i^t(\tau)\sum_{j \in N} B_j R_{jj}^{-1}B_j'M_j^t(\tau) = 0,$$

$$M_i^t(1) = 0, \quad i \in N.$$

(8)

Then, the differential game with continuous updating admits an open-loop-based Nash equilibrium with continuous updating solution given by

$$u_i^{NE}(t, x) = -R_{ii}^{-1}B_i'M_i^t(0)\overline{T}x(t), \quad i \in N.$$

Proof. In order to prove the Theorem we introduce the following change of variables

$$s = t + \overline{T}\tau,$$
$$y^t(\tau) = x^t(t + \overline{T}\tau), \qquad (9)$$
$$v_i^t(\tau, y) = u_i(t + \overline{T}\tau, x), \ i \in N.$$

By substituting (9) to the motion Eqs. (3), payoff function (4) we obtain

$$\dot{y}^t(\tau) = \overline{T}Ay^t(\tau) + \sum_{i=1}^N \overline{T}B_i v_i^t(\tau, y) \qquad (10)$$

and

$$K_i^t(y^t, \tau; v^t) = \int_0^1 (y^t(s))' Q_i y^t(s) + \sum_{j=1}^N (v_j^t(s, y))' R_{ij}v_j^t(s, y)ds, \ i \in N. \quad (11)$$

The Theorem 6.12 from [1] and existence of solution for the system of differential equations (8) lead to open-loop-based Nash equilibrium strategies in the subgame $\Gamma(x, t, \overline{T})$ have the form

$$v_i^{t,NE}(\tau, y_0) = -R_{ii}^{-1}B_i'M_i^t(\tau)\overline{T}\varPhi^t(\tau)y_0,$$

where

$$\frac{d\varPhi^t}{d\tau} = \left(A - \sum_{i \in N} B_i R_{ii}^{-1}B_i'\right)\varPhi^t(\tau),$$

$$\varPhi^t(0) = E.$$

Returning to original variables we obtain the following strategies

$$u_i^t(s, x) = -R_{ii}^{-1} B_i' M_i^t \left(\frac{s-t}{\overline{T}}\right) \overline{T} \varPhi^t \left(\frac{s-t}{\overline{T}}\right) x.$$

Then a generalized open-loop Nash equilibrium in the game with continuous updating has the form

$$\widetilde{u}_i^{NE}(t, s, x) = -R_{ii}^{-1} B_i' M_i^t \left(\frac{s-t}{\overline{T}}\right) \overline{T} \varPhi^t \left(\frac{s-t}{\overline{T}}\right) x. \tag{12}$$

Apply the procedure (7) to determine Nash equilibrium with continuous updating using generalized Nash equilibrium (12), $s = t$:

$$u_i^{NE}(t, x) = -R_{ii}^{-1} B_i' M_i^t(0) \overline{T} x, \ t \in [t_0, +\infty), \ i \in N. \tag{13}$$

This proves the theorem.

Remark 1. Notice, open-loop-based solution with continuous updating has a feedback form, i. e. open-loop-based Nash equilibrium with continuous updating explicitly depends on the current state. This happens because of the way the solution is constructed, as a value of solution generalized open-loop Nash equilibrium.

3.3 Concept of Cooperative Strategy for Games with Continuous Updating

In contrast to the non-cooperative case, here all players minimize one functional

$$K^t(x^t, t, \overline{T}; u^t) = \sum_{i \in N} K_i^t(x^t, t, \overline{T}; u^t)$$

$$= \sum_{i \in N} \int_t^{t+\overline{T}} \left((x^t(s))' Q_i x^t(s) + \sum_{j=1}^n (u_j^t(s, x^t))' R_{ij} u_j^t(s, x^t) \right) ds,$$
$$\tag{14}$$

Define concepts of cooperative solution for a game with continuous updating.

Definition 3. *Strategy profile* $\widetilde{u}^*(t, s, x) = (\widetilde{u}_1^*(t, s, x), \ldots, \widetilde{u}_n^*(t, s, x))$, $t \geqslant t_0$, $s \in [t, t+\overline{T}]$ *is a generalized cooperative solution in the game with continuous updating, if for any fixed* $t \in [t_0, +\infty)$ *strategy profile* $\widetilde{u}^*(t, s, x)$ *is cooperative solution in the game* $\Gamma(x, t, \overline{T})$, $0 < \overline{T} < \infty$.

Using a generalized cooperative solution it is possible to define a solution concept for a game model with continuous updating.

Definition 4. *Strategy profile* $u^*(t, x)$ *is called the cooperative solution with continuous updating, if it is defined in the following way:*

$$u^*(t, x) = \widetilde{u}^*(t, s, x)|_{s=t} = (\widetilde{u}_1^*(t, s, x)|_{s=t}, \ldots, \widetilde{u}_n^*(t, s, x)|_{s=t}), \tag{15}$$

where $t \in [t_0, +\infty)$, $\widetilde{u}^*(t, s, x)$ *is the generalized cooperative solution defined above.*

Strategy profile $u^*(t, x)$ will be used as a solution concept in the cooperative game with continuous updating. Sufficient conditions for the existence of a cooperative solution with continuous updating are presented below.

Theorem 2. *For an N-person linear-quadratic differential game with $Q_i \geq 0$, $R_{ij} \geq 0$ $(i, j \in N, i \neq j)$, let there exist a solution set $\{Z^t, i \in N, t \geqslant t_0\}$ to the matrix Riccati differential quations*

$$\dot{Z}^t(\tau) = -\overline{T}A'Z^t(\tau) - \overline{T}Z^t(\tau)A + Z^t(\tau)SZ^t(\tau) - Q,$$
$$Z^t(1) = 0, \tag{16}$$

where $S = \overline{T}^2 BR^{-1}B'$, $Q = \sum\limits_{i \in N} Q_i$, and $B = [B_1, \ldots, B_n]$, $R = \{R_{ij}\}_{i,j=1}^n$ — block matrices. Then, the differential game with continuous updating admits an cooperative solution with continuous updating given by

$$u^*(t, x) = -R^{-1}B'Z^t(0)\overline{T}x.$$

Proof. To prove the Theorem we introduce the change of variables. By substituting (9) to the motion Eqs. (3), payoff function (14) we obtain

$$\dot{y}^t(\tau) = \overline{T}Ay^t(\tau) + \sum_{i=1}^{N} \overline{T}B_i v_i^t(\tau, y) \tag{17}$$

and

$$K^t(y^t, \tau; v^t) = \sum_{i \in N} \int_0^1 \left((y^t(s))' Q_i y^t(s) + \sum_{j=1}^{N} (v_j^t(s, y))' R_{ij} v_j^t(s, y) \right) ds, \ i \in N. \tag{18}$$

The Theorem 5.1 from [4] and existence of solution for the system of differential equations (16) lead to cooperative solution in the subgame $\Gamma(x, t, \overline{T})$ have the form

$$v^{t,*}(\tau, y_0) = -R^{-1}B'Z^t(\tau)\overline{T}\Phi^t(\tau)y_0,$$

where

$$\frac{d\Phi^t}{d\tau} = \left(A - SZ^t(\tau) \right) \Phi^t(\tau),$$

$$\Phi^t(0) = E.$$

Returning to original variables we obtain the following strategies

$$u^t(s, x) = -R^{-1}B'Z^t\left(\frac{s-t}{\overline{T}}\right)\overline{T}\Phi^t\left(\frac{s-t}{\overline{T}}\right)x.$$

Then a generalized cooperative solution in the game with continuous updating has the form

$$\tilde{u}^*(t, s, x) = -R^{-1}B'Z^t\left(\frac{s-t}{\overline{T}}\right)\overline{T}\Phi^t\left(\frac{s-t}{\overline{T}}\right)x. \tag{19}$$

Apply the procedure (15) to determine Nash equilibrium with continuous updating using generalized Nash equilibrium (19), $s = t$:

$$u^*(t, x) = -R^{-1}B'Z^t(0)\overline{T}x, \ t \in [t_0, +\infty), \ i \in N. \tag{20}$$

This proves the theorem.

Remark 2. Notice, open-loop-based cooperative strategies with continuous updating explicitly depend on current state.

4   Cooperative Solution with Continuous Updating

To determine how to allocate joint payoff among the players it is necessary to define how the overall game evolves, how players forecast their behavior at every current time instant $t \in [t_0, +\infty]$ for the future interval $[t, t + \overline{T}]$:

1. forecasted trajectory
2. how the characteristic function is calculated along the forecasted trajectory
3. how the imputation and the cooperative solution is chosen along the forecasted trajectory

and finally how the cooperative solution with continuous updating is constructed and what are the properties.

4.1   Characteristic Function for Subgame on Interval $[t, t + \overline{T}]$

Consider coalition S in n-player differential game $\Gamma(x, t, \overline{T})$ (3) (4). The characteristic function is defined as the total payoff of the coalition S in the Nash equilibrium $u^{NE} = (u_1^{NE}, \ldots, u_{n_S}^{NE})$ in a game $\Gamma^S(x, t, \overline{T})$ with the following set of players: a coalition S acting as one player and players from the set $N \setminus S$ i. e., in the game of $n_S = |N \setminus S| + 1$ players.

Describe the building of the auxiliary game $\Gamma^S(x, t, \overline{T})$. Let the first player of this game be player associated with coalition S for convenience and let other players have been renumbered in some way players from $N \setminus S$. Relabel matrices for $N \setminus S$ players $A^S = A$, $B_i^S = B_{k_i}$, $Q_i^S = Q_{k_i}$, $R_{i,j} = R_{k_i, k_j}$, where $i, j = \overline{2, n_S}$ and i—new index in $\Gamma^S(x, t, \overline{T})$ of k_i-th player from $\Gamma(x, t, \overline{T})$. Some matrices for coalition player have a block structure: $B_1^S = [B_{m_1} \ldots B_{m_c}]$, $R_{1,1}^S = \text{diag}(R_{m_1,m_1}, \ldots, R_{m_c,m_c})$, $R_{i,1}^S = \text{diag}(R_{k_i,m_1}, \ldots, R_{k_i,m_c})$; other—are sum of corresponding matrices from $\Gamma(x, t, \overline{T})$: $Q_1^S = \sum_{m \in S} Q_m$, $R_{1,i}^S = \sum_{m \in S} R_{m,k_i}$, where $m_1, \ldots, m_s \in S$, $i = \overline{2, n_S}$.

Thus motion equations of $\Gamma^S(x, t, \overline{T})$ have the form

$$\dot{x}^t(s) = A^S x^t(s) + B_1^S u_1^t(s, x^t) + \ldots + B_n^S u_n^t(s, x^t),$$
$$x^t(t) = x.$$

Payoff function of player $i \in N^S$ in the game $\Gamma^S(x, t, \overline{T})$ is defined as

$$K_i^{S,t}(x^t, t, \overline{T}; u^t) = \int\limits_{t}^{t+\overline{T}} \left((x^t(s))' Q_i^S x^t(s) + \sum_{j=1}^{n} (u_j^t(s, x^t))' R_{ij}^S u_j^t(s, x^t) \right) ds,$$

where $x^t(s)$, $u^t(s, x)$ are trajectory and strategies in the game $\Gamma^S(x, t, \overline{T})$.

Lemma 1. *For an N-person linear-quadratic differential game $\Gamma(x, t, \overline{T})$ with $Q_i^S \geq 0$, $R_{ij}^S \geq 0$ $(i, j \in N, i \neq j)$, let there exist a solution set $\{M_i^S, i \in N_S, t \geq t_0\}$ to the matrix Riccati differential equations*

$$\frac{dM_i^S(\tau)}{d\tau} + \overline{T} M_i^S(\tau) A^S + \overline{T} \left(A^S\right)' M_i^S(\tau) + Q_i^S$$
$$- \overline{T}^2 M_i^S(\tau) \sum_{j \in N_S} B_j^S \left(R_{jj}^S\right)^{-1} \left(B_j^S\right)' M_j^S(\tau) = 0, \qquad (21)$$
$$M_i^S(1) = 0, \quad i \in N_S.$$

Then, the characteristic function for game $\Gamma(x, t, \overline{T})$ has form

$$V^t(S, x, \xi, t + \overline{T}) = \int\limits_{\xi}^{t+\overline{T}} \left((x^*(s, t, x))' Q_1^S x^*(s, t, x) \right.$$
$$\left. + \sum_{j=1}^{n_S} (u_j^t(s, x))' R_{1j}^S u_j^t(s, x) \right) ds, \qquad (22)$$

where

$$u_i^t(s, x) = - \left(R_{ii}^S\right)^{-1} \left(B_i^S\right)' M_i^S \left(\frac{s-t}{\overline{T}}\right) \overline{T} \Phi^S \left(\frac{s-t}{\overline{T}}\right) x, \qquad (23)$$

$\Phi^S(\tau)$—*solution of system*

$$\frac{d\Phi^S}{d\tau} = \left(A^S - \sum_{i \in N_S} B_i^S \left(R_{ii}^S\right)^{-1} \left(B_i^S\right)'\right) \Phi^S(\tau),$$
$$\Phi^S(0) = E,$$

$x^*(s, t, x)$—*solution of system*

$$\dot{x}^t(s) = A^S x^t(s) + B_1^S u_1^t(s, x) + \ldots + B_n^S u_n^t(s, x), \qquad (24)$$
$$x^t(t) = x.$$

Proof. We defined the characteristic function as the total payoff of the coalition S in the Nash equilibrium in a game $\Gamma^S(x, t, \overline{T})$.

Similar to Theorem 1 proof we use the change of variables (9), the Theorem 6.12 from [1] and existence of solution for the system of differential equations (21).

It implies open-loop-based Nash equilibrium strategies in the subgame $\Gamma^S(x,t,\overline{T})$ have the form (23).

According to building auxiliary game $\Gamma^S(x,t,\overline{T})$ the total payoff of the coalition S could calculate as $K_1^{S,t}(x^t,t,\overline{T},u^t)$. Then function (22) is characteristic and system dynamics evolves by (24).

4.2   Characteristic Function for Games with Continuous Updating

Suppose that the function $\widetilde{V}^t(S;\widetilde{x}_t^*(s),s,t+\overline{T})$, $S \subseteq N$ is continuously differentiable by $s \in [t,\overline{T}]$ and integrable by $t \in [t_0,+\infty)$. Define characteristic function in game model with continuous updating $V(S;x^*(t),t)$ in the following way:

Definition 5. *Function $V(S;x^*(t),t)$, $t \in [t_0,+\infty)$, $S \subseteq N$ is a characteristic function with continuous updating, if it is defined as the following integral:*

$$V(S;x^*(t),t) = \int\limits_{t}^{+\infty} -\frac{d}{ds}\widetilde{V}^\tau(S;\widetilde{x}_\tau^*(s),s,\tau+\overline{T})|_{s=\tau}d\tau, \ t \in [t_0,+\infty), \ S \subseteq N,$$

(25)

where $\widetilde{V}_\tau(S;\widetilde{x}_\tau^*(s),s,t+\overline{T})$, $s \in [\tau,\tau+\overline{T}]$, $\tau \in [t,+\infty)$, $S \subseteq N$ is a characteristic function in the game $\Gamma(\widetilde{x}_\tau^*(s),s,\tau+\overline{T})$ defined on the interval $[s,t+\overline{T}]$.

Notice that the integral in (25) can be infinite. Therefore it is necessary to define cooperative game model on finite time interval, i.e. $t \in [t_0,T]$:

$$V(S;x^*(t),t,T) = \int\limits_{t}^{T} -\frac{d}{ds}\widetilde{V}^\tau(S;\widetilde{x}_\tau^*(s),s,\tau+\overline{T})|_{s=\tau}d\tau, \ t \in [t_0,T], \ S \subseteq N. \quad (26)$$

Theorem 3. *For a coalitions S in N-person linear-quadratic differential game with $Q_i^S \geq 0$, $R_{ij}^S \geq 0$ $(i,j \in N, i \neq j)$, let there exist a solution set $\{M_i^S, i \in N_S, t \geqslant t_0\}$ to the matrix Riccati differential equations (21). Then, the characteristic function for game with continuous updating has form*

$$V(S,x,t,T) = \int\limits_{t}^{T} (x^*(s,t,x))' \left(Q_1^S - \overline{T}^2 \sum_{j=1}^{n_S} P_j' R_{1j}^S P_j\right) x^*(s,t,x)ds, \quad (27)$$

where

$$P_j = \left(R_{jj}^S\right)^{-1} \left(B_j^S\right)' M_j^S(0), \quad (28)$$

$x^*(s,t,x)$—solution of system

$$\dot{x}^t(s) = \left(A^S - \sum_{i=1}^{n_S} B_i^S P_i\right) x^t(s),$$

$$x^t(t) = x.$$

(29)

Proof. According to definition we have general form (26) of characteristic function for game with continuous updating. Lemma 1 gives the characteristic function (22) for subgame $\Gamma(x, t, \overline{T})$. By substituting (22) into (26) we get

$$
V(S, x, t, T) = \int_t^T -\frac{d}{ds} V^\tau(S; \widetilde{x}_\tau^*(s), s, \tau + \overline{T})|_{s=\tau} d\tau
$$

$$
= \int_t^T \Bigg((x^*(s, t, x))' Q_1^S x^*(s, t, x) \tag{30}
$$

$$
+ \sum_{j=1}^{n_S} (u_j^s(s, x^*(s, t, x)))' R_{1j}^S u_j^s(s, x^*(s, t, x)) \Bigg) ds,
$$

Taking into account $\Phi^S(0) = E$ we have

$$
u_i^s(s, x) = - \left(R_{ii}^S \right)^{-1} \left(B_i^S \right)' M_i^S(0) \overline{T} x. \tag{31}
$$

By substituting (31) into (30) with (28) we get

$$
V(S, x, t, T) = \int_t^T \Bigg((x^*(s, t, x))' Q_1^S x^*(s, t, x)
$$

$$
- \overline{T}^2 \sum_{j=1}^{n_S} (x^*(s, t, x))' P_j' R_{1j}^S P_j x^*(s, t, x) \Bigg) ds. \tag{32}
$$

Give (32) in similar terms to (27). Taking into account (31), (24) we can describe system dynamics as (29).

4.3 Shapley Value with Continuous Updating

Suppose that all the players united in a coalition of N, then, moving along the cooperative trajectory $x^*(t)$, they can secure a total payoff of $V(N, x^*(t), t, T)$. To determine the payoffs of each player $i \in N$, we introduce the concept of imputation $\xi(x^*(t), t, T) = (\xi_1(x^*(t), t, T), \ldots, \xi_n(x^*(t), t, T))$, i.e., e. the payoff that the player i will receive after the redistribution of the maximum total win $V(N, x^*(t), t, T)$ between all players.

For imputation two conditions must be met:

$$
\xi_i(x^*(t), t, T) \geq V(\{i\}, t, T, x^*(t)), \quad i \in N,
$$

$$
\sum_{i \in N} \xi_i(x^*(t), t, T) = V(N, t, T, x^*(t)).
$$

As a imputation in the game with continuous updating, we will use the Shapley vector:

$$Sh_i(x^*(t), t, T) = \sum_{\substack{S \subseteq N: \\ i \in S}} \frac{(k-1)!\,(n-k)!}{n!}$$

$$\times (V(S, t, T, x^*(t)) - V(S \setminus \{i\}, t, T, x^*(t)),\ i = 1, \ldots, n,),$$

(33)

where $k = |S|$.

Let there exist a solution set $\{M_i^S, i \in N_S, t \geqslant t_0\}$ to the matrix Riccati differential equations (21) for every coalition $S \subset N$. Then we can calculate Shapley vector (33) where $V(S, t, T, x^*(t))$ is calculated according to (27).

5   Example Model

5.1   Common Description

Consider the model in which there are two individuals investing in a public stock of knowledge (see also Dockner et al. [3]). Let $x(t)$ be the stock of knowledge at time t and $u_i(t)$ – the investment of player i in public knowledge at time t. Assume that the stock of knowledge evolves according to the accumulation equation

$$\dot{x}(t) = -\beta x(t) + u_1(t, x_0) + u_2(t, x_0), \quad x(0) = x_0,$$

(34)

where β is the depreciation rate. Assume that each player derives quadratic utility from the consumption of the stock of knowledge and that the cost of investment increases quadratically with the investment effort. That is, the cost function of both players is given by

$$K_i(x_0, t_0, T; u) = \int_0^T \left(-q_i x^2(t) + r_i u_i^2(t, x_0) \right) dt, \ i = 1, 2.$$

5.2   Game Model with Continuous Updating

Now consider the case of continuous updating. Here we suppose that two individuals at each time instant $t \in [t_0, +\infty)$ use information about motion equations and payoff functions on the interval $[t, t + \overline{T}]$. As the current time t evolves the interval, which defines the information shifts as well. Motion equations for the game model with continuous updating have the form

$$\dot{x}^t(s) = -\beta x^t(s) + u_1^t(s, x) + u_2^t(s, x), \quad x^t(t) = x, \quad t \in [t_0, +\infty).$$

Non-cooperative Case. Payoff function of player $i \in N$ for the game model with continuous updating is defined as

$$K_i^t(x^t, t, \overline{T}; u^t) = \int_t^{t+\overline{T}} \left(-\left(x^t(s)\right)^2 q_i + \left(u_i^t(s, x)\right)^2 r_i \right) ds, \quad i = 1, 2.$$

As an example consider the symmetric case $r_1 = r_2 = r$, $q_1 = q_2 = q$ here and thereafter. According to the Theorem 1 defining the form of open-loop Nash equilibrium with continuous updating on the first step we need to solve the following differential equation:

$$\begin{cases} \dot{k}(\tau) = 2\beta \overline{T} k(\tau) + \frac{2\overline{T} k^2(\tau)}{r} + q, \\ k(1) = 0. \end{cases} \tag{35}$$

The solution of (35) is

$$k(\tau) = \frac{r(\beta - v)}{2\overline{T}} \left(\frac{2v}{v - \beta + (v + \beta)e^{2v\overline{T}(1-\tau)}} - 1 \right), \tag{36}$$

where $v = \sqrt{\beta^2 - \frac{2q}{r}}$. According to Theorem 1 open-loop-based Nash equilibrium with continuous updating has the form:

$$\tilde{u}_i^{NE}(t, x) = -\frac{k(0)x\overline{T}}{r}. \tag{37}$$

By substituting (36) in (37) we obtain:

$$\tilde{u}_i^{NE}(t, x) = \frac{\beta - v}{2} \left(\frac{2v}{v - \beta + (v + \beta)e^{2v\overline{T}(1-\tau)}} - 1 \right) x, \tag{38}$$

by substituting (38) in (34) we obtain $\tilde{x}^{NE}(t)$ as solution of equation

$$\dot{\tilde{x}}^{NE}(t) = -\beta \tilde{x}^{NE}(t) + \tilde{u}_1^{NE}(t, x) + \tilde{u}_2^{NE}(t, x), \quad \tilde{x}^{NE}(0) = x_0. \tag{39}$$

Cooperative Case. Payoff function of player for the cooperative game model with continuous updating is defined as

$$K^t(x^t, t, \overline{T}; u^t) = \sum_{i=1}^{2} \int_t^{t+\overline{T}} \left(-\left(x^t(s)\right)^2 q_i + \left(u_i^t(s, x)\right)^2 r_i \right) ds.$$

Consider the symmetric case $r_1 = r_2 = r$, $q_1 = q_2 = q$ again. According to the Theorem 2 defining the form of cooperative strategies with continuous updating on the first step we need to solve the following differential equation:

$$\begin{cases} \dot{k}(\tau) = 2\beta \overline{T} k(\tau) + \frac{2\overline{T} k^2(\tau)}{r} + 2q, \\ k(1) = 0. \end{cases} \tag{40}$$

The solution of (40) is

$$k(\tau) = \frac{r(\beta - v_1)}{2\overline{T}}\left(\frac{2v_1}{v_1 - \beta + (v_1 + \beta)e^{2v_1\overline{T}(1-\tau)}} - 1\right), \tag{41}$$

where $v_1 = \sqrt{\beta^2 - \frac{4q}{r}}$. According to Theorem 2 cooperative strategies with continuous updating has the form:

$$\tilde{u}_i^*(t, x) = -\frac{k(0)x\overline{T}}{r}. \tag{42}$$

By substituting (41) in (42) we obtain:

$$\tilde{u}_i^*(t, x) = \frac{\beta - v_1}{2}\left(\frac{2v_1}{v_1 - \beta + (v_1 + \beta)e^{2v_1\overline{T}(1-\tau)}} - 1\right)x, \tag{43}$$

by substituting (43) in (34) we obtain $\tilde{x}^*(t)$ as solution of equation

$$\dot{\tilde{x}}^*(t) = -\beta\tilde{x}^*(t) + \tilde{u}_1^*(t, x) + \tilde{u}_2^*(t, x), \quad \tilde{x}^*(0) = x_0. \tag{44}$$

We have characteristic function as

$$\tilde{V}(S, x, t, T) = \sum_{i \in S}\int_t^T\left(-(\tilde{x}^*(s))^2 q_i + (\tilde{u}_i^*(s, x))^2 r_i\right)ds, \tag{45}$$

taking into account (43) (44). By substituting (45) in (33) we obtain Shapley value with continuous updating $\tilde{Sh}_i(\tilde{x}^*(t), t, T)$ for this example.

5.3   Game Model on Infinite Interval

Consider classic approach for Nash equilibrium for the game on infinite interval $[0, +\infty)$. Motion equations have the form

$$\dot{x}(t) = -\beta x(t) + u_1(t, x) + u_2(t, x), \quad x(0) = x_0. \tag{46}$$

Payoff function of player $i \in N$ is defined as

$$K_i(x_0; u) = \lim_{T \to \infty}\int_0^T\left(-q_i x^2(t) + r_i u_i^2(t, x)\right)dt, \quad i = 1, 2.$$

Non-cooperative Case. According to [4] open-loop Nash equilibrium strategies have the form

$$u_i^*(t, x_0) = -\frac{kx_0}{r}e^{-(\beta + \frac{2k}{r})t} \tag{47}$$

in our symmetric case ($r_1 = r_2 = r$, $q_1 = q_2 = q$), where k is solution of

$$\frac{2k^2}{r} + 2\beta k + q = 0.$$

By substituting (47) in (46) we obtain $x^{NE}(t)$ as solution of equation

$$\dot{x}^{NE}(t) = -\beta x^{NE}(t) - \frac{2kx_0}{r}e^{-(\beta + \frac{2k}{r})t}, \quad x^{NE}(0) = x_0. \tag{48}$$

Cooperative Case. Payoff function in cooperative case is defined as

$$K(x_0; u) = \sum_{i=1}^{2} \lim_{T \to \infty} \int_0^T \left(-q_i x^2(t) + r_i u_i^2(t, x) \right) dt.$$

According to [4] open-loop Nash equilibrium strategies have the form

$$u_i^*(t, x_0) = -\frac{kx_0}{r} e^{-\left(\beta + \frac{2k}{r}\right)t} \tag{49}$$

in our symmetric case ($r_1 = r_2 = r$, $q_1 = q_2 = q$), where k is solution of

$$\frac{2k^2}{r} + 2\beta k + 2q = 0.$$

By substituting (49) in (46) we obtain $x^*(t)$ as solution of equation

$$\dot{x}^*(t) = -\beta x^*(t) - \frac{2kx_0}{r} e^{-\left(\beta + \frac{2k}{r}\right)t}, \quad x^*(0) = x_0. \tag{50}$$

We have characteristic function as

$$V(S, x, t, T) = \sum_{i \in S} \int_t^T \left(-(x^*(s))^2 q_i + (u_i^*(s, x))^2 r_i \right) ds, \tag{51}$$

taking into account (49) (50). By substituting (51) in (33) we obtain Shapley value on infinity interval $Sh_i(x^*(t), t, T)$ for this example.

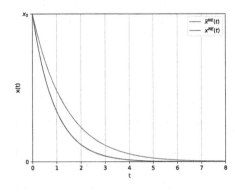

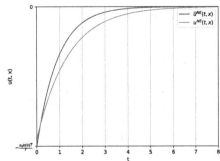

Fig. 1. $\tilde{x}^{NE}(t)$ (39) - red lower line, $x^{NE}(t)$ (48) - green upper line. (Color figure online)

Fig. 2. $\tilde{u}^{NE}(t)$ (38) - red upper line, $u^{NE}(t)$ (47) - green lower line. (Color figure online)

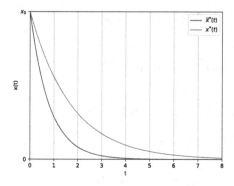

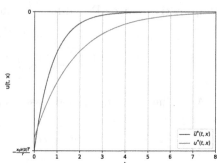

Fig. 3. $\tilde{x}^*(t)$ (44) - red lower line, $x^*(t)$ (50) - green upper line. (Color figure online)

Fig. 4. $\tilde{u}^*(t)$ (43) - red upper line, $u^*(t)$ (49) - green lower line. (Color figure online)

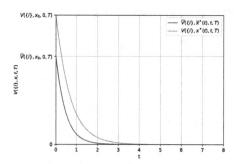

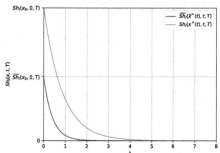

Fig. 5. Payoff function with continuous updating $\tilde{V}(\{i\}, \tilde{x}^*(t), t, T)$ - red lower line, payoff function on infinity interval $V(\{i\}, x^*(t), t, T)$ - green upper line. (Color figure online)

Fig. 6. Shapley value with continuous updating $\widetilde{Sh}_i(\tilde{x}^*(t), t, T)$ - red lower line, Shapley value on infinity interval $Sh_i(x^*(t), t, T)$ - green upper line. (Color figure online)

5.4   Numerical Simulation

Consider the results of numerical simulation for the game model presented above on the interval $[0, 8]$, i.e. $t_0 = 0$, $T = 8$. At the initial instant $t_0 = 0$ the stock of knowledge is 100, i.e. $x_0 = 100$. The other parameters of models: $\beta = 0.9$, $r = 6$, $q = -1$, $\overline{T} = 3$. In Fig. 1 the comparison of Nash equilibrium with continuous updating (red lines) and Nash equilibrium in game on infinite interval is presented. In Fig. 2 similar results are presented for the strategies. In Figs. 3 and 4 the similar comparisons of cooperative solutions are presented. In Fig. 5 the comparison of payoff function for a non-cooperative case is presented. In Fig. 6 the comparison of Shapley value for these two cases is presented.

6 Conclusion

The concept of open-loop based strategies for the class of linear-quadratic differential games with continuous updating is presented. Open-loop based Nash equilibrium and cooperative strategies in this class of differential games are constructed and the corresponding Theorems are presented. The characteristic function for a cooperative case is constructed, the path of possible solutions (e.g. Shapley value) based on it is indicated. The results are demonstrated using the differential game model of knowledge stock. Obtained results are both fundamental and applied in nature since they allow specialists from the applied field to use a new mathematical tool for more realistic modeling of engineering systems describing human-machine interaction.

References

1. Basar, T., Olsder, G.: Dynamic Noncooperative Game Theory. Academic Press, London (1995)
2. Bellman, R.: Dynamic Programming. Princeton University Press, Princeton (1957)
3. Dockner, E., Jorgensen, S., Long, N., Sorger, G.: Differential Games in Economics and Management Science. Cambridge University Press, Cambridge (2000)
4. Engwerda, J.: LQ Dynamic Optimization and Differential Games. Willey, New York (2005)
5. Gao, H., Petrosyan, L., Qiao, H., Sedakov, A.: Cooperation in two-stage games on undirected networks. J. Syst. Sci. Complexity **30**(3), 680–693 (2017). https://doi.org/10.1007/s11424-016-5164-7
6. Gromova, E., Petrosian, O.: Control of information horizon for cooperative differential game of pollution control. In: 2016 International Conference Stability and Oscillations of Nonlinear Control Systems (Pyatnitskiy's Conference) (2016)
7. Kleimenov, A.: Non-antagonistic Positional Differential Games. Science, Ekaterinburg (1993)
8. Kuchkarov, I., Petrosian, O.: On class of linear quadratic non-cooperative differential games with continuous updating. In: Khachay, M., Kochetov, Y., Pardalos, P. (eds.) MOTOR 2019. LNCS, vol. 11548, pp. 635–650. Springer, Cham (2019). https://doi.org/10.1007/978-3-030-22629-9_45
9. Petrosian, O., Kuchkarov, I.: About the looking forward approach in cooperative differential games with transferable utility. In: Petrosyan, L.A., Mazalov, V.V., Zenkevich, N.A. (eds.) Frontiers of Dynamic Games. SDGTFA, pp. 175–208. Springer, Cham (2019). https://doi.org/10.1007/978-3-030-23699-1_10
10. Petrosian, O., Shi, L., Li, Y., Gao, H.: Moving information horizon approach for dynamic game models. Mathematics **7**(12) (2019). https://doi.org/10.3390/math7121239
11. Petrosian, O., Tur, A.: Hamilton-Jacobi-Bellman equations for non-cooperative differential games with continuous updating. In: Bykadorov, I., Strusevich, V., Tchemisova, T. (eds.) MOTOR 2019. CCIS, vol. 1090, pp. 178–191. Springer, Cham (2019). https://doi.org/10.1007/978-3-030-33394-2_14
12. Petrosian, O.: Looking forward approach in cooperative differential games. Int. Game Theory Rev. **18**(2), 1–14 (2016). https://doi.org/10.1142/S0219198916400077

13. Petrosian, O.: Looking forward approach in cooperative differential games with infinite-horizon. Vestnik S.-Petersburg Univ. Ser. 10. Prikl. Mat. Inform. Prots. Upr. **4**, 18–30 (2016)
14. Petrosian, O., Barabanov, A.: Looking forward approach in cooperative differential games with uncertain-stochastic dynamics. J. Optim. Theory Appl. **172**, 328–347 (2017)
15. Petrosian, O., Nastych, M., Volf, D.: Non-cooperative differential game model of oil market with looking forward approach. In: Petrosyan, L.A., Mazalov, V.V., Zenkevich, N.A. (eds.) Frontiers of Dynamic Games. SDGTFA, pp. 189–202. Springer, Cham (2018). https://doi.org/10.1007/978-3-319-92988-0_11
16. Petrosian, O., Nastych, M., Volf, D.: Differential game of oil market with moving informational horizon and non-transferable utility. In: 2017 Constructive Nonsmooth Analysis and Related Topics (dedicated to the memory of V.F. Demyanov) (2017)
17. Petrosyan, L., Murzov, N.: Game-theoretic problems in mechanics. Lith. Math. Collect. **6**, 423–433 (1966)
18. Pontryagin, L.: On the theory of differential games. Successes Math. Sci. **26**(4(130)), 219–274 (1966)
19. Shevkoplyas, E.: Optimal solutions in differential games with random duration. J. Math. Sci. **199**(6), 715–722 (2014)
20. Vasin, A., Divtsova, A.: A game-theoretic model of agreement on limitation of transboundary air pollution. Autom. Remote Control **80**, 1164–1176 (2019)
21. Yeung, D., Petrosian, O.: Cooperative stochastic differential games with information adaptation. In: International Conference on Communication and Electronic Information Engineering (2017)

On Iterative Methods for Searching Equilibrium in Pure Exchange Economy with Multiplicative Utilities of Its Agents

Leonid D. Popov[1,2]([⊠]) [ID]

[1] Krasovskii Institute of Mathematics and Mechanics UB RAS,
Yekaterinburg, Russia
popld@imm.uran.ru
[2] Ural Federal University, Yekaterinburg, Russia

Abstract. We consider the classical Arrow–Debreu model for a pure exchange economy with multiplicative utilities of its agents. To calculate its equilibrium prices, we present a new iterative algorithm that simulates the simplest intuitive forms of the economic behavior of market agents. It converges under very weak assumptions. The algorithm relies on increasing prices for scarce products only. Moderate inflation, accompanying the computational process, plays a positive role in establishing an equilibrium between commodity supply and demand. Schemes have a meaningful economic interpretation. The convergence theorems are proved, and the results of numerical experiments are presented, including other types of economies.

Keywords: Arrow–Debreu model · Cobb–Douglas utility · Economic equilibrium · Tâtonnement

1 Introduction

The investigation of market mechanisms, which create a balance between the commodities demand and their supply, is the central issue of modern economic theory. Fundamental ideas and hypotheses in this area were stated by Leon Walras in 1874 [1]. The first mathematical formulations describing the relationship between the changing of the prices and the changing of market excess demand can be found in the works of P. Samuelson, K. Arrow, L. Hurvicz, J. Debreu, G. Scarf, H. Uzawa [2–6]. Some later, an extensive series of mathematical models of general and particular economic equilibrium was proposed as well as various algorithms for its search (see also [7–13] and many others).

Very soon, it became clear that the relatively simple assumptions, using by the founders of the theory to prove the existence of the competitive equilibrium and convergence of continuous schemes for its search, were not enough for convergence of discrete analogs of these algorithms [10–17]. Besides, the discrete schemes themselves gradually became more and more complicated and went away further and further from modeling simple business operations. It has

© Springer Nature Switzerland AG 2020
A. Kononov et al. (Eds.): MOTOR 2020, LNCS 12095, pp. 231–245, 2020.
https://doi.org/10.1007/978-3-030-49988-4_16

aroused a feeling of the incompleteness of the initial premises in this area and created a motivation to find additional ways and ideas.

One of such ideas is to implement instinctive forms of behavior of the economic agents under conditions of their weak economic awareness. In the work, we propose a new algorithm which corrects only prices for scarce commodities at each exchange cycle. This implies some inflation which plays a positive role in equilibrium search. Maybe, being not very efficient, the new iterative schemes nevertheless converge under the most relaxed assumptions. As a testing base to verify these ideas, the classic Arrow–Debreu exchange model with the multiplicative utilities of its participants is chosen.

The work is organized as follows. Sect. 2 describes the Arrow–Debreu model of pure exchange and provides a brief overview of known approaches to its analysis. Sect. 3 describes new algorithms and assumptions under which its convergence will be proved. Sect. 4 contains the proof of convergence itself. Sect. 5 presents the results of numerical experiments, including for other types of economies. A conclusion and bibliography complete the work.

2   The Arrow–Debreu Model

The Arrow–Debreu model [3] describes the behavior of m market participants (agents) who exchange of n types of commodities on the base of prevailing prices. We analyze only the Cobb–Douglas economy, where the preferences of agents are described by multiplicative utilities. Data on these functions and the initial distribution of commodities (endowments) can be gathered in two non-negative matrices $A = (a_{ij})_{m \times n}$ and $B = (b_{ij})_{m \times n}$. Each row of these matrices corresponds to one of the market participants and each column corresponds to one of the commodity types.

Matrix A consists of elasticity coefficients of the utility functions of each of the market agents; these functions are of the form

$$u_i(x_{i1}, x_{i2}, \ldots, x_{in}) = \prod_{j=1}^{n} x_{ij}^{a_{ij}} \qquad (i \in \overline{1, m}).$$

Matrix B consists of the initial stocks of commodities held by each of the participants before exchange. As usual, we assume that each of the participants has a non-zero supply of at least one type of product and needs at least one other type of commodity. Also, each type of commodity is available in a non-zero total quantity. Without loss of generality, we can assume that all row sums of the matrix A and all column sums of the matrix B are equal to 1.

Let $p = (p_1, \ldots, p_n) > 0$ be the vector of current prices for the commodities under consideration. Following these prices, each market participant (agent) sells its commodities on the market and, using the proceeds, acquires other commodities in such a way as to maximize the value of its utility function. Individual choice of participants (their demand for commodities) can be organized in a non-negative matrix $X(p) = (x_{ij}(p))_{m \times n}$. Each row of this matrix corresponds to one of the market participants, and each column corresponds to one type of commodities.

Essentially, every row $x_i(p)$ of the matrix $X(p)$ is a solution to one of the optimization problems

$$(\mathcal{P}_i) \quad \max\left\{ \prod_{j=1}^{n} x_{ij}^{a_{ij}} : \sum_{j=1}^{n} p_j x_{ij} \leq \sum_{j=1}^{n} p_j b_{ij}, \text{ all } x_{ij} \geq 0 \right\}.$$

The corresponding solution has the form

$$x_{ij}(p) = \frac{a_{ij}}{p_j} \sum_{s=1}^{n} b_{is} p_s \qquad (i \in \overline{1,m}, \ j \in \overline{1,n}).$$

Given these formulas, the components of the vector of market excess demand, which we denote $\mathbf{E}(p)$, are equal to

$$\mathbf{E}_j(p) = \frac{1}{p_j} \sum_{i=1}^{m} \sum_{s=1}^{n} a_{ij} b_{is} p_s - b_j \qquad (j \in \overline{1,n}). \tag{1}$$

Using vectors and matrix denoting, we can rewrite it as

$$\mathbf{E}(p) = \operatorname{diag}(p)^{-1} A^\top B p - b = \operatorname{diag}(p)^{-1} C p - b,$$

where $C = A^\top B$ and $\operatorname{diag}(w)$ is the diagonal matrix with the vector w on the diagonal.

We introduce the notation $\mathcal{E}(C, b)$ for the exchange economy under consideration. The auxiliary matrix C incorporates the properties of both matrices A and B and describes the characteristics of the interaction between the market agents. Recall that market excess demand is always (and not only in Cobb–Douglas economy) satisfies the homogeneity condition $\mathbf{E}(\lambda p) = \mathbf{E}(p)$ $(\forall \lambda > 0)$ and the Walras law, according to which $p^\top \mathbf{E}(p) = 0$.

Definition. The price vector $\bar{p} > 0$ is called the *equilibrium point* in the Arrow–Debreu model if the market clearing condition holds, i.e.

$$\mathbf{E}(\bar{p}) = 0. \tag{2}$$

For the Cobb–Douglas economy, the equilibrium conditions (2) can also be written as a system of homogeneous linear algebraic equations

$$\sum_{i=1}^{m} \sum_{s=1}^{n} a_{ij} b_{is} p_s = b_j p_j \qquad (j \in \overline{1,n}),$$

or, in matrix denoting,

$$C p - \operatorname{diag}(b) p = 0, \qquad p > 0. \tag{3}$$

To date, a wide range of algorithms has been developed to find the equilibrium, including ones for general types of utility functions of market agents. Almost all of these algorithms, somehow or other, came from a continuous dynamic Samuelson's system

$$\frac{dp(t)}{dt} = \mathbf{E}(p(t)), \qquad t \in [0, +\infty). \tag{4}$$

In the above system, $E(p)$ is a single-valued continuous function of prices. Prices are constantly changing over time. The system itself formalizes Walras's argument that commodity prices rise with increasing demand for them and fall along with its fall.

The stability of the system (4), i. e., its convergence to an equilibrium point, rests upon that equilibrium prices $\bar{p} > 0$ exist and that the inequality $\bar{p}^\top \mathbf{E}(p) > 0$ holds for any vector of out-of-equilibrium prices $p > 0$. In particular, such stability was earlier established for the cases when the market excess demand satisfies the Property of Global Gross Substitutability (PGGS)[1] or Weak Axiom of Revealed Preferences (WARP)[2]. However, such conditions are not enough when we pass from continuous trajectories $p(t)$ of the system (4) to their discrete approximations $\{p^k\}_{k=0}^\infty$. Moreover, as shown in [11–13, 15], many discrete algorithms exhibit chaotic behavior for some economies, including the economy with multiplicative utility functions of participants.

Of cause, it is possible to find an equilibrium of the Cobb–Douglas economy through alternative methods based on the resolving of the system of linear equations (3). It may be finite methods [9] which are similar to the simplex method as well as the infinite methods [17] liking to the regularized method of simple iteration

$$p^{k+1} = \frac{k+1}{k+2} C p^k + \frac{1}{k+2} p^0 \qquad (k = 0, 1, 2, \dots). \tag{5}$$

However, these algorithms use the specificity of Cobb–Douglas model too explicitly and are difficult for economic interpretation.

Below we propose a new algorithm simulating the simplest forms of behavior of economic agents. The last usually have no complete information about other people's preferences and spending plans neither about the initial distribution of the commodities nor can execute any complex mathematical calculations. The only ability of the agents we have in mind is to determine whether a given particular type of commodity is in short supply or not.

3   Algorithm Description and Initial Assumptions

New algorithm has an iterative nature. At each iteration (cycle), agents exchange their commodities according to their budgets and preferences. Some times one

---

[1] It means that the partial derivatives $d\mathbf{E}_i(p)/dp_j > 0$ for all $i \neq j$.

[2] This axiom states that for any price vectors p', $p'' > 0$, such that $\mathbf{E}(p') \neq \mathbf{E}(p'')$, the implication is true $\left(\mathbf{E}(p'')^\top p' \leq 0\right) \Rightarrow \left(\mathbf{E}(p')^\top p'' > 0\right)$.

or other commodities disappear from the market and pass into the category of scarce ones. Then agents correct the prices. But the mechanisms for such correction will differ from using usually. Namely, only prices corresponding to scarce commodities grow by a bit. Other ones do not change at all. Because the total price level grows, we can observe weak inflation (which is custom in any real economy). This inflation forces the demand for scarce commodities to decrease and switch to other groups of commodities according to current norms of their substitutability. Though in any cycle the increase in the prices is fixed, their not absolute but relative changes fade from cycle to cycle, providing a generalized convergence of the iterative process. To get a converging sequence, we can divide current prices by the total cost of commodities sold. In the real economy, such money denomination occurs from time to time.

Consider the formal scheme of algorithm proposed:

Algorithm 1. (checking of group demand)
Step 1. Choose an arbitrary initial prices $p^0 > 0$. Put $k = 0$.
Step 2. Determine the current set of scarce commodities $I_+(k)$ by the rule

$$i \in I_+(k) \iff \mathbf{E}_i(p^k) = \frac{c_{i1}p_1^k + c_{i2}p_2^k + \cdots + c_{in}p_n^k}{p_i^k} - 1 > 0.$$

Step 3. Correct prices of scarce commodities as below

$$p_i^{k+1} = \begin{cases} p_i^k + \delta_i^k, & \delta_i^k > 0, \text{ iff } i \in I_+(k), \\ p_i^k & \text{otherwise.} \end{cases}$$

Step 4. Increase $k := k + 1$.
Step 5. Return to step 2.

Let us discuss some assumptions for Cobb–Douglas economy in question.

Consider the matrix $C = A^\top B$. All its elements are non-negative, and all its column-wise sums are equal to 1, since by assumption $C^\top e = B^\top Ae = B^\top e = e$.

The arrangement structure of nonzero elements of the matrix C carries significant economic information and can also be described in terms of graph theory. Namely, since all elements of the matrices A and B are not negative, the element $c_{ij} = \sum_{s=1}^n a_{si}b_{sj}$ of the matrix C is nonzero if and only if at least one market participant s owns the commodity j and needs the commodity i, i. e.

$$\exists s : (a_{si} > 0)\&(b_{sj} > 0).$$

Definition. In the economy $\mathcal{E}(C, b)$, the commodity i is defined to access the commodity j directly if the element $c_{ij} > 0$. The commodity i is defined to access the commodity j if there exists a chain of numbers $j_1, j_2, \ldots, j_r$ such that $i = j_1$, $j_r = j$ and in intermediate pairs the commodity j_{s-1} accesses the commodity j_s directly.

Consider the directed graph $\Gamma(C) = (U, V)$, where U is the set of its nodes, and V is the set of its arcs. The nodes $u \in U$ correspond to commodities circulating within the economy $\mathcal{E}(C, b)$, and their pairs (i, j) form arcs if and only if $c_{ij} > 0$. Thus, the item i accesses the item j if and only if there exists a path in the graph $\Gamma(C)$ leading from the node i to the node j. If all the commodities in the economy are accessing each other, the graph $\Gamma(C)$ is strongly connected.

Following to [9], without loss of generality, we may assume that the following assumptions is fulfilled.

Assumption 1. The economy in question does not contain sub-markets, i.e. graph $\Gamma(C)$ is strongly connected.

Assumption 2. All price correction constants in algorithms 1 lie in some limited range of admissible values $0 < \underline{\delta} < \delta_i^k < \bar{\delta} < +\infty$, which is separated from zero.

To illustrate fulfilling Assumption 1, recall the well-known example with multiplicative (3×3)-economy defined by the matrices:

$$A = \begin{pmatrix} 0\,0\,1 \\ 1\,0\,0 \\ 0\,1\,0 \end{pmatrix}, \quad B = \begin{pmatrix} 1\,0\,0 \\ 0\,1\,0 \\ 0\,0\,1 \end{pmatrix}, \quad C = A^\top B = \begin{pmatrix} 0\,1\,0 \\ 0\,0\,1 \\ 1\,0\,0 \end{pmatrix}.$$

As shown in [17], many popular calculating schemas can not work with this example.

4   Convergence Analysis

Let us prove the convergence of algorithm 1. We consider three cases.

Case 1.   Suppose that at the current iteration of the algorithm, the set of indices of scarce commodities is full $I_+(k) = \{1, 2, ..., n\}$. Then all $\mathbf{E}_i(p^k) > 0$. It implies the inequality $\mathbf{E}(p^k)^\top p^k > 0$ that contradicts the Walras law. The case is impossible.

Case 2.   Next suppose that at the current iteration of the algorithm, $I_+(k) = \emptyset$, that is, all $\mathbf{E}_i(p^k) \leq 0$. It means that all prices $\bar{p}_i = p_i^k > 0$ do not change further, and all $\mathbf{E}_i(\bar{p})\bar{p}_i \leq 0$. But according to the Walras law, their sum must be equal to zero, which is possible only if all $\mathbf{E}_i(\bar{p}) = 0$, that is, the prices $\bar{p}_i$ form an equilibrium point. The work finishes.

Case 3.   Finally, let the sets $I_+(k) \neq \emptyset$ and $I_+(k) \neq \{1, 2, ..., n\}$ for all k. In other words, we assume that at each iteration, part of the prices rises, and other ones remain their values unchanged. Since the assortment of commodities is finite, at least one of them belongs to the list of scarce commodities infinitely many times, and the price of it grows unlimitedly. It is just the case that will be studied in detail below. Recall that we work under assumptions 1–2.

Lemma 1. *Let the sequence* $\{p_j^k\}_{k=0}^\infty$, *generated by the algorithm 1, grow unlimitedly, and the commodity i accesses the commodity j, that is $c_{ij} > 0$. Then the sequence* $\{p_i^k\}_{k=0}^\infty$ *also grows unlimitedly.*

Proof. From the unlimited increasing sequence $\{p_j^k\}_{k=0}^\infty$, let us single out a suitable subsequence, also, of course, unlimited from above, with the indexes $k_0, k_1, k_2, \ldots$. We make it as follows. Firstly, we put $k_0 = 0$. Then we choose the smallest k_1 for which the inequality holds

$$p_j^{k_1-1} \geq p_j^{k_0} + \frac{p_i^{k_0}}{c_{ij}}. \tag{6}$$

This can be done because the sequence $\{p_j^k\}_{k=0}^\infty$ grows unlimitedly and the difference between its successive elements does not exceed a finite number $\bar{\delta} > 0$.

Obviously, $k_1 > k_0 + 1$. Let show that $p_i^{k_1} \geq p_i^{k_0} + \underline{\delta}$, that is, during the time between iterations with indexes from k_0 to k_1 the price of the i-th commodity must increase at least once.

Indeed, if, during all iterations from k_0 to $k_1 - 1$, the price of the ith commodity has never been adjusted and kept the same, then by (6) we have inequality

$$\mathbf{E}_i(p^{k_1-1}) = \frac{1}{p_i^{k_1-1}} \sum_{s=1}^n c_{is} p_s^{k_1-1} - 1 = \frac{1}{p_i^{k_0}} \sum_{s=1}^n c_{is} p_s^{k_1-1} - 1 \geq \frac{c_{ij} p_j^{k_1-1}}{p_i^{k_0}} - 1 > 0. \tag{7}$$

But then, by the rules of Algorithm 1, the price of the ith commodity will be adjusted at the iteration of k_1.

So, in any case, $p_i^{k_1} \geq p_i^{k_0} + \underline{\delta}$.

Next, choose the smallest k_2 such that

$$p_j^{k_2-1} \geq p_j^{k_1} + \frac{p_i^{k_1}}{c_{ij}}, \tag{8}$$

This inequality is similar to (6) from the previous step. Obviously, $k_2 > k_1 + 1$. And again, we have an estimate of the price increase $p_i^{k_2} \geq p_i^{k_1} + \underline{\delta}$, since during the time between iterations from k_1 to k_2 the price of the j-th commodity, by virtue of the formulas of algorithm 1, must increase at least once.

Indeed, if, during all iterations from k_1 to $k_2 - 1$, the price of the j-th commodity would never have been adjusted, then according to (8) the inequality holds (compare with (7))

$$\mathbf{E}_i(p^{k_2-1}) = \frac{1}{p_i^{k_2-1}} \sum_{s=1}^n c_{is} p_s^{k_2-1} - 1 = \frac{1}{p_i^{k_1}} \sum_{s=1}^n c_{is} p_s^{k_2-1} - 1 \geq \frac{c_{ij} p_j^{k_2-1}}{p_i^{k_1}} - 1 > 0.$$

But then, by the rules of Algorithm 1, the price of the i-th commodity will be adjusted at the iteration of k_2.

So, in any case, $p_i^{k_2} \geq p_i^{k_1} + \underline{\delta}$.

Continuing the above reasoning, we construct the next indexes $k_3, k_4, \ldots$, and so on, and each time it turns out that

$$p_i^{k_s} \geq p_i^{k_{s-1}} + \underline{\delta} \geq p_i^{k_{s-2}} + 2\underline{\delta} \geq \ldots \geq p_i^0 + s\underline{\delta} \to \infty,$$

so the sequence $\{p_i^{k_s}\}_{s=0}^\infty$, and therefore the sequence $\{p_i^k\}_{k=0}^\infty$ (being monotonous) really grows unlimitedly, q.e.d.

Corollary 1. *Let at least one of the price sequences, for example, $\{p_i^k\}_{k=0}^{\infty}$, grow unlimitedly. Then all the other price sequences also grow unlimitedly under the Assumption 1.*

Proof. By Assumption 1, the exchange graph $\Gamma(C)$ is strongly connected. Thus, any pair of its nodes are connected by a path containing at most $(n-1)$ arcs. Consider a path connecting the initial node i, for which the corresponding price grows unlimitedly, and an arbitrary node j that does not coincide with it. Let this path be consist of intermediate nodes $i = l_0, l_1, ..., l_N = j$ so that all $c_{l_s l_{s+1}} > 0$, $s = 0, 1, ..., N-1$. By the Lemma 1, an unlimited increase in the price of commodity $l_0 = i$ forces an unlimited increase in the price of commodity l_1. In its turn, an unlimited increase in the price of commodity l_1 leads to an unlimited increase in the price of commodity l_2. And so on along the chain, an unlimited rise in the price of commodity l_{N-1} will ultimately determine an unlimited rise in the price of commodity $l_N = j$, q.e.d.

Now analyze the discrete trajectories of the components of market excess demand. If current prices are out-of-equilibrium, then, by the Walras law, some of the components $\mathbf{E}_i(p^k)$ must be positive, and some another must be negative. In reality, these components make a wave-like motion. For example, let $\mathbf{E}_i(p^k) < 0$ at some iteration k. Because the price of ith commodity is temporary unchanged and the prices for some other commodities keep on to increase, $\mathbf{E}_i(p^k)$ gradually grows and after several iterations become positive. But for i with positive $\mathbf{E}_i(p^{k'})$, the price of ith commodity starts to grow. That is why, step by step, $\mathbf{E}_i(p^k)$ returns to the area of negative values. The price of ith commodity is temporarily stabilized. Then everything repeats once more.

Let us evaluate the depth of immersion of an arbitrary trajectory in the negative and positive regions.

Lemma 2. *Suppose that $\mathbf{E}_i(p^k) \leq 0$. Then, $\mathbf{E}_i(p^k) \leq \mathbf{E}_i(p^{k+1}) \leq n\bar{\delta}/p_i^k \to +0$.*

Proof. When $\mathbf{E}_i(p^k) \leq 0$, then, according to the rules of the algorithm 1, the price of the i-th commodity remains unchanged, that is, $p_i^{k+1} = p_i^k > 0$. Since all other prices do not decrease, we have

$$\mathbf{E}_i(p^{k+1}) = \frac{1}{p_i^{k+1}} \sum_{s=1}^{n} c_{is} p_s^{k+1} - 1 = \frac{1}{p_i^k} \sum_{s=1}^{n} c_{is} p_s^{k+1} - 1 \geq \frac{1}{p_i^k} \sum_{s=1}^{n} c_{is} p_s^k - 1 = \mathbf{E}_i(p^k).$$

It gives us the left side of proving inequality. As for the right-hand side of this inequality, then, by Assumption 2, the price increase is bounded from above, and even when these prices rise all together, the new value of the ith component of excess market demand will be bounded from above

$$\mathbf{E}_i(p^{k+1}) = \frac{1}{p_i^k} \sum_{s=1}^{n} c_{is} p_s^{k+1} - 1 \leq \frac{1}{p_i^k} \sum_{s=1}^{n} c_{is}(p_s^k + \bar{\delta}) - 1 \leq \mathbf{E}_i(p^k) + \frac{n\bar{\delta}}{p_i^k} \leq \frac{n\bar{\delta}}{p_i^k},$$

q.e.d.

Thus, the negative component of market excess being growing can come into a positive area. But the altitude of such a jump will be less and less in absolute value.

Consider now the case when the current component of market excess demand is positive.

Lemma 3. *Let* $\mathbf{E}_i(p^k) > 0$. *Then* $\mathbf{E}_i(p^{k+1}) > -\bar{\delta}/p_i^k \to -0$.

Proof. If $\mathbf{E}_i(p^k) > 0$, then, according to the rules of algorithm 1, the price for the i-th commodity increases by $\delta_i^k > 0$, so

$$
\mathbf{E}_i(p^{k+1}) = \frac{1}{p_i^k + \delta_i^k} \cdot \sum_{s=1}^{n} c_{is} p_s^{k+1} - 1 \geq \frac{1}{p_i^k + \delta_i^k} \cdot \sum_{s=1}^{n} c_{is} p_s^k - 1
$$

$$
= \frac{p_i^k}{p_i^k + \delta_i^k} \cdot \mathbf{E}_i(p^k) - \frac{\delta_i^k}{p_i^k + \delta_i^k} > -\frac{\delta_i^k}{p_i^k + \delta_i^k} > -\frac{\bar{\delta}}{p_i^k} \to -0,
$$

q.e.d.

Thus, the possible depth of immersion of each of the positive components of market excess demand in the negative area occurs at ever smaller and smaller.

Corollary 2. *Let all the price sequences* $\{p_i^k\}_{k=0}^{\infty}$ *generated by the algorithm* 1, *grow unlimitedly. Then*

$$
\lim_{k \to \infty} \inf \mathbf{E}_i(p^k) \geq 0 \qquad (\forall i \in \overline{1, n}).
$$

It remains to study the possible behavior of the trajectory of the components of market excess demand within the positive range of its values.

Lemma 4. *Let* $\mathbf{E}_i(p^{k_1-1}) \leq 0$ *and* $\mathbf{E}_i(p^k) > 0$ *for* $k_1 \leq k < k_2$. *Then*

$$
\gamma(k_1, k_2) = \max_{k_1 \leq k < k_2} \{\mathbf{E}_i(p^k)\} \leq n \max\{\bar{\delta}/p_i^0; \ \bar{\delta}/\underline{\delta}\} = Const.
$$

Proof. Two cases are possible. In the first one, the maximum under consideration is reached at the left end of the interval, and, taking into account the Lemma 2, we conclude that

$$
\gamma(k_1, k_2) = \max_{k_1 \leq k < k_2} \{\mathbf{E}_i(p^k)\} = \mathbf{E}_i(p^{k_1}) \leq n\bar{\delta}/p_i^{k_1} \leq n\bar{\delta}/p_i^0.
$$

In the second one, the same maximum is reached inside this interval, i.e.

$$
\gamma(k_1, k_2) = \max_{k_1 \leq k < k_2} \{\mathbf{E}_i(p^k)\} = \mathbf{E}_i(p^{k_0}), \quad k_1 < k_0 < k_2.
$$

But since for all $k_1 \leq k < k_2$ the inequality holds

$$
\mathbf{E}_i(p^{k+1}) = \frac{1}{p_i^{k+1}} \sum_{s=1}^{n} c_{is} p_s^{k+1} - 1 \leq \frac{1}{p_i^{k+1}} \sum_{s=1}^{n} c_{is}(p_s^k + \bar{\delta}) - 1
$$

$$\leq \frac{p_i^k}{p_i^{k+1}} \cdot \frac{\sum_{s=1}^n c_{is} p_s^k}{p_i^k} + \frac{n\bar{\delta}}{p_i^{k+1}} - 1 \leq \frac{p_i^k}{p_i^{k+1}} \mathbf{E}_i(p^k) + \frac{n\bar{\delta}}{p_i^{k+1}},$$

which implies

$$\mathbf{E}_i(p^{k_0}) \leq \frac{p_i^{k_0-1}}{p_i^{k_0}} \mathbf{E}_i(p^{k_0-1}) + \frac{n\bar{\delta}}{p_i^{k_0}} \leq \frac{p_i^{k_0-1}}{p_i^{k_0}} \mathbf{E}_i(p^{k_0}) + \frac{n\bar{\delta}}{p_i^{k_0}}.$$

After multiplying both sides of the last inequality by $p_i^{k_0} > 0$ and rearranging the terms, we have

$$p_i^{k_0} \mathbf{E}_i(p^{k_0}) - p_i^{k_0-1} \mathbf{E}_i(p^{k_0}) \leq n\bar{\delta},$$

so, finally,

$$\underline{\delta} \, \mathbf{E}_i(p^{k_0}) \leq (p_i^{k_0} - p_i^{k_0-1}) \, \mathbf{E}_i(p^{k_0}) \leq n\bar{\delta},$$

or

$$\gamma(k_1, k_2) = \max_{k_1 \leq k < k_2} \{\mathbf{E}_i(p^k)\} = \mathbf{E}_i(p^{k_0}) \leq n\bar{\delta}/\underline{\delta}.$$

Combining the both cases above, we get

$$\gamma(k_1, k_2) = \max_{k_1 \leq k < k_2} \{\mathbf{E}_i(p^k)\} \leq \max \left\{ n\bar{\delta}/p_i^0; \ n\bar{\delta}/\underline{\delta} \right\}.$$

q.e.d.

Thus, we show that excess market demand is bounded from above by a certain constant, and it is so despite the fact that the prices of various commodities may increase with different rates. This allows us to formulate the next statement.

Lemma 5. *Let the sequence $\{p^k\}_{k=0}^\infty$ generated by the algorithm 1, be unlimited. Then an arbitrary limit point $\bar{p}$ of the normalized sequence $\{p^k/\|p^k\|_1\}_{k=0}^\infty$ is necessarily positive, that is, $\bar{p} > 0$.*

Proof. All components of the normalized price sequence are positive, and due to the homogeneity of market excess demand, the equality $\mathbf{E}(p^k) = \mathbf{E}(\breve{p}^k)$ holds, where $\breve{p}^k = p^k/\|p^k\|_1$. Suppose that for some commodity i the limit normalized price $\bar{p}_i$ is equal to zero. Then the convergence of some subsequence of $\breve{p}_i^k$ to $+0$ takes place (we preserve the original numbering for it). But by virtue of the Lemma 4 excess demand for the i-th commodity is bounded from above by some constant

$$\mathbf{E}_i(p^k) = \mathbf{E}_i(\breve{p}^k) = \frac{1}{\breve{p}_i^k} \sum_{s=1}^n c_{is} \breve{p}_s^k - 1 \leq Const.$$

It implies

$$\sum_{s=1}^n c_{is} \breve{p}_s^k \leq (Const + 1) \breve{p}_i^k \to +0,$$

that is, within the framework of the chosen subsequence, the normalized prices $\breve{p}_s^k$ for all those commodities for which the commodity i is exchanged directly also converge to zero (since for them $c_{is} > 0$).

Using the same arguments to the normalized prices of commodities which are connected by arcs with commodities connected by arcs with the commodity i, we obtain the convergence to zero of the same subsequences made up of their normalized prices. But by the Assumption 1, the graph $\Gamma(C)$ is strongly connected. So, we conclude that this property is true for all commodities, so a subsequence of normalized prices is found that converges to zero simultaneously for all its components. This, however, contradicts the assumption that all $\|\check{p}^k\|_1 = 1$. The revealed contradiction proves the strict inequality $\bar{p} > 0$. The proof is complete.

Now we can formulate the final convergence statement.

Theorem 1. *Let the economy $\mathcal{E}(C, b)$ satisfy the Assumption 1, and the step parameters of the algorithm 1 satisfy the Assumption 2. Then only the following two cases are possible:*

(a)   *price sequence $\{p^k\}_{k=0}^{\infty}$ reaches equilibrium in a finite number of steps,*

(b)   *the sequence of prices $\{p^k\}_{k=0}^{\infty}$ grows unlimitedly and*

$$\lim_{k \to \infty} \mathbf{E}_i(p^k) = 0 \quad (\forall i \in \overline{1, n}).$$

Moreover, any limit point $\bar{p}$ of the normalized sequence $\{p^k\|p^k\|_1^{-1}\}_{k=0}^{\infty}$ is positive and gives the equilibrium point of the economy in the sense of (2) .

Proof. Let us combine the above lemmas into a joint assertion. Firstly, if at some step of the iterative process 1 the price system has stabilized, then it represents an equilibrium point (this fact was considered at the beginning of the section, see case 2). Secondly, if price stabilization does not occur, then we can apply successively all the lemmas from case 3. We conclude that all prices rise unlimitedly (Lemma 1). Moreover, due to Corollary 2 and the fact that the function of market excess demand is homogeneous, we can write

$$\liminf_{k \to \infty} \mathbf{E}(p^k\|p^k\|_1^{-1}) = \liminf_{k \to \infty} \mathbf{E}(p^k) \geq 0. \tag{9}$$

Next, market excess demand is continuous function of prices and all the limit points $\bar{p}$ of the sequence $\{p^k\|p^k\|_1^{-1}\}_{k=0}^{\infty}$ are positive. Therefore (see Corollary 2) $\mathbf{E}(\bar{p}) \geq 0$. It implies $\bar{p}^{\top}\mathbf{E}(\bar{p}) \geq 0$ and we conclude that all $\mathbf{E}_i(\bar{p}) = 0$. Indeed, if the inequality $\mathbf{E}_i(\bar{p}) > 0$ holds at least for one i, then we have the inequality $\bar{p}^{\top}\mathbf{E}(\bar{p}) > 0$, which contradicts the Walras law.

Therefore, all limit points of the sequence $\{p^k\|p^k\|_1^{-1}\}_{k=0}^{\infty}$ are equilibrium points, and (to strengthen the property (9))

$$\lim_{k \to \infty} \mathbf{E}_i(p^k) = 0 \quad (\forall i \in \overline{1, n}).$$

The proof is complete.

It remains to estimate the rate of convergence of algorithm 1 for the case where the elements of the sequence $p^k > 0$ do not repeat.

Lemma 6. *Suppose that at each iteration of algorithm 1, at least one of the components of current market excess demand is positive. Then a small $\varepsilon > 0$ exists such that $p_i^k > (\varepsilon \underline{\delta}/n)k$ for all i and sufficiently large k.*

Proof. We can assert that there are $\varepsilon > 0$ and number K such that for all i, j and $k > K$ the inequalities hold $p_i^k \geq \varepsilon p_j^k$. Otherwise, one could single out a pair of numbers i and j such that $\lim_{k \to \infty} \inf\{p_i^k/p_j^k\} = 0$, which means that there would be the limit point $\bar{p}$ of the sequence $\{p^k\}_{k=0}^{\infty}$ lying on the boundary of a non-negative orthant. But it contradicts the Lemma 5.

Next we denote $\Delta p_i^k = p_i^{k+1} - p_i^k$. By assumption, $\sum_{i=1}^n \Delta p_i^k \geq \underline{\delta}$ for all k. As we already establish,

$$p_i^k \geq \varepsilon p_j^k = \varepsilon \left(p_j^0 + \sum_{s=0}^k \Delta p_j^s \right)$$

for all sufficiently large k. Summing up this inequality over all j, we obtain

$$n p_i^k \geq \varepsilon \sum_{j=1}^n \left(p_j^0 + \sum_{s=0}^{k-1} \Delta p_j^s \right) = \varepsilon \|p^0\|_1 + \varepsilon \sum_{s=0}^{k-1} \sum_{j=1}^n \Delta p_j^s \geq \varepsilon \left(\|p^0\|_1 + k\underline{\delta} \right) > \varepsilon \underline{\delta} k.$$

The desired conclusion obviously follows from this.

Corollary 3. *Suppose, that at each iteration of algorithm 1, at least one of the components of market excess demand is positive. Then there exists $\varepsilon > 0$ such that*

$$|\mathbf{E}_i(p^k)| < 2n^2(\bar{\delta}/\varepsilon\underline{\delta}) \cdot (k-1)^{-1}$$

for all sufficiently large k.

Proof. By the Lemma 3 all $\mathbf{E}_i(p^k) > -\bar{\delta}/p_i^{k-1}$, so, by the Lemma 6, for all non-positive $\mathbf{E}_i(p^k)$ the inequality holds

$$|\mathbf{E}_i(p^k)| < \bar{\delta}/p_i^{k-1} < n\bar{\delta}/[\varepsilon\underline{\delta}(k-1)].$$

According to the Walras law,

$$p^k \cdot \mathbf{E}(p^k) = \sum_{i \in I_+(k)} p_i^k \mathbf{E}_i(p^k) + \sum_{i \notin I_+(k)} p_i^k \mathbf{E}_i(p^k) = 0,$$

so the last lemma gives us the inequality

$$\frac{\varepsilon\underline{\delta}k}{n} \max_{i \in I_+(k)} \left\{ \mathbf{E}_i(p^k) \right\} \leq \sum_{i \in I_+(k)} p_i^k \mathbf{E}_i(p^k) = - \sum_{i \notin I_+(k)} p_i^k \mathbf{E}_i(p^k) < \bar{\delta} \sum_{i \notin I_+(k)} \frac{p_i^k}{p_i^{k-1}}.$$

Whence for all positive $\mathbf{E}_i(p^k)$ and all sufficiently large k we have an upper bound

$$|\mathbf{E}_i(p^k)| \leq \max_{i \in I_+(k)} \left\{ \mathbf{E}_i(p^k) \right\} < \frac{n\bar{\delta}}{\varepsilon\underline{\delta}k} \sum_{i \notin I_+(k)} \frac{p_i^k}{p_i^{k-1}} < \frac{n\bar{\delta}}{\varepsilon\underline{\delta}k} \sum_{i \notin I_+(k)} \frac{p_i^{k-1} + \bar{\delta}}{p_i^{k-1}} < \frac{n\bar{\delta}}{\varepsilon\underline{\delta}k} 2n.$$

Combining both obtained estimates, we complete the proof.

5   Computational Experiments

For a computational experiment, we choose two methods: our algorithm 1, and algorithm 2 from [17] (see (5), it also converges under weak assumptions). Tests were selected from a well-known list of "bad" problems. Some of them were constructed using a random generating. The starting price vectors for all examples are chosen the same, and only the control rules for step parameters in (5) are varied (version 2.1: $\alpha_{k+1} = 1/(k+1)$, version 2.2: $\alpha_{k+1} = \alpha_k(1-\alpha_k)$, $\alpha_1 = 1/2$). For calculations, we use the MATLAB system.

Table 1 demonstrates the results for the cyclical economy [17] already mentioned above. For this economy, only the necessary conditions for the existence of equilibrium are satisfied. Besides, many well-known methods demonstrate chaotic behavior for it. Below, the data in the first column indicates the number of iterations. The data in the rest columns show the equilibrium accuracy achieved to these iterations and estimated by the 1-norm of the market excess demand vector. The initial prices are equal, $\|\mathbf{E}(p^0)\|_1 = 4.091$.

As can be seen from Table 1, in the beginning, Algorithm 2 is better than Algorithm 1. But then, with increasing accuracy requirements, both algorithms demonstrate similar results. Nevertheless, Algorithm 1 requires less information to choose the next direction of descent. As for the choice of step parameters, it affects the convergence rate in the beginning. But with a sufficiently large number of iterations, these differences disappear.

Table 2 demonstrates the analogous results for Cobb–Douglas economy of dimension 10×10, randomly constructed. In this case, the matrix A is tridiagonal, and the matrix B is the identity matrix. The substantive meaning of the data given in Table 2 is the same as in Table 1. The initial prices in each method are the same too, $\|\mathbf{E}(p^0)\|_1 = 15.42$. We see that the experiments confirm the previous conclusions.

In Table 3, Algorithm 1 is applied to G. Scarf's economy from [10] and to an economy with consumer preferences expressed by C.E.S. utility functions. Here, starting prices $p^0 > 0$ are chosen randomly from the range $(0.1; 10)$. For them, $\|\mathbf{E}(p^0)\|_1 = 2.21$ for C.E.S. economics, and $\|\mathbf{E}(p^0)\|_1 = 0.61$ for G. Scarf's example. Although there is no theoretical analysis for these economies, the calculation results are encouraging.

Recall that, in C.E.S. economy, the utility functions are of the form

$$u(x) = \left(\sum_{j=1}^{n} a_j^{1/r} x_j^{(r-1)/r} \right)^{r/(r-1)},$$

where all $a_j > 0$, $r < 1$. It means that

$$\mathbf{E}_j(p) = \frac{a_j}{p_j^r} \sum_{s=1}^{n} p_s b_s \left(\sum_{s=1}^{n} p_s^{1-r} a_s \right)^{-1} - b_j \quad (j = 1, ..., n).$$

For the experiments, we take $m = n = 3$.

As for G. Scarf's example, its market excess demand is as

$$\mathbf{E}(p_1, p_2, p_3) = \left(\frac{p_3}{p_1 + p_3} - \frac{p_2}{p_1 + p_2}; \ \frac{p_1}{p_1 + p_2} - \frac{p_3}{p_2 + p_3}; \ \frac{p_2}{p_2 + p_3} - \frac{p_1}{p_1 + p_3} \right).$$

This economy is a popular counterexample to many tâtonnement algorithms, though has a single equilibrium (it is the ray $p_1 = p_2 = p_3 > 0$). Nevertheless, there is neither the property of global gross substitutability nor the fulfillment of the weak axiom of revealed preferences nor the property $\bar{p} \cdot \mathbf{E}(p) > 0$. As we see, for G. Scarf's example converges is especially slow. But it is not surprising.

Table 1. The accuracy of the solution for problem 1

Iteration number	Algorithm 1			Algorithm 2	
	$\delta = 1.0$	$\delta = 0.1$	$\delta = 0.01$	vers. 2.1	vers. 2.2
100	0.051	0.135	3.075	0.039	0.040
500	0.197	0.012	1.120	0.048	0.037
1000	0.005	0.048	0.115	0.036	0.043
5000	0.002	0.001	0.001	0.0005	0.0004
10000	0.0005	0.0005	0.0005	0.0004	0.0004
100000	5.e−5	5.e−5	5.e−5	4.e−5	4.e−5

Table 2. The accuracy of the solution for problem 2

Iteration number	Algorithm 1			Algorithm 2	
	$\delta = 1.0$	$\delta = 0.1$	$\delta = 0.01$	vers. 2.1	vers. 2.2
100	0.425	5.56	12.7	0.034	0.033
500	0.007	1.55	7.9	0.007	0.007
1000	0.007	0.414	5.56	0.004	0.004
10000	0.0003	0.0007	0.414	0.0003	0.0003
100000	2.e−5	1.e−5	3.e−5	3.e−5	3.e−5

Table 3. Behavior of algorithm 1 for other types of economies

Iteration number	C.E.S Economy			Economy by G.Scarf		
	$\delta = 1.0$	$\delta = 0.1$	$\delta = 0.01$	$\delta = 1.0$	$\delta = 0.1$	$\delta = 0.01$
500	0.006	0.127	0.245	0.043	0.18	0.24
1000	0.003	0.015	0.0721	0.021	0.09	0.10
5000	0.0007	0.0016	0.0022	0.0044	0.012	0.022
10000	0.0005	0.0008	0.0011	0.0023	0.006	0.009
100000	4.e−5	4.e−5	5.e−5	0.0002	0.002	0.03
1000000				2.e−5	0.0002	0.002

6 Conclusion

To find an equilibrium in the classical Arrow–Debreu model of pure exchange with the multiplicative utilities of its agents, we construct new iterative method simulating the simplest and instinctive forms of behavior of economic agents. The method use increases in prices for scarce commodities only and converge under very weak assumptions. The convergence theorems are given and the results of computing experiments are presented, including ones for other types of economies. It turned out that new method converges for these economies too.

References

1. Walras, L.: Elements d'economie politique pure. Corbaz, Lausanne (1874)
2. Samuelson, P.A.: The stability of equilibrium: comparative statics and dynamics. Econometrica **9**(2), 97–120 (1941)
3. Arrow, K.J., Debreu, G.: Existence of equilibrium for a competitive economy. Econometrica **25**, 265–290 (1954)
4. Arrow, K.J., Hurwicz, L.: On the stability of the competitive equilibrium. Econometrica **26**, 522–552 (1958)
5. Uzawa, H.: Walras' tatonnenment in the theory of exchange. Rev. Econ. Stud. **27**, 182–194 (1960)
6. Arrow, K.J., Hahn, F.H.: General Competitive Analysis. North-Holland, Amsterdam (1971)
7. Nicaido, H.: Convex Structures and Economic Theory. Academic Press, New York (1968)
8. Shafer, W.J., Sonnenschein, H.F.: Some theorems on the existence of competitive equilibrium. J. Econ.Theory **11**, 83–93 (1975)
9. Eaves, B.C.: Finite solution of pure trade markets with Cobb – Douglas utilities. Math. Program. Study **23**, 226–239 (1985)
10. Scarf, H.: Some examples of global instability of the competitive equilibrium. Internat Econ. Rev. **1**, 157–172 (1960)
11. Bala, V., Majumdar, M.: Chaotic Tatonnement. Econ. Theory **2**(4), 437–445 (1992)
12. Mukherji, A.: A simple example of complex dynamics. Econ. Theory **14**, 741–749 (1999)
13. Tuinstra, J.: A discrete and symmetric price adjustment process on the simplex. J. Econ. Dyn. Control **24**(5–7), 881–907 (2000)
14. Antipin, A.S.: Extra-proximal approach to calculating equilibriums in pure exchange models. Comput. Math. Math. Phys. **46**(10), 1687–1698 (2006)
15. Cole, R., Fleischer, I.: Fast-converging tatonnement algorithms for one-time and ongoing market problems. In: Proceedings of the Fortieth Annual ACM Symposium on Theory of Computing, pp. 315–324. Association for Computing Machinery, New York (2008)
16. Kitti, M.: Convergence of iterative tatonnement without price normalization. J. Econ. Dyn. Control **34**, 1077–1091 (2010)
17. Shikhman, V., Nesterov, Y., Ginsburg, V.: Power method tatonnements for Cobb-Douglas economies. J. Math. Econ. **75**, 84–92 (2018)

The Continuous Hotelling Pure Location Game with Elastic Demand Revisited

Pierre von Mouche[(✉)]

Wageningen Universiteit, Wageningen, The Netherlands
pvmouche@deds.nl

Abstract. The Hotelling pure location game has been revisited. It is assumed that there are two identical players, strategy sets are one-dimensional, and demand as a function of distance is constant or strictly decreasing. Besides qualitative properties of conditional payoff functions, attention is given to the structure of the equilibrium set, best-response correspondences and the existence of potentials.

Keywords: Hotelling game · Potential game · Pure Nash equilibrium existence · Principle of Minimum Differentiation

1 Introduction

In mathematics, a strategy to make progress is by studying concrete examples and thereby trying to find out what drives the results. This in particular holds in game theory for the topic of Nash equilibria of games in strategic form. An important example in this context is Cournot oligopoly games. In the present article we consider another one: Hotelling games.

By 'Hotelling games', one understands a variety of games that appeared in the literature after the seminal article of Hotelling [1].[1] The focus of interest of the present article is pure location Hotelling games. In fact we consider the pure location part of the model in [1] dealing with two sellers of a homogeneous product locating a single plant on a finite one-dimensional geographic market.

The aim of our article is to further develop the theory for the Hotelling pure location game with elastic demand. The bulk of articles presupposes inelastic demand, meaning that the demand function f is constant; the elastic case was first treated by [2]. Our article deals with the more difficult elastic case. The game we consider is a generalisation of that in [7] in the sense that more general demand functions are allowed.

The article is organized as follows. In Sect. 2, we fix the setting. Section 3 makes some useful observations about Nash equilibria of games with location and player symmetry. Section 4 reviews the inelastic case. Before proceeding in Sect. 6 to the equilibrium structure of the elastic case, Sect. 5 establishes properties of game-theoretic fundamental objects for this case. Section 7 investigates in which sense the game is a potential game. Finally, Sect. 8 provides some concluding remarks.

---

[1] For an overview and discussion of the literature we refer to [5] and [6].

© Springer Nature Switzerland AG 2020
A. Kononov et al. (Eds.): MOTOR 2020, LNCS 12095, pp. 246–262, 2020.
https://doi.org/10.1007/978-3-030-49988-4_17

2 Setting

Below we fix the setting for the Hotelling game that we are going to consider. Well, strategy sets are one-dimensional, there are two identical players and demand may be inelastic. We will allow for a non-continuous demand function. In order to distinguish the game in the present article from discrete variants (see [14] and references therein), we simply refer to it as the cHg ('continuous Hotelling game').

Throughout the whole article, S denotes a real interval $[0, L]$ with $L > 0$ and $f : S \to \mathbb{R}$ is a positive function which is constant or is strictly decreasing; without loss of generality we assume $f = 1$ in the case f is constant. The case of constant f is referred to as the inelastic case and the other one as elastic case.

In this article by a continuous Hotelling game (cHg), we understand a two-person game in strategic form with player set $N = \{1, 2\}$, common strategy set S and defining the function $\mathcal{L} : S \to \mathbb{R}$ by

$$\mathcal{L}(x) := \int_0^x f(z)\, dz$$

payoff functions $u_1, u_2 \colon S \times S \to \mathbb{R}$ given by,

$$u_i(x_1, x_2) = \begin{cases} \mathcal{L}(x_i) + \mathcal{L}(\frac{|x_1 - x_2|}{2}) & \text{if } x_i < x_j, \\ \mathcal{L}(L - x_i) + \mathcal{L}(\frac{|x_1 - x_2|}{2}) & \text{if } x_i > x_j, \\ \frac{1}{2}(\mathcal{L}(x_i) + \mathcal{L}(L - x_i)) & \text{if } x_i = x_j. \end{cases}$$

In the case where f even is continuously differentiable, the cHg becomes the game in [7]. We refer to f as a demand function.[2]

The functions $f_- : [0, L[\to \mathbb{R}$ and $f_+ : [0, L[\to \mathbb{R}$ are well-defined by

$$f_-(z) := \lim_{w \uparrow z} f(w), \quad f_+(z) := \lim_{w \downarrow z} f(w).$$

So f_- and f_+ are decreasing, $f_- \geq f \geq f_+$. One also knows that f_- is right continuous and that f_+ is left continuous.

$\mathcal{L}$ has the following simple properties:

A. $\mathcal{L} \geq 0$ and $\mathcal{L}(0) = 0$.
B. $\mathcal{L}$ is strictly increasing.
C. $\mathcal{L}$ is continuous.
D. $\mathcal{L}$ is linear if f is constant and $\mathcal{L}$ is strictly concave if f is strictly decreasing.
E. f is semidifferentiable: $D_-\mathcal{L}(x) = f_-(x)$ and $D_+\mathcal{L}(x) = f_+(x)$. And if f is continuous at x, then $\mathcal{L}$ is differentiable at x and $D\mathcal{L}(x) = f(x)$.

---

[2] Its standard interpretation in location theory concerns two competing vendors on a beach. The vendors simultaneously and independently select a position. Customers go to the closest vendor and split themselves evenly if the vendors choose an identical position. Each vendor wants to maximize his number of customers. One can reframe the interpretation as two candidates placing themselves along an ideological spectrum, with citizens voting for whichever one is closest (see e.g. [4]).

F. For $x' > x$ the inequalities $(x' - x)f(x') \leq \mathcal{L}(x') - \mathcal{L}(x) \leq (x' - x)f(x)$ hold and these inequalities are strict if f is strictly decreasing.

There is an interesting principle for the cHg, the so called Principle of Minimum Differentiation. For our (interpretation of the) cHg, this principle, coined by Boulding [3], comes down to that firms liking[3] to locate together. We formalize this principle for the cHg as follows: the <u>Principle of Minimum Differentiation</u> holds if the game has $(\frac{L}{2}, \frac{L}{2})$ as unique (pure) Nash equilibrium.

3   Games with Player and Location Symmetry

The content of this section is borrowed from [14].

In this section we consider a game in strategic form with two players 1, and 2, with common strategy set $S = [0, L]$ with $L > 0$, and with payoff functions $g_1, g_2 : S \times S \to \mathbb{R}$. Assume, player symmetry, i.e.

$$g_2(x_1, x_2) = g_1(x_2, x_1) \quad (x_1, x_2 \in S).$$

Also assume,

$$g_i(x_1, x_2) = g_i(L - x_1, L - x_2) \quad (x_1, x_2 \in S),$$

i.e. location symmetry. The cHg is an example of such a game.

We denote the conditional payoff function of player i where his opponent plays $x_j \in S$ by $g_i^{(x_j)}$; so $g_1^{(x_2)} : S \to \mathbb{R}$ is defined by $g_1^{(x_2)}(x_1) = g_1(x_1, x_2)$ and $g_2^{(x_1)} : S \to \mathbb{R}$ of player 2 is defined by $g_2^{(x_1)}(x_2) = g_2(x_1, x_2)$. With B_i we denote the best-response correspondence of player i; so $B_i : S \multimap S$.

The location symmetry implies the formulas[4]

$$g_i^{(L-z)}(x) = g_i^{(z)}(L - x) \text{ and } B_i(x) = \{L\} - B_i(L - x).$$

And player symmetry implies

$$g_1^{(z)} = g_2^{(z)} \text{ and } B_1 = B_2 =: B.$$

Denoting the Nash equilibrium set by E, player symmetry also implies for every $(e_1, e_2) \in E$ that $\{(e_1, e_2), (e_2, e_1)\} \subseteq E$ and location symmetry implies that $\{(e_1, e_2), (L - e_1, L - e_2)\} \subseteq E$. Thus for every $(e_1, e_2) \in E$

$$\{(e_1, e_2), (e_2, e_1), (L - e_1, L - e_2), (L - e_2, L - e_1)\} \subseteq E. \tag{1}$$

Having this, we like to see $(e_1, e_2), (e_2, e_1), (L - e_1, L - e_2), (L - e_2, L - e_1)$ as the same equilibrium. We formalize this by defining on E the relation $\sim$ by

$$(e_1, e_2) \sim (e'_1, e'_2) \text{ means:}$$

---

[3] However, see concluding remark 3 in Sect. 8.

[4] Here $\{L\} - B_i(L - x)$ is the Minkowski sum of the sets $\{L\}$ and $-B_i(L - x)$.

$$(e_1, e_2) \in \{(e_1', e_2'), (e_2', e_1'), (L - e_1', L - e_2'), (L - e_2', L - e_1')\}.$$

It is straightforward to check that this relation is an equivalence relation. Denote by $[E]$ the set of its equivalence classes, to be called equilibrium classes, and by $[(e_1, e_2)]$ the equilibrium class of $(e_1, e_2) \in E$. We have

$$[(e_1, e_2)] = \{(e_1, e_2), (e_2, e_1), (L - e_1, L - e_2), (L - e_2, L - e_1)\}.$$

We define the multiplicity of an equilibrium as the number of elements of its equilibrium class. Of course, if the game has a unique equilibrium (e_1, e_2), then there is just one equilibrium class consisting of this equilibrium and (e_1, e_2) has multiplicity 1. Note that with the action distance of an action profile (x_1, x_2) defined by $|x_2 - x_1|$, each element of a given equilibrium class has the same action distance. Also note that (1) implies:

$$\#E = 1 \Rightarrow E = \{(\frac{L}{2}, \frac{L}{2})\}.$$

Theorem 1. *If $(e_1, e_2) \in E$, then (e_1, e_2) has multiplicity $1, 2$ or 4 and*

$$\#[(e_1, e_2)] = 1 \Leftrightarrow e_1 = e_2 \wedge e_1 + e_2 = L \Leftrightarrow e_1 = e_2 = \frac{L}{2};$$

$$\#[(e_1, e_2)] = 2 \Leftrightarrow [e_1 = e_2 \wedge e_1 + e_2 \neq L] \vee [e_1 \neq e_2 \wedge e_1 + e_2 = L];$$

$$\#[(e_1, e_2)] = 4 \Leftrightarrow [e_1 \neq e_2 \wedge e_1 + e_2 \neq L]. \qquad \diamond$$

Proof. It is easy to prove the three displayed statements. They in turn imply, as desired, that $\#[(e_1, e_2)] \neq 3$. Q.E.D.

We shall freely use all the results in this section together with the results A–F for the function $\mathcal{L}$ in Sect. 2. Because of player symmetry, we often only present results for player 1.

4   Inelastic Case

Let us start our investigation of the continuous Hotelling game by considering the well-known inelastic case, i.e. the case where f is constant. Without loss of generality, we assume $f = 1$.

First it may good to have a look to the simple results in Lemma 1 and Proposition 1 below. In addition to these results, the following simple results hold for the inelastic case (but not with exception of those in parts 2a and 2b for the elastic case):

Theorem 2. *1. (a) $u_1^{(x_2)}$ is on $[0, x_2[$ strictly increasing and on $]x_2, L]$ strictly decreasing.*

(b) If $x_2 \leq \frac{L}{2}$, then $u_1^{(x_2)}$ is on $[0, x_2]$ strictly increasing.

(c) If $x_2 \geq \frac{L}{2}$, then $u_1^{(x_2)}$ is on $[x_2, L]$ strictly decreasing.

2. (a) If $x_2 < \frac{L}{2}$, then $u_1^{(x_2)}$ is on $[0, x_2[$ concave and on $[x_2, L]$ concave.

   (b) If $x_2 > \frac{L}{2}$, then $u_1^{(x_2)}$ is on $[0, x_2]$ concave and on $]x_2, L]$ concave.

   (c) $u_1^{(\frac{L}{2})}$ is concave.

   (d) $u_1^{(x_2)}$ is strictly quasi-concave.

3. $B(x) = \begin{cases} \emptyset & \text{if } x \neq \frac{L}{2}, \\ \{\frac{L}{2}\} & \text{if } x = \frac{L}{2}. \end{cases}$

4. $E = \{(\frac{L}{2}, \frac{L}{2})\}$. Thus there is one equilibrium class, this class contains one element and the Principle of Minimum Differentiation holds.   ◇

5 Properties of Fundamental Objects

5.1 Smoothness

Lemma 1. 1. If $0 < x_2 < \frac{L}{2}$, then $\lim_{x_1 \uparrow x_2} u_1^{(x_2)}(x_1) = \mathcal{L}(x_2) < u_1^{(x_2)}(x_2) < \mathcal{L}(L - x_2) = \lim_{x_1 \downarrow x_2} u_1^{(x_2)}(x_1)$. And $u_1^{(0)}(0) < \mathcal{L}(L) = \lim_{x_1 \downarrow 0} u_1^{(0)}(x_1)$.

2. If $L > x_2 > \frac{L}{2}$, then $\lim_{x_1 \uparrow x_2} u_1^{(x_2)}(x_1) = \mathcal{L}(x_2) > u_1^{(x_2)}(x_2) > \mathcal{L}(L - x_2) = \lim_{x_1 \downarrow x_2} u_1^{(x_2)}(x_1)$. And $u_1^{(L)}(L) < \mathcal{L}(L) = \lim_{x_1 \uparrow L} u_1^{(0)}(x_1)$.

3. $\lim_{x_1 \to \frac{L}{2}} u_1^{(\frac{L}{2})}(x_1) = u_1^{(\frac{L}{2})}(\frac{L}{2}) = \mathcal{L}(\frac{L}{2})$.   ◇

Proof. 1. $\lim_{x_1 \uparrow x_2} u_1^{(x_2)}(x_1) = \lim_{x_1 \uparrow x_2}(\mathcal{L}(x_1) + \mathcal{L}(\frac{|x_1 - x_2|}{2})) = \mathcal{L}(x_2) + \mathcal{L}(0) = \mathcal{L}(x_2) < \frac{\mathcal{L}(x_2) + \mathcal{L}(L - x_2)}{2} = u_1^{(x_2)}(x_2)$ and $\lim_{x_1 \downarrow x_2} u_1^{(x_2)}(x_1) = \lim_{x_1 \uparrow x_2}(\mathcal{L}(L - x_1) + \mathcal{L}(\frac{|x_1 - x_2|}{2})) = \mathcal{L}(L - x_2) + \mathcal{L}(0) = \mathcal{L}(L - x_2) > \frac{\mathcal{L}(x_2) + \mathcal{L}(L - x_2)}{2} = u_1^{(x_2)}(x_2)$.

2. Analogous to part 1.

3. $\lim_{x_1 \downarrow \frac{L}{2}} u_1^{(\frac{L}{2})}(x_1) = \lim_{x_1 \downarrow \frac{L}{2}}(\mathcal{L}(L - x_1) + \mathcal{L}(\frac{|x_1 - \frac{L}{2}|}{2})) = \mathcal{L}(\frac{L}{2}) = u_1^{(\frac{L}{2})}(\frac{L}{2})$.

With this $\lim_{x_1 \uparrow \frac{L}{2}} u_1^{(\frac{L}{2})}(x_1) = \lim_{x_1 \uparrow \frac{L}{2}} u_1^{(\frac{L}{2})}(L - x_1) = \lim_{x_1 \downarrow \frac{L}{2}} u_1^{(\frac{L}{2})}(x_1) = u_1^{(\frac{L}{2})}(\frac{L}{2})$. So the desired result follows. Q.E.D.

Proposition 1. 1. $u_1^{(x_2)}$ is continuous at every $x_1 \neq x_2$ and discontinuous at every $x_1 = x_2 \neq \frac{L}{2}$.

2. If $x_2 \notin \{0, \frac{L}{2}, L\}$, then $u_1^{(x_2)}$ is at $x_1 = x_2$ neither upper-semicontinuous nor lower-semicontinuous. $u_1^{(0)}$ is at 0 lower-semicontinuous, but not upper-semicontinuous. $u_1^{(L)}$ is at L upper-semicontinuous, but not lower-semicontinuous.

3. For $x_2 < \frac{L}{2}$, $u_1^{(x_2)}$ is left-upper-semicontinuous at x_2 and right-lower-semicontinuous at x_2. For $x_2 > \frac{L}{2}$, $u_1^{(x_2)}$ is right-upper-semicontinuous at x_2 and left-lower-semicontinuous at x_2.

4. $u_1^{(\frac{L}{2})}$ is continuous.

5. $u_1^{(x_2)}$ is semidifferentiable at each $x_1 \neq x_2$. If f is continuous, then $u_1^{(x_2)}$ even is differentiable at each $x_1 \neq x_2$.   ◇

Proof. 2. This follows from Lemma 1(1, 2).

1. First statement: clear. Second statement: from part 2.
3. By Lemma 1(1, 2).
4. By Lemma 1(3) together with part 1.
5. Clear. Q.E.D.

Here is an improvement of Proposition 1(5).

Proposition 2. *1.* $D_\pm u_1^{(x_2)}(x_1) = f_\pm(x_1) - \frac{1}{2}f_\mp(\frac{x_2-x_1}{2})$ $(x_1 < x_2)$ and

$D_\pm u_1^{(x_2)}(x_1) = -f_\mp(L-x_1) + \frac{1}{2}f_\pm(\frac{x_1-x_2}{2})$ $(x_1 > x_2)$.[5]

2. $u_1^{(\frac{L}{2})}$ *is semidifferentiable at* $\frac{L}{2}$ *and* $D_\pm u_1^{(\frac{L}{2})}(\frac{L}{2}) = \mp f_-(\frac{L}{2}) \pm \frac{1}{2}f_+(0)$. ◇

Proof. 1. First statement: for $x_1 \in [0, x_2[$, we have $u_1^{(x_2)}(x_1) = \mathcal{L}(x_1) + \mathcal{L}(\frac{x_2-x_1}{2})$.
This implies $D_\pm u_1^{(x_2)}(x_1) = \mathcal{L}_\pm(x_1) + D_\pm(\mathcal{L}(\frac{x_2-x_1}{2})) = f_\pm(x_1) - \frac{1}{2}f_\mp(\frac{x_2-x_1}{2})$.
Second statement: in the same way.
2. Suppose f is continuous at 0 and $\frac{L}{2}$. We have

$$D_+ u_1^{(\frac{L}{2})}\left(\frac{L}{2}\right) = \lim_{h\downarrow 0} \frac{u_1^{(\frac{L}{2})}(\frac{L}{2}+h) - u_1^{(\frac{L}{2})}(\frac{L}{2})}{h} = \lim_{h\downarrow 0} \frac{\mathcal{L}(\frac{L}{2}-h) + \mathcal{L}(\frac{h}{2}) - \mathcal{L}(\frac{L}{2})}{h}$$

$$= \lim_{h\downarrow 0} \frac{\mathcal{L}(\frac{L}{2}-h) - \mathcal{L}(\frac{L}{2})}{h} + \lim_{h\downarrow 0} \frac{\mathcal{L}(\frac{h}{2})}{h} = \lim_{h\uparrow 0} -\frac{\mathcal{L}(\frac{L}{2}+h) - \mathcal{L}(\frac{L}{2})}{h} + \lim_{h\downarrow 0} \frac{\mathcal{L}(h) - \mathcal{L}(0)}{2h}$$

$$= -D_-\mathcal{L}\left(\frac{L}{2}\right) + \frac{1}{2}D_+\mathcal{L}(0) = -f_-\left(\frac{L}{2}\right) + \frac{1}{2}f_+(0).$$

From this $D_- u_1^{(\frac{L}{2})}(\frac{L}{2}) = \lim_{h\uparrow 0} \frac{u_1^{(\frac{L}{2})}(\frac{L}{2}+h) - u_1^{(\frac{L}{2})}(\frac{L}{2})}{h} = \lim_{h\downarrow 0} \frac{u_1^{(\frac{L}{2})}(\frac{L}{2}-h) - u_1^{(\frac{L}{2})}(\frac{L}{2})}{-h} =$

$\lim_{h\downarrow 0} -\frac{u_1^{(\frac{L}{2})}(\frac{L}{2}+h) - u_1^{(\frac{L}{2})}(\frac{L}{2})}{h} = -D_+ u_1^{(\frac{L}{2})}(\frac{L}{2})$. Q.E.D.

5.2   Monotonicity

Lemma 2. *1. For all* $x_2, x_1, x_1' \in S$ *with* $x_2 < x_1 < x_1'$

$$u_1^{(x_2)}(x_1) - u_1^{(x_2)}(x_1') \begin{cases} \geq (x_1' - x_1)(f(L-x_1) - \frac{1}{2}f(\frac{x_1-x_2}{2})), \\ \leq (x_1' - x_1)(f(L-x_1') - \frac{1}{2}f(\frac{x_1'-x_2}{2})). \end{cases}$$

2. For all $x_2, x_1, x_1' \in S$ *with* $x_1 < x_1' < x_2$

$$u_1^{(x_2)}(x_1) - u_1^{(x_2)}(x_1') \begin{cases} \geq (x_1' - x_1)(\frac{1}{2}f(\frac{x_2-x_1}{2}) - f(x_1)), \\ \leq (x_1' - x_1)(\frac{1}{2}f(\frac{x_2-x_1'}{2}) - f(x_1')). \end{cases}$$ ◇

---

[5] So, if f is continuous, then by Proposition 1(5), these formulas become $D u_1^{(x_2)}(x_1) = f(x_1) - \frac{1}{2}f(\frac{x_2-x_1}{2})$ $(x_1 < x_2)$ and $D u_1^{(x_2)}(x_1) = -f(L-x_1) + \frac{1}{2}f(\frac{x_1-x_2}{2})$ $(x_1 > x_2)$.

Proof. 1. We have $u_1^{(x_2)}(x_1) - u_1^{(x_2)}(x_1')$

$$= \left(\mathcal{L}(L - x_1) - \mathcal{L}(L - x_1')\right) - \left(\mathcal{L}(\frac{x_1' - x_2}{2}) - \mathcal{L}(\frac{x_1 - x_2}{2})\right).$$

From this, as desired, $u_1^{(x_2)}(x_1) - u_1^{(x_2)}(x_1') \geq (x_1' - x_1)f(L - x_1) - \frac{x_1' - x_1}{2}f(\frac{x_1 - x_2}{2})$
and $u_1^{(x_2)}(x_1) - u_1^{(x_2)}(x_1') \leq (x_1' - x_1)f(L - x_1') - \frac{x_1' - x_1}{2}f(\frac{x_1' - x_2}{2})$.

2. Analogous to part 1. Q.E.D.
   Notation:

$$V := \{x_1 \in S \mid f(L - x_1) \geq \frac{1}{2}f(0)\}. \tag{2}$$

Note that V is a real interval containing L.

Lemma 3. $u_1^{(x_2)}$ *is strictly decreasing on* $V \cap \,]x_2, L]$. ◇

Proof. Suppose $x_1, x_1' \in V \cap \,]x_2, L]$ with $x_1 < x_1'$. We prove that $u_1^{(x_2)}(x_1) > u_1^{(x_2)}(x_1')$. Well, as $f(L - x_1) \geq \frac{1}{2}f(0)$, we obtain with Lemma 2(1), as desired,

$$u_1^{(x_2)}(x_1) - u_1^{(x_2)}(x_1') \geq (x_1' - x_1)(f(L - x_1) - \frac{1}{2}f(\frac{x_1 - x_2}{2})) \geq$$

$$(x_1' - x_1)(f(L - x_1) - \frac{1}{2}f(0)) \geq 0.$$

Finally, note that here the last inequality is strict if f is constant and otherwise the second inequality is strict. Q.E.D.

Lemma 4. *Suppose f is upper-semicontinuous at $L - x_2$ and $f(L - x_2) < \frac{1}{2}f(0)$. Then there exists a punctured right neighbourhood of x_2 on which $u_1^{(x_2)}$ is strictly increasing.* ◇

Proof. Note that $x_2 < L$. Let $d = \frac{1}{2}f(0) - f(L - x_2)$. As f is upper-semicontinuous at $L - x_2$, we can fix x_1' with $L > x_1' > x_2$ such that

$$f(L - x_1') < f(L - x_2) + \frac{d}{2} \ (x_2 < x_1' < x_1'').$$

We have $\frac{1}{2}f(\frac{x_1'' - x_2}{2}) > 0 > \frac{d}{2} - \frac{1}{2}f(0)$. With Lemma 2(1), we obtain for $x_1, x_1' \in \,]x_2, x_1'[$ with $x_1 < x_1'$, as desired that $u_1^{(x_2)}(x_1) - u_1^{(x_2)}(x_1') \leq (x_1' - x_1)(f(L - x_1') - \frac{1}{2}f(\frac{x_1' - x_2}{2})) < (x_1' - x_1)(f(L - x_2) + \frac{d}{2} + (\frac{d}{2} - \frac{1}{2}f(0))) = 0$. Q.E.D.

Proposition 3. *Suppose $f(\frac{L}{2}) \geq \frac{1}{2}f(0)$.*

1. *If $x_2 \leq \frac{L}{2}$, then $u_1^{(x_2)}$ is strictly increasing on $[0, x_2]$.*
2. *If $x_2 \geq \frac{L}{2}$, then $u_1^{(x_2)}$ is strictly decreasing on $[x_2, L]$.*
3. *$u_1^{(\frac{L}{2})}$ has $\frac{L}{2}$ as unique maximiser.* ◇

Proof. 1, 2. By location symmetry, it is sufficient to prove part 2. Fix $x_2 \geq \frac{L}{2}$. We prove that $u_1^{(x_2)}$ is strictly decreasing on $]x_2, L]$; together with Lemma 1(2) and Proposition 1(4) the desired result follows. Well, with V as in (2), $[\frac{L}{2}, L] \subseteq V$. This implies $]x_2, L] \subseteq V \cap]x_2, L]$. Finally, apply Lemma 3.

3. By parts 1 and 2. Q.E.D.

The statements in Proposition 3 are no longer valid for the situation $f(\frac{L}{2}) < \frac{1}{2}f(0)$. For example, Lemma 4 shows (by taking $x_2 = \frac{L}{2}$), that then Proposition 3(2) is no longer valid.

Proposition 4. *Suppose f is continuous.*

1. *If $x_2 < \frac{L}{2}$, then $u_1^{(x_2)}$ is strictly decreasing on $[\frac{5}{6}L, L]$.*
2. *If $x_2 > \frac{L}{2}$, then $u_1^{(x_2)}$ is strictly increasing on $[0, \frac{1}{6}L]$.*   ◇

Proof. By location symmetry, it is sufficient to prove part 1. So suppose $x_2 < \frac{L}{2}$. Note that $\frac{x_1 - x_2}{2} > \frac{2x_1 - L}{4}$ and for $x_1 \geq \frac{5}{6}L$ that $L - x_1 \leq \frac{2L - x_1}{4}$. Finally, note that, together with Proposition 2(1), for $x_1 \geq \frac{5}{6}L$

$$Du_1^{(x_2)}(x_1) = -f(L - x_1) + \frac{1}{2}f(\frac{x_1 - x_2}{2}) \leq -f(\frac{2L - x_1}{4}) + \frac{1}{2}f(\frac{2L - x_1}{4}) < 0. \quad \text{Q.E.D.}$$

5.3   Concavity

Proposition 5. *Suppose f is strictly decreasing.*

1. *If $x_2 < \frac{L}{2}$, then $u_1^{(x_2)}$ is on $[0, x_2[$ strictly concave and on $[x_2, L]$ strictly concave.*
2. *If $x_2 > \frac{L}{2}$, then $u_1^{(x_2)}$ is on $[0, x_2]$ strictly concave and on $]x_2, L]$ strictly concave.*
3. *$u_1^{(\frac{L}{2})}$ is on $[0, \frac{L}{2}]$ strictly concave and on $[\frac{L}{2}, L]$ strictly concave.*
4. *If $f(\frac{L}{2}) \geq \frac{1}{2}f(0)$, then $u_1^{(x_2)}$ is strictly quasi-concave.*   ◇

Proof. 1. First statement: for the function $u_1^{(x_2)} : [0, x_2[\rightarrow \mathbb{R}$ we have $u_1^{(x_2)}(x_1) = \mathcal{L}(x_1) + \mathcal{L}(\frac{x_2 - x_1}{2})$. So this function is a sum of strictly concave functions and therefore strictly concave.

Second statement: for the function $u_1^{(x_2)} :]x_2, L] \rightarrow \mathbb{R}$ we have $u_1^{(x_2)}(x_1) = \mathcal{L}(L - x_1) + \mathcal{L}(\frac{x_1 - x_2}{2})$. So this function is a sum of strictly concave functions and therefore strictly concave. As the function $u_1^{(x_2)} : [x_2, L] \rightarrow \mathbb{R}$ is, by Proposition 1(1, 3) right lower-semicontinuous, it follows, as desired, that also this function is strictly concave.

2. Analogous to part 1.

3. For the function $u_1^{(\frac{L}{2})} : [0, \frac{L}{2}[\rightarrow \mathbb{R}$ we have $u_1^{(\frac{L}{2})}(x_1) = \mathcal{L}(x_1) + \mathcal{L}(\frac{\frac{L}{2} - x_1}{2})$. So this function is a sum of strictly concave functions and therefore strictly concave. As, by Proposition 1(2), $u_1^{(\frac{L}{2})}$ is continuous, it follows that $u_1^{(\frac{L}{2})}$ is on

$[0, \frac{L}{2}]$ strictly concave. By location symmetry, it follows that $u_1^{(\frac{L}{2})}$ is on $[\frac{L}{2}, L]$ strictly concave.

4. Suppose $x_2 \geq \frac{L}{2}$. By Proposition 3(2), $u_1^{(x_2)}$ is strictly decreasing on $[x_2, L]$. By parts 2 and 3, $u_1^{(x_2)}$ is on $[0, x_2]$ strictly concave. If $x_2 = \frac{L}{2}$, then $u_1^{(x_2)}$ is continuous by Proposition 1(4) and it follows that $u_1^{(x_2)}$ is strictly concave. If $x_2 > \frac{L}{2}$, then by Lemma 1(2), $\lim_{x_1 \uparrow x_2} u_1^{(x_2)}(x_1) > u_1^{(x_2)}(x_2) > \lim_{x_1 \downarrow x_2} u_1^{(x_2)}(x_1)$ and it follows that $u_1^{(x_2)}$ is strictly quasi-concave.

So the statement holds for $x_2 \geq \frac{L}{2}$. Noting that $u_1^{(x_2)}(x_1) = u_1^{(L-x_2)}(L - x_1)$ the statement now also holds for $x_2 \leq \frac{L}{2}$. Q.E.D.

5.4   Best-Response Correspondences

Notation: by fix(B) we denote the set of fixed points of the best-response correspondence B, i.e. the set $\{x \in S \mid x \in B(x)\}$.

Proposition 6. $0 \notin B(x_2)$ $(x_2 \in S)$ and $L \notin B(x_2)$ $(x_2 \in S)$. ◇

Proof. By location symmetry, it is sufficient to prove the first statement. By Lemma 1(1), this statement holds for $x_2 = 0$. Now suppose $0 < x_2 < \frac{L}{2}$. We have $u_1^{(x_2)}(0) = \mathcal{L}(0) + \mathcal{L}(\frac{x_2}{2}) < \mathcal{L}(\frac{3}{4}x_2) < \mathcal{L}(\frac{3}{4}x_2) + \mathcal{L}(\frac{x_2}{4}) = u_1^{(x_2)}(\frac{3}{4}x_2)$. Thus $0 \notin B(x_2)$. Q.E.D.

Lemma 5. $B(x_2) \subseteq \begin{cases} \text{argmax}_{x_1 \in]x_2, L]} u_1^{(x_2)}(x_1) \ (0 \leq x_2 < \frac{L}{2}), \\ \text{argmax}_{x_1 \in [0, x_2[} u_1^{(x_2)}(x_1) \ (\frac{L}{2} < x_2 \leq L). \end{cases}$ ◇

Proof. We prove the statement for $0 \leq x_2 < \frac{L}{2}$); then the other statement follows by location symmetry. Well, part 1 implies that the statement is true for $x_2 = 0$. Now suppose $0 < x_2 < \frac{L}{2}$. For every $h > 0$ with $0 \leq x_2 - h < x_2 < x_2 + h \leq L$ we have $u_1^{(x_2)}(x_2 - h) = \mathcal{L}(x_2 - h) + \mathcal{L}(\frac{h}{2}) < \mathcal{L}(L - x_2 - h) + \mathcal{L}(\frac{h}{2}) = u_1^{(x_2)}(x_2 + h)$. This implies $B(x_2) \subseteq [x_2, L]$. By Lemma 1(1), $u_1^{(x_2)}(x_2) < \lim_{x_1 \downarrow x_2} u_1^{(x_2)}(x_1)$. Therefore $B(x_2) \subseteq]x_2, L]$. The desired result now follows. Q.E.D.

Proposition 7. 1. $x_2 \neq \frac{L}{2} \Rightarrow x_2 \notin B(x_2)$. Thus fix$(B) \subseteq \{\frac{L}{2}\}$.
2. If f is continuous at 0 and $\frac{L}{2}$ and $f(\frac{L}{2}) < \frac{1}{2}f(0)$, then $\frac{L}{2} \notin$ fix(B). ◇

Proof. 1. By Lemma 5.
2. For $x_1 \in]0, \frac{L}{2}[$ we obtain

$$u_1^{(\frac{L}{2})}(x_1) = \mathcal{L}(x_1) + \mathcal{L}(\frac{L}{4} - \frac{x_1}{2}) = \mathcal{L}(\frac{L}{2}) + \mathcal{L}(x_1) - \mathcal{L}(\frac{L}{2}) + \mathcal{L}(\frac{L}{4} - \frac{x_1}{2})$$

$$= u_1^{(\frac{L}{2})}(\frac{L}{2}) - (\mathcal{L}(\frac{L}{2}) - \mathcal{L}(x_1)) + \mathcal{L}(\frac{L}{4} - \frac{x_1}{2})$$

$$\geq u_1^{(\frac{L}{2})}(\frac{L}{2}) - (\frac{L}{2} - x_1)f(x_1) + (\frac{L}{4} - \frac{x_1}{2})f(\frac{L}{4} - \frac{x_1}{2})$$

$$= u_1^{(\frac{L}{2})}(\frac{L}{2}) + \frac{1}{2}(\frac{L}{2} - x_1)(-2f(x_1) + f(\frac{L - 2x_1}{4})).$$

As $-2f(\frac{L}{2}) + f(\frac{L-2\frac{L}{2}}{4}) = -2f(\frac{L}{2}) + f(0) > 0$ and f is continuous at 0 and $\frac{L}{2}$, there exists $\delta > 0$ such that $-2f(x_1) + f(\frac{L-2x_1}{4}) > 0$ for every $x_1 \in]\frac{L}{2} - \delta, \frac{L}{2}[$. So for these x_1 we obtain $u_1^{(\frac{L}{2})}(x_1) > u_1^{(\frac{L}{2})}(\frac{L}{2})$. It follows that $\frac{L}{2} \notin B(\frac{L}{2})$. Q.E.D.

Terminology: given a correspondence $F : A \multimap B$, we call F <u>proper</u> if $F(a) \neq \emptyset$ $(a \in A)$ and call F <u>at most single-valued</u> if $\#F(a) \leq 1$ $(a \in A)$.

Proposition 8. *1. Suppose $f(\frac{L}{2}) > \frac{1}{2}f(0)$.*
  (a) B is at most single-valued.
  (b) $B(\frac{L}{2}) = \{\frac{L}{2}\}$.
  (c) Suppose f is lower-semicontinuous at $\frac{L}{2}$. Then there exists a punctured open interval around $\frac{L}{2}$ on which B is empty-valued, thus B is not proper.
2. Suppose $f(\frac{L}{2}) = \frac{1}{2}f(0)$.
  (a) B is single-valued.
  (b) $B(\frac{L}{2}) = \{\frac{L}{2}\}$.
3. Suppose $f(\frac{L}{2}) < \frac{1}{2}f(0)$.
  (a) Suppose f is upper-semicontinuous. Then B is proper and B is on $S\backslash\{\frac{L}{2}\}$ single-valued.
  (b) Suppose f is continuous. Then $B(\frac{L}{2}) = \{x_1, L - x_1\}$ with x_1 the unique solution $y \in]0, \frac{L}{2}[$ of the equation $f(y) = \frac{1}{2}f(\frac{\frac{L}{2}-y}{2})$.
4. Suppose $f(\frac{L}{2}) \leq \frac{1}{2}f(0)$, f continuous and $x_2 \in]\frac{L}{2}, L]$. Then $B(x_2) = \{x_1\}$ with x_1 the unique solution $y \in]0, x_2[$ of the equation $f(y) = \frac{1}{2}f(\frac{x_2-y}{2})$. ◇

Proof. 1b, 2b. By Proposition 3(3).

1a. As $u_1^{(x_2)}$ is, by Proposition 5, strictly quasi-concave.

1c. By location symmetry, it is sufficient to prove that there exists $\delta > 0$ such that $B(x_2) = \emptyset$ for all $x_2 \in]\frac{L}{2} - \delta, \frac{L}{2}[$. Well, as f is lower-semicontinuous at $\frac{L}{2}$, we can take $\delta > 0$ such that $f(\frac{L}{2} + \delta) > \frac{1}{2}f(0)$. Next fix $x_2 \in]\frac{L}{2} - \delta, \frac{L}{2}[$. As $\frac{L}{2} - \delta < x_2$, we have $L - x_2 < \frac{L}{2} + \delta$ and therefore for $x_1 \in]x_2, L]$ it follows that $f(L - x_1) \geq f(L - x_2) \geq f(\frac{L}{2} + \delta) > \frac{1}{2}f(0)$. So, with V as in (2), $]x_2, L] \subseteq V \cap]x_2, L]$. By Lemma 3, $u_1^{(x_2)}$ is strictly decreasing on $]x_2, L]$. As $x_2 < \frac{L}{2}$, Proposition 7(1) guarantees $B(x_2) \subseteq]x_2, L]$. It follows that $B(x_2) = \emptyset$.

2a. As $u_1^{(x_2)}$ is, by Proposition 5, strictly quasi-concave, we have $\#B(x_2) \leq 1$ $(x_2 \in S)$. So we still need to prove that $B(x_2) \neq \emptyset$ $(x_2 \in S)$. By part 2b and location symmetry, it is sufficient to show that $B(x_2) \neq \emptyset$ for $x_2 < \frac{L}{2}$.

Fix $x_2 < \frac{L}{2}$. It is sufficient to show that $u_1^{(x_2)}$ has a maximiser on $[0, x_2]$ and on $]x_2, L]$. Well, by Proposition 1(1, 3, 4), $u_1^{(x_2)}$ is on $[0, x_2]$ upper-semicontinuous and therefore, by the Lemma of Weierstrass-Lebesgue, has a maximiser on this segment. As $f(L - x_2) < f(\frac{L}{2}) = \frac{1}{2}f(0)$, Lemma 4 guarantees that there exists $\delta > 0$ such that $u_1^{(x_2)}$ is strictly increasing on $]x_2, x_2 + \delta]$. Also $u_1^{(x_2)}$ is continuous on $[x_2 + \delta, L]$. It follows that $u_1^{(x_2)}$ has a maximiser on $]x_2, L]$.

3a. First statement: by location symmetry, it is sufficient to show that $B(x_2) \neq \emptyset$ for $x_2 \leq \frac{L}{2}$. Well, for $x_2 = \frac{L}{2}$, this follows from the Weierstrass Theorem, as $u_1^{(\frac{L}{2})}$ is continuous by Proposition 1(4). Now fix $x_2 < \frac{L}{2}$. The rest of the proof is the same as that in part 3a after 'Fix $x_2 < \frac{L}{2}$'.

Second statement: by the above is sufficient to prove that $\#B(x_2) \leq 1$ for all $x_2 \neq \frac{L}{2}$. By location symmetry it is sufficient to prove this inequality for $x_2 < \frac{L}{2}$. So suppose $x_2 < \frac{L}{2}$. By Proposition 7(1), $B(x_2) \subseteq]x_2, L]$. As, by Proposition 5(1), $u_1^{(x_2)} :]x_2, L] \to \mathbb{R}$ is strictly concave, this function has at most one maximiser. This implies $\#B(x_2) \leq 1$.

3b. By Propositions 5(3) and 1(4), the function $u_1^{(\frac{L}{2})} : [0, \frac{L}{2}] \to \mathbb{R}$ is strictly concave and continuous. It follows that this function has a unique maximiser, say x_1. Noting that $u_1^{(\frac{L}{2})}(x) = u_1^{(\frac{L}{2})}(L - x)$ $(x \in S)$, it follows that $B(\frac{L}{2}) = \{x_1, L - x_1\}$. By Propositions 6 and 7(2) we have $0 < x_1 < \frac{L}{2}$. As f is continuous, the function $u_1^{(\frac{L}{2})} : [0, \frac{L}{2}] \to \mathbb{R}$ is by Proposition 1(5) differentiable at its interior maximiser x_1, Fermat's theorem gives $Du_1^{(\frac{L}{2})}(x_1) = 0$. Proposition 2(1) implies $f(x_1) = \frac{1}{2}f(\frac{\frac{L}{2}-x_1}{2})$. As the function $y \mapsto f(y) - \frac{1}{2}f(\frac{\frac{L}{2}-y}{2})$ is strictly increasing on $]0, \frac{L}{2}[$, the proof is complete.

4. By parts 2b and 3c, $\#B(x_2) = 1$. Let $B(x_2) = \{x_1\}$. Now, $0 < x_1 < x_2$ by Lemma 5. As $u_1^{(x_2)}$ is differentiable at its interior maximiser x_1, Fermat's theorem gives $Du_1^{(x_2)}(x_1) = 0$. Proposition 2(1) implies $f(x_1) = \frac{1}{2}f(\frac{x_2-x_1}{2})$. As the function $y \mapsto f(y) - \frac{1}{2}f(\frac{x_2-y}{2})$ is strictly increasing on $]0, x_2[$, the proof is complete. Q.E.D.

For the inelastic case $B(x) = \emptyset$ holds for all $x \neq \frac{L}{2}$. Proposition 8(1c) shows that this property continues to hold in case of a continuous demand function f with $f(\frac{L}{2}) > \frac{1}{2}f(0)$ for x in a punctured neighbourhood of $\frac{L}{2}$.

Proposition 9. *Suppose f is upper-semicontinuous and $f(\frac{L}{2}) \leq \frac{1}{2}f(0)$. Then*

$$B(x_2) = \begin{cases} \mathrm{argmax}_{x_1 \in]x_2, L]} \, u_1^{(x_2)}(x_1) \ (0 \leq x_2 < \frac{L}{2}), \\ \mathrm{argmax}_{x_1 \in [0, x_2[} \, u_1^{(x_2)}(x_1) \ (\frac{L}{2} < x_2 \leq L). \end{cases}$$ ◇

Proof. By Lemma 5, we still have to prove '⊇'. We prove the statement for $0 \leq x_2 < \frac{L}{2}$; then the other statement follows by location symmetry. So fix $x_2 \in [0, \frac{L}{2}[$ and suppose $\tilde{x}_1 \in \mathrm{argmax}_{x_1 \in]x_2, L]} u_1^{(x_2)}(x_1)$. By Proposition 8(2a, 3a), $u_1^{(x_2)}$ has a maximiser, say $\overline{x}_1$. By Lemma 5, $\overline{x}_1 \in \mathrm{argmax}_{x_1 \in]x_2, L]} u_1^{(x_2)}(x_1)$. It follows that $u_1^{(x_2)}(\tilde{x}_1) = u_1^{(x_2)}(\overline{x}_1)$. This implies $\tilde{x}_1 \in B(x_2)$. Q.E.D.

Concerning the statement in the next theorem, note that we have Proposition 8(2a, 3a) on single-valuedness of the correspondence B.

Theorem 3. *Suppose $f(\frac{L}{2}) \leq \frac{1}{2}f(0)$ and f is continuously differentiable with $Df < 0$. Then the functions $B : [0, \frac{L}{2}[\to \mathbb{R}$ and $B :]\frac{L}{2}, L] \to \mathbb{R}$ are continuously differentiable and strictly increasing.*   ◇

Proof. By location symmetry, it is sufficient to prove the second statement. Well, by Proposition 8(4), $B(x_2)$ is the unique solution y of the equation $f(y) = \frac{1}{2}f(\frac{x_2-y}{2})$ $(0 < y < x_2)$. So we have $f(B(x_2)) = \frac{1}{2}f(\frac{x_2-B(x_2)}{2})$. The implicit function theorem applies and implies that $B : [0, \frac{L}{2}[\rightarrow \mathbb{R}$ is continuously differentiable with $DB(x_2) = \frac{Df(\frac{x_2-B(x_2)}{2})}{4Df(B(x_2))+Df(\frac{x_2-B(x_2)}{2})} < 0$. Q.E.D.

6   Equilibria

In this section we provide results for the Nash equilibrium set E.

Proposition 10. *If f is continuous at 0 and $\frac{L}{2}$ and $f(\frac{L}{2}) < \frac{1}{2}f(0)$, then $(\frac{L}{2}, \frac{L}{2}) \notin E$.*   ⋄

Proof. By Proposition 7(2). Q.E.D.

Theorem 4. *If f is continuous and $(e_1, e_2) \in E$, then $e_1 + e_2 = L$.*   ⋄

Proof. Suppose f is continuous. If f is constant, then, by Theorem 2(3), $E = \{(\frac{L}{2}, \frac{L}{2})\}$ and so the statement is true. Now assume that f is strictly decreasing. First we prove by contradiction that $e_1 + e_2 \geq L$ for each equilibrium (e_1, e_2). So suppose (e_1, e_2) is an equilibrium with $e_1 + e_2 < L$. By player symmetry, we may assume that $e_2 \leq e_1$. Proposition 7(1) and $e_1 + e_2 < L$ imply $e_2 \neq e_1$. So $e_2 < e_1$ holds. By Proposition 6, $0 < e_2 < e_1 < L$. As $u_1^{(e_2)}$ is differentiable at e_1 and $u_2^{(e_1)}$ is differentiable at e_2, it follows $Du_1^{(e_2)}(e_1) = Du_2^{(e_1)}(e_2) = 0$. So, by Proposition 2(1), noting that $Du_2^{(e_1)}(e_2) = Du_1^{(e_1)}(e_2)$

$$0 = -f(L - e_1) + \frac{1}{2}f(\frac{e_1 - e_2}{2}), \; 0 = f(e_2) - \frac{1}{2}f(\frac{e_1 - e_2}{2}).$$

We obtain $f(L - e_1) = f(e_2)$. As $L - e_1 \neq e_2$, this contradicts the strict decreasingness of f. Thus $e_1 + e_2 \geq L$ for each equilibrium (e_1, e_2). By location symmetry, $(L - e_1, L - e_2)$ is also an equilibrium. Therefore $(L - e_1) + (L - e_2) \geq L$. Hence $e_1 + e_2 = L$ follows. Q.E.D.

The next example shows that Theorem 4 no longer holds if we allow f therein to be discontinuous.

Example 1. This example is taken from [11]. Consider the case $L = 1$ with the following discontinuous demand function

$$f(z) = \begin{cases} 2 - z & \text{if } 0 \leq z \leq \frac{5}{24}, \\ \frac{1}{2} - \frac{1}{4}z & \text{if } \frac{5}{24} < z \leq 1. \end{cases}$$

Note that f is upper-semicontinuous and that $f(\frac{L}{2}) < \frac{1}{2}f(0)$.
    We now prove that $(\frac{2}{3}, \frac{1}{4})$ is an equilibrium.

By Proposition 9, $B(\frac{1}{4}) = \text{argmax}_{x_1 \in]\frac{1}{4}, 1]} u_1^{(\frac{1}{4})}(x_1)$ and by Proposition 5(1), the function $u_1^{(\frac{1}{4})}$ is on $[\frac{1}{4}, 1]$ strictly concave. By Proposition 2(1), we have $D_- u_1^{(\frac{1}{4})}(\frac{2}{3}) = -f_+(\frac{1}{3}) + \frac{1}{2}f_-(\frac{5}{24}) = -\frac{5}{12} + \frac{43}{48} = \frac{23}{48} > 0$. And $D_+ u_1^{(\frac{1}{4})}(\frac{2}{3}) = -f_-(\frac{1}{3}) + \frac{1}{2}f_+(\frac{5}{24}) = -\frac{5}{12} + \frac{43}{192} = -\frac{37}{192} < 0$. This implies $\frac{2}{3} \in B(\frac{1}{4})$.

By Proposition 9, $B(\frac{2}{3}) = \text{argmax}_{x_1 \in [0, \frac{2}{3}[} u_1^{(\frac{2}{3})}(x_1)$ and by Proposition 5(2), the function $u_1^{(\frac{2}{3})}$ is on $[0, \frac{2}{3}]$ strictly concave. By Proposition 2(1), we have $D_- u_1^{(\frac{2}{3})}(\frac{1}{4}) = f_-(\frac{1}{4}) - \frac{1}{2}f_+(\frac{5}{24}) = \frac{42}{96} - \frac{43}{192} = \frac{41}{192} > 0$. And $D_+ u_1^{(\frac{2}{3})}(\frac{1}{4}) = f_+(\frac{1}{4}) - \frac{1}{2}f_-(\frac{5}{24}) = \frac{42}{96} - \frac{43}{48} = -\frac{44}{96} < 0$. This implies $\frac{1}{4} \in B(\frac{2}{3})$. ◇

Define the function $H : [0, \frac{L}{2}] \to \mathbb{R}$ by

$$H(x_1) := f(x_1) - \frac{1}{2}f(\frac{L}{2} - x_1).$$

Note that $H(0) > 0$, H is decreasing, and strictly decreasing if f is not constant. Thus H has at most one zero. If f is continuous and $f(\frac{L}{2}) \leq \frac{1}{2}f(0)$, then H has a unique zero; we denote this zero by

$$x^\star.$$

As $H(\frac{L}{4}) = \frac{1}{2}f(\frac{L}{4}) > 0$, we obtain

$$x^\star \in \begin{cases}]\frac{L}{4}, \frac{L}{2}[\text{ if } f(\frac{L}{2}) < \frac{1}{2}f(0), \\ = \frac{L}{2} \text{ if } f(\frac{L}{2}) = \frac{1}{2}f(0). \end{cases} \tag{3}$$

Theorem 5. *Suppose f is continuous. Then the game has a Nash equilibrium. Even:*

1. *if $f(\frac{L}{2}) \geq \frac{1}{2}f(0)$, then $E = \{(\frac{L}{2}, \frac{L}{2})\}$.*
2. *if $f(\frac{L}{2}) < \frac{1}{2}f(0)$, then $E = \{(x^\star, L - x^\star), (L - x^\star, x^\star)\}$.* ◇

Proof. 1. Suppose $f(\frac{L}{2}) \geq \frac{1}{2}f(0)$. By Theorem 2(3), we may suppose that f is strictly decreasing. By Proposition 8(1b, 2b), we have $\{(\frac{L}{2}, \frac{L}{2})\} \subseteq E$. Now suppose $(e_1, e_2) \in E$. We have to prove that $(e_1, e_2) = (\frac{L}{2}, \frac{L}{2})$. This we do by contradiction. So suppose $(e_1, e_2) \neq (\frac{L}{2}, \frac{L}{2})$. By player symmetry, we may suppose $e_1 \leq e_2$. By Theorem 4, $e_2 - L = e_1$. By Proposition 6, $e_1 \neq 0$ and $e_2 \neq L$. It follows that $0 < e_1 < \frac{L}{2} < e_2 < L$. As $(e_1, L - e_1) \in E$, we have $Du_1^{(L-e_1)}(e_1) = 0$. By Proposition 2(1), $f(e_1) - \frac{1}{2}f(\frac{L}{2} - e_1) = 0$. Thus $f(\frac{L}{2}) < f(e_1) = \frac{1}{2}f(\frac{L}{2} - e_1) \leq \frac{1}{2}f(0)$, a contradiction.

2. Suppose $f(\frac{L}{2}) < \frac{1}{2}f(0)$. '⊆': suppose $(e_1, e_2) \in E$. By location symmetry, we may suppose $e_1 \leq e_2$. By Theorem 4, $e_1 + e_2 = L$. Propositions 6 and 10 now imply $0 < e_1 < \frac{L}{2} < e_2 < L$. By Proposition 8(3a), $e_1 = B(e_2)$. By Proposition 8(4), $f(e_1) = \frac{1}{2}f(\frac{e_2 - e_1}{2}) = \frac{1}{2}f(\frac{L}{2} - e_1)$. Thus e_1 is a zero of H, and therefore $e_1 = x^\star$. We see $(e_1, e_2) = (x^\star, L - x^\star) \in \{(x^\star, L - x^\star), (L - x^\star, x^\star)\}$.

'$\supseteq$': by location symmetry, it is sufficient to prove that $(x^\star, L - x^\star) \in E$. By definition of $x^\star$ we have $0 = f(x^\star) - \frac{1}{2}f(\frac{L}{2} - x^\star) = f(x^\star) - \frac{1}{2}f(\frac{(L-x^\star)-x^\star}{2})$. Therefore Proposition 8(4) guarantees that $x^\star = B(L - x^\star)$. This implies $L - x^\star = B(x^\star)$. It follows that $(x^\star, L - x^\star) \in E$. Q.E.D.

Thus for a continuous demand function, the Principle of Minimum Differentiation holds if and only if $f(\frac{L}{2}) \geq \frac{1}{2}f(0)$.

Corollary 1. *Suppose f is continuous. Then the game has one equilibrium class and this class contains one or two elements.* ◇

In the next section we shall prove by a completely different approach that each cHg has a Nash equilibrium (even if f is not continuous).

Proposition 11. *Suppose f is continuous. Then for all $(e_1, e_2) \in E$ it holds that $e_1, e_2 \in \,]\frac{L}{4}, \frac{3}{4}L\,[$.* ◇

Proof. By Theorem 5 and (3). Q.E.D.

7 Potentials

In this section we review the results in [12] on potentials for the cHg. We shall encounter the notions of generalized ordinal potential, best-response potential, a weak quasi-potential and quasi-potential; again we denote with E the Nash equilibrium set.[6] We note that each generalized ordinal potential game and each best-response potential game is a weak quasi potential game.

Define the function $P^\bullet : S \times S \to \mathbb{R}$ by

$$P^\bullet(x_1, x_2) := \mathcal{L}(\min\{x_1, x_2\}) + \mathcal{L}(L - \max\{x_1, x_2\}) + \mathcal{L}(\frac{|x_2 - x_1|}{2}). \quad (4)$$

Note that $P^\bullet$ is continuous irrespective of the continuity of f. In the case of $f = 1$, i.e. elastic demand, we have

$$P^\bullet(x_1, x_2) = \begin{cases} L - \frac{x_2 - x_1}{2} & \text{if } x_1 < x_2, \\ L & \text{if } x_1 = x_2, \\ L - \frac{x_1 - x_2}{2} & \text{if } x_1 > x_2. \end{cases}$$

Theorem 6. *1. Suppose f is continuous.*
  (a) If $f(\frac{L}{2}) \leq \frac{1}{2}f(0)$, then $P^\bullet$ is a continuous best-response potential.

---

[6] For the cHg, a function $P : S \times S \to \mathbb{R}$ is (1) a generalized ordinal potential if for every $a_1, b_1, z \in [0, L]$ it holds that $u_1(a_1, z) < u_1(b_1; z) \Rightarrow P(a_1, z) < P(b_1, z)$ and for every $a_2, b_2, z \in [0, L]$ it holds that $u_2(z, a_1) < u_2(z, b_1) \Rightarrow P(z, a_1) < P(z, b_1)$; (2) a best-response potential if $B_1(x_2) = \text{argmax}_{x_1 \in S} P(x_1, x_2)$ $(x_2 \in S)$ and $B_2(x_1) = \text{argmax}_{x_2 \in S} P(x_1, x_2)$ $(x_1 \in S)$; (3) a quasi potential if $\text{argmax}\, P = E$. (4) a weak quasi potential if $\text{argmax}\, P \subseteq E$. In this case, one calls the game a 'generalized ordinal potential game' (etc.).

(b) If $f(\frac{L}{2}) > \frac{1}{2}f(0)$, then there does not exist a continuous best-response potential.

(c) If f is strictly decreasing, then $P^\bullet$ is a continuous quasi potential.

2. If f is strictly decreasing, then $P^\bullet$ is a continuous weak quasi potential.

3. If f is constant, then $P(x_1, x_2) = -(|\frac{L}{2} - x_1| + |\frac{L}{2} - x_2|)$ is a continuous quasi potential.

4. The game may not have a generalized ordinal potential, even if f is continuous. ◇

Proof. 1a. See Proposition 3.1 in [12].

1b. Suppose $f(\frac{L}{2}) > \frac{1}{2}f(0)$. By contradiction, suppose P is a continuous best-response potential. By the Weierstrass theorem this would imply that B is proper. But, by Proposition 8(1c), B is not proper.

1c. See Proposition 3.2 in [12].

2. See concluding remark 3 in [12].

3. By Theorem 2(3), $E = \{(\frac{L}{2}, \frac{L}{2})\}$. Thus, as desired, $\mathrm{argmax}(P) = E$.

4. See Proposition 3.3 in [12]. Q.E.D.

Theorem 6(2, 3) implies:

Corollary 2. *Each cHg has a Nash equilibrium.* ◇

In addition to Theorem 6(1b), we have:

Proposition 12. *In the case f is constant, $P : [0, L] \times [0, L] \to \mathbb{R}$ defined by*

$$P(x_1, x_2) := \begin{cases} -|x_1 - x_2| & \text{if } x_1 = \frac{L}{2} \vee x_2 = \frac{L}{2}, \\ \frac{1}{|x_1 - \frac{L}{2}|} + \frac{1}{|x_2 - \frac{L}{2}|} & \text{if } x_1 \neq \frac{L}{2} \wedge x_2 \neq \frac{L}{2} \end{cases}$$

is a (discontinuous) best-response potential. ◇

Proof. As P is symmetric, P being a quasi-potential comes down to

$$\text{for all } x_2 \in [0, L]: \quad B(x_2) = \mathrm{argmax}_{x_1 \in [0, L]} P(x_1, x_2).$$

We have $\mathrm{argmax}_{x_1 \in [0, L]} P(x_1, x_2) = \begin{cases} \emptyset & \text{if } x_2 \neq \frac{L}{2}, \\ \{\frac{L}{2}\} & \text{if } x_2 = \frac{L}{2}. \end{cases}$ So $P(x_1, \frac{L}{2}) = -|x_1 - \frac{L}{2}|$; thus, as desired, $\mathrm{argmax}_{x_1 \in [0, L]} P(x_1, \frac{L}{2}) = \{\frac{L}{2}\}$. Now fix $x_2 \neq \frac{L}{2}$. We have $P(x_1, x_2) = \begin{cases} \frac{1}{|x_1 - \frac{L}{2}|} + \frac{1}{|x_2 - \frac{L}{2}|} & \text{if } x_1 \neq \frac{L}{2}, \\ -|\frac{L}{2} - x_2| & \text{if } x_1 = \frac{L}{2}. \end{cases}$ From this formula, one sees, as desired, that $\mathrm{argmax}_{x_1 \in [0, L]} P(x_1, x_2) = \emptyset$. Q.E.D.

8   Concluding Remarks

1. We presented results for the Hotelling pure location game with two identical players, one-dimensional strategy sets and a demand function f which is constant (inelastic case) or strictly decreasing (elastic case). The elastic case has been poorly studied in the literature. For the elastic case we tried to derive the results without further smoothness assumptions on f.

2. Although the inelastic case for our one-dimensional case of the cHg is simple to analyse, this is no longer true for the two-dimensional case (see [9]).
3. We have shown that for a continuous f, the Principle of Minimum Differentiation holds if and only if $f(\frac{L}{2}) \geq \frac{1}{2}f(0)$. This is in accordance with the observations in the literature (e.g. [5]) that this principle is not so robust.
4. In Corollary 1 we have shown that in the case of a continuous f, the game has at most one equilibrium class and this class contains one or two elements. In Example 1 we have shown for a specific cHg with a discontinuous demand function, that it has $(\frac{2}{3}, \frac{1}{4})$ as equilibrium. Therefore also $(\frac{1}{4}, \frac{2}{3})$, $(\frac{1}{3}, \frac{3}{4})$ and $(\frac{3}{4}, \frac{1}{3})$ are equilibria and this game has an equilibrium class with four elements. An interesting question is whether there exists a cHg with more than one equilibrium class.
5. We have shown 'by hand' that the cHg has a Nash equilibrium in the case of a continuous demand function. One might want to have a deeper reason for this existence. Concerning this Theorem 6 shows that each cHg admits a continuous weak quasi potential (and therefore has an equilibrium).
6. As the cHg is a game with discontinuous payoff functions, it may be interesting to find out in which sense general equilibrium existence results for games in strategic form with discontinuous payoff functions apply. We here only mention that the result in [8] does not apply as it assumes quasi-concave conditional payoff functions.
7. The type of strict quasi-concavity in Proposition 5(5) has been studied in more detail in [10], where it was called 'semi-strict demi-concavity'.
8. A direction for further research concerns the comparison of the results in the present article with those for the, also poorly studied, discrete variant of the cHg [13,14].

References

1. Hotelling, H.: Stability in competition. Econ. J. **39**(153), 41–57 (1929)
2. Smithies, A.: Optimum location in spatial competition. J. Polit. Econ. **44**, 423–439 (1941)
3. Boulding, K.: Economics Analysis: Microeconomics. Harper & Row, New York (1955)
4. Davis, O., Hinich, M.J., Ordeshook, P.C.: An expository development of a mathematical model of the electoral process. Am. Polit. Sci. Rev. **44**, 426–448 (1970)
5. Eaton, B.C., Lipsey, R.G.: The principle of minimum differentiation reconsidered: some new developments in the theory of spatial competition. Rev. Econ. Stud. **42**(1), 27–49 (1975)
6. Graitson, D.: Spatial competition à la hotelling: a selective survey. J. Ind. Econ. **31**(1/2), 11–25 (1982)
7. Anderson, S.P., de Palma, A., Thisse, J.-F.: Discrete Choice Theory of Product Differentiation. MIT Press, Cambridge (1992)
8. Reny, P.: On the existence of pure and mixed strategy Nash equilibria in discontinuous games. Econometrica **67**, 1029–1056 (1999)
9. Mazalov, V., Sakaguchi, M.: Location game on the plain. Int. Game Theory Rev. **5**(1), 13–25 (2003)

10. von Mouche, P.H.M., Quartieri, F.: Cournot equilibrium uniqueness via demi-concavity. Optimization **67**(4), 41–455 (2017)
11. Iimura, T.: Private Communication. Tokyo Metropolitan University, Tokyo, Japan (2017)
12. Iimura, T., von Mouche, P.H.M., Watanabe, T.: Best-response potential for hotelling pure location games. Econ. Lett. **160**, 73–77 (2017)
13. von Mouche, P., Pijnappel, W.: The hotelling bi-matrix game. Optim. Lett. **12**(1), 187–202 (2015). https://doi.org/10.1007/s11590-015-0964-6
14. Iimura, T., von Mouche, P.H.M.: Discrete hotelling pure location games: potentials and equilibria. Working Paper, Wageningen Universiteit (2020)

Scheduling Problem

An Improved Approximation Algorithm for the Coupled-Task Scheduling Problem with Equal Exact Delays

Alexander Ageev[1]([⊠]) [iD] and Mikhail Ivanov[2]

[1] Sobolev Institute of Mathematics, pr. Koptyuga 4, Novosibirsk, Russia
ageev@math.nsc.ru
[2] Novosibirsk State University, Pirogova 2, Novosibirsk, Russia
mi337@mail.ru

Abstract. We study the coupled-task single machine scheduling problem with equal exact delays and makespan as the objective function. It is known that the problem cannot be approximated with a factor better than 1.25 unless P = NP. In this paper, we present a 2.5-approximation algorithm for this problem, which improves the best previously known approximation bound of 3. The algorithm runs in time $O(n \log n)$ where n is the number of jobs.

Keywords: Coupled-task scheduling · Inapproximability lower bound · Approximation algorithm · Worst-case analysis

1 Introduction

We consider the single-machine coupled-task scheduling problem with exact delays. In the problem, a set $J = \{1, \ldots, n\}$ of independent jobs is given. Each job $j \in J$ is composed of two operations with processing times a_j and b_j separated by a given *exact* delay l_j, which means that the second operation of job j must start processing exactly l_j time units after the completion of first operation of job j. It is assumed that at any time the machine can process at most one operation and no preemption is allowed. The objective is to minimize the makespan (the schedule length). In the standard three-field notation scheme introduced by Graham et al. [12] (see also [14]) this single machine problem is denoted by $1 \mid \text{exact } l_j \mid C_{\max}$.

In this paper, we consider the case of the single machine problem when all delays are equal, i.e., $l_j = L$ for all $j \in J$. We refer to this case as $1 \mid \text{exact } l_j = L \mid C_{\max}$.

The scheduling problems with exact delays spring from command-and-control applications where an administrator gives away a set of orders (associated with the first operations) and must wait to get responses (corresponding to the second operations) that do not collide with each other (for more detailed discussion,

The work was supported by the program of fundamental scientific researches of the SB RAS, project N 0314-2019-0014.

see [10,16]). Investigations on problem $1 \mid$ exact $l_j \mid C_{\max}$ are mostly inspired by applications in pulsed radar systems, where the machine is a multifunctional radar whose goal is to simultaneously keep track of numerous targets by transmitting a pulse and accepting its reflection some time later [8,10,11,15,16]. Coupled-task scheduling problems with exact delays also have applications in chemistry manufacturing where there may be an exact technological delay between the completion time of some operation and the starting time of the next operation.

1.1 Related Work

Coupled-task scheduling problems have been investigated for decades. Quite a few various results related to these problem are surveyed by Blazewicz et al. in [6] (for later results see [7], [13]). We cite here only previously known approximation results as well as those related to the case of equal delays.

Orman and Potts [15] establish that the problem is strongly NP-hard even in some special cases. In particular, they prove this fact for $1 \mid$ exact $l_j = L, b_j = b \mid C_{\max}$, i.e., in the case when $l_j = L, b_j = b$ for all $j \in J$. Baptiste [4] presents an algorithm with running time $O(\log n)$ for the very special case when $a_j = a, b_j = b, l_j = L$ for all jobs j provided that a, b, and L are fixed. The complexity status of the case, when a, b, and L are part of the input, remains open [15],[5].

Ageev and Baburin [1] present non-trivial constant-factor approximation algorithms for the single and the two machine problems subject to unit processing times. More specifically, it is shown in [1] that problem $1 \mid$ exact $l_j, a_j = b_j = 1 \mid C_{\max}$ is approximable within a factor of 7/4.

Ageev and Kononov [3] present a 3.5-approximation algorithm for the general case of $1 \mid$ exact $l_j \mid C_{\max}$ and 3-approximation algorithms for the cases when either $a_j \leq b_j$, or $a_j \geq b_j$ for all $j \in J$. They also show that the last two algorithms provide a 2.5-approximation for the case when $a_j = b_j$ for all $j \in J$. Moreover, they prove that problem $1 \mid$ exact $l_j \mid C_{\max}$ is not $(2-\varepsilon)$-approximable in polynomial time unless P = NP even in the case of $a_j = b_j$ for all $j \in J$.

Ageev and Ivanov [2] consider the coupled-task single machine scheduling problem with equal exact delays. They show that the existence of a polynomial-time $(1.25 - \varepsilon)$-approximation for $1 \mid$ exact $l_j = L \mid C_{\max}$ even in the case $a_j = b_j$ for all jobs $j = 1, \ldots n$ implies P = NP. On the positive side, they design a 3-approximation for $1 \mid$ exact $l_j = L \mid C_{\max}$. For the cases of $1 \mid$ exact $l_j = L \mid C_{\max}$ when either $a_j \leq b_j$, or $a_j = b_j$ for all jobs $j = 1, \ldots n$ they present 2- and 1.5-approximations, respectively. All approximation algorithms mentioned above are polynomial-time.

1.2 Our Results

In this paper we design and analyze a 2.5-approximation algorithm for the general case of $1 \mid$ exact $l_j = L \mid C_{\max}$ 2.5, which improves the best previously known approximation bound of 3 established in the previous paper by the authors [3].

Table 1. A summary of the approximability results.

Problem	Appr. factor	Inappr. bound
$1 \mid$ exact $l_j, \mid C_{\max}$	3.5 [3]	$2 - \varepsilon$ [3]
$1 \mid$ exact $l_j, a_j \leq b_j \mid C_{\max}$	3 [3]	$2 - \varepsilon$ [3]
$1 \mid$ exact $l_j, a_j = b_j \mid C_{\max}$	2.5 [3]	$2 - \varepsilon$ [3]
$1 \mid$ exact $l_j, a_j = b_j = 1 \mid C_{\max}$	1.75 [1]	
$1 \mid$ exact $l_j = L \mid C_{\max}$	2.5 (this paper)	$1.25 - \varepsilon$ [2]
$1 \mid$ exact $l_j = L, a_j \leq b_j \mid C_{\max}$	2 [2]	$1.25 - \varepsilon$ [2]
$1 \mid$ exact $l_j = L, a_j = b_j \mid C_{\max}$	1.5 [2]	$1.25 - \varepsilon$ [2]

The algorithm runs in time $O(n \log n)$ and has an interesting property: its approximation factor tends to 2 when the number of blocks it constructs tends to infinity.

Our result compared with the previously known approximation results is shown in Table 1.

1.3 Basic Notation

For the problems under consideration an instance will be represented as a collection of triples $\{(a_j, l_j, b_j) : j \in J\}$ where $J = \{1, \ldots, n\}$ is the set of jobs, a_j and b_j are the lengths of the first and the second operations of job j, respectively and l_j is the given delay between these operations. As usual, we assume that all input data are nonnegative integers. For a schedule σ and any $j \in J$, we denote by $\sigma(j)$ the starting time of the first operation of job j. Since the starting times of the first operations uniquely determine the starting times of the second operations, any feasible schedule is uniquely specified by the collection of starting times of the first operations $\{\sigma(1), \ldots, \sigma(n)\}$. For a schedule σ and any $j \in J$, denote by $C_j(\sigma)$ the completion time of job j in σ; note that $C_j(\sigma) = \sigma(j) + l_j + a_j + b_j$ for all $j \in J$. The length of a schedule σ is denoted by $C_{\max}(\sigma)$ and thus $C_{\max}(\sigma) = \max_{j \in J} C_j(\sigma)$. The length of a shortest schedule is denoted by $C_{\max}^*$.

2 Preliminaries

Our algorithm uses as a procedure algorithm $A^{\leq}$ for $1 \mid$ exact $l_j = L, a_j \leq b_j \mid C_{\max}$ described in [2,3]. Since the general problem is symmetric with respect to the time axis an evident modification of this algorithm is also applicable to the case $1 \mid$ exact $l_j = L, a_j \geq b_j \mid C_{\max}$. We will denote this modification by $A^{\geq}$.

In what follows we will use structural properties of the schedules constructed by algorithm $A^{\leq}$ (and its counterpart $A^{\geq}$). To make the paper self-contained we give informal and formal descriptions of the algorithm.

Informally, algorithm $A^{\leq}$ does the following. First it numbers the jobs in non-increasing order of the durations of the first operations. Then it scans the list of the jobs in this order and successively constructs blocks B_s $(s = 1, \ldots r)$ which are some bundles of jobs $j_s, \ldots, j_{s+1} - 1$ $(j_1 = 1)$. In each block B_s, the second operations of job $j_s = 1, \ldots, j_s - 1$ are processed one after the other without idle times (see Fig. 1(a)). At that, both the first operations and the second operations are scheduled in each block in increasing order of their indices. A block becomes complete when it cannot be augmented in this way by the current job. Then the algorithm starts constructing the next block. Finally, the algorithm outputs a schedule which consists in the successive execution of blocks $B_1, \ldots B_r$ (see Fig. 1(b) with $r = 4$). Note that block B_r may be incomplete.

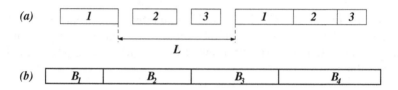

Fig. 1. (a) a block consisting of three jobs; (b) a schedule consisting of four blocks.

Now we present a formal description of the algorithm.

ALGORITHM $A^{\leq}$.

PHASE I *(jobs ordering)*. Number the jobs in the following way:

$$a_1 \geq a_2 \geq \ldots \geq a_n .$$

PHASE II *(constructing blocks $1, \ldots, r$)*. By scanning the set of jobs in the order $j = 1, \ldots, n$ calculate the indices $j_1 < j_2 < \ldots < j_r \leq n$ in the following way.

Step 1. Set $j_1 = 1$. If $\sum_{s=1}^{n-1} b_s \leq L$, then retrieve $r = 1$ and go to Phase III. Otherwise go to Step 2.

Step $k(k \geq 2)$. Set j_k to be equal to the minimum index among indices t such that $n + 1 > t > j_{k-1}$ and $\sum_{s=j_{k-1}}^{t-1} b_s > L$. If $j_k = n$ or $\sum_{s=j_k}^{t-1} b_s \leq L$ for all $t = j_k + 1, \ldots, n$, then set $r = k$ and go to Phase III. Otherwise go to Step $k+1$.

PHASE III *(constructing the schedule)*. Set $\sigma(j_1) = \sigma(1) = 0$. If $r > 1$, then for $s = 2, \ldots, r$ set

$$\sigma(j_s) = \sigma(j_{s-1}) + a_{j_{s-1}} + L + \sum_{k=j_{s-1}}^{j_s - 1} b_k .$$

For every $j \in J \setminus \{j_1, \ldots, j_r\}$, set

$$\sigma(j) = \sigma(j_s) + a_{j_s} - a_j + \sum_{k=j_s}^{j-1} b_k$$

where s is the maximum index such that $j_s < j$.

The correctness of the algorithm is established by Lemma 1 in [3]. The running time is $O(n \log n)$. The worst-case analysis is based on exploiting the lower bounds W_1 and W_2 that will be also crucial in analyzing our algorithm. It is easy to observe that $C^*_{\max} \geq \max\{W_1, W_2\}$ where

$$W_1 = \sum_{j=1}^{n} (a_j + b_j) \tag{1}$$

and

$$W_2 = L + \max\{\sum_{j \in J} a_j, \sum_{j \in J} b_j\}. \tag{2}$$

The lower bound W_1 (machine load) is evident. The bound W_2 follows from the fact that in any feasible schedule all first or all second operations are executed outside the delay $(=L)$ of the last or the first executed job, respectively.

A block B consisting of jobs $j_1, \ldots, j_t$ is called *complete* if $\sum_{s=1}^{k} b_{j_s} \geq L$. Otherwise the block is *incomplete*. The same term we will use for the blocks retrieved by Algorithm $A^{\geq}$. In this case the block B is complete if $\sum_{s=1}^{k} a_{j_s} \geq L$.

Assume that σ consists of $k+1$ blocks. Since the first k blocks are complete we have

$$C_{\max}(\sigma) \leq \sum_{j=1}^{n}(a_j + b_j) + kL + L \leq \sum_{j=1}^{n}(a_j + b_j) + \sum_{j=1}^{n} b_j + L \leq W_1 + W_2 \leq 2C^*_{\max}.$$

3   A 2.5-Approximation

Assume that Algorithm $A^{\leq}$ (or Algorithm $A^{\geq}$) finds a schedule consisting of $k+1$ blocks. Then by the definition of complete block $kL \leq W_1$.

We say that two blocks are *combinable* if they can be combined as in Fig. 2 (a) otherwise they are *non-combinable*. If two incomplete blocks are combinable then they can be combined as shown in Fig. 2 (b) by a mutual shift. We call the configuration (b) in Fig. 2 (or its symmetric counterpart) the *combined block* of two blocks.

Observe that if two blocks are combinable then they are both incomplete. Furthermore, if at least one of them is complete then they are non-combinable.

3.1   Algorithm CombineBlocks

Divide the set of jobs J into two subsets J_1, J_2 such that $J_1 = \{j \in J | a_j \leq b_j\}$, $J_2 = \{j \in J | a_j > b_j\}$. Let $J_1 = \{(a_j, L, b_j) : j = 1, \ldots, r\}$, $J_2 = \{(a_j, L, b_j) :$

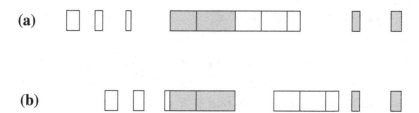

Fig. 2. Two combinable blocks (a) and the combined block of these blocks (b).

$j = t + 1,\ldots,n\}$. If $t = 0$ or $t = n$, i.e., either J_1 or J_2 is empty, apply algorithm $A^{\geq}$ or $A^{\leq}$, respectively. Otherwise for the set J_1 apply algorithm $A^{\leq}$ and for the set J_2, algorithm $A^{\geq}$. Assume that the retrieved schedule σ_1 for J_1 consists of blocks $B_1, B_2, \ldots B_{k_1+1}$ and the retrieved schedule σ_2 for J_2 consists of blocks $D_{k_2+1}, D_{k_2}, \ldots D_1$. The blocks B_{k_1+1} and D_{k_2+1} may be incomplete, the remaining $k = k_1 + k_2$ blocks are necessarily complete. By arranging the schedules σ_1 and σ_2 one after another we get a feasible schedule $\widetilde{\sigma} = (B_1, B_2, \ldots B_{k_1+1}, D_{k_2+1}, D_{k_2}, \ldots D_1)$ for the set of jobs J (see Fig. 3). If blocks B_{k_1+1} and D_{k_2+1} are non-combinable or $k = 0$, set $\sigma = \widetilde{\sigma}$. Otherwise combine blocks B_{k_1+1} and D_{k_2+1} into a block F and define σ as the schedule $(B_1, B_2, \ldots B_{k_2} F, D_{k_2}, D_{k_2-1}, \ldots, D_1)$ (see Fig. 4). Output σ.

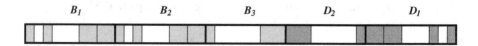

Fig. 3. A schedule $\widetilde{\sigma}$ with $k_1 = 2$ and $k_2 = 1$. The block boundaries are shown in bold lines. Blocks B_1, B_2, D_1 are complete, blocks B_3 and D_2 are incomplete.

Fig. 4. A schedule σ with $k_1 = 2$ and $k_2 = 1$ and the combined block F. The combined block boundary is shown in bold lines.

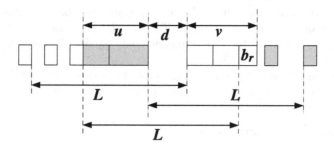

Fig. 5. A combined block.

3.2   Analysis

Note that $C_{\max}(\sigma) \leq C_{\max}(\tilde{\sigma})$. For the length of $\tilde{\sigma}$ we have the following straightforward bound:

$$C_{\max}(\tilde{\sigma}) \leq \sum_{j=1}^{n} a_j + \sum_{i=1}^{n} b_j + 2L + kL,$$

which does not imply the claimed approximation factor. Fortunately, the schedule σ has some helpful properties that will lead us to the desired goal.

Lemma 1. *If the schedule σ contains at least one complete block, then $2L \leq C_{\max}^*$, i.e., $2L$ is a lower bound.*

Proof. If the schedule σ contains a complete block, then either $L \leq \sum_{j \in J} a_j$, or $L \leq \sum_{j \in J} b_j$. However by (2) we have the lower bound

$$W_2 = L + \max\{\sum_{j \in J} a_j, \sum_{j \in J} b_j\}.$$

Thus

$$2L \leq L + \max\{\sum_{j \in J} a_j, \sum_{j \in J} b_j\} \leq C_{\max}^*.$$

$\square$

Lemma 2. *If the incomplete blocks B_{k_1+1} and D_{k_2+1} are combinable, then the idle time in the combined block does not exceed L.*

Proof. Let x denote the idle time in the combined block and let b_r be the length of the last operation in the block B_{k_1+1}. Then (see Fig. 5)

$$x \leq (L - u) + (L - v) - d = (L - u) + (L - v) - (L + b_r - u - v) = L - b_r \leq L.$$

$\square$

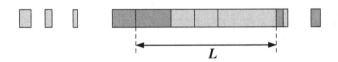

Fig. 6. Non-combinable blocks

Lemma 3. *If blocks B_{k_1+1} and D_{k_2+1} are non-combinable, then the total machine load in the two blocks is at least L.*

Proof. If blocks B_{k_1+1} and D_{k_2+1} are non-combinable, then the total machine load in the blocks is at least $u + v$, with $u + v \geq L$ where u and v denote the total lengths of the first and second operations in blocks D_{k_2+1} and B_{k_1+1}, respectively (see Fig. 6). $\square$

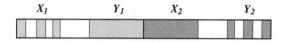

X_1 Y_1 X_2 Y_2

Fig. 7. The case of $k = 0$.

Now we consider the following four cases.

Case 1: $k = 0$. If $\tilde{\sigma}$ consists of a single block then its length is clearly at most $W_1 + W_2 \leq 2C^*_{\max}$. Assume that $\tilde{\sigma} = \{B_1, D_1\}$. Let X_i and Y_i denote the total lengths of the first and the second operations in blocks B_1 and D_1, respectively (see Fig. 7). Then $C_{\max}(\sigma)(\sigma) \leq (L + X_1 + X_2) + (L + Y_1 + Y_2) \leq 2W_2 \leq$, which is at most $2C^*_{\max}$.

Case 2: $k = 1$, B_{k_1+1} and D_{k_2+1} are combinable. W.l.o.g. we may assume that $\sigma = (B_1, F)$ where F is the combined block. By Lemma 1, $2L \leq C^*_{\max}$. On the other hand, by Lemma 2 the idle time in F is at most L. So by (1)

$$C_{\max}(\sigma) \leq 2L + \sum_{j \in J}(a_j + b_j) \leq C^*_{\max} + W_1 \leq 2C^*_{\max}.$$

Case 3: $k \geq 2$, B_{k_1+1} and D_{k_2+1} are combinable. Note that $W_1 \geq kL$. By Lemma 2, $C_{\max}(\sigma) \leq \sum_{j \in J}(a_j + b_j) + kL + L$ and we have

$$\frac{C_{\max}(\sigma)}{W_1} \leq \frac{W_1 + L + kL}{W_1} = 1 + \frac{L + kL}{W_1} \leq 1 + \frac{L + kL}{kL} = 2 + \frac{1}{k},$$

which gives $C_{\max}(\sigma) \leq (2 + 1/k)C^*_{\max}$.

Case 4: $k \geq 1$, B_{k_1+1} and D_{k_2+1} are non-combinable. Then by Lemma 3, $W_1 \geq kL + L$ and therefore

$$\frac{C_{\max}(\sigma)}{W_1} \leq \frac{W_1 + 2L + kL}{W_1} \leq 1 + \frac{2L + kL}{kL + L} = 2 + \frac{1}{k+1},$$

which implies $C_{\max}(\sigma) \leq (2 + \frac{1}{k+1})C^*_{\max}$.

Thus we arrive at

Theorem 1. *Algorithm* CombineBlocks *runs in time* $O(n \log n)$ *and outputs a schedule of length within a factor of* 2.5 *of the length of the optimal schedule. Moreover, when the number of blocks* CombineBlocks *constructs tends to infinity the approximation factor tends to* 2. □

Acknowledgments. The authors thank the anonymous referees for their helpful comments and suggestions.

References

1. Ageev, A.A., Baburin, A.E.: Approximation algorithms for UET scheduling problems with exact delays. Oper. Res. Lett. **35**, 533–540 (2007)
2. Ageev, A., Ivanov, M.: Approximating coupled-task scheduling problems with equal exact delays. In: Kochetov, Y., Khachay, M., Beresnev, V., Nurminski, E., Pardalos, P. (eds.) DOOR 2016. LNCS, vol. 9869, pp. 259–271. Springer, Cham (2016). https://doi.org/10.1007/978-3-319-44914-2_21
3. Ageev, A.A., Kononov, A.V.: Approximation algorithms for scheduling problems with exact delays. In: Erlebach, T., Kaklamanis, C. (eds.) WAOA 2006. LNCS, vol. 4368, pp. 1–14. Springer, Heidelberg (2007). https://doi.org/10.1007/11970125_1
4. Baptiste, P.: A note on scheduling identical coupled tasks in logarithmic time. Disc. Appl. Math. **158**, 583–587 (2010)
5. Békési, J., Galambos, G., Jung, M.N., Oswald, M., Reinelt, G.: A branch-and-bound algorithm for the coupled task problem. Math. Methods Oper. Res. **80**(1), 47–81 (2014). https://doi.org/10.1007/s00186-014-0469-6
6. Blazewicz, J., Pawlak, G., Tanas, M., Wojciechowicz, W.: New algorithms for coupled tasks scheduling – a survey. RAIRO - Operations Research - Recherche Operationnelle **46**, 335–353 (2012)
7. Condotta, A., Shakhlevich, N.V.: Scheduling coupled-operation jobs with exact time-lags. Discrete Appl. Math. **160**, 2370–2388 (2012)
8. Farina, A., Neri, P.: Multitarget interleaved tracking for phased array radar. IEEE Proc. Part F Comm. Radar Signal Process. **127**, 312–318 (1980)
9. Garey, M.R., Johnson, D.S.: Computers and Intractability: A Guide to the Theory of NP-Completeness. Freeman, San Francisco (1979)
10. Elshafei, M., Sherali, H.D., Smith, J.C.: Radar pulse interleaving for multi-target tracking. Naval Res. Logist. **51**, 79–94 (2004)
11. Izquierdo-Fuente, A., Casar-Corredera, J.R.: Optimal radar pulse scheduling using neural networks. In: IEEE International Conference on Neural Networks, vol. 7, pp. 4588–4591 (1994)
12. Graham, R.L., Lawler, E.L., Lenstra, J.K., Rinnooy Kan, A.H.G.: Optimization and approximation in deterministic sequencing and scheduling: a survey. Ann. Discrete Math. **5**, 287–326 (1979)
13. Hwang, F.J., Lin, B.M.T.: Coupled-task scheduling on a single machine subject to a fixed-job-sequence. J. Comput. Ind. Eng. **60**, 690–698 (2011)
14. Lawler, E., Lenstra, J., Rinnooy Kan, A., Shmoys, D.: Sequencing and scheduling: algorithms and complexity, In: Handbooks in Operations Research and Management Science, Logistics of Production and Inventory, 4, North Holland, Amsterdam, pp. 445–522 (1993)
15. Orman, A.J., Potts, C.N.: On the complexity of coupled-task scheduling. Discrete Appl. Math. **72**, 141–154 (1997)
16. Sherali, H.D., Smith, J.C.: Interleaving two-phased jobs on a single machine. Discrete Optim. **2**, 348–361 (2005)

On the Optima Localization
for the Three-Machine Routing
Open Shop

Ilya Chernykh[1]([⊠]) [iD] and Olga Krivonogova[2]

[1] Sobolev Institute of Mathematics, Koptyug ave. 4, Novosibirsk 630090, Russia
idchern@math.nsc.ru
[2] Novosibirsk State University, Pirogova str. 2, Novosibirsk 630090, Russia
olga.krivonogova@gmail.com

Abstract. A tight optima localization interval for the classical open shop scheduling problem with three machines was established by S. Sevastyanov and I. Chernykh in 1998. It was proved that for any problem instance its optimal makespan does not exceed $\frac{4}{3}$ times the standard lower bound. The process of proof involved massive computer-aided enumeration of the subsets of instances of the problem considered and took about 200 h of the running time to complete. This makes it seemingly impossible to use the same approach for more complicated problems, i.e. the four machine open shop for which the optima localization interval is still unknown. In this paper we apply that computer-aided approach to the three-machine routing open shop problem on a two-node transportation network. For this generalization of the plain open shop problem we derive some extreme instance properties and prove that the optimal makespan does not exceed $\frac{4}{3}$ times the standard lower bound, thus generalizing the result previously known for the three-machine open shop.

Keywords: Open shop · Routing open shop · Optima localization · Computer-aided proof · Approximation algorithm

1 Introduction

The main direction of this research is the search for the *tight optima localization interval* (OL-interval for short) for a special case of one generalization of scheduling and routing problems. To describe that interval, consider some optimization problem $F(x) \to \min$ over some class of instances $\mathcal{K}$. Let $LB(I)$ be some lower bound on the optimum, defined for every instance I from $\mathcal{K}$. Then the OL-interval for the problem $F(x) \to \min$ other $\mathcal{K}$ with respect to LB is defined as the tightest interval

This research was supported by the program of fundamental scientific researches of the SB RAS No. I.5.1., project No. 0314-2019-0014, and by the Russian Foundation for Basic Research, projects 20-01-00045 and 18-01-00747.

A. Kononov et al. (Eds.): MOTOR 2020, LNCS 12095, pp. 274–288, 2020.
https://doi.org/10.1007/978-3-030-49988-4_19

of type $[LB, \rho^* LB]$ such that for any $I \in \mathcal{K}$ we have $F^*(I) \in [LB(I), \rho^* LB(I)]$. In other words, the goal of this research is to find

$$\rho^* = \sup_{I \in \mathcal{K}} \frac{F^*(I)}{LB(I)},$$

assuming that a lower bound is strictly positive. Note that the value of ρ^* does not depend on the instance, but is depends on the optimization problem, a lower bound and class $\mathcal{K}$.

This direction of research is important due to the following reasons. First, it helps to estimate the quality of the lower bound for class $\mathcal{K}$: the tighter is the interval, the less is the gap between the optimum and LB. Second, it describes some property of the optimal solution in form $F^* \leqslant \rho^* LB$. And at last (but not the least), it often helps to describe some good approximation algorithm for the optimization problem under consideration, due to the following observation. The tightness of the localization interval is proved by the construction of a *critical* instance I^* such that $F^*(I^*) = \rho^* LB(I^*)$ (or by presenting an infinite series of instances I_k^* such that $\lim_{k \to \infty} \dfrac{F^*(I_k^*)}{LB(I_k^*)} = \rho^*$). On the other hand, the fact that for any $I \in \mathcal{K}$ the inequality $F^*(I) \leqslant \rho^* LB$ is usually established constructively, by presenting an efficient algorithm that obtains an approximate solution x_I with $F(x_I) \leqslant \rho^* LB(I)$. Such an algorithm is obviously a ρ^*-approximation, moreover it has an approximation ratio ρ^* not only with respect to the optimum, but to the lower bound LB. By definition of the OL-interval, this approximation is as good as theoretically possible with respect to LB. This means, that in order to improve the approximation ratio we need to improve the lower bound, or to compare the solution obtained with the optimum itself.

The first (to the best of our knowledge) OL-interval for a scheduling problem—namely the *open shop*—was found in [12]. The open shop problem ([8]) can be described as follows. Given a set of *machines* $\mathcal{M} = \{M_1, \ldots, M_m\}$, set of *jobs* $\mathcal{J} = \{J_1, \ldots, J_n\}$ and a matrix of *processing times* (p_{ji}), one needs to construct a schedule of processing each of the jobs on each machine (operation O_{ji} of machine M_i on job J_j takes p_{ji} time units) such that operations of each job are performed consecutively (in any order) and no machine operates two jobs at any time. The goal is to minimize the *makespan* $C_{\max}$, *i.e.* the completion time of the latest operation. Following the traditional three-field notation for scheduling problems (see [11] for instance) the open shop problem with m machines is denoted by $Om||C_{\max}$. It is known to be NP-hard for $m \geqslant 3$ and polynomially solvable in the two-machine case [8].

The *standard lower bound* for any shop scheduling problem (open shop included) is the following combination of the maximum *machine load* $\ell_{\max} = \max_i \ell_i = \max_i \sum_j p_{ji}$ and the maximum *job length* $d_{\max} = \max_j d_j = \max_j \sum_i p_{ji}$:

$$\bar{C} = \max\{\ell_{\max}, d_{\max}\}. \tag{1}$$

It is shown in [8] that for any instance of $O2||C_{\max}$ the optimal makespan of I coincides with the lower bound $\bar{C}$, therefore the respective OL-interval consists of a single point. This is not the case for $m \geqslant 3$. The OL-interval for $O3||C_{max}$—$[\bar{C}, \frac{4}{3}\bar{C}]$—was established in [12]. That research required a computer-aided approach, developed by Sevastyanov and Chernykh. The idea behind this approach is an intelligent branch-and-bound-style enumeration of subsets of instances. We utilize the same approach in this paper, it is described in detail in Sect. 5. Another application of this approach can be found in [9] for a problem with multiprocessor tasks, what suggests that the method is viable. It actually helped to describe a bunch of polynomially solvable cases, as well as establish OL-intervals for other subproblems with respect to a simple lower bound ℓ_{max} [9].

Recently the OL result for $O3||C_{max}$ problem was improved for a special case—so-called *proportionate open shop*—in which for every job its processing times are equal: $p_{ji} = p_j$. The OL-interval for this case is $[\bar{C}, \frac{10}{9}\bar{C}]$, and the proof didn't require a computer-aided approach [13]. As for the general $O||C_{max}$ problem (when the number of machines is a part of an input), we have only partial knowledge of the OL-interval $[\bar{C}, \rho^*\bar{C}]$: the value of ρ^* belongs to $[\frac{3}{2}, 2)$. The lower bound is supported by a well-known series of instances (see [12] for instance), and the upper bound follows from the properties of *dense schedules* (see [1]).

Other known OL results concern the routing open shop problem, which is a subject of our investigation. Routing open shop was introduced in [2,3]. It extends the classical open shop in the following way. Jobs are located at the nodes of a transportation network described by an edge-weighted graph $G = \langle V; E \rangle$. Each node contains at least one job. A weight function on E represents the travel time $\mathbf{dist}(v, u)$ of mobile machines over the edge $[v, u] \in E$. Initially machines are located at the predefined *depot* $v_0 \in V$, and have to return there after performing all the operations. No restrictions on the traveling are in place: any number of machines can travel simultaneously over the same edge in any direction, machines are allowed to visit any node multiple times. The goal is to construct a feasible schedule for each machine to travel and perform operations of jobs with respect to the constraints from the classical open shop, and to return to the depot minimizing the makespan, which is in this case the completion time of last machine's activity (either traveling to the depot or performing an operation of the job from the depot). To distinguish this makespan from the objective function of the classical open shop, we use notation R_{max} for the routing open shop problem. Following the three-field notation we use $ROm||R_{max}$ for the m-machine routing open shop. An optional piece of notation $G = X$ in the second field is used if we want to specify the structure of the transportation network. Optimal makespan of a problem instance I is denoted by $R^*_{max}(I)$.

The routing open shop problem includes the metric traveling salesman problem as a special case even for $m = 1$, therefore it is strongly NP-hard in general. Moreover, its very special case $RO2|G = K_2|R_{max}$ is still NP-hard in the ordinary sense [3]. An FPTAS for this special case is described in [10]. An OL result for $RO2|G = K_2|R_{max}$ was presented in [2] with respect to the following lower bound $\bar{R}$. Let T^* denotes the total weight of the optimal TSP solution on graph G, $\mathcal{J}(v)$ is the set of jobs located at v, and $d_{max}(v) = \max\limits_{J_j \in \mathcal{J}(v)} d_j$. Then

$$\bar{R} = \max\left\{\ell_{\max} + T^*, \max_v\left(d_{\max}(v) + 2\mathbf{dist}(v_0, v)\right)\right\} \tag{2}$$

is the *standard lower bound* for the routing open shop problem. Note that it coincides with $\bar{C}$ in case when all travel distances are zero or the transportation network consists of a single node (in this case we have a plain open shop problem).

The known OL results for the two-machine routing open shop are shown in Table 1. The OL-interval for the general $RO2||R_{\max}$ is still unknown, although we do not have any evidence that it differs from the common interval for the known cases. On the other hand, a $\frac{4}{3}$-approximation algorithm for $RO2|easy - TSP|R_{\max}$ (with a known optimal solution of the underlying TSP) [4] suggests that such an interval is not wider than $[\bar{R}, \frac{4}{3}\bar{R}]$.

Table 1. OL-intervals for special cases of $RO2||R_{\max}$.

Problem:	Interval:	Reference:
$RO2|G = K_2|R_{\max}$	$[\bar{R}, \frac{6}{5}\bar{R}]$	[2]
$RO2|G = K_3|R_{\max}$		[6]
$RO2|G = tree|R_{\max}$		[5]

The OL result for the $RO2|G = K_2|R_{\max}$ problem was detalized in [7], where the tight upper bound of the OL-interval is described as a function of distribution of the total processing time between the nodes.

For a larger number of machines, the problem is still open. The only result (mentioned above) concerns $O3||C_{\max}$ (which we can denote as $RO3|G = K_1|R_{\max}$ for consistency) [12]. The OL-interval for $O4||C_{\max}$ is still unknown, and we have no evidence that it is wider than the one for the $O3||C_{\max}$ problem. The research for the case $m \geqslant 3$ is complicated and the only known result achieved required massive, though intelligent, computer-aided enumeration of subclasses of instances. The main goal of this paper is to establish the OL-interval for at least some special cases of $RO3||R_{\max}$ problem. Naturally, we could not avoid that computer-aided approach, moreover, in order to make it work in a reasonable time we had to derive a number of reduction techniques which allowed the approach to work faster, which in turn helped us to verify the surprising result, that the OL-interval for $RO3|G = K_2|R_{\max}$ is in fact $[\bar{R}, \frac{4}{3}\bar{R}]$—the same as for the classical $O3||C_{\max}$ problem. Therefore we generalized the known result from [12]. As a by-product, we describe a linear-time $\frac{4}{3}$-approximation algorithm for the $RO3|G = K_2|R_{\max}$ problem.

The structure of the paper is the following. Section 2 contains preliminary notes, definitions and known results we utilize in our research. In Sect. 3 the properties of critical instances for $ROm||R_{\max}$ are investigated. We show the OL result for one special case of $RO3|G = K_2|R_{\max}$ in Sect. 4. Section 5 contains the detailed description of the computer-aided approach for other special cases, followed by conclusive remarks in Sect. 6.

2   Instance Reduction Procedures

Together with notation p_{ji}, G, T^*, $\bar{R}$ and $\mathcal{J}(v)$, introduced in the previous section, we use $p_{ji}(I)$, $G(I)$, $T^*(I)$, $\bar{R}(I)$ and $\mathcal{J}(I; v)$ if we want to refer to a specific problem instance I. We denote the *node load* of v by $\Delta(v) = \sum\limits_{J_j \in \mathcal{J}(v)} d_j$.

The latter notation will also be used in the form $\Delta(I; v)$ for a specific instance I. Note that due to (2)

$$\Delta = \sum_v \Delta(v) \leqslant m\ell_{\max} \leqslant m(\bar{R} - T^*). \tag{3}$$

By a reversible simplification procedure of a class of instances $\mathcal{K}$ we understand an instance transformation $\varphi : \mathcal{K} \to \mathcal{K}$ with the following properties:

1. Procedure *simplifies* the instance (reduces the number of jobs/machines/ nodes, or structure of the transportation network), unless it is simple already.
2. Transformation is *reversible*: any feasible schedule of instance I' can be treated as a feasible schedule of instance I with the same makespan.

On order to use such a procedure for the research of OL-intervals we need another property.

Definition 1. *An instance transformation φ on $\mathcal{K}$ is referred to as* valid *if $\bar{R}(\varphi(I)) = \bar{R}(I)$ for any $I \in \mathcal{K}$.*

For any valid reversible transformation φ on $\mathcal{K}$ we have

$$\frac{R^*_{\max}(\varphi(I))}{\bar{R}(\varphi(I))} \geqslant \frac{R^*_{\max}(I)}{\bar{R}(I)},$$

therefore it is sufficient to investigate the OL for the image $\varphi(\mathcal{K})$.

In our research we use simplification procedure based on the two known simplification operations: *job aggregation* and *terminal edge contraction*.

The job aggregation operation (also known as *grouping*) utilizes a simple idea of replacing a number of jobs with a single *aggregated* or *composite* job with processing times of each equal to the total processing time of the operations of jobs combined (see [6, 12] for example).

Definition 2. *For problem instance I, let $K \subseteq \mathcal{J}(I; v)$ for some node v. Then by* job aggregation *of set K we understand the following instance transformation $I \to I'$:*

$$G(I') = G(I), \; \mathcal{J}(I'; v) = \mathcal{J}(I; v) \setminus K \cup \{J_K\}, \; p_{Ki} = \sum_{J_j \in K} p_{ji}.$$

The job aggregation is clearly a reversible transformation: any schedule of operation of a composite job J_K can be treated as a schedule of respective operations of jobs from set K processed without any idle time in an arbitrary sequence.

Note that the machine loads and the node loads are preserved by any job aggregation operation. However, it is possible that $d_{\max}(I'; v) = d_K > d_{\max}(I; v)$, and there is a possibility that $\bar{R}(I') > \bar{R}(I)$ so that the job aggregation is not valid. The sufficient condition of the validity of an aggregation of set $K \subseteq \mathcal{J}(v)$ is the following inequality:

$$\sum_{J_i \in K} d_j \leqslant \bar{R} - 2\mathbf{dist}(v_0, v). \tag{4}$$

In particular we have the following

Definition 3. *A node v from $G(I)$ of some problem instance I is* overloaded *if*

$$\Delta(I; v) > \bar{R}(I) - 2\mathbf{dist}(I; v_0, v). \tag{5}$$

Otherwise the node v is referred to as underloaded.

By this definition and (4), the aggregation of $\mathcal{J}(I; v)$ is valid if and only if the node v is underloaded.

We use the following

Definition 4. *An instance I is called* irreducible, *if no valid job aggregation is possible for I.*

Note that there is a linear time valid transformation procedure converting any instance into an irreducible one [6,12]. According to the earlier observation it is sufficient to establish the OL-interval for irreducible instances only.

Terminal edge contraction can be used in case some terminal v node contains a single job and is based on the following idea: transfer the single job from node v to an adjacent one u, modifying its processing times to include the travel times between v and u.

Definition 5. *Let I be a problem instance, $v \neq v_0$ is a terminal node in $G(I)$ and $\mathcal{J}(I; v) = \{J_j\}$. Let $e = [u, v]$ be an edge incident to v. Then by the contraction of edge e we understand the following instance transformation $I \to I'$:*

$$\mathcal{J}(I'; u) = \mathcal{J}(I; u) \cup \{J_j\}, \ G(I') = G(I) \setminus \{v\}, \ p_{ji}(I') = p_{ji}(I) + 2\mathbf{dist}(u, v).$$

Again, we want to perform an edge contraction operation only if it is valid. The exact condition of invalidity of an edge contraction operation is described in the following

Definition 6. *In the settings of Definition 5, the edge e is called* overloaded *if*

$$d_j + 2m\mathbf{dist}(u, v) + 2\mathbf{dist}(v_0, u) > \bar{R}(I), \tag{6}$$

and underloaded *otherwise.*

Overloaded elements make the instance somehow problematic. Fortunately, the number of such elements is rather small. It was proved in [6] that any irreducible instance I of the $RO2||R_{\max}$ problem contains at most one overloaded node, and the only overloaded node (if any) contains at most three jobs. Recently that result was generalized in [5].

Lemma 1 ([5]). *Any instance of the $ROm\|R_{\max}$ problem contains at most $m - 1$ overloaded elements.*

The upper bound on the number of jobs in overloaded nodes is generalized for the case of m machines in the next section.

3   Properties of Irreducible Instances

Proposition 1. *Let I be an irreducible instance of the $ROm\|R_{\max}$ problem such that $|\mathcal{J}(v)| = k > 1$. Then*

$$\Delta(v) > \frac{k}{2}(\bar{R} - T^*). \tag{7}$$

Proof. Let $\mathcal{J}(I; v) = \{J_1, \ldots, J_k\}$. Since I is irreducible, (4) implies

$$\left.\begin{array}{l} d_1 + d_2 > \bar{R} - 2\mathbf{dist}(v_0, v) \geqslant \bar{R} - T^* \\ d_1 + d_3 > \bar{R} - 2\mathbf{dist}(v_0, v) \geqslant \bar{R} - T^* \\ \qquad \cdots \\ d_{k-1} + d_k > \bar{R} - 2\mathbf{dist}(v_0, v) \geqslant \bar{R} - T^* \end{array}\right\} \ \frac{k(k-1)}{2} \ \text{inequalities}$$

Each d_j occurs $k - 1$ times in the left parts. Adding up these inequalities we have

$$(k - 1)(d_1 + d_2 + \ldots + d_k) > \frac{k(k-1)}{2}(\bar{R} - T^*),$$

and (7) follows. □

Theorem 1. *Let I be an irreducible instance of $ROm\|R_{\max}$. Then every underloaded node in I contains exactly one job, and all the overloaded nodes (if any) contain at most $2m - 1$ jobs.*

We prove this theorem by contradiction to the existence of a counterexample with at least $2m$ jobs in all overloaded nodes together. To that end, we use the following

Definition 7. *Let m and R be positive integers, collection $D_1, \ldots, D_h$ is a partition of the set of positive numbers $\{\delta_1, \ldots, \delta_{2m}\}$, and*

$$\sum_{j=1}^{2m} \delta_j \leq mR, \tag{8}$$

$$\forall k \in \{1, \ldots, h\} \forall \delta_i, \delta_j \in D_k \ \ \delta_i + \delta_j > R, \tag{9}$$

$$\forall i |D_i| \geq 2. \tag{10}$$

Then a tuple $\langle m; R; D_1, \ldots, D_h \rangle$ is called a counter-structure.

Note that for any counter-structure we have

$$h \leqslant m - 1, \qquad (11)$$

and for $|D_k| = x$

$$\sum_{D_k} \delta_j > \frac{x}{2} R. \qquad (12)$$

(The proof is similar to that of Proposition 1 and Lemma 1).

Lemma 2. *A counterexample I to Theorem 1 with m machines, h overloaded nodes and exactly $2m$ jobs in the overloaded nodes all together exists if an only if there exists a counter-structure $\langle m; R; D_1, \ldots, D_h \rangle$ with $R \geqslant \bar{R}(I) - T^*(I)$.*

Proof. Let I be an counterexample with overloaded nodes $v_1, \ldots, v_h$, and $D_k = \{d_j | J_j \in \mathcal{J}(v_k)\}$, $k = 1, \ldots, h$. Then the tuple $\langle m; \bar{R} - T^*; D_1, \ldots, D_h \rangle$ is a counter-structure. Indeed, due to the irreducibility of I we have $\forall k \forall d_i, d_j \in D_k$ $d_i + d_j > \bar{R} - T^*$, and properties (8)–(10) follow from (3).

Now let us have a counter-structure $\langle m; R; D_1, \ldots, D_h \rangle$. We can construct the corresponding counter-example as follows. Let G be a star with h terminal nodes $v_1, \ldots, v_h$ and zero weight of edges. Node v_0 contains a dummy job with zero processing times, and each $\delta_j \in D_k$ corresponds to a job $J_j \in \mathcal{J}(v_k)$ with equal processing times δ_j / m. Then the irreducibility of the instance created follows from (9) and the fact that $\bar{R} = \ell_{\max} = \sum_{j=1}^{2m} \delta_j / m \leqslant R$. □

Lemma 3. *For any $m > 0$ there is no counter-structure $\langle m; R; D_1, \ldots, D_h \rangle$.*

Proof. Assume there exists a counter-structure $\mathcal{D} = \langle m; R; D_1, \ldots, D_h \rangle$ for some minimal value of m.

Case 1. $h = 1$ or $h = 2$.
Assume $|D_1| = m + r$ and $|D_2| = m - r$ (in case $h = 1$ the second set is empty and $r = m$). From (12) we have

$$\sum_{D_1} \delta_j > \frac{m+r}{2} R, \quad \sum_{D_2} \delta_j > \frac{m-r}{2} R,$$

therefore

$$\sum_{j=1}^{2m} \delta_j = \sum_{D_1 \cup D_2} \delta_j > mR,$$

which contradicts (8).

Case 2. $h > 2$.
In this case from (11) we have $m - 3 > 0$. Suppose for some of sets $D_1, \ldots, D_h$ its cardinality is not equal to 3. Without loss of generality let it be D_h, and $\delta_{2m-1}, \delta_{2m} \in D_h$. Now we reduce our counter-structure in the following manner:

remove elements $\delta_{2m-1}, \delta_{2m}$ from D_h, and if D_h becomes empty, remove it from $\mathcal{D}$. As a result we have

$$\sum_{j=1}^{2(m-1)} \delta_j < (m-1)R,$$

$$\forall D_k \in \mathcal{D} \forall \delta_x, \delta_y \in D_k, \quad \delta_x + \delta_y > R,$$

$$\forall D_k \in \mathcal{D} \ |D_k| \geq 2,$$

$$\sum_k |D_k| = 2(m-1).$$

That means that we obtained a counter-structure with a smaller m, which contradicts the choice of $\mathcal{D}$.

By contradiction we have $\forall D_k \in \mathcal{D} \ |D_k| = 3$, and from (12)

$$\sum_{D_k} \delta_j > \frac{3}{2}R.$$

Consider a structure $\mathcal{D}' = \langle m-3; R; D_1, \ldots, D_{h-2} \rangle$. Then $\mathcal{D}'$ is also a counter-structure. Indeed, properties (9) and (10) are preserved. On the other hand,

$$\sum_{k=1}^{h-2} \sum_{D_k} \delta_j \leqslant mR - \sum_{D_{h-1}} \delta_j - \sum_{D_h} \delta_j < mR - 3R = (m-3)R.$$

Hence property (8) holds, and lemma is proved by contradiction to the choice of $\mathcal{D}$.   □

To derive the proof of Theorem 1 from Lemmas 2 and 3 it is sufficient to prove that any counterexample with $2m + x$ jobs in overloaded nodes can be reduced to a counterexample with exactly $2m$ jobs in overloaded nodes. This can be easily done by the following observation. Consider any overloaded node v with at least three jobs (such a node exists due to Lemma 1). Introduce a new dummy terminal node u adjacent to v with zero weight of the incident edge. Transform one of the jobs from v to u. (This can be seen as an inverse of the terminal edge contraction operation.) New node u is not overloaded (as it contains a single job), and we can continue in this manner until we have exactly $2m$ jobs in overloaded nodes.   □

4   Optima Localization for $RO3|G = K_2|R_{\max}$

We use notation a_j, b_j and c_j instead of p_{j1}, p_{j2} and p_{j3} for each job, moreover, same notation is used to represent the operations themselves and not only their processing times. The node v, different from the depot, will be referred to as *distant node*, and the weight of the only edge is denoted by τ. Notation $\mathcal{I}_{3,n_0+n_1}$ is used for the set of irreducible instances with exactly n_0 jobs at v_0 and n_1 jobs at v.

At first we show that even for *proportionate* instances from $\mathcal{I}_{3,1+1}$ the optimal makespan can be as large as $\frac{4}{3}\bar{R}$.

Lemma 4. *There exists a proportionate instance* $\hat{I} \in \mathcal{I}_{3,1+1}$ *such that* $R_{\max}(S) = \frac{4}{3}\bar{R}$.

Proof. The instance $\hat{I}$ consists of two jobs with $a_0 = b_0 = c_0 = 1$ and $a_1 = b_1 = c_1 = 0$, $\mathcal{J}(v_0) = \{J_0\}$ and $\mathcal{J}(v_1) = \{J_1\}$. The distance $\tau = 1$. Note that $\bar{R}(\hat{I}) = 3$.

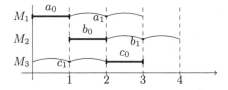

Fig. 1. An optimal schedule for instance $\hat{I}$.

A schedule of makespan 4 for $\hat{I}$ is shown in Fig. 1. That schedule is optimal due to the following reasoning. Let S be some schedule for $\hat{I}$, without loss of generality the operations of J_0 are performed in order $a_0 \prec b_0 \prec c_0$ ($O_1 \prec O_2$ means that operation O_1 precedes O_2 in the schedule under consideration). If $b_0 \prec b_1$ in S then $R_{\max}(S) \geqslant a_0 + b_0 + 2\tau = 4$. The case $b_1 \prec b_0$ is considered similarly. □

The main goal of the rest of the paper is to describe the proof of the following

Theorem 2. *There exists a linear time algorithm that obtains a feasible schedule* S *for each instance of the problem* $RO3|G = K_2|R_{\max}$ *such that*

$$R_{\max}(S) \leqslant \frac{4}{3}\bar{R}(I).$$

The algorithm consists of three steps: transforming the initial instance I to the irreducible one I' (and contracting the edge if it is underloaded), building a schedule S' for I' with a desired makespan not greater than $\frac{4}{3}\bar{R}$, and restoring the schedule S for the initial I from S'. The first and the last steps can be done in linear time, and the second step requires constant time (because instance I' has a constant number of operations). The difficult part is to guarantee the performance of the algorithm, and to that end we need to establish the OL-interval for the problem under consideration.

Further we assume that the only edge is overloaded, otherwise the problem can be reduced to the plain $O3||C_{\max}$ for which the OL-interval is known. The following lemma describes a simple sufficient condition for that.

Lemma 5. *Let I be such an irreducible instance of $ROm|G = K_2|R_{\max}$ problem that*

$$\Delta(v_0) \geq (m-1)\bar{R}. \tag{13}$$

Then the only edge is underloaded.

Proof. Due to (13) and (3) the node v is underloaded and therefore contains a single job J_α. Suppose that the edge is overloaded. By Definition 6 we have $d_\alpha + 2m\tau > \bar{R}$. On the other hand, from (3) and (13) we have $d_\alpha + 2m\tau \leq m\bar{R} - \Delta(v_0) < \bar{R}$. The lemma is proved by contradiction. $\square$

Corollary 1. *Let an irreducible instance I of $RO3|G = K_2|R_{\max}$ problem contain at least 4 jobs at the depot. Then the edge is underloaded.*

Proof. Straightforward from Lemma 5 and Proposition 1.

By Theorem 1 and Corollary 1 it is sufficient to consider instances from the following collection of sets:

$$\left(\bigcup_{n_1 \leqslant 5} \mathcal{I}_{3,1+n_1} \right) \cup \left(\bigcup_{n_0 \leqslant 3} \mathcal{I}_{3,n_0+1} \right) \cup \left(\bigcup_{\substack{n_0+n_1 \leqslant 5 \\ n_0,n_1 > 1}} \mathcal{I}_{3,n_0+n_1} \right). \tag{14}$$

In this section we present a detailed proof for the set $\mathcal{I}_{3,1+1}$. Another part of the proof is done by the computer-aided approach, and described in detail in the next section.

We use the branch-and-bounds approach to prove the main result. A similar approach to the proofs of the OL was used, for example, in [12], [6]. To describe a schedule for an arbitrary instance, we use the schedule *templates* defined by digraphs, which determine a linear order of operations for each job and for each machine. Nodes of the template represent operations and have weights of respective processing times, while arcs represent precedence constraints on the operations, and some of them also have weights representing travel times. We will use two dummy nodes S and F of zero weight, denoting the start and the finish of the schedule, respectively. By $S_{\mathcal{H}}(I)$ we denote the early (or *active*) schedule built according to template $\mathcal{H}$ for instance I. We can estimate the makespan of that schedule as a weight of a *critical path* in graph $\mathcal{H}$. The weight of a path is a total weight of all its elements, and a critical path is a path of the maximum weight.

However we cannot tell which path is critical until we know the instance, therefore to complete the proof we consider all possible complete paths. We exclude from consideration so-called trivial paths whose length does not exceed $\bar{R}$ regardless of the instance. The same approach lies in the foundation of computer-aided proofs and is described in greater detail in Sect. 5.

Lemma 6. *Let $I \in \mathcal{I}_{3,1+1}$. Then one can in linear time build a feasible schedule S for I such that $R_{\max}(S) \leq \frac{4}{3}\bar{R}$.*

Proof. Let us have set of jobs $\mathcal{J}(v_0) = \{J_0\}$ and $\mathcal{J}(v_1) = \{J_1\}$. Without loss of generality assume $a_0 = \max\{a_0, b_0, c_0\}$, therefore

$$b_0 + c_0 \leq \frac{2}{3}\bar{R}. \tag{15}$$

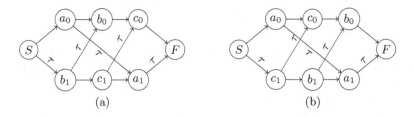

Fig. 2. Templates $\mathcal{H}_1$ and $\mathcal{H}_2$.

Consider schedule $S_1 = S_{\mathcal{H}_1}(I)$ (see Fig. 2, a).
Assuming $R_{\max}(S_1) > \bar{R}$ (otherwise the lemma is proved), we have

$$R_{\max}(S_1) \leq 2\tau + b_1 + c_0 + \max\{b_0, c_1\}.$$

Consider schedule $S_2 = S_{\mathcal{H}_2}(I)$ (Fig. 2, b).
By similar reasoning, assuming $R_{\max}(S_2) > \bar{R}$,

$$R_{\max}(S_2) = 2\tau + c_1 + b_0 + \max\{c_0, b_1\}.$$

By (2) and (15) we have

$$R_{\max}(S_1) + R_{\max}(S_2) = \underbrace{2\tau + b_1 + c_1}_{\leq \bar{R}} + \underbrace{b_0 + c_0}_{\leq \frac{2}{3}\bar{R}} + 2\tau + \max\{b_0, c_1\} + \max\{c_0, b_1\}.$$

Note that $2\tau + \max\{b_0 + b_1, c_0 + c_1, b_1 + c_1\} \leqslant \bar{R}$, therefore, unless $b_0 > c_1$ and $c_0 > b_1$,

$$R_{\max}(S_1) + R_{\max}(S_2) \leq \frac{8}{3}.$$

For the remaining case $b_0 > c_1$ and $c_0 > b_1$ regroup the sum in another way:

$$R_{\max}(S_1) + R_{\max}(S_2) = \underbrace{2\tau + b_1 + b_0}_{\leq \bar{R}} + \underbrace{2\tau + c_1 + c_0}_{\leq \bar{R}} + \underbrace{b_0 + c_0}_{\leq \frac{2}{3}\bar{R}} \leq \frac{8}{3}\bar{R}.$$

Therefore $\min\{R_{\max}(S_1), R_{\max}(S_2)\} \leqslant \frac{4}{3}\bar{R}$ and the lemma is proved.   □

5   Computer-Aided Approach and Approximation Algorithm

Let $\overline{\mathcal{I}}_{3,n_1+n_2} = \{I \in \mathcal{I}_{3,n_1+n_2} | \bar{R}(I) = 1\}$ is the set of the *normalized* instances. The processing and the travel times are considered to be rational numbers. Obviously, any instance (except the trivial ones with zero standard lower bound) can be normalized by dividing every processing and travel time by $\bar{R}$, and an optimal schedule for normalized instance is a scaled version of an optimal schedule for initial instance. Therefore, in order to justify the OL-interval for $RO3|G = K_2|R_{\max}$ it is sufficient to prove theorems of the following type (similar to Lemma 6):

Theorem. *For any instance* $I \in \overline{\mathcal{I}}_{3,n_1+n_2}$ *a schedule S with $R_{\max}(S) \leqslant \frac{4}{3}$ can be built in polynomial time.*

Sometimes such theorems can be proved in traditional ways (Lemma 6 is an example). For the others there is the computer-aided approach to search for the OL-intervals for irreducible instances, developed by Sevastyanov and Chernykh [12]. It can be described as follows.

The goal is to construct a *tree of proof* (which is actually an *outtree, i.e.* a directed tree where all the arcs are oriented away from the root). Each vertex of the tree (except for the terminal ones) corresponds to a *template* of a schedule, feasible for the set of instances under consideration. Outgoing arcs represent *non-trivial* complete paths in the corresponding template. The length of each of those paths can be described by a linear expression on the variables p_{ji} and τ. A sample template $\mathcal{H}_1$ (with jobs $J_1, J_2 \in \mathcal{J}(v_0)$ and $\mathcal{J}(v_1) = \{J_3\}$) is presented in Fig. 3 (a), below is the list of its non-trivial paths in the form of linear expressions:

- $R_1 = p_{11} + p_{21} + p_{22}$
- $R_3 = p_{32} + p_{12} + p_{13} + 2\tau$
- $R_5 = p_{23} + p_{21} + p_{31} + 2\tau$

- $R_2 = p_{11} + p_{12} + p_{22}$
- $R_4 = p_{32} + p_{33} + p_{13} + 2\tau$
- $R_6 = p_{23} + p_{33} + p_{31} + 2\tau$

A fragment of an example structure of a tree of proof is depicted in Fig. 3 (b). Each terminal vertex of the tree represents a subset of instances, *e.g.* for terminal vertex T_1 this subset $\mathcal{I}(T_1)$ is defined as the following intersection:

$$\mathcal{I}(T_1) = \{I | \text{path } R_1 \text{ is critical in } S_{\mathcal{H}_1}(I)\} \cup \{I | \text{path } R_{11} \text{ is critical in } S_{\mathcal{H}_{21}}(I)\}.$$

Naturally, the number of subsets in the intersection coincides with the depth of the terminal vertex in the tree of proof.

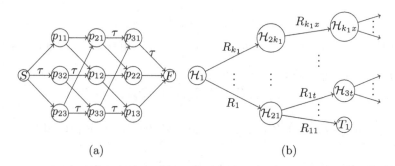

(a)   (b)

Fig. 3. (a) Sample template for the set $\overline{\mathcal{I}}_{3,2+1}$, and (b) an example fragment of the tree of proof.

Now we can use this information to derive an *upper bound* on optima of all instances from $\mathcal{I}(T_1)$ in the following way. The idea is to find a *critical instance* from $\mathcal{I}(T_1)$, *i.e.* such an instance $I^{crit}(T_1)$ for which the best among the schedules $S_{\mathcal{H}_1}(I^{crit}(T_1))$ and $S_{\mathcal{H}_{21}}(I^{crit}(T_1))$ has the greatest makespan:

$$\min\{R_{\max}(S_{\mathcal{H}_1}(I)), R_{\max}(S_{\mathcal{H}_{21}}(I))\} \to \max_{I \in \mathcal{I}(T_1)}.$$

This can be done by means of a linear program. Consider the lower bound defined as $LB = \max\{LB_1, \ldots, LB_l\}$, there LB_k is a linear expression on the operations processing times and travel times. For instance, for set $\overline{\mathcal{I}}_{3,2+1}$ we have $\bar{R} = \max\{\ell_1 + 2\tau, \ell_2 + 2\tau, \ell_3 + 2\tau, d_1, d_2, d_3 + 2\tau\}$, assuming $\mathcal{J}(v_0) = \{J_1, J_2\}$ and $\mathcal{J}(v) = \{J_3\}$. The critical instance and the upper bound on the worst-case optimal makespan can be found with the following LP (the general form is on the left, with τ_s being parameters of an instance in addition to the processing times, *i.e.* travel times, and the example form is on the right):

$$\rho \rightarrow \max$$
$$s.t.$$
$$\begin{cases} LB_1 \leqslant 1, \\ \vdots \\ LB_l \leqslant 1, \\ R_{t_1} \geqslant \rho, \\ R_{t_1 t_2} \geqslant \rho, \\ \vdots \\ R_{t_1 t_2 \ldots t_k} \geqslant \rho, \\ p_{ji}, \tau_s, \rho \geqslant 0; \end{cases}$$

$$\rho \rightarrow \max$$
$$s.t.$$
$$\begin{cases} \ell_1 + 2\tau \leqslant 1, \\ \ell_2 + 2\tau \leqslant 1, \\ \ell_3 + 2\tau \leqslant 1, \\ d_1 \leqslant 1, \\ d_2 \leqslant 1, \\ d_3 + 2\tau \leqslant 1, \\ R_1 \geqslant \rho, \\ R_{11} \geqslant \rho \\ p_{ji}, \tau, \rho \geqslant 0. \end{cases} \quad (16)$$

Thus, if the optimum value ρ^* of LP (16) does not exceed the upper bound we want to prove, we can discard the whole subset of instances (which becomes a terminal node of the tree of proof). Otherwise, we obtain a critical instance I', for which the upper bound ρ^* is achieved. Further, we can use this instance to select the next template for the current subset of instances, which works the best for the instance I'. That is the core idea of the computer-aided way to build and verify a tree of proof. Note that for verification purposes one needs an exact solution of LP (16), so it has to be solved in rational numbers instead of using the floating point arithmetic.

We were able to construct a tree of proof in about 28 h of the running time. For that end we discovered a few techniques to reduce the running time. The description of those ideas is omitted from this paper due to the volume limitation.

6 Conclusion

The purpose of this paper is twofold. The first part is a theoretical foundation for further research of OL intervals for $ROm||R_{\max}$ (Theorem 1). Another one is a mostly computer-aided proof of Theorem 2. We only provide a description of that proof though, so how can one verify the result? Our strong belief is that the best way to do it is not to use the program we have written (which is probable inefficient), but to create an alternative one using our description. Indeed, it is usually more difficult to figure out someone else's code than to write one yourself.

References

1. Aksyonov, V.: An approximation polynomial time algorithm for one scheduling problem. Upravlyaemye systemy **28**, 8–11 (1988). (in Russian)
2. Averbakh, I., Berman, O., Chernykh, I.: A 6/5-approximation algorithm for the two-machine routing open-shop problem on a two-node network. Eur. J. Oper. Res. **166**(1), 3–24 (2005). https://doi.org/10.1016/j.ejor.2003.06.050
3. Averbakh, I., Berman, O., Chernykh, I.: The routing open-shop problem on a network: complexity and approximation. Eur. J. Oper. Res. **173**(2), 531–539 (2006). https://doi.org/10.1016/j.ejor.2005.01.034
4. Chernykh, I., Kononov, A.V., Sevastyanov, S.: Efficient approximation algorithms for the routing open shop problem. Comput. Oper. Res. **40**(3), 841–847 (2013). https://doi.org/10.1016/j.cor.2012.01.006
5. Chernykh, I., Krivonogiva, O.: Optima localization for the two-machine routing open shop on a tree, submitted to Diskretnyj Analiz i Issledovanie Operacij (2019). (in Russian)
6. Chernykh, I., Lgotina, E.: The 2-machine routing open shop on a triangular transportation network. In: Kochetov, Y., Khachay, M., Beresnev, V., Nurminski, E., Pardalos, P. (eds.) DOOR 2016. LNCS, vol. 9869, pp. 284–297. Springer, Cham (2016). https://doi.org/10.1007/978-3-319-44914-2_23
7. Chernykh, I., Pyatkin, A.: Refinement of the optima localization for the two-machine routing open shop. In: OPTIMA 2017 Proceedings, vol. 1987, pp. 131–138 (2017)
8. Gonzalez, T.F., Sahni, S.: Open shop scheduling to minimize finish time. J. ACM **23**(4), 665–679 (1976). https://doi.org/10.1145/321978.321985
9. Kononov, A., Kononova, P., Gordeev, A.: Branch-and-bound approach for optima localization in scheduling multiprocessor jobs. Int. Trans. Oper. Res. **27**(1), 381–393 (2017). https://doi.org/10.1111/itor.12503
10. Kononov, A.: On the routing open shop problem with two machines on a two-vertex network. J. Appl. Ind. Math. **6**(3), 318–331 (2012). https://doi.org/10.1134/s1990478912030064
11. Lawler, E.L., Lenstra, J.K., Rinnooy Kan, A.H.G., Shmoys, G.B.: Sequencing and scheduling: algorithms and complexity. In: Logistics of Production and Inventory. Elsevier (1993)
12. Sevastianov, S.V., Tchernykh, I.D.: Computer-aided way to prove theorems in scheduling. In: Bilardi, G., Italiano, G.F., Pietracaprina, A., Pucci, G. (eds.) ESA 1998. LNCS, vol. 1461, pp. 502–513. Springer, Heidelberg (1998). https://doi.org/10.1007/3-540-68530-8_42
13. Sevastyanov, S.V.: Some positive news on the proportionate open shop problem. Sibirskie Elektronnye Matematicheskie Izvestiya **16**, 406–426 (2018). https://doi.org/10.33048/semi.2019.16.023

Makespan Minimization for Parallel Jobs with Energy Constraint

Alexander Kononov[1] and Yulia Kovalenko[2,3]([✉])

[1] Sobolev Institute of Mathematics SB RAS,
4, Akad. Koptyug Avenue, 630090 Novosibirsk, Russia
`alvenko@math.nsc.ru`
[2] Sobolev Institute of Mathematics SB RAS, Omsk Department,
13, Pevtsov Street, 644043 Omsk, Russia
`julia.kovalenko.ya@yandex.ru`
[3] Dostoevsky Omsk State University,
55a, Mira Prospect, 644077 Omsk, Russia

Abstract. We are given a set of parallel jobs that have to be executed on a set of speed-scalable processors varying their speeds dynamically. Running a job at a slower speed is more energy efficient, however it takes longer time and affects the performance. Every job is characterized by the processing volume and the number of the required processors. Our objective is to minimize the maximum completion time so that the energy consumption is not greater than a given energy budget. For various particular cases we propose polynomial-time approximation algorithms, consisting of two stages. At the first stage, we give an auxiliary convex program. By solving this problem in polynomial time, we find processing times of jobs and a lower bound on the makespan. Then, at the second stage, we transform our problem to the classical problem without speed scaling and construct a feasible schedule.

Keywords: Parallel job · Speed scaling · Scheduling · Approximation algorithm

1 Introduction

We consider the problem of scheduling a set of jobs $\mathcal{J} = \{1, \ldots, n\}$ on m speed scalable parallel processors. Each job $j \in \mathcal{J}$ is characterized by processing volume (work) W_j and number of required processors $size_j$. Note that parameter $size_j$ for job $j \in \mathcal{J}$ indicates that the job can be processed on any subset of parallel processors of the given size. Such jobs are called rigid jobs. It is assumed that all jobs arrive at time step 0 unless stated otherwise. Job preemption, migration and precedence constraint might or might not be allowed in the exploring of scheduling in this paper.

The reported study was funded by RFBR, project number 20-07-00458.

The standard homogeneous model in speed-scaling is considered. When a processor runs at a speed s, then the rate with which the energy is consumed (the *power*) is s^α, where $\alpha > 1$ is a constant. Each of m processors may operate at variable speed. However, we assume that if processors execute the same job simultaneously then all these processors run at the same speed. It is supposed that a continuous spectrum of processor speeds is available.

The aim is to find a feasible schedule with the smallest value of the maximum completion time (makespan) so that the energy consumption is not greater than a given energy budget E. This is a natural assumption in the case when the energy of a battery is fixed, i.e. the problem finds applications in computer devices whose lifetime depends on a limited battery efficiency (for example, multi-core laptops). Moreover, the bicriteria problems of minimizing energy consumption and a scheduling metric arise in real practice. The most obvious approach is to bound one of the objective functions and optimize the other. The energy of the battery may reasonably be estimated, so we bound the energy used, and optimize the makespan.

Note that there is an optimal solution where each job j is processed with a fixed speed s_j due to the convexity of the speed-to-power function. Later we consider only feasible schedules with constant speeds of jobs. In this case, the energy consumption during the execution of job j depends on the duration $p_j = W_j/s_j$ of its performing and is equal to $size_j W_j^\alpha p_j^{1-\alpha}$. The preemptive and nonpreemptive variants of the speed-scaling scheduling of rigid jobs subject to the bound on energy consumption are denoted by $P|size_j, pmtn, energy|C_{\max}$ ($P|size_j, pmtn*, energy|C_{\max}$ if migrations are disallowed) and $P|size_j, energy|C_{\max}$, respectively. We denote the special case with precedence constraints on the set of jobs by placing "prec" in the second field of the three-field notation.

2   Previous Research

The makespan scheduling with single-processor jobs has been widely investigated. For the single-processor problem there is an optimal non-preemptive schedule where all jobs are executed with the same speed. This schedule may be computed via Karush-Kuhn-Tuker conditions in linear time (see, e.g., [13]). Bunde [3] developed an exact constructive algorithm for the uniprocessor setting with arbitrary release times of jobs. The implementation of the algorithm has also running time $O(n)$.

The makespan multiprocessor problem without migration is NP-hard even in the case of two processors [3]. The non-preemptive problem is strongly NP-hard. Pruhs et al. [13] proposed a PTAS for non-migrative independent jobs with zero release times. The makespan minimization is reduced to minimizing the l_α norm of the loads on processors, and in the formed schedule each processor finishes at the same time. They also proposed $O(\log^{1+2/\alpha} m)$-approximation algorithm for non-migrative jobs with precedence constraints. In the constructed schedule the sum of the instantaneous powers at which the processors run is constant over time. In [2] the latter result was improved. Bampis et al. [2] proposed an $\left(2 - \frac{1}{m}\right)$-approximation algorithm, which computes durations of jobs and the lower bound on the energy consumption. This allows reducing a speed scaling problem to the classical makespan minimization problem with fixed speeds of jobs. A fast approximation algorithm was proposed in [3] for multiprocessor non-migrative instances with independent equal-work jobs having arbitrary release times. The algorithm assigns jobs to processors using Round Robin strategy, and finds a schedule for each processor separately.

Shabtay et al. [14] analyzed a closely related problem of scheduling single-processor jobs on identical parallel processors, where job-processing times p_j are controllable through the allocation of a nonrenewable common limited resource as $p_j(R_j) = \left(\frac{W_j}{R_j}\right)^\kappa$. Here W_j is the workload of job j, R_j is the amount of resource allocated to processing job j and $\kappa \leq 1$ is a positive constant. Exact polynomial time algorithms were proposed for the following cases:

- uniprocessor makespan setting with precedence constraints;
- multiprocessor makespan instances with independent preemptive jobs;
- multiprocessor non-preemptive instances of minimizing the sum of completion times.

Note that the algorithms from [14] can be adopted to solving the presented three cases in the context of the speed scaling scheduling of single-processor jobs.

The speed scaling scheduling is important in computational systems. In addition to the time criteria, the energy consumption minimization under deadline constraints is widely investigated. For details we refer the reader to the surveys [1,5].

Our Results. Reductions from 2-Partition and 3-Partition imply that problem $P|size_j, energy, pmtn|C_{\max}$ is weakly NP-hard and problem $P|size_j, energy|C_{\max}$ is strongly NP-hard even if all jobs have a common release time and unit processing volumes.

We propose approximation algorithms, which consist of two stages. At the first stage, we obtain a lower bound on the makespan and calculate processing times of jobs using an auxiliary convex program. Then, at the second stage, we transform our problem to the classical problem without speed scaling, and we use "list-scheduling" algorithms to obtain schedules of constant factor approximation. We consider the following makespan minimization cases:

1. rigid jobs without preemption ($\left(2 - \frac{1}{m}\right)$ -approximation algorithm);
2. non-preemptive rigid jobs with precedence constraints $\left(\left(\frac{2-q}{1-q}\right)\right.$- approximation algorithm, where $size_j \leq qm$, $0 < q < 1$);
3. rigid jobs with preemptions and release times ($\left(2 - \frac{1}{m}\right)$ -approximation algorithm).

3   Rigid Jobs Without Preemption

In this section, we present a constant factor approximation algorithm for the case of non-preemptive rigid jobs.

The First Stage. We relax the condition that the processors should execute the job simultaneously. More precisely, we replace job $j \in \mathcal{J}$ with $size_j$ single-processor jobs of volume W_j, provided that all these copies have the same processing time p_j. Then the relaxed problem can be formulated as the following convex program with the variables p_j and T, where p_j is understood as an actual processing time of each copy of job j.

$$T \rightarrow \min, \tag{1}$$

$$\max_{j \in \mathcal{J}} p_j \leq T, \tag{2}$$

$$\frac{1}{m} \sum_{j \in \mathcal{J}} size_j p_j \leq T, \tag{3}$$

$$\sum_{j \in \mathcal{J}} size_j W_j^\alpha p_j^{1-\alpha} \leq E, \tag{4}$$

$$p_j \geq 0, \; j \in \mathcal{J}. \tag{5}$$

The constraints (2)–(3) guarantee that the duration of any job does not exceed T, and the total load of all processors is no more than Tm. Inequality (4) ensures that the total energy consumption is not greater than the budget E.

Problem (1)–(5) can be solved using KKT (Karush-Kuhn-Tucker) conditions (see, e.g., [9]). The time complexity is $O(n^2)$ if we have oracles to raise to a power and extract a root. Let $(p_1^*, \ldots, p_n^*, T^*)$ denote the optimal solution to (1)–(5). As a result, we obtain the lower bound T^* on the optimal makespan, C_{max}^*.

The Second Stage. The non-preemptive schedule S is constructed using "non-preemptive list-scheduling" algorithm [11]: Whenever a subset of processors falls idle, the "non-preemptive list-scheduling" algorithm schedules a rigid job j of duration p_j^* that does not require more processors than are available (if it is possible). The time complexity of the algorithm is $O(n^2)$.

Lemma 1. *"Non-preemptive list-scheduling" algorithm generates a feasible schedule S with length at most $\left(2 - \frac{1}{m}\right) T^*$ for problem $P|size_j, energy|C_{\max}$.*

Proof. Let T' denote the length of schedule S. If at least $\left\lceil \frac{m+1}{2} \right\rceil$ processors are used at any time step in S, we have

$$T^* \geq \frac{1}{m} \sum_{j \in \mathcal{J}} p_j^* size_j \geq \left\lceil \frac{m+1}{2} \right\rceil \frac{T'}{m} \geq \frac{T'}{2 - 1/m}.$$

Otherwise, assume that I is the last time interval of schedule S with $m_I < \lceil \frac{m+1}{2} \rceil$ processors are used during I. By the construction of S there is a job j that is performed during the whole interval I. Let C_j be the completion time of j in S. It is easy to see that at every point in time during interval $[0, C_j - p_j^*)$ schedule S uses at least $m - size_j + 1 \geq m - m_I + 1$ processors (otherwise job j should be started earlier). Moreover, at least m_I processors are utilized in interval $[C_j - p_j, C_j)$, therefore, each job executed in interval $[C_j, T')$ requires no less than $m - m_I + 1$ processors. Thus, the total load of all processors $\sum_{j \in \mathcal{J}} p_j^* size_j$ is at least

$$m_I p_j^* + (T' - p_j^*)(m - m_I + 1) \leq T^* m. \tag{6}$$

As $1 \leq m_I < \lceil \frac{m+1}{2} \rceil$, $m - 2m_I + 1 > 0$ and $p_j^* \leq T^*$ for all jobs $j \in \mathcal{J}$, we have from (6)

$$T' \leq \frac{mLB + p_j^*(m - 2m_I + 1)}{m - m_I + 1} \leq \frac{T^*(2m - 2m_I + 1)}{m - m_I + 1}$$

$$= 2T^* - T^* \frac{1}{m - m_I + 1} \leq T^* \left(2 - \frac{1}{m} \right).$$

$\square$

Using the results from [6], we conclude that the approximation ratio of $\left(2 - \frac{1}{m} \right)$ for the "non-preemptive list-scheduling" algorithm is tight even if all jobs have single-processor type. Indeed, consider an instance with $(m^2 - m)$ jobs which have sizes $size_j = 1$ and works $W_j = 1$, and one single-processor job with work m, i. e. $n = m^2 - m + 1$. The energy budget E is m^2. In the optimal schedule the speed of each processor is equal to 1, the job of length m is executed in interval $[0, m]$, for example, on the first processor, and all other jobs are executed in arbitrary order in interval $[0, m]$ on processors $\{2, \ldots, m\}$. From lower-bound model (1)–(5) we have $T^* = m$, $p_1^* = \ldots = p_{n-1}^* = 1$ and $p_n^* = m$. Suppose that jobs are sorted in order of nondecreasing processing times p_j^*. Then the resulting schedule obtained by "non-preemptive list-scheduling" algorithm has makespan $2m - 1$, since $(m^2 - m)$ jobs of durations $p_j^* = 1$ are executed firstly.

Theorem 1. *A* $\left(2 - \frac{1}{m} \right)$*-approximate schedule can be found in* $O(n^2)$ *time for problems* $P|size_j, pmtn, energy|C_{\max}$, $P|size_j, pmtn*, energy|C_{\max}$ *and* $P|size_j, energy|C_{\max}$.

4   Rigid Jobs with Precedence Constraints

In this section, we consider the case of non-preemptive rigid jobs and precedence relations between the jobs. If job j precedes job j' (we write $j \prec j'$), then j' cannot start until j is completed. The precedence constraints are represented in the form of a directed acyclic graph $G = (J, A)$, where arc (j, j') belongs to set A if and only if job j is constrained to precede job j'. Additionally, we assume that each job j requires $size_j \leq qm$ processors, where $0 < q < 1$.

In [4], the strong NP-hardness of problem $P2|size_j, chain|C_{max}$ has been proven, when the partial order consists of a set of chains. Using the approach from [8] and a polynomial-time reduction from [4], we can prove that problem $P2|size_j, chain, energy|C_{max}$ is also strongly NP-hard. Now we propose a two-stage approximation algorithm for problem $P|size_j, prec, energy|C_{max}$.

The First Stage. We formulate the following convex problem in order to obtain a lower bound on the optimal makespan:

$$T \to \min, \tag{7}$$

$$\frac{1}{m} \sum_{j \in \mathcal{J}} p_j size_j \le T, \tag{8}$$

$$C_j \le T, \ j \in \mathcal{J}, \tag{9}$$

$$p_j \le C_j, \ j \in \mathcal{J}, \tag{10}$$

$$C_j + p_{j'} \le C_{j'}, \ (j, j') \in A, \tag{11}$$

$$\sum_{j \in \mathcal{J}} size_j W_j^\alpha p_j^{1-\alpha} \le E, \tag{12}$$

$$C_j \ge 0, \ p_j \ge 0, \ j \in \mathcal{J}. \tag{13}$$

Here variable p_j (C_j) represents the processing time (the completion time) of job $j \in \mathcal{J}$. Inequalities (10)–(11) allow to calculate the completion times of jobs taking into account the processing times and precedence constraints. Problem (7)–(13) can be solved in polynomial time using the Ellipsoid algorithm (see, e.g., [7]).

To apply the ellipsoid method in polynomial time, we need to check two additional technical conditions. The first condition is that the values of all variables are upper bounded by some number R. The second condition is that for the convex program there is a feasible point (or solution) and every point in a radius r is feasible. Then the running time of the ellipsoid method will be polynomial in $\log \frac{R}{r}$. The first condition and the bound on R can be derived from the fact that the value of an optimal solution is bounded ($p_j \le \left(\frac{size_j W_j^\alpha}{E} \right)^{\frac{1}{\alpha-1}}$, $C_j \le T$ for all $j \in \mathcal{J}$, and $T \le \sum_{i \in \mathcal{J}} \left(\frac{size_i W_i^\alpha}{E} \right)^{\frac{1}{\alpha-1}}$). Therefore, R is a polynomial involving various input parameters. The second condition is satisfied for the point (p_j', C_j', T') defined as follows: $p_j' := W_j \left(\frac{\sum_i W_i size_i}{E} \right)^{\frac{1}{\alpha-1}} + 1$, $j \in \mathcal{J}$, $C_1' := (p_1' + 1) + 1$, $C_j' := C_{j-1}' + (p_j' + 1) + 1$ for $j = 2, \ldots, n$ (we suppose that jobs are ordered in accordance with the partial order) and T' is large enough such that

$$T' - 1 \ge \max \left\{ C_n' + 1; \frac{1}{m} \sum_{j \in \mathcal{J}} ((p_j' + 1) size_j) \right\}.$$

Hence, the inequalities are satisfied in the ball of radius 1 around (p'_j, C'_j, T'), that is $r = 1$ (for details see, e.g., Paragraph 3.2.6 in the book [12]).

The Second Stage. Let $(p_1^*, \ldots, p_n^*, C_1^*, \ldots, C_n^*, T^*)$ denote the optimal solution to (7)–(13). As a result, we obtain the lower bound T^* on the optimal makespan, C_{max}^*. Using the found durations of jobs, we construct a feasible schedule S by "precedence-dependent list-scheduling" algorithm: Whenever a subset of processors falls idle, the algorithm schedules a rigid job that does not require more processors than are available, for which all the predecessors have been completed (if it is possible). The running time is $O(n^3)$. We state

Lemma 2. *"Precedence-dependent list-scheduling" algorithm generates a feasible schedule S with length at most* $\left(\frac{(2-q)m}{(1-q)m+1}\right) T^* < \frac{2-q}{1-q} T^*$ *for problem* $P|size_j \leq qm, prec, energy|C_{\max}$.

Proof. Our proof uses some ideas from [10]. Let $t_1 < t_2 < \cdots < t_\mu < t_{\mu+1}$ be the events corresponded to the completion times of jobs in schedule S sorted according to the increasing time, $\mu \leq n$. We define sub-intervals $I_{i-1} = (t_{i-1}, t_i]$, $i = 1, \ldots, \mu + 1$, where $t_0 := 0$ is the starting time of the first job, and $t_{\mu+1} := L_S$ is the completion time of the last job in schedule S.

We consider the sub-intervals in the reverse order $I_\mu, I_{\mu-1}, \ldots, I_2, I_1, I_0$, and partition them into two disjoint subsets $\Im_1$ and $\Im_2$ as follows. Let j_1 be the job that completes at time L_S and starts at time t_{k_1}. We put sub-intervals $I_{k_1}, I_{k_1+1}, \ldots, I_\mu$ into set $\Im_1$, and call them *covered* by job j_1.

There are two reasons by which job j_1 does not start at time t_{k_1-1}:

(C1) a job j_2 is executed in interval I_{k_1-1}, and this job precedes job j_1;
(C2) job j_1 is ready at time t_{k_1-1}, but interval I_{k_1-1} contains less than $size_{j_1}$ available processors.

In the first case (C1), we find time instant t_{k_2}, when job j_2 starts. The sub-intervals $I_{k_2}, I_{k_1+1}, \ldots, I_{k_1-1}$ are put into set $\Im_1$, and called covered by job j_2. Then we check why job j_2 does not start at time t_{k_2-1} and so on. In the second case (C2), we put sub-interval I_{k_1-1} into set $\Im_2$. Note that the number of busy processors in interval I_{k_1-1} is larger than $(1-q)m$. Then we go to sub-interval I_{k_1-2} and check why job j_1 does not start at time t_{k_1-2} and so on.

Continuing the above reasoning, we divide all sub-intervals I_i, $i = 0, \ldots, \mu$, into two disjoint subsets $\Im_1$ and $\Im_2$, and find a chain of jobs $j_k \prec j_{k-1} \prec j_{k-2} \prec \cdots \prec j_1$ such that each sub-interval in $\Im_1$ is covered by one of the jobs in the chain. Let T_1 and T_2 be the total length of all sub-intervals in $\Im_1$ and $\Im_2$, respectively. Then, the schedule length

$$L_S = T_1 + T_2. \tag{14}$$

From model (7)–(13) we have

$$T^* \geq C_{j_1}^* \geq C_{j_1}^* - C_{j_k}^* + p_{j_k}^* \geq \sum_{l=1}^{k-1}(C_{j_l}^* - C_{j_{l+1}}^*) + p_{j_k}^* \geq \sum_{l=1}^{k} p_{j_l}^* = T_1. \tag{15}$$

Moreover, inequality (8) guarantees that

$$T^* \geq \frac{1}{m} \sum_{j \in \mathcal{J}} p_j^* size_j. \tag{16}$$

The number of busy processors in a sub-interval $I_i \in \Im_2$ is larger than $(1-q)m$. Otherwise, more ready jobs would be executed in this sub-interval as all jobs $j \in \mathcal{J}$ have $size_j \leq qm$. Also, at least one processor is used during each sub-interval $I_i \in \Im_1$. Therefore, the total load

$$\sum_{j \in \mathcal{J}} p_j^* size_j \geq T_1 + ((1-q)m + 1) T_2 \tag{17}$$

Summing up the relations (14), (15), (16), and (17) we obtain the following bound on the ratio of schedule length L_S to lower bound T^*

$$\frac{L_S}{T^*} \leq \frac{T_1 + T_2}{\max\{T_1; \; T_1/m + (1 - q + 1/m)T_2\}} =: \rho. \tag{18}$$

When $T_1 \geq \frac{T_1}{m} + \left(1 - q + \frac{1}{m}\right) T_2$, i.e.

$$\frac{T_2}{T_1} \leq \frac{1 - 1/m}{1 - q + 1/m},$$

we have

$$\rho = \frac{T_1 + T_2}{T_1} \leq 1 + \frac{1 - 1/m}{1 - q + 1/m}.$$

When $T_1 \leq \frac{T_1}{m} + \left(1 - q + \frac{1}{m}\right) T_2$, i.e.

$$T_2 \geq T_1 \frac{1 - 1/m}{1 - q + 1/m},$$

we have

$$\rho = \frac{T_1 + T_2}{T_1/m + (1 - q + 1/m)T_2} \leq \frac{T_1 + T_1 \frac{1-1/m}{1-q+1/m}}{T_1/m + (1 - 1/m)T_1} = 1 + \frac{1 - 1/m}{1 - q + 1/m}.$$

This completes the proof. □

Note that the approximation ratio of $\left(\frac{(2-q)m}{(1-q)m+1}\right)$ is tight for single-processor jobs with empty partial order as we can see from the example in Sect. 3.

Theorem 2. *A $\left(\frac{2-q}{1-q}\right)$-approximate schedule can be found in polynomial time for problem $P|size_j, prec, energy|C_{\max}$ with $size_j \leq qm$, $0 < q < 1$.*

5   Rigid Jobs with Preemptions and Release Times

In this section, we consider the problem with preemptions, where jobs have arbitrary release times $r_j \geq 0$, $j \in \mathcal{J}$, and construct a polynomial time approximation algorithm.

The First Stage. We formulate the following program:

$$T \to \min, \tag{19}$$

$$r_j + \sum_{i \in R_j} p_i \leq T, \ j \in \mathcal{J}_{\lfloor \frac{m}{2} \rfloor}, \tag{20}$$

$$r_j + \frac{1}{m} \sum_{i \in \mathcal{J}: \ r_i \geq r_j} p_i size_i \leq T, \ j \in \mathcal{J}, \tag{21}$$

$$r_j + p_j \leq T, \ j \in \mathcal{J}, \tag{22}$$

$$\sum_{j \in \mathcal{J}} size_j p_j \left(\frac{W_j}{p_j}\right)^\alpha \leq E, \ j \in \mathcal{J}, \tag{23}$$

$$p_j \geq 0, \ j \in \mathcal{J}, \tag{24}$$

where $\mathcal{J}_{\lfloor \frac{m}{2} \rfloor} = \{j \in \mathcal{J} : size_j > \lfloor \frac{m}{2} \rfloor\}$, $R_j = \{i \in \mathcal{J}_{\lfloor \frac{m}{2} \rfloor} : r_i \geq r_j\}$ for $j \in \mathcal{J}_{\lfloor \frac{m}{2} \rfloor}$. So, we have the convex program with a polynomial number of constraints. The motivation to include constraint (20) is the fact that at most one job from $\mathcal{J}_{\lfloor \frac{m}{2} \rfloor}$ can be performed at the same time. Constraint (21) indicates that the total load of jobs with release times no less than r_j does not exceed $(T - r_j)m$.

Program (19)–(24) can be solved by the Ellipsoid method in polynomial time. Here we also need to check two technical conditions as in the previous section. The first condition follows from the polynomial bound on the value of an optimal solution: $p_j \leq \left(\frac{size_j W_j^\alpha}{E}\right)^{\frac{1}{\alpha-1}}$ for all $j \in \mathcal{J}$ and $T \leq \max_{j \in \mathcal{J}} r_j + \sum_{j \in \mathcal{J}} \left(\frac{size_j W_j^\alpha}{E}\right)^{\frac{1}{\alpha-1}}$. The second condition is satisfied for the point (p_j', T') defined as $p_j' := W_j \left(\frac{\sum_i W_i size_i}{E}\right)^{\frac{1}{\alpha-1}} + 1$ for all $j \in \mathcal{J}$, and T' is large enough such that

$$T' - 1 \geq \max \left\{ \max_{j \in \mathcal{J}_{\lfloor \frac{m}{2} \rfloor}} \left(r_j + \sum_{i \in R_j} (p_i' + 1)\right); \right.$$

$$\left. \max_{j \in \mathcal{J}} \left(r_j + \frac{1}{m} \sum_{i \in \mathcal{J}: \ r_i \geq r_j} (p_i' + 1) size_i\right); \max_{j \in \mathcal{J}} \left(r_j + (p_j' + 1)\right) \right\}.$$

The second stage. Let $(p_1^*, \ldots, p_n^*, T^*)$ denote the optimal solution to (19)–(24). We set the duration of each job j equal p_j^*, and apply the "preemptive $size_j$-list-scheduling" algorithm to construct a $\left(2 - \frac{1}{m}\right)$-approximate schedule. At every decision point (i.e., the release time or completion time of a job) all currently running jobs are interrupted. Then not yet completed jobs are considered in order of nonincreasing $size_j$-values, and as many of them are greedily assigned to the processors as feasibly possible. The time complexity of the algorithm is $O(n^2)$.

Lemma 3. "Preemptive $size_j$-list-scheduling" algorithm generates a feasible schedule S with length at most $\left(2 - \frac{1}{m}\right) T^*$ for problem $P|r_j, size_j, pmtn, energy| C_{\max}$.

Proof. Let l be the job that determines the makespan in the constructed schedule S, and let C_l be its completion time. We distinguish two cases: (I) $size_l > \left\lfloor \frac{m}{2} \right\rfloor$ and (II) $size_l \leq \left\lfloor \frac{m}{2} \right\rfloor$.

(I) $size_l > \left\lfloor \frac{m}{2} \right\rfloor$. We consider the last time interval, where at every time-moment a job with $size_j > \left\lfloor \frac{m}{2} \right\rfloor$ is executed. It follows from the definition of "preemptive $size_j$-list-scheduling" algorithm that this interval begins with the minimal release time r_z of all jobs with $size_j > \left\lfloor \frac{m}{2} \right\rfloor$ that are scheduled in this last contiguous interval. So, we have $C_l = r_z + (C_l - r_z) \leq r_z + \sum_{i \in R_z} p_i^* \leq T^*$ due to constraint (20).

(II) $size_l \leq \left\lfloor \frac{m}{2} \right\rfloor$. We consider the last time interval, where at every time-moment a job with $size_j \geq size_l$ is performed. By definition, the start time of this interval is the minimal release time r_z of all jobs that are scheduled in this last contiguous interval. Thus, the total load of jobs from the last considered interval is

$$(r_l - r_z)size_l + (C_l - r_l - p_l^*)(m - size_l + 1) + p_l^* size_l \leq \Delta m, \qquad (25)$$

where Δm is an estimation of the total load for jobs with $r_j \geq r_z$. Now we define the value of Δ. From constraints (21) and (22) the following inequalities hold:

$$r_l + p_l^* \leq T^*,$$

$$r_z + \frac{1}{m} \sum_{i \in \mathcal{J}:\, r_i \geq r_z} p_i^* size_i \leq T^*.$$

If $\frac{1}{m} \sum_{i \in \mathcal{J}:\, r_i \geq r_z} p_i^* size_i \leq (r_l - r_z) + p_l^*$, then we put $\Delta := (r_l - r_z) + p_l^*$ (in this case $p_l^* + r_l = \Delta + r_z$ and $\Delta + r_z \leq T^*$). Otherwise, we set $\Delta := \frac{1}{m} \sum_{i \in \mathcal{J}:\, r_i \geq r_z} p_i^* size_i$ (in this case $p_l^* + r_l < \Delta + r_z$ and $\Delta + r_z \leq T^*$).

Next we suppose $C_l > \left(2 - \frac{1}{m}\right)(r_z + \Delta)$, and prove that the total load (25) will be greater than Δm. Indeed,

$$(r_l - r_z)size_l + (C_l - r_l - p_l^*)(m - size_l + 1) + p_l^* size_l >$$
$$(r_l - r_z)size_l + \left(\left(2 - \frac{1}{m}\right)(r_z + \Delta) - r_l - p_l^*\right)(m - size_l + 1) + p_l^* size_l =$$
$$\Delta m + (\Delta + r_z - p_l^* - r_l)(m - 2size_l + 1) + r_z(m - size_l) + \left(\frac{r_z}{m} + \frac{\Delta}{m}\right)(size_l - 1) \geq \Delta m.$$

Here we use the following properties: $\Delta + r_z \geq p_l^* + r_l$, $size_l \leq \lfloor \frac{m}{2} \rfloor$, and $size_l \geq 1$. Therefore, $C_l \leq \left(2 - \frac{1}{m}\right)(r_z + \Delta) \leq \left(2 - \frac{1}{m}\right)T^*$.   □

Using the results from [6], we conclude that the approximation ratio of $\left(2 - \frac{1}{m}\right)$ is tight even if all jobs have single-processor type (see the example in Sect. 3).

Theorem 3. A $\left(2 - \frac{1}{m}\right)$-approximate schedule can be found in polynomial time for problem $P|r_j, size_j, pmtn, energy|C_{\max}$.

6 Non-uniform Partition of Work

Theorems 1, 2 and 3 can be generalized to the case of rigid jobs with non-uniform partition of the work between processors. Let a job $j \in \mathcal{J}$ consists of $size_j$ operations with processing volume W_{jl}, $l = 1, \ldots, size_j$, and each operation must be executed on an individual processor.

We only need to slightly correct the first stage of the presented algorithms. Namely, in the lower-bound models, the constraint on the energy consumption is modified as follows

$$\sum_{j \in \mathcal{J}} \sum_{l=1}^{size_j} W_{jl}^{\alpha} \cdot p_j^{1-\alpha} \leq E.$$

7 Conclusion

We have studied the makespan minimization under the energy consumption constraint. Strongly polynomial-time approximation algorithms were constructed for the case of rigid jobs. Our algorithms have constant factor approximation guarantees. Further research might address the approaches to the problems with more complex structure, where processors are heterogeneous and jobs have alternative execution modes with various characteristics.

References

1. Albers, S., Müller, F., Schmelzer, S.: Speed scaling on parallel processors. Algorithmica **68**(2), 404–425 (2014)
2. Bampis, E., Letsios, D., Lucarelli, G.: A note on multiprocessor speed scaling with precedence constraints. In: 26th ACM symposium on Parallelism in algorithms and architectures, SPAA 2014, pp. 138–142. ACM (2014)
3. Bunde, D.: Power-aware scheduling for makespan and flow. J. Sched. **12**, 489–500 (2009). https://doi.org/10.1007/s10951-009-0123-y
4. Du, J., Leung, J.T.: Complexity of scheduling parallel task systems. SIAM J. Discrete Math. **2**(4), 472–478 (1989)
5. Gerards, M.E.T., Hurink, J.L., Hölzenspies, P.K.F.: A survey of offline algorithms for energy minimization under deadline constraints. J. Sched. **19**(1), 3–19 (2016). https://doi.org/10.1007/s10951-015-0463-8

6. Graham, R.L.: Bounds for certain multiprocessor anomalies. SIAM J. Appl. Math. **17**(2), 416–429 (1966)
7. Grötschel, M., Lovász, L., Schrijver, A.: Geometric Algorithms and Combinatorial Optimizations, 2nd edn. Springer, Heidelberg (1993). https://doi.org/10.1007/978-3-642-78240-4
8. Kononov, A., Kovalenko, Y.: On speed scaling scheduling of parallel jobs with preemption. In: Kochetov, Y., Khachay, M., Beresnev, V., Nurminski, E., Pardalos, P. (eds.) DOOR 2016. LNCS, vol. 9869, pp. 309–321. Springer, Cham (2016). https://doi.org/10.1007/978-3-319-44914-2_25
9. Kuhn, H., Tucker, A.: Nonlinear programming. In: The Second Berkeley Symposium on Mathematical Statistics and Probability, pp. 481–492. University of California Press, Berkeley (1951)
10. Li, K.: Analysis of the list scheduling algorithm for precedence constrained parallel tasks. J. Comb. Opt. **3**, 73–88 (1999). https://doi.org/10.1023/A:1009817206440
11. Naroska, E., Schwiegelshohn, U.: On an on-line scheduling problem for parallel jobs. Inform. Process. Lett. **81**(6), 297–304 (2002)
12. Nesterov, E.: Methods of Convex Optimization. Moscow (2010). (in Russian)
13. Pruhs, K., van Stee, R.: Speed scaling of tasks with precedence constraints. Theory Comput. Syst. **43**, 67–80 (2007). https://doi.org/10.1007/s00224-007-9070-1
14. Shabtay, D., Kaspi, M.: Parallel machine scheduling with a convex resource consumption function. Eur. J. Oper. Res. **173**, 92–107 (2006)

A Polynomial-Time Algorithm for the Routing Flow Shop Problem with Two Machines: An Asymmetric Network with a Fixed Number of Nodes

Ilya Chernykh[1,2], Alexander Kononov[1], and Sergey Sevastyanov[1(✉)]

[1] Sobolev Institute of Mathematics, Koptyug ave. 4, Novosibirsk 630090, Russia
{idchern,alvenko,seva}@math.nsc.ru
[2] Novosibirsk State University, Pirogov str. 2, Novosibirsk 630090, Russia

Abstract. We consider the routing flow shop problem with two machines on an asymmetric network. For this problem we discuss properties of an optimal schedule and present a polynomial time algorithm assuming the number of nodes of the network to be bounded by a constant. To the best of our knowledge, this is the first positive result on the complexity of the routing flow shop problem with an arbitrary structure of the transportation network, even in the case of a symmetric network. This result stands in contrast with the complexity of the two-machine routing open shop problem, which was shown to be NP-hard even on the two-node network.

Keywords: Scheduling · Flow shop · Routing flow shop · Polynomially-solvable case · Dynamic programming

1 Introduction

A flow shop problem to minimize the makespan (also known as *Johnson's problem*) is probably the first machine scheduling problem described in the literature [7]. It can be set as follows.

Flow Shop Problem. Sets $\mathcal{M}$ of machines and $\mathcal{J}$ of jobs are given, each machine $M_i \in \mathcal{M}$ has to process each job $J_j \in \mathcal{J}$; such an operation takes p_{ji} time units. Each job has to be processed by machines in the same order: first by machine M_1, then M_2 and so on. No machine can process two jobs simultaneously. The goal is to construct a feasible schedule of processing all the jobs within the minimum *makespan* (which means, with the minimum completion time of the last operation). According to the traditional three-field notation of scheduling problems (see [9]), Johnson's problem with a fixed number m of machines is denoted as $Fm||C_{\max}$.

This research was supported by the program of fundamental scientific researches of the SB RAS No I.5.1., project No 0314-2019-0014, and by the Russian Foundation for Basic Research, projects 20-07-00458 and 18-01-00747.

© Springer Nature Switzerland AG 2020
A. Kononov et al. (Eds.): MOTOR 2020, LNCS 12095, pp. 301–312, 2020.
https://doi.org/10.1007/978-3-030-49988-4_21

Problem $F2||C_{\max}$ can be solved to the optimum by the well-known Johnson's algorithm, which basically is a sorting of the set of jobs according to Johnson's rule [7]. On the other hand, problem $F3||C_{\max}$ is NP-hard in the strong sense [6].

In classical scheduling problems (including flow shop), it is assumed that the location of each machine is fixed, and either there is no pre-specified delay between the processing of two consecutive operations of a job or such a delay depends on the distance between the corresponding machines. However, this assumption often diverges from real-life situations. Imagine that the company is engaged in the construction or maintenance of country houses, cottages or chalets. The company has several crews which, for example, specialize either in preparing the site for construction, or filling the foundation, or building a house, or landscaping the site. The facilities are located in a suburban area, and each team must move from place to place to carry out their work. The sequence of jobs performed by various crews is fixed, e.g., you cannot start to build a house before filling the foundation[1].

To take into account the situation described above, we consider a natural combination of $Fm||C_{\max}$ with the well-known traveling salesman problem, a so-called *routing flow shop problem* introduced in [1]. In this model, jobs are located at nodes of a transportation network G, while machines have to travel over the edges of the network to visit each job and perform their operation in the flowshop environment. All machines start from the same location (the *depot*) and have to return to the depot after performing all the operations. The completion time of the last machine *action* (either traveling or processing an operation of some job in the depot) is considered to be the *makespan* of the schedule ($C_{\max}$) and has to be minimized. (See Sect. 2 for the detailed formulation of the problem.)

We denote the m-machine routing flow shop problem as $RFm||C_{\max}$ or $RFm|G = W|C_{\max}$, when we want to specify a certain structure W of the transportation network.

The routing-scheduling problems can simulate many problems in real-world applications. Examples of applications where machines have to travel between jobs include situations where parts are too big or heavy to be moved between machines (e.g., engine casings of ships), or scheduling of robots that perform daily maintenance operations on immovable machines located in different places of a workshop [2]. Another interesting application is related to the routing and scheduling of museum visitors traveling as homogeneous groups [10]. The model is embedded in a prototype wireless context-aware museum tour guide system developed for the National Palace Museum of Taiwan, one of the top five museums in the world.

The routing flow shop problem is still understudied. Averbakh and Berman [1] considered $RF2||C_{\max}$ with exactly one job at each node, under the following restriction: each machine has to follow some **shortest route** through the set of nodes of the network (not necessarily the same for both machines). This will be

---

[1] The relationship between the routing flow shop problem and the problem of planning the construction of country houses in England was proposed by Natalia Shakhlevich in a private communication.

referred to as an *AB-restriction*. They proved that for the two-machine problem the AB-restriction affects the optimal makespan by a factor of at most $\frac{3}{2}$, and this bound is tight. They also showed that, under this restriction, there always exists a *permutation* optimal schedule, in which machines process jobs in the same order (a *permutation property*). Using this property, they presented $O(n \log n)$ algorithms for solving $RF2|AB$-restriction, $G = W|C_{\max}$ to the optimum, where W is a tree or a cactus, n is the number of jobs. These algorithms, therefore, provide a $\frac{3}{2}$-approximation for the problem without the AB-restriction on a tree or on a cactus with a single job at each node. Later on ([2]), they extended these results to the case of an arbitrary graph G and an arbitrary number of machines m by presenting a $\frac{m+1}{2}$-approximation algorithm for the $RFm||C_{\max}$ problem. Yu and Znang [12] improved on the latter result and presented an $O(m^{\frac{2}{3}})$-approximation algorithm based on a reduction of the original problem to the permutation flow shop problem.

A generalized routing flow shop problem with buffers and release dates of jobs was also considered in [8]. The authors present a heuristic based on solving the corresponding multiple TSP.

Yu *et al.* [11] investigated the $RF2||C_{\max}$ problem with a single job at each node farther. They obtained the following results:

1. The permutation property also holds for the problem without the AB-restriction.
2. The problem is ordinary NP-hard, even if G is a tree (moreover, if G is a spider of diameter 4 with the depot in the center).
3. There is a $\frac{10}{7}$-approximation algorithm that solves the $RF2|G = tree|C_{\max}$ problem in $O(n)$ time.

Finally, the possibility of designing a polynomial-time algorithm for the special case of our problem, when the transportation network is symmetric, was claimed in [5] (although, without any proof).

In the present paper, we investigate the generalization of $RF2||C_{\max}$ problem to the case of asymmetric travel times and of an arbitrary number of jobs at any node. Thus, we have to consider a **directed network** G in which the travel times through an edge may be different in the opposite directions. (We will denote such a problem by $\overrightarrow{RF2}||C_{\max}$.) We prove that the permutation property holds for this version of the problem, as well. We also establish another important property: there exists an optimal permutation schedule (with the same job processing order π on both machines) such that for each node v, sub-sequence π_v of π consisting of all jobs from node v obeys Johnson's rule. These two properties allow us to design a dynamic programming algorithm which solves this problem in time $O(n^{g^2+1})$, where g is the number of nodes in G. Thereby, we have established a polynomial-time solvability of the asymmetric two-machine routing flow shop problem with a constant number of network nodes. This result stands in contrast with the complexity result for the two-machine routing open shop problem, which is known to be ordinary NP-hard even if G consists of only two nodes (including the depot) [3].

The structure of the paper is as follows. Section 2 contains a formal description of the problem under investigation, as well as some notation and definitions. Properties of an optimal schedule are established at the beginning of Sect. 3 which also contains a description of the exact algorithm for solving the problem. The analysis of its qualities follows in Sect. 4. Section 5 concludes the paper with some open questions for further investigation.

2   Problem Setting, Definitions and Notation

Farther, throughout the paper, an expression of the form $x \in [\alpha, \beta]$ (where α and β are integers, and x is an integer variable, by definition) means that x takes any integral values from this interval; $[\beta] \doteq \{1, 2, \ldots, \beta\}$. In this paper we will consider the following problem.

Problem $\overrightarrow{RF2}||C_{\max}$. We are given n jobs $\{J_1, \ldots, J_n\}$ that are to be processed by two dedicated machines denoted as A and B. For each $j \in [n]$, job J_j consists of two operations that should be performed in the given order: first the operation on machine A, and then on machine B. Processing times of the operations are equal to a_j and b_j, respectively. All jobs are located at nodes of a transportation network; the machines move between those nodes along the arcs of that network. At the beginning of the process, both machines are located at a node called a *depot*, and they must return to that very node after completing all the jobs.

Without loss of generality of the problem (and for the sake of convenience of the further description and analysis of the algorithm presented in Sect. 3), we will assume that a *reduced network* $G = (V, E)$ ($|V| = g + 2$) is given, in which: (1) only *active nodes* are retained, i.e., the nodes containing jobs (they will be referred to as *job nodes*) and two *node-depots*: the *start-depot* and the *finish-depot*; (2) there are no jobs in both depots (otherwise, we split the original depot into three copies, the distances between which are equal to zero; one of those copies is treated as a job node, while the other two are job-free); the start-depot and the finish-depot get indices 0 and $g + 1$, respectively, while all job nodes get indices $i \in [g]$ (g is the number of job nodes); thus, starting from the start-depot, each machine will travel among the job nodes, and only after completing all the jobs it may arrive at the finish-depot; (3) G is a **complete directed graph** in which each arc $e = (v_i, v_j) \in E$ is assigned a non-negative weight $\rho(e) = \rho_{i,j}$ representing the shortest distance between the nodes corresponding to i and j in the source network **in the given direction**; therefore, the weights of arcs satisfy the triangle inequalities; at that, the **symmetry of the weights is not assumed**, i.e., the weights of the forward and the backward arcs may not coincide. The objective function $C(S)$ is the time, when machine B arrives at the finish-depot in schedule S, and this time should be minimized.

Other designations: $\mathbf{N} \doteq (n_1, \ldots, n_g)$, where n_i denotes the number of jobs located at job node $i \in [g]$. $\|K\|_1 \doteq \sum_{i \in [g]} |k_i|$ denotes the *1-norm* of vector $K = (k_1, \ldots, k_g)$.

Given an integer $d > 0$, we define a partial order $\prec$ on the set R^d of d-dimensional real-valued vectors, such that for any two vectors $x' = (x'_1, \ldots, x'_d)$,

$x'' = (x_1'', \ldots, x_d'') \in R^d$ the relation $x' \ll x''$ holds, if and only if $x_i' \le x_i''$, $\forall\, i \in [d]$. By $\mathcal{J}(v)$, we will denote the set of indices of jobs located at node $v \in V$.

By a *schedule*, we will mean, as usual, the set of starting and the completion times of all operations. Since, however, such a schedule model admits a continuum set of admissible values of its parameters, it will be more convenient for us to switch to a discrete model in which any schedule is determined by a pair of permutations $\{\pi', \pi''\}$ specifying the orders of processing the jobs by machines A and B, respectively. Each pair (π', π'') uniquely defines both the routes of the machines through the nodes of network G and an *active schedule* $S(\pi', \pi'')$ of job processing which is defined as follows.

A schedule $S(\pi', \pi'')$ is called *active, iff*: (1) it is feasible for the given instance of problem $\overrightarrow{RF2}||C_{\max}$; (2) it meets the precedence constraints imposed by permutations $\{\pi', \pi''\}$; (3) the starting time of no operation in this schedule can be decreased without violating the above mentioned requirements.

An active schedule $S(\pi', \pi'')$ is called a *permutation* one, if $\pi' = \pi''$.

Definition 1. For each $j \in [n]$, we define a *priority vector* $\chi_j = (\chi_j', \chi_j'', j)$ of job J_j, where $(\chi_j' = 1,\ \chi_j'' = a_j)$, if $a_j \le b_j$, and $(\chi_j' = 2,\ \chi_j'' = -b_j)$, otherwise. We next define a strict linear order $\prec$ on the set of jobs: for two jobs J_j, J_k $(j, k \in [n])$ the relation $J_j \prec J_k$ holds, *iff* $\chi_j <_{\text{lex}} \chi_k$ (i.e., vector χ_j is lexicographically less than χ_k). Clearly, for any two jobs J_j, J_k $(j \ne k)$, one and only one of two relations holds: either $J_j \prec J_k$ or $J_k \prec J_j$.

We will say that a permutation of jobs π and the corresponding permutation schedule meet the *Johnson local property*, if for each node $v \in V$ the jobs from $\mathcal{J}(v)$ are sequenced in permutation π *properly*, which means: in the lexicographically increasing order of their priority vectors. (Johnson [7] showed that in the case of the networkless two-machine flow shop problem, such a job order π provides the optimality of the corresponding permutation schedule.)

3   Properties of the Optimal Schedule and an Algorithm for the Exact Solution of Problem $\overrightarrow{RF2}||C_{\max}$

The algorithm described in this section is based on two important properties of the optimal schedule established in the following theorems.

Theorem 1. *For any instance I of problem $\overrightarrow{RF2}||C_{\max}$ there exists an optimal schedule which is a permutation one.*

Theorem 2. *For any instance I of problem $\overrightarrow{RF2}||C_{\max}$ there exists a permutation schedule which meets the Johnson local property and provides the minimum makespan on the set of all permutation schedules.*

The proofs of these theorems are omitted due to the volume limitations. They can be found in arXiv [4]. Two theorems above imply the following

Corollary 1. *For any instance I of problem $\overrightarrow{R}F2||C_{\max}$ there exists an optimal schedule which is a permutation one and meets the Johnson local property.*

The algorithm for computing the exact solution of problem $\overrightarrow{R}F2||C_{\max}$ is based on the idea of Dynamic Programming and on the two properties of optimal solutions mentioned in Corollary 1 (and, thus, enabling us to restrict the set of schedules under consideration by job sequences which meet these properties). So, from now on, we will consider only permutation schedules which meet the Johnson local property.

Let us number the jobs at each node v_i *properly*, i.e., in the ascending order of the relation $\prec$ (see Definition 1, p. 5). Then, due to Theorem 2, jobs at each node v_i ($i \in [g]$) should be processed in the order $\pi_i = (1, 2, \ldots, n_i)$. According to this order, the jobs at node v_i will be numbered by two indices: J_{ij} ($j \in [n_i]$).

In the schedule under construction, we will highlight the time moments when a machine $M \in \{A, B\}$ completes a portion of jobs at node v_i and is preparing to move to another node. Each such moment will be called an *intermediate finish point* of machine M or, in short, an *if-point* of machine M. It follows from Theorem 2 that at each *if*-point t' of machine A the set of jobs already completed by the machine is a collection of some **initial segments** $[1, \ldots, k_i]$ of sequences $\{\pi_i \mid i \in [g]\}$. This collection can be specified by a g-dimensional integral vector $K = (k_1, \ldots, k_g)$ (and will be denoted as $\mathcal{J}(K)$), where k_i denotes the number of jobs performed by machine A at node i by time t'.

By Theorem 1, machine B completely reproduces the route of machine A through network nodes (as well as the order of processing the jobs by that machine) and, at some later point in time $t'' \geq t'$, it also finds itself at its *if*-point with **the same set** $\mathcal{J}(K)$ of completed jobs, defined by vector K. Thus, a natural correspondence is established between the *if*-points of machines A and B: they are combined into pairs (t_s', t_s'') of *if*-points at which the sets of jobs completed by machines A and B coincide and are defined by the same vector $K_s = (k_1^s, \ldots, k_g^s)$. The pairs of *if*-points divide the whole process of performing the jobs by machines A and B into *steps* ($s = 1, 2, \ldots, \bar{s}$), each step s being defined by two parameters: the node index (i_s) and the number of jobs (d_s) performed in this step at node i_s.

The tuple $\widehat{K} \doteq (K, i^*)$ consisting of a value of vector $K = (k_1, \ldots, k_g)$ and a value of a node index i^* determines a *configuration* of a partial schedule of processing the subset of jobs $\mathcal{J}(K)$, with the final job at node i^*. The set of *admissible configurations* is defined as the set including all *basic configurations* (with values $K = (k_1, \ldots, k_g) \in [0, n_1] \times \cdots \times [0, n_g]$, $i^* \in [g]$, such that $k_{i^*} > 0$), as well as two *special configurations*: the *initial* one $\widehat{K}_S = (\mathbf{0}, 0)$ and the *final* one $\widehat{K}_F = (\mathbf{N}, 0)$.

Algorithm $\mathcal{A}_{DP}$ for constructing the optimal schedule makes two things: 1) it enumerates all possible configurations of partial schedules, and 2) for each of them, it accumulates the maximum possible set of pairwise incomparable solutions (characterized by pairwise incomparable pairs (t', t'') of *if*-points with respect to the relation $\lessdot$). In other words, given a configuration $\widehat{K} \doteq (K, i^*)$, we consider a "partial" bi-criteria problem $\mathcal{P}(\widehat{K})$ of processing the jobs from $\mathcal{J}(K)$, with the final job at node i^*. The objective is to minimize the two-dimensional vector-function $\bar{F} \doteq (F_1, F_2)$, where F_1, F_2 are the completion times of jobs from $\mathcal{J}(K)$ by machines A and B, respectively. We compute the complete set $\mathcal{F}(\widehat{K})$ of representatives of Pareto-optimal solutions of this problem.

For each solution $\bar{F} = (F_1, F_2) \in \mathcal{F}(\widehat{K})$, let us define the parameter $\Delta(\bar{F}) = F_2 - F_1$. The set $\mathcal{F}(\widehat{K})$ for each configuration $\widehat{K}$ will be stored as the list sorted in the ascending order of component F_1. (At that, the values of F_2 and $\Delta(\bar{F})$ strictly decrease). The first element of each list $\mathcal{F}(\widehat{K})$ will be a solution with the value $F_1 = 0$. This is either a *dummy solution* $\tilde{F} = (0, \infty)$ (added to each list $\mathcal{F}(\widehat{K})$ at the beginning of its formation), or a real solution with the value $F_1 = 0$ (if it is found).

We create lists $\mathcal{F}(\widehat{K})$ successively for all configurations $\{\widehat{K} = (K, i^*)\}$. At that, the whole algorithm is divided into three stages: the *initial*, the *main*, and the *final* one. Configurations with $i^* = 0$ are considered in the initial and the final stages only.

List $\mathcal{F}(\widehat{K}_S)$ for the *initial configuration* $\widehat{K}_S = (\mathbf{0}, 0)$ is created in the **initial stage** and consists of the single solution $(0, 0)$.

In the **main stage**, we enumerate configurations $\{\widehat{K} = (K, i^*)\}$ in non-decreasing order of the norm $\|K\|_1 = \sum k_i$ of vectors K (which varies from 1 to n); configurations with the same $\|K\|_1$ are enumerated in the lexicographically ascending order of vectors K. Next, for each given vector $K = (k_1, \ldots, k_g)$, only those values of $i^* \in [g]$ are enumerated for which $k_{i^*} > 0$ holds.

In the **final stage**, for the *final configuration* $\widehat{K}_F = (\mathbf{N}, 0)$, we find its optimal solution by comparing g variants of solutions obtained from the optimal solutions of configurations $\{(\mathbf{N}, i) \mid i \in [g]\}$. For each configuration $\widehat{K}_i = (\mathbf{N}, i)$, its optimal solution $\bar{F}_i^* = (F_1^*, F_2^*)$ (with the minimum value of the component F_2) is located at the very end of list $\mathcal{F}(\widehat{K}_i)$. Having added to F_2^* the distance $\rho_{i,0}$ from node v_i to the depot, we obtain the value of the objective function $C(S)$ of our problem for the given variant of schedule S. Having chosen (from g variants) the variant with the minimum value of the objective function, we find the optimum.

In order to create list $\mathcal{F}(\widehat{K})$ for a given configuration $\widehat{K} = (K, i^*)$ in the **main stage**, we enumerate all possible values $\widehat{K}' = (K', i')$ of *pre-configurations* ("p-c", for short), i.e., such configurations that could be obtained in the previous step of the algorithm. They should meet the inequality $i' \neq i^*$, and their vectors K and K' should differ in exactly one (i^*th) component, so as $k'_{i^*} < k_{i^*}$. At that, if k_{i^*} is the only non-zero component of vector K, then $\widehat{K}'$ can only be the *initial configuration*. Alternatively, if $K' \neq \mathbf{0}$, then should be $k'_{i'} > 0$. (Clearly, there is no need for a machine to come to node $v_{i'}$ without doing any job at it).

We note that for each configuration $\widehat{K} = (K, i^*)$ in the main stage, each variant of its p-c $\widehat{K}' = (K', i')$ can be uniquely defined by the pair (d, i'), where $i' \in [0, g] \setminus \{i^*\}$, and $d \in [k_{i^*}]$ is the number of jobs being processed in this step at node v_{i*}. The pairs (d, i') are enumerated so as the loop on d is an exterior one with respect to the loop on i'.

For each given value of d, we construct an **optimal schedule** $S_d = S(\widehat{K}, d)$ in problem $F2||C_{\max}$ for the jobs from $\mathcal{J}(\widehat{K}, d) \doteq \{J_{i*,j} \mid j \in [k_{i*} - d + 1, k_{i*}]\}$, and then compute three characteristics of that schedule: $L_1(\widehat{K}, d)$ and $L_2(\widehat{K}, d)$, which are the total workloads of machines A and B on the set of jobs $\mathcal{J}(\widehat{K}, d)$, and also $\delta(\widehat{K}, d) = C^*_{\max}(\widehat{K}, d) - L_2(\widehat{K}, d)$, where $C^*_{\max}(\widehat{K}, d)$ is the length of schedule S_d.

After that, we start the loop on i' which will be farther referred to as a *c-loop* (which means, "the loop on configurations and pre-configurations"). Indeed, after specifying the value of i', a configuration and its pre-configuration are completely defined. At each step of the c-loop, we adjust the current list $\mathcal{F}(\widehat{K})$ of solutions of configuration $\widehat{K}$. (Before starting the loop on d, the list consists of the single dummy solution $\tilde{F} = (0, \infty)$).

For each value of i', we enumerate all Pareto-optimal solutions $\bar{F}' = (F_1', F_2') \in \mathcal{F}(\widehat{K}')$ of the p-c $\widehat{K}' = (K', i')$ in the ascending order of F_1' (and the descending order of $\Delta(\bar{F}') = F_2' - F_1'$). Given a solution $\bar{F}'$ and schedule S_d, we form a solution $\bar{F}'' = (F_1'', F_2'')$ of configuration $\widehat{K}$ as follows.

$$F_1'' := F_1' + \rho_{i',i*} + L_1(\widehat{K}, d).$$

$$F_2'' := \begin{cases} F_2' + \rho_{i',i*} + L_2(\widehat{K}, d), & \text{if } \Delta(\bar{F}') \geq \delta(\widehat{K}, d) \ (\textit{a solution of type (a)}); \\ F_1' + \rho_{i',i*} + C^*_{\max}(\widehat{K}, d), & \text{if } \Delta(\bar{F}') < \delta(\widehat{K}, d) \ (\textit{a solution of type (b)}). \end{cases}$$

Case (b) means that the component F_2' does not affect the parameters of the resulting solution $\bar{F}''$ any more, and so, considering further solutions $\bar{F}' \in \mathcal{F}(\widehat{K}')$ (with greater values of F_1' and smaller values of $\Delta(\bar{F}')$) makes no sense, since it is accompanied by a monotonous increasing of both F_1'' and F_2'' (between which, a constant difference is established equal to $C^*_{\max}(\widehat{K}, d) - L_1(\widehat{K}, d)$). Thus, for any given p-c $\widehat{K}'$, a solution of "type (b)" can be obtained at most once.

For each solution $\bar{F}''$ obtained, we immediately try to understand whether it should be **added** to the current list $\mathcal{F}(\widehat{K})$, and if so, whether we should remove some solutions from list $\mathcal{F}(\widehat{K})$ (majorized by the new solution $\bar{F}''$).

To get answers to these questions, we find a solution $\bar{F}^\ell = (F_1^\ell, F_2^\ell)$ in list $\mathcal{F}(\widehat{K})$ with the maximum value of the component F_1^ℓ such that $F_1^\ell \leq F_1''$. Such a solution always exists (we call it a *control element* of list $\mathcal{F}(\widehat{K})$). Since in the loop on $\bar{F}' \in \mathcal{F}(\widehat{K}')$, component F_1'' monotonously increases, the search for the control element matching $\bar{F}''$ can be performed not from the beginning of list $\mathcal{F}(\widehat{K})$, but from the current control element. Before starting the loop on $\bar{F}'$, we assign the first item of list $\mathcal{F}(\widehat{K})$ to be the current control element.

If the inequality $F_2^\ell \leq F_2''$ holds, the current step of the loop on $\bar{F}' \in \mathcal{F}(\widehat{K}')$ ends **without including** the solution $\bar{F}''$ in list $\mathcal{F}(\widehat{K})$ (we pass on to the next solution $\bar{F}' \in \mathcal{F}(\widehat{K}')$). Otherwise, if $F_2'' < F_2^\ell$, we look through list $\mathcal{F}(\widehat{K})$ (starting from the control element $\bar{F}^\ell$) and remove from the list all solutions $\bar{F} = (F_1, F_2)$ majorized by the new solution $\bar{F}''$ (which is expressed by the relations $F_1'' \leq F_1$, $F_2'' \leq F_2$). At that, the condition $F_1'' = F_1$ is sufficient for removing the current control element, while the inequality $F_2'' \leq F_2$ is sufficient for removing subsequent elements. The scanning of list $\mathcal{F}(\widehat{K})$ stops as soon as either the first non-majorized list item is found distinct from the control element (for this item and for all subsequent items, the relations $F_2 < F_2''$ hold), or if the list has been scanned till the end. Include solution $\bar{F}''$ in list $\mathcal{F}(\widehat{K})$ and assign it to be a new *control element*, which completes the current step of the loop on $\bar{F}' \in \mathcal{F}(\widehat{K}')$.

4   The Analysis of Algorithm $\mathcal{A}_{DP}$

Theorem 3. *Algorithm $\mathcal{A}_{DP}$ finds an optimal solution of problem $\overrightarrow{R}F2||C_{\max}$ in time $O(n^{g^2+1})$.*

Proof. Since the optimality of the solution found by algorithm $\mathcal{A}_{DP}$ follows explicitly from the properties of the optimal solution proved in Theorems 1 and 2, to complete the proof of Theorem 3, it remains to show the validity of bounds on the running time of the algorithm; to that end, it is sufficient to estimate the running time (T_{BS}) of the *Main stage* of the algorithm.

In the Main stage, for each basic configuration $\widehat{K}$, the set $\mathcal{F}(\widehat{K})$ of all its Pareto-optimal solutions is found. Since this set is formed from the solutions obtained in the previous steps of the algorithm for various pre-configurations of configuration $\widehat{K}$, the obvious upper bound on the value of T_{BS} is the **product** of the number of configurations (N_C), of the number of pre-configurations (N_{PC}) for a given configuration, and of the bound (T_{step}) on the running time of any step of the loop on configurations and pre-configurations (a *c-loop*).

In each step of the c-loop, list of solutions $\mathcal{F}(\widehat{K}')$ of a given p-c $\widehat{K}'$ is scanned. From each such solution, a solution for configuration $\widehat{K}$ is generated which is then either included or not included in list $\mathcal{F}(\widehat{K})$. The solutions included in the list **in this step** of the c-loop will be called "new" ones; other solutions, included in $\mathcal{F}(\widehat{K})$ **before starting this step** will be called "old".

While estimating a new solution claiming to be included in $\mathcal{F}(\widehat{K})$, we scan some "old" solutions of list $\mathcal{F}(\widehat{K})$, which is performed in two stages. In the first stage, we look through the elements from $\mathcal{F}(\widehat{K})$, starting from the current *control element*, in order to find a **new control element** immediately preceding the applicant. In the second stage (in the case of the **positive decision** on including the applicant in the list), we check the (new) control element and the subsequent elements from $\mathcal{F}(\widehat{K})$ subject to their **removal from the list** (if they are majorized by the applicant). We continue this process until we find either the first undeletable element or the end of the list. We would like to know: **how many views** of items of list $\mathcal{F}(\widehat{K})$ will be required in total in one step of the c-loop? It is stated that no more than $O(Z)$, where Z is the maximum possible size of list $\mathcal{F}(\widehat{K})$ in any step of the algorithm for all possible configurations $\widehat{K}$.

To prove this statement, we first note that none of the "new" elements included in list $\mathcal{F}(\widehat{K})$ in this step of the c-loop will be deleted in this step,

since all "new" solutions included in the list are incomparable by the relation
$\lessdot$. This follows from the facts that: 1) all applicants formed by type (a) are
incomparable; 2) if the last solution is formed by type (b), then it is either
incomparable with the previous applicant, or majorized by it (and therefore, is
not included in the list). Thus, only "old" elements will be deleted from the list,
and the total (in the c-loop step) number of such deletions does not exceed Z.

In addition, the viewing of an element from $\mathcal{F}(\widehat{K})$, when it receives the status
of a "control element", occurs at most once during each c-loop step, and so, the
total number of such views in one step does not exceed Z. There may be also "idle
views" of elements subject to assigning them the status of a "control element".
Such an idle view may happen only once for each applicant, and so, the number
of such idle views during one step of the c-loop does not exceed $|\mathcal{F}(\widehat{K}')| \leq Z$.

Next, the total (over a step of the c-loop) number of views of elements from
$\mathcal{F}(\widehat{K})$ subject to their removal from the list does not exceed $O(Z)$, as well.
Indeed, viewing an element of $\mathcal{F}(\widehat{K})$ **with its removal** occurs, obviously, for
each element at most once (or, in total over the whole step, at most Z times).
Possible "idle view" of an element from $\mathcal{F}(\widehat{K})$ (without its deleting) happens
at most once for each applicant, which totally amounts (over the current step
of the c-loop) at most $|\mathcal{F}(\widehat{K}')| \leq Z$. Thus, the total number of views of items
from $\mathcal{F}(\widehat{K})$, as well as the total running time of the c-loop step (T_{step}), does not
exceed $O(Z)$. Let us estimate now number Z itself.

We know that for any given configuration $\widehat{K}$ the solutions $\bar{F} = (F_1, F_2)$ from
list $\mathcal{F}(\widehat{K})$ are incomparable with respect to relation $\lessdot$. Thus, the number of
elements in list $\mathcal{F}(\widehat{K})$ does not exceed the number of different values of the
component F_1. The value of the component F_1 is the sum of the workload of
machine A and the total duration of its movement. (There are no idle times of
machine A in the optimal schedule.) Since the workload of machine A (for a
fixed configuration $\widehat{K}$) is fixed, the number of different values of the component
F_1 can be bounded above by the number of different values that the length of a
machine route along the nodes of network G can take. As we know, each passage
of the machine along the arc (v_i, v_j) is associated with the performance of at
least one job located at node v_j. Thus, any machine route contains $x \leq k_j \leq n_j$
arcs entering node v_j, and the same number of arcs (x) leaving the node.

Let us define a *configuration of a machine route* as a matrix $H = (h_{ij})$ of size
$g \times g$, where h_{ij} $(i \neq j)$ specifies the multiplicity of passage of an arc $(v_i, v_j) \in G$
in the route; $h_{jj} = n_j - \sum_{i \neq j} h_{ij}$. Thus, for any $j \in [g]$, the equality holds:

$$\sum_{j=1}^{g} h_{ij} = n_i. \tag{1}$$

Clearly, for any closed route the following equalities are also valid:

$$\sum_{i=1}^{g} h_{ij} = n_j, \quad j \in [g]. \tag{2}$$

Hence, it follows that the number of different values of the route length of a machine does not exceed the number of configurations of a closed route. The latter does not exceed the number of different matrices H with properties (1) and (2). Let us (roughly) estimate from above the number (Z') of such matrices without taking into account property (2).

The number of variants of the ith row of matrix H does not exceed the number of partitions of the number n_i into g parts, i.e., is not greater than

$$C_{n_i+g-1}^{g-1} = \frac{1}{(g-1)!}(n_i+1)(n_i+2)\dots(n_i+g-1)$$

$$= \frac{n_i^{g-1}}{(g-1)!}\left(1+\frac{1}{n_i}\right)\left(1+\frac{2}{n_i}\right)\dots\left(1+\frac{g-1}{n_i}\right) \le \frac{n_i^{g-1}}{(g-1)!}\exp\frac{(g-1)g}{2n_i}.$$

Since the value of $\sup_{n_i\in[1,\infty)}\exp\frac{(g-1)g}{2n_i}$ depends only on g, a function $f(g)$ can be defined such that

$$C_{n_i+g-1}^{g-1} \le f(g)n_i^{g-1}.$$

Let $\Pi \doteq n_1 n_2 \dots n_g$. Then $Z \le Z' \le (f(g))^g \cdot \Pi^{g-1}$, and the number of configurations (N_C) can be bounded above by $O(g\Pi)$. Finally, the number of pre-configurations is bounded by $N_{PC} \le O(gn)$. Taking into account the above bounds, the bound $\Pi \le n^g/g^g$, and the inequality $g \le O(1)$, we obtain (for some function $\varphi(g)$) the final bound on the running time of the algorithm:

$$T_{\mathcal{A}} \approx T_{BS} \approx N_C N_{PC} T_{step} \le \varphi(g) \cdot O(\Pi^g n) \le O(n^{g^2+1}).$$

Theorem 3 is proved. □

5 Conclusion

We have considered the two-machine routing flow shop problem on an asymmetric network $(\overrightarrow{RF2}||C_{\max})$. We have improved the result by Yu et al. [11] by showing that for a more general problem (the problem with an arbitrary asymmetric network) the property of existing an optimal permutation schedule also holds. Next, we have presented a polynomial time algorithm for the problem with a fixed number of nodes, which is the first positive result on the computational complexity of the general $\overrightarrow{RF2}||C_{\max}$ problem.

We now propose a few open questions for future investigation.

Question 1. What is the parametrized complexity of problem $\overrightarrow{RF2}||C_{\max}$ with respect to the parameter g?

Question 2. Are there any subcases of problem $\overrightarrow{RF2}||C_{\max}$ with unbounded g (e.g., G is a chain, or a cycle, or a tree of diameter 3, or a tree with a constant maximum degree, etc.) solvable in polynomial time?

Question 3. Are there any strongly NP-hard subcases of problem $\overrightarrow{RF2}||C_{\max}$ for which NP-hardness is not based on the underlying TSP? In other words, is it possible that for some graph structure $G = W$ the TSP on W is easy, but problem $\overrightarrow{RF2}|G = W|C_{\max}$ is strongly NP-hard?

References

1. Averbakh, I., Berman, O.: Routing two-machine flowshop problems on networks with special structure. Transp. Sci. **30**(4), 303–314 (1996). https://doi.org/10.1287/trsc.30.4.303

2. Averbakh, I., Berman, O.: A simple heuristic for m-machine flow-shop and its applications in routing-scheduling problems. Oper. Res. **47**(1), 165–170 (1999). https://doi.org/10.1287/opre.47.1.165

3. Averbakh, I., Berman, O., Chernykh, I.: The routing open-shop problem on a network: complexity and approximation. Eur. J. Oper. Res. **173**(2), 531–539 (2006). https://doi.org/10.1016/j.ejor.2005.01.034

4. Chernykh, I., Kononov, A., Sevastyanov, S.: A polynomial-time algorithm for the routing flow shop problem with two machines: an asymmetric network with a fixed number of nodes. http://arxiv.org/abs/2004.03942

5. Chernykh, I., Kononov, A., Sevastyanov, S.: Exact polynomial-time algorithm for the two-machine routing flow shop problem with a restricted transportation network. In: Optimization problems and their applications (OPTA-2018), Abstracts of the VII International Conference, Omsk, Russia, 8–14 July 2018, pp. 37–37. Omsk State University (2018)

6. Garey, M.R., Johnson, D.S., Sethi, R.: The complexity of flowshop and job-shop scheduling. Math. Oper. Res. **1**(2), 117–129 (1976). https://www.jstor.org/stable/3689278

7. Johnson, S.M.: Optimal two- and three-stage production schedules with setup times included. Rand Corporation (1953). https://www.rand.org/pubs/papers/P402.html

8. Józefczyk, J., Markowski, M.: Heuristic solution algorithm for routing flow shop with buffers and ready times. In: Swiątek, J., Grzech, A., Swiątek, P., Tomczak, J.M. (eds.) Advances in Systems Science. AISC, vol. 240, pp. 531–541. Springer, Cham (2014). https://doi.org/10.1007/978-3-319-01857-7_52

9. Lawler, E.L., Lenstra, J.K., Rinnooy Kan, A.H.G., Shmoys, D.B.: Sequencing and scheduling: algorithms and complexity. In: Logistics of Production and Inventory, Handbooks in Operations Research and Management Science, vol. 4, pp. 445–522. Elsevier (1993). https://doi.org/10.1016/S0927-0507(05)80189-6

10. Yu, V.F., Lin, S., Chou, S.: The museum visitor routing problem. Appl. Math. Comput. **216**(3), 719–729 (2010). https://doi.org/10.1016/j.amc.2010.01.066

11. Yu, W., Liu, Z., Wang, L., Fan, T.: Routing open shop and flow shop scheduling problems. Eur. J. Oper. Res. **213**(1), 24–36 (2011). https://doi.org/10.1016/j.ejor.2011.02.028

12. Yu, W., Zhang, G.: Improved approximation algorithms for routing shop scheduling. In: Asano, T., Nakano, S., Okamoto, Y., Watanabe, O. (eds.) ISAAC 2011. LNCS, vol. 7074, pp. 30–39. Springer, Heidelberg (2011). https://doi.org/10.1007/978-3-642-25591-5_5

Heuristics and Metaheuristics

Optimal Location of Welds on the Vehicle Wiring Harness: P-Median Based Exact and Heuristic Approaches

Maurizio Boccia[1], Adriano Masone[1], Antonio Sforza[1], and Claudio Sterle[1,2(✉)]

[1] Department of Electrical Engineering and Information Technology, University Federico II of Naples, Via Claudio 21, 80125 Naples, Italy
{maurizio.boccia,adriano.masone,antonio.sforza,claudio.sterle}@unina.it
[2] Istituto di Analisi dei Sistemi ed Informatica A. Ruberti, IASI-CNR, Via dei Taurini, 19, 00185 Rome, Italy

Abstract. Nowadays vehicles are highly customizable products. Indeed, they can be equipped with a great number of options directly chosen by the customers. This situation provides several harness design problems to automotive companies, where by harness we mean the set of conducting wires (*cables*), positioned within the vehicle frame (*chassis*), which transmit information and electrical power to the options to make them operative. In this context we focus on an optimization problem arising in the construction and assembly phase of the harness within a vehicle. The options selected by customers have to be connected through a harness shaped in a tree structure within the vehicle chassis. In particular, the wiring has to connect subsets composed of two or more options. The total length of the connecting cables could be very large if a dedicated cable would be used for each couple of options in each subset. This length can be significantly reduced by realizing the connection through the usage of cable weldings. This work introduces for the first time the problem of the optimal placement of the weldings on the wiring harness tree of a vehicle, aimed at minimizing the total length and/or the cost of the cables, weighted by their gauge. The problem can be schematized as a p-median problem (*PMP*) on a tree in a continuous and discrete domain, with additional technological constraints related to the welding positions and mutual distance. This work proposes an integer linear programming model and a matheuristic aimed at finding exact and/or heuristic solutions for this constrained *PMP*. The efficiency and the effectiveness of the proposed methodologies have been proved through the solution of test instances built from real data provided by an automotive company.

Keywords: Harness design · Optimal diversity management · P-median with mutual distance · ILP formulation and matheuristic

© Springer Nature Switzerland AG 2020
A. Kononov et al. (Eds.): MOTOR 2020, LNCS 12095, pp. 315–328, 2020.
https://doi.org/10.1007/978-3-030-49988-4_22

1   Introduction

Automobile Engineering also known an Automotive Engineering is a field concerned with the activity of designing, developing, constructing, manufacturing, operating and safety testing of vehicles (automobiles, buses and trucks) and related subsystems. Its main objective is represented by the improvement of vehicle technical performance, aesthetics and software. Given the wide range of elements present in a vehicle (mechanical, electrical, electronic, software and safety elements), Automotive Engineering generally tackles complex decision problems whose solution requires the involvement of different technical skills and expertise. Among the others, several optimization problems arise for which Operations Research methodologies can represent a valuable decision support tool.

In this work we focus on the manufacturing activities dealing with the creation and the assembly of the whole parts of a vehicle. One of the main criticalities in this field is represented by the fact that vehicles are highly customizable products. Indeed, nowadays, in order to respond to the market needs, vehicles can be equipped with a great number of electrical components (50 or more), generically referred to as *options*. These options are made operative by the harness of the vehicle, i.e., the set of conducting wires (*cables*), positioned within the vehicle frame (*chassis*), which transmit information and electrical power, so allowing the interconnection among the options and the connection to the power and control unit.

A customer can configure a vehicle with the desired combination of options (*demand*). Even if one cannot require all the possible option combinations, because of technical and functional constraints, the number of different admissible demands can be very high. Each demand is enabled by the corresponding *wiring configuration*, containing all the needed cables. In this context, for a company, satisfying all the admissible demands installing exactly the related cables, would mean to have a specific wiring configuration for each of them. This is obviously impossible, since it would mean that an automotive company should have to produce in advance and manage at the assembly line a great number of wiring configurations, each of them characterized by its own installation requirements.

From this brief discussion about the harness of a vehicle, it is easy to understand that an effective and efficient management of the wiring configuration production and assembly activities can represent a key factor for automotive companies. In this context, three main optimization problems arise:

1. *Production problem*: it concerns the selection of the wiring configurations to be produced in order to meet the forecast demands at the minimum cost.
2. *Data representation problem*: it is related to the development of data clustering algorithms for an effective and efficient demand database management.
3. *Design problem*: it concerns the determination of the most effective and efficient deployment of the cables within the vehicle chassis, taking into account the technical constraints at the production and assembly line.

This work is focused on the third problem. However, in the following we give some hints to clarify the first problem, since its comprehension is preparatory for a better understanding of the third problem. Regarding the second problem, we just address the interested reader to [8] and [9].

The first problem is known in literature as the *Optimal Diversity Management Problem, ODMP*. In order to explain this problem we have to introduce the extra-cost concept. Automotive companies produce in advance a limited number of opportunely chosen wiring configurations, instead of producing a specific wiring configuration after receiving the demand of each customer. Then, if a wiring configuration containing just the options of a demand is not produced, the company substitutes it with a compatible (dominating) one. This substitution allows the company to overcome the drawbacks related to the product customization, but it also imposes that it has to sustain an additional wiring cost (extra-cost), since it is giving the cables for one or more options not demanded by the customer. Hence the choice of the number and the kind of wiring configurations to produce in order to meet the forecast demand should be done in order to minimize the extra-cost. *ODMP* was introduced as a particular variant of the p-median problem (*PMP*) in [7], where it was solved by a Lagrangian relaxation-based algorithm. Later, an exact solution approach exploiting the decomposition of the *ODMP* in several reduced size p-median sub-problems has been proposed in [6]. Exploiting the same idea, several decomposition approaches have been proposed in [1,2,4] and [3]. The most recent contribution on *ODMP* is presented in [14]. This work starts from several findings of [15] and it proposes an algorithm structured in three stages, where Lagrangian relaxation-based techniques, variable fixing and reduction tests, and a dynamic programming algorithm are effectively combined. This method represents the state of the art for the problem.

The above introduced design problem arises downstream the (*ODMP*). Indeed, the set options of the configurations selected through the *ODMP* solution have to be connected and made operative through a wiring harness, shaped in a tree structure within the vehicle frame (*chassis*). The total length of the connecting cables could be very large if a dedicated wire should be used for each couple of options in each subset. This reflects also in the cost of the wiring configuration (*connection cost*), which depends on the length and the gauge of the used cables. Hence, it is necessary to use cable weldings to avoid expensive dedicated connections among the options and reduce the overall length.

This work describes the design problem of the optimal placement of the weldings (*Optimal Welding Location Problem, OWLP*) on the wiring harness tree of a vehicle, aimed at minimizing the weighted length/cost of the cables. This problem can be schematized as a p-Median Problem (*PMP*) on a tree in a continuous domain. However, exploiting the property imposing the correspondence between vertices and medians of a graph [10,13], using an ad-hoc discretization procedure, we can formulate and tackle the problem as a *PMP* in a discrete domain. This allows to pass from an uncountable to a countable set of welding locations along the tree, for which it is easier to impose additional technological constraints related to the mutual distance between weldings or to the minimum distance between the weldings and the fixing hooks/pegs. To the best of authors knowledge, no contribution in literature deals with this distance constrained variant of the *PMP*, as can be concluded from the comprehensive review on the topic present in the following surveys and recent papers: [5,11,12,16,17].

In the following we present an original integer linear programming (*ILP*) formulation and a matheuristic aimed at finding exact and/or heuristic solutions for the *OWLP*. The proposed methods are then validated on test instances built from real data coming from a company operating in the field and a discussion about the obtained results is provided.

The paper is structured as follows: in Sect. 2, we provide a detailed description of the *OWLP*; in Sect. 3, we introduce the problem setting and present a p-median based *ILP* formulation and a matheuristic approach for the *OWLP*; Sect. 4 is devoted to the presentation of computational results of the proposed methods; finally, Sect. 5 provide conclusions and future work perspectives.

2   Problem Description

The harness of a vehicle, also known as wiring harness is the set of electrical cables/wires which transmit signals or electrical power to the electrical components. These masses of loose cables, if unrolled and stretched, could reach the extension of several hundred metres. Hence, they are generally jacketed together by a non-flexible, durable and insulating material, such as rubber, vinyl, tapes, etc., or a combination of them (binding process). This solution gathering together in common paths a large number of cables, has a twofold target: on one side, it allows to secure the harness against potential damages and short; on the other side, it provides an optimization of the used space and of the installation time, since the worker has to install just a single 'wire' within the vehicle chassis. Before performing the binding process, first the cables are cut to the desired length, and, then, they are assembled and clamped together on a special workbench, or onto a pin board (assembly board), according to the design specification, to form the cable harness. After this, the ends of the cables are stripped to expose the metal (or core) of the cables and fitted with terminals and connectors.

The harness design problem that we tackle in this work arises before the binding process, when the cables are cut on the basis of the geometric and electrical requirements in order to be positioned within the vehicle according to a tree like topology. In the following, in order to facilitate the explanation and the comprehension of the optimal welding location problem (*OWLP*), we will use some graphical representations and samples.

Let us consider the harness tree network sketched in Fig. 1a, composed of 6 options coinciding with the leaves of the tree. Each option has to be connected with one or more other options to form a functional package. Hence, in other words, a functional package is a sub-set or an *n-tuple* of options, e.g.: couple $(3,6)$, triplets $(1,2,5)$, etc. For the sake of the completeness, it is important to underline that each option can be composed of more pins and each pin or subset of pins can be part of a functional package. A pin or a subset of pins can be part of just one *n-tuple*, but an option can be involved in more *n-tuples*. In the following we will focus just on the options, since it is easy to extend our approach to the pins by duplicating each leaf of the tree as many times as are the pins of each option.

In order to make operative a functional package, we should connect each couple of options forming it. Hence, for example, to connect the triplet $(1, 2, 5)$ we should construct three connections: $(1–2)$, $(2–5)$ and $(1–5)$. Figure 1b shows the sub-tree network deriving from the construction of all the connections among the interested options. This kind of interconnection among the options provide a total length of the harness which is equal to the sum of the lengths of each connection. This length, and consequently the deriving connection cost, could be reduced if all the options are connected through a welding performed in any point along the sub-tree network identified by the functional package, as shown in Fig. 1c.

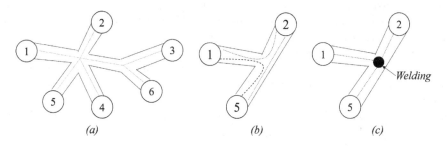

Fig. 1. a) Harness tree network with 6 options. b) Harness sub-tree network connecting options 1, 2 and 5 by dedicated cables. c) Harness sub-tree network connecting options 1, 2 and 5 by the usage of a welding.

The *OWLP* consists in determining the welding locations of all the functional packages along related sub-tree networks, minimizing the total length/cost of the used cables. By this description, we can easily understand that *OWLP* can be schematized as a p-Median Problem (*PMP*) on a tree in a continuous domain, where each median corresponds to a welding and just one welding has to be performed for each functional package along the links or in the nodes constituting the related sub-tree network. However, exploiting the above cited property about the correspondence between vertices and absolute medians of a graph, using an ad-hoc discretization procedure along the links of the tree, we can formulate and tackle the *OWLP* as a *PMP* in a discrete domain, where weldings are located only in the vertices of the tree, so passing from an uncountable to a countable set of welding locations [10,13].

In next section we will provide the details to pass from the continuous to the discrete domain. Here, we use the discretization of the welding locations to explain other features and restrictions of the problem under investigation, related to the diameter (gauge) of the cables and the welding positions, respectively.

The cables can be characterized by different diameters (gauge), depending on the electrical requirement of the option to be supplied. The gauge obviously reflects on the weight and cost of each cable. Moreover, the cables forming the harness are grasped to the chassis by the usage of fixing hooks and pegs. Technological constraints impose that no welding can be made at the leaves of the tree and in correspondence with hooks and pegs, and the distance of the weldings with respect to these points has to be higher than or equal to a predefined value denoted as d. This distance restriction has to be satisfied also for each couple of welds and it is motivated by the fact that the soldering of wire ends and the plugging of wires into connector housings require enough space to be made without causing flaws to the system. This restriction is the one which makes the *OWLP* a complex problem, since otherwise it could be solved to optimality just solving several 1-median problems, one for each functional package.

Figure 2 and Fig. 3 show two *OWLPs* in a discrete domain, characterized by 6 options and 2 hooks and 9 options and 2 hooks, respectively. The case of Fig. 2 considers two functional packages: $p_1 = (1, 2, 3)$ and $p_2 = (4, 5, 6)$. The case of Fig. 3 considers also a third package $p_3 = (7, 8, 9)$. The discretized potential welding locations for the three packages p_1, p_2 and p_3 are highlighted in red, blue and green, respectively. These locations have been obtained discretizing the links with a step size equal to $25\,cm$. Let us assume also that $d = 25\,cm$. For the sake of the simplicity, the length associated to each link, whose value is reported in the figure, is a multiple of the step size. Let us assume that the options of the packages have to be connected using cables with three different unitary costs (depending on their gauge): $C_1 = C_4 = C_8 = 0.05€/cm$, $C_2 = C_5 = C_7 = 0.025€/cm$, $C_3 = C_6 = C_9 = 0.1€/cm$. The total connection cost of each package with respect to all the potential welding locations can be easily computed multiplying the length of each connection by the corresponding unitary cost. Then, the optimal welding location for each package is the one corresponding to the minimum connection cost. The welding locations for the two cases reported in the figures are identified by large coloured circles. It is easy to note that the solution reported in Fig. 2 is optimal and feasible, whereas the one reported in Fig. 3 is optimal but unfeasible because the mutual distance constraints are not satisfied, since two weldings are in the same positions. This simple example explains the need of a solution method which does not consider the weldings one-by-one but optimize all of them at the same time.

3   *OWLP* Solution Approaches

In this section we introduce the problem setting of the *OWLP* detailing also the discretization procedure to be implemented in order to pass from a *PMP* in a continuous domain to a *PMP* in a discrete domain. Then we present an original *ILP* formulation for the *OWLP*, achieved by the modification of the classical *PMP* formulation. Finally, a matheuristic approach exploiting the proposed formulation will be presented.

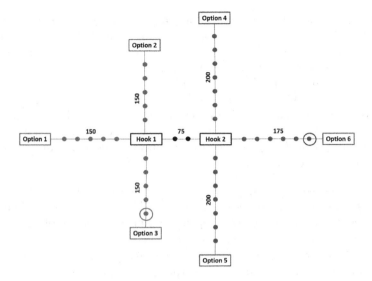

Fig. 2. Optimal solution of a 2-welding case. (Color figure online)

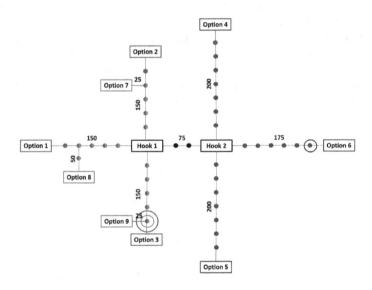

Fig. 3. Infeasible solution of a 3-welding case. (Color figure online)

3.1 Problem Setting

Let $G(V, A)$ be the harness tree network to be installed within the vehicle, where V is the set of vertices and A is the set of links, each of them indicated with its extreme vertices (e.g., (k_1, k_2), $k_1, k_2 \in V$). The set of vertices V is the union of three disjoint subsets: L, the set of leaves of the tree; I, the set of internal vertices of the tree; J, the set of potential welding locations along the links. Each generic leaf $l, l \in L$, corresponds with an option to be connected; each intermediate point $i, i \in I$, corresponds with an anchor point (fixing hook or peg) within the vehicle chassis; each potential welding location $j, j \in J$, is an intermediate point between any couple of point k_1 and k_2, $k_1, k_2 \in L \cup I$, generated by the discretization procedure.

We used a simple discretization procedure. Let us denote by d the minimum mutual distance between two weldings or between a welding and a leaf/anchor point. Moreover, let $e(k_1, k_2)$ be the distance between vertices k_1 and k_2, $k_1, k_2 \in L \cup I$. Two cases are possible:

- If $n = \frac{e(k_1,k_2)-2d}{d}$ is an integer value, then we have exactly $n + 2$ admissible welding locations between nodes k_1 and k_2. This means that we should find the same discretization points independently of the direction of the dicretization along the link (from k_1 to k_2 or viceversa).
- If $n = \frac{e(k_1,k_2)-2d}{d}$ is a real value, then the discretization is affected by the direction of the discretization along the link (from k_1 to k_2 or viceversa). In this case, we can use $n' = \lfloor \frac{e(k_1,k_2)-2d}{d} \rfloor + 1$ welding points, spaced out with distance d, along each discretization directions. This case is represented in Fig. 4, where the green points $(1, 2, 3)$ and the red points $(4, 5, 6)$ are the welding potential locations obtained discretizing from k_1 to k_2 and viceversa, respectively. It is important to say that, as demonstrated in [13], given a link (k_1, k_2), the position of a median is always left or right shifted towards one of the two extreme vertices. Hence, by a pre-processing procedure which evaluates this situation for each welding and for each link, we could halve the potential locations of each welding.

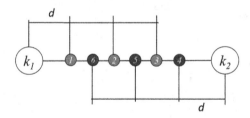

Fig. 4. Discretization procedure along a link with length not multiple of d. (Color figure online)

Let us now define K as the set of weldings (corresponding to the functional packages) to be performed and $L_k \subset K$ as the subset of options that have to be connected by the generic welding $k, k \in K$ in order to make operative a package. We recall that $\cup_{k \in K} L_k = L$ and $\cap_{k \in K} L_k = \emptyset$, since each option can be part of just one functional package. Similarly, let us define as J_k the set of potential welding locations for the generic welding $k, k \in K$, obtained by the previously described discretization procedure along the sub-tree network identified by the functional package. We highlight that $\cup_{k \in K} J_k = J$ whereas the intersection among the subsets J_k can be empty or not. Moreover, let G be the set of gauge values of the cables to be used and let H_g be the unitary cost associated with the generic gauge $g, g \in G$. We denote by LG_l the unitary cost of the cable used to connect the option $l, l \in L$ to the other options of the functional package.

Given the previous problem setting, for each generic welding $k, k \in K$, we can easily compute the minimum distance D_{lk} from each generic option $l, l \in L_k$, to the generic welding potential location $j, j \in J_k$. Indeed, since the harness has a tree topology, there is a unique path connecting an option l with a potential welding location j. Thus, the value of D_{lk} is just the sum of the lengths of the links composing the unique path connecting an option with a welding location. On this basis, given a welding $k, k \in K$, we can easily obtain the connection cost from each option $l, l \in L$, to each potential welding location $j, j \in K$, as: $C_{lj} = \sum_{l \in L_k} LG_l \, D_{lj}$. Obviously, for all the $l, l \notin L_k$, the connection cost is infinite.

Finally, let us indicate as $\hat{j}_k$ a specific potential location of a welding $k, k \in K$. Then we define $J_{\hat{j}_k} = \{j \in J | D_{j\hat{j}_k} \leq d\}$, i.e., the subset containing all the welding potential locations whose distance with respect to $\hat{j}_k$ is lower than d. Figure 5 reports a graphical representation of the set $J_{\hat{j}_k}$ for a generic welding position k_1 with respect to another welding k_2.

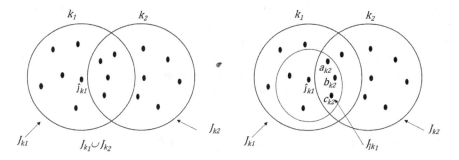

Fig. 5. Representation of set $J_{\hat{j}_k}$ for a generic welding.

3.2   A P-Median Based Formulation for the *OWLP*

On the basis of previous problem setting, the following binary variables have to be defined in order to model the *OWLP*.

- $y_{jk} = \{0, 1\}$: binary variable equal to 1 if the vertex $j, j \in J_k$, is selected as location of the welding $k, k \in K$, 0 otherwise;
- $x_{ljk} = \{0, 1\}$: binary variable equal to 1 if the option $l, l \in L$ is connected to the vertex $j, j \in J_k$, selected as location of the welding $k, k \in K$, 0 otherwise.

$$(OWLP) \qquad \min z = \sum_{k \in K} \sum_{j \in J_k} \sum_{l \in L} C_{lj} x_{ljk} \qquad (1)$$

subject to:

$$\sum_{j \in J_k} x_{ljk} = 1 \qquad\qquad \forall l \in L, k \in K \qquad (2)$$

$$x_{ljk} \le y_{jk} z \qquad\qquad \forall k \in K, l \in L, j \in J_k \qquad (3)$$

$$\sum_{j \in J} \sum_{k \in K} y_{jk} = |K| \qquad (4)$$

$$y_{\hat{j}_k} + \sum_{k' \in K, k' \neq k} \sum_{j \in J | j \in J_{\hat{j}_k}} y_{jk'} \le 1 \qquad \forall k \in K, \hat{j}_k \in J_k \qquad (5)$$

$$x_{ljk} \in \{0,1\}, \qquad\qquad \forall k \in K, l \in L, j \in J_k \qquad (6)$$

$$y_{jk} \in \{0,1\}, \qquad\qquad \forall k \in K, j \in J_k \qquad (7)$$

The objective function (1) and the constraints (2), (3), (4) configure a classical *PMP* formulation. Indeed, the objective function (1) minimizes total connection cost of the options to the weldings. Constraints (2) impose that each option of a functional package has to be connected to just one welding performed along the related sub-tree network. Constraints (3) are consistency constraints among the variables. Constraints (4) impose that the number of weldings (medians) to be located is equal to the cardinality of K, i.e., to the number of functional packages. Finally, constraints (5) are the additional constraints imposing the mutual distance restrictions among the weldings positions. Indeed, they impose that no more than one welding location is selected between $\hat{j}_k$ and the ones in $J_{\hat{j}_k}$. It is important to highlight that the usage of a step size equal to d in the discretization procedure guarantees by construction the satisfaction of the distance constraints between all the potential welding locations and the leaves/anchor points.

3.3   A Matheuristic Approach for the *OWLP*

In this section we propose a matheuristic approach for the *OWLP*, based on the exploitation of the proposed *ILP* formulation. The key idea is to generate only a subset of all the potential welding locations used in the model, trying then to add new locations just if they can provide an improvement of the objective function.

- STEP 0. Generate the set of potential welding locations performing the discretization procedure along all the links of the harness tree, considering only the welding points which are at distance d and $2d$ from the extreme vertices of each link. In other words, if we indicate by m a generic integer value, we perform the discretization procedure considering all the points at distance md, with $m \le 2$, from the vertices of each link. We denote this set as $\overline{J} \subset J$ and we denote as $\overline{J_k} \subset \overline{J}$ the set of potential locations of the generic welding k, $k \in K$. Compute the C_{lj} values forall k, $k \in K$, $l \in L$, $j \in \overline{J_k}$.

– STEP 1. Solve the *OWLP* formulation to optimality considering sets L, K and $\overline{J}$. Two situations can occur:

- In the obtained optimal solution all the weldings are located in positions with distance d from the extreme vertices of the links. This means that the solution is optimal also for the problem considering the whole set J, since all the weldings are shifted right or left towards the extreme vertices of each link, as shown in [13]. Hence, the matheuristic STOPS.
- In the obtained optimal solution at least one welding is located in a position with distance md from the extreme vertices of a link. This means that, because of the mutual distance restriction, one or more weldings instead of being shifted towards the extreme vertices of a link, have been moved towards the internal side. Hence, this is a feasible solution, but not the optimal one, since it could be possible to find a better solution allowing the relocation of a welding on a different link. Hence, the matherustic proceeds to STEP 2.

– STEP 2. Integrate the set $\overline{J}$ with additional welding locations. If $\overline{m}$ is the current value of m, the set $\overline{J}$ is increased with all the welding locations at distance md, with $m = \overline{m} + 1$. After this integration the matheuristic returns to STEP 1.

As explained above, at each iteration, the proposed matheuristic either provides the optimal solution or a feasible solution, i.e., an upper bound, for the *OWLP*.

4   Computational Results

In this section we summarize the results obtained on four test instances built from real data coming from a company operating in the automotive field. The instances differ in number of options (leaves of harness tree), weldings to be performed, potential welding locations and structure of the harness tree. The settings of the instances, using the notation of Subsect. 3.1, are reported in Table 1.

Table 1. Settings of the four test instances

$|L|$	$|I|$	$|K|$	$|J|$
221	438	70	2500
203	438	65	2460
181	438	60	2310
162	438	55	2140

The sets J have been obtained using a discretization step $d = 5$ cm, corresponding to the minimum distance value between the weldings and the leaves/anchor points and to the mutual distance among the weldings. In all the instances 6 or 4 different gauge values, each of them characterized by its own unitary cost. In all the instances the unitary costs are related to the weight of a cable and are expressed as gr/m.

The *ILP* formulations have been solved by the off-the-shelf software FICO Xpress-MP 8.2 and the experimentation was run an Intel(R) Core(TM) i5-3210M CPU @ 2.50 Ghz Processor and 4,00 GB RAM.

All the instances have been solved to optimality by both the *ILP* formulation and the proposed matheuristic, within a computation time limit of 30 min and

Table 2. Solutions of the four test instances

70 weldings				
Gauge	Unitary cost (gr/m)	Number of cables	Total length (m)	Total weight (Kg)
0.35	4.5	116	255.57	1.150
0.5	6.6	85	158.84	1.046
0.75	9	14	34.52	0.311
1	11	7	16.05	0.177
1.5	16	19	43.36	0.694
4	42	4	6.58	0.276
6	61	1	0.05	0.003
65 weldings				
Gauge	Unitary cost (gr/m)	Number of cables	Total length (m)	Total weight (Kg)
0.35	4.5	108	240.03	1.080
0.5	6.6	85	158.84	1.046
0.75	9	14	34.52	0.311
1.5	16	19	43.36	0.694
4	42	1	3.43	0.143
6	61	1	0.05	0.003
60 weldings				
Gauge	Unitary cost (gr/m)	Number of cables	Total length (m)	Total weight (Kg)
0.35	4.5	97	212.45	0.956
0.5	6.6	77	141.74	0.934
0.75	9	13	31.51	0.284
1.5	16	19	43.36	0.694
55 weldings				
Gauge	Unitary cost (gr/m)	Number of cables	Total length (m)	Total weight (Kg)
0.35	4.5	85	193.56	0.871
0.5	6.6	74	137.87	0.909
0.75	9	13	31.51	0.284
1.5	16	15	36.64	0.587

5 min, respectively. It is important to highlight that for these instances the proposed mathheuristic determines the optimal solution at the first step for the two smaller instances and with a value of m equal to 5 for the other larger ones. For each instance, Table 2 reports the following information for each cable gauge: the unitary cost (gr/m); the number of cables used in the harness; the total length of the cables and the corresponding weight. It is interesting to note that the total length of the cables, corresponding to the loose unbounded cables unrolled and stretched, can be higher than several hundred metres. This gives an idea of the saving that a company can achieve by the optimization of this harness design problem, where dedicated connections among the options are substituted by the weldings.

5   Conclusions

In this work we presented a particular location problem arising in the vehicle manufacturing field. It consists in the determination of the optimal positions of the weldings needed to minimize the harness (connection) costs within a vehicle. The problem has been tackled by both an exact and a mathheuristic approach. The performed experimentation on instances derived from real data confirm the applicability and the effectiveness of the proposed methodologies. We want to remark that even if in terms of cost of a single wiring harness the obtained saving can be very small (few euros or less for few cable metres), the total saving for a company has to be evaluated considering all the vehicle production, which is around several millions of vehicles.

Future work will be aimed at providing some managerial insights about the harness optimization of a vehicle, taking also into account some operative aspects of the problem. Indeed, in spite of increasing automation, hand manufacture continues to be the primary method of vehicle harness installation, because it involves several activities which are not easy to be mechanized: routing wires through sleeves; taping with fabric tape; clamps or cable ties; etc. Hence, it could be interesting to investigate the possibility of minimizing the differences between the welding locations for two or more harnesses, in order to standardise the installation procedure performed by the workers.

References

1. Agra, A., Cardoso, D.M., Cerdeira, J.O., Miranda, M., Rocha, E.: Solving huge size instances of the optimal diversity management problem. J. Math. Sci. **161**(6), 956–960 (2009)
2. Agra, A., Cerdeira, J.O., Raquejo, C.: Using decomposition to improve greedy solutions of the optimal diversity management. Int. Trans. Oper. Res. **20**(6), 617–625 (2013)
3. Agra, A., Cerdeira, J.O., Requejo, C.: A decomposition approach for the p-median problem on disconnected graphs. Comput. Oper. Res. **86**, 79–85 (2017)
4. Agra, A., Requejo, C.: The linking set problem: a polynomial special case of the multiple-choice knapsack problem. J. Math. Sci. **161**(6), 919–929 (2009)

5. An, H.-C., Svensson, O.: Recent developments in approximation algorithms for facility location and clustering problems. In: Fukunaga, T., Kawarabayashi, K. (eds.) Combinatorial Optimization and Graph Algorithms, pp. 1–19. Springer, Singapore (2017). https://doi.org/10.1007/978-981-10-6147-9_1

6. Avella, P., Boccia, M., Martino, C.D., Oliviero, G., Sforza, A., Vasilyev, I.: A decomposition approach for a very large scale optimal diversity management problem. 4OR **3**(1), 23–37 (2005). https://doi.org/10.1007/s10288-004-0059-1

7. Briant, O., Naddef, D.: The optimal diversity management problem. Oper. Res. **52**(4), 515–526 (2004)

8. Boccia, M., Masone, A., Sforza, A., Sterle, C.: A partitioning based heuristic for a variant of the simple pattern minimality problem. In: Sforza, A., Sterle, C. (eds.) ODS 2017. SPMS, vol. 217, pp. 93–102. Springer, Cham (2017). https://doi.org/10.1007/978-3-319-67308-0_10

9. Boccia, M., Sforza, A., Sterle, C.: Simple pattern minimality problems: integer linear programming formulations and covering-based heuristic solving approaches. Informs J. Comput., 93–102 (2020). https://doi.org/10.1287/ijoc.2019.0940. to appear

10. Cheng, Y., Kang, L., Lu, C.: The pos/neg-weighted 1-median problem on tree graphs with subtree-shaped customers. Theor. Comput. Sci. **411**(7–9), 1038–1044 (2010)

11. Daskin, M.S., Maass, K.L.: The p-Median Problem. In: Laporte, G., Nickel, S., da Gama, F.S. (eds.) Location Science, pp. 21–45. Springer, Cham (2015). https://doi.org/10.1007/978-3-319-13111-5_2

12. Drezner, Z., Hamacher, H.W.: Facility Location: Applications and Theory. Springer, Heidelberg (2004)

13. Hakimi, S.L.: Optimum locations of switching centers and the absolute centers and medians of a graph. Oper. Res. **12**(3), 450–459 (1964)

14. Masone, A., Sterle, C., Vasilyev, I., Ushakov, A.: A three-stage p-median based exact method for the optimal diversity management problem. Networks **74**(2), 174–189 (2019)

15. Masone, A., Sforza, A., Sterle, C., Vasiliev, I.: A graph clustering based decomposition approach for large scale p-median problems. Int. J. Artif. Intell. **16**(1), 116–129 (2018)

16. Mladenovic, N., Brimberg, J., Hansen, P., Moreno-Pérez, J.A.: The p-median problem: a survey of metaheuristic approaches. Eur. J. Oper. Res. **179**(3), 927–939 (2007)

17. Reese, J.: Solution methods for the p-median problem: an annotated bibliography. Networks **28**(3), 125–142 (2006)

On Non-elitist Evolutionary Algorithms Optimizing Fitness Functions with a Plateau

Anton V. Eremeev$^{(\boxtimes)}$ (iD)

Sobolev Institute of Mathematics, Omsk, Russia
eremeev@ofim.oscsbras.ru

Abstract. We consider the expected runtime of non-elitist evolutionary algorithms (EAs), when they are applied to a family of fitness functions PLATEAU$_r$ with a plateau of second-best fitness in a Hamming ball of radius r around a unique global optimum. On one hand, using the level-based theorems, we obtain polynomial upper bounds on the expected runtime for some modes of non-elitist EA based on unbiased mutation and the bitwise mutation in particular. On the other hand, we show that the EA with fitness proportionate selection is inefficient if the bitwise mutation is used with the standard settings of mutation probability.

Keywords: Evolutionary algorithm · Selection · Runtime · Plateau · Unbiased mutation

1 Introduction

Realising the potential and usefulness of each operator that can constitute evolutionary algorithms (EAs) and their interplay is an important step towards the efficient design of these algorithms for practical applications. The proofs showing how and when the population size, recombination operators, mutation operators or self-adaptation techniques are essential in EAs can be found in [5,6,13,16,20] and other works.

In the present paper, we study the efficiency of non-elitist EAs without recombination, applied to optimization problems with a single plateau of constant values of objective function around the unique global optimum. Significance of plateaus analysis is associated with several reasons. Plateaus often occur in combinatorial optimization problems, especially in the unweighted problems, such as Maximum Satisfiability Problem [11,19]. As a measure of efficiency, we consider the expected runtime, i.e., the expected number of objective (or fitness) function evaluations until the optimal solution is reached. We study the EAs without elite individuals, based on bitwise mutation, when they are applied to optimize fitness functions with plateaus of constant fitness. To this end, we consider the PLATEAU$_r$ function with a plateau of second-best fitness in a ball of radius r around the unique optimum. The goal of this paper is to study the expected runtime of non-elitist EAs,

© Springer Nature Switzerland AG 2020
A. Kononov et al. (Eds.): MOTOR 2020, LNCS 12095, pp. 329–342, 2020.
https://doi.org/10.1007/978-3-030-49988-4_23

optimizing PLATEAU$_r$, asymptotically for unbounded increasing number of binary variables n, assuming constant parameter r.

It is shown in [1] that the $(1 + 1)$ EA, which is one of the simplest mutation-based evolutionary algorithms, using an unbiased mutation operator (e.g., the bitwise mutation or the one-point mutation) optimizes PLATEAU$_r$ function with expected runtime $\frac{n^r(1+o(1))}{r!\Pr(1\leq\xi\leq r)}$, where ξ is a random variable, equal to the number of bits flipped in an application of the mutation operator. This is proved under the condition that mutation flips exactly one bit with probability $\omega(n^{-\frac{1}{2r-2}})$. The most natural special case when this condition is satisfied is when exactly one bit is flipped with probability $\Omega(1)$.

In the present paper, with the similar conditions on unbiased mutation we obtain polynomial upper bounds on the expected runtime of non-elitist EAs, using tournament selection, (μ, λ)-selection and, in the case of bitwise mutation with low mutation probability of order $1/n^2$, using fitness proportionate selection. The bounds are obtained using the level-based theorems [2,9] and [7].

Taking into account the similarity of function PLATEAU$_r$ to the well-known ONEMAX function, we derive an exponential lower bound on the expected runtime of the EAs with the proportionate selection and standard mutation probability $1/n$, and more generally, with mutation probability χ/n, where χ is a constant greater than $\ln 2$. It is assumed that population size $\lambda = \Omega(n^{2+\delta})$ for some constant $\delta > 0$. In these conditions, we also show that finding an approximate solution within some constant approximation ratio also requires an exponential time in expectation. The lower bounds for the case of proportionate selection are based on the proof outlines suggested for linear functions in [3] and coincide with those results in the special case of ONEMAX function.

2   Preliminaries

We use the same notation as in [2,4,15]. For any $n \in \mathbb{N}$, define $[n] := \{1, 2, \ldots, n\}$. The natural logarithm and logarithm to the base 2 are denoted by $\ln(\cdot)$ and $\log(\cdot)$ respectively. For $x \in \{0,1\}^n$, we write x_i for the ith bit value. The Hamming distance is denoted by $H(\cdot, \cdot)$ and the Iverson bracket by $[\cdot]$. Throughout the paper the maximisation of a *fitness function* $f: \mathcal{X} \to \mathbb{R}$ over a finite *search space* $\mathcal{X} := \{0,1\}^n$ is considered. Given a partition of $\mathcal{X}$ into m ordered subsets/*levels* $(A_1, \ldots, A_m)$, let $A_{\geq j} := \cup_{i=j}^m A_i$. Note that by this definition, $A_{\geq 1} = \mathcal{X}$. A *population* is a vector $P \in \mathcal{X}^\lambda$, where the ith element $P(i)$ is called the ith *individual*. For $A \subseteq \mathcal{X}$, define $|P \cap A| := |\{i \mid P(i) \in A\}|$, i.e., the count of individuals of P in A.

2.1   The Objective Function

We are specifically interested in two fitness functions defined on $\mathcal{X} = \{0,1\}^n$:

– The most well-known benchmark function

$$\text{ONEMAX}(x) := \sum_{i=1}^{n} x_i,$$

it is deeply studied in the literature on the theory of EAs, and will be referred here several times.

- A function from [1] with a single plateau of the second-best fitness in a ball of radius r around the unique optimum

$$\text{PLATEAU}_r := \begin{cases} \text{ONEMAX}(x) & \text{if } \text{ONEMAX}(x) \leq n - r, \\ n - r & \text{if } n - r < \text{ONEMAX}(x) < n, \\ n & \text{if } \text{ONEMAX}(x) = n, \end{cases}$$

parametrized by an integer r, assumed to be a constant greater than one.

Note that our results will also hold for the generalised classes of such functions (see, e.g., [8]), where the meaning of 0-bit and 1-bit in each position can be exchanged, and/or x is rearranged according to a fixed permutation before each evaluation.

2.2 Non-elitist Evolutionary Algorithm and Its Operators

The non-elitist EAs considered in this paper fall into the framework of Algorithm 1, see, e.g., [4,15]. Suppose that the fitness function $f(x)$ should be maximized. Starting with some P_0 which is sampled uniformly from $\mathcal{X}^\lambda$, in each iteration t of the outer loop a new population P_{t+1} is generated by independently sampling λ individuals from the existing population P_t using two operators: *selection* `select`: $\mathcal{X}^\lambda \to [\lambda]$ and *mutation* `mutate`: $\mathcal{X} \to \mathcal{X}$. Here, `select` takes a vector of λ individuals as input, then implicitly makes use of the function f, i.e., through *fitness evaluations*, to return the index of the individual to be selected.

Algorithm 1. Non-Elitist Evolutionary Algorithm

Require: Finite state space $\mathcal{X}$, and initial population $P_0 \in \mathcal{X}^\lambda$
1: **for** $t = 0, 1, 2, \ldots$ until termination condition met **do**
2:   **for** $i = 0, 1, 2, \ldots, \lambda$ **do**
3:     Sample $I_t(i) :=$ `select`(P_t), and set $x := P_t(I_t(i))$
4:     Sample $P_{t+1}(i) :=$ `mutate`(x)

The function is optimised when an optimum x^*, i.e., $f(x^*) = \max_{x \in \mathcal{X}} \{f(x)\}$, appears in P_t for the first time, i.e., x^* is sampled by `mutate`, and the optimisation time (or runtime) is the number of fitness evaluations made until that time.

In this paper, we assume that the termination condition is never satisfied and the algorithm produces an infinite sequence of iterations. This simplifyng assumption is frequently used in the theoretical analysis of EAs.

Formally, select is represented by a probability distribution over $[\lambda]$, and we use $p_{\text{sel}}(i \mid P)$ to denote the probability of selecting the ith individual $P(i)$ of P. The well-known *fitness-proportionate selection* is an implementation of select with

$$\forall P \in \mathcal{X}^\lambda, \forall i \in [\lambda]: p_{\text{sel}}(i \mid P) = \frac{f(P(i))}{\sum_{j=1}^{\lambda} f(P(j))}$$

(if $\sum_{j=1}^{\lambda} f(P(j)) = 0$, then one can assume that select has the uniform distribution). By definition, in the *k-tournament selection*, k individuals are sampled uniformly at random with replacement from the population, and a fittest of these individuals is returned. In (μ, λ)-*selection*, parents are sampled uniformly at random among the fittest μ individuals in the population. The ties in terms of fitness function are resolved arbitrarily.

We say that select is *f-monotone* if for all $P \in \mathcal{X}^\lambda$ and all $i, j \in [\lambda]$ it holds that $p_{\text{sel}}(i \mid P) \geq p_{\text{sel}}(j \mid P) \Leftrightarrow f(P(i)) \geq f(P(j))$. It is easy to see that all three selection mechanisms mentioned above are *f-monotone*.

The *cumulative selection probability* β of select (P) for any $\gamma \in (0, 1]$ is

$$\beta(\gamma, P) := \sum_{i=1}^{\lambda} p_{\text{sel}}(i \mid P) \cdot \left[f(P(i)) \geq f_{\lceil \gamma \lambda \rceil} \right], \quad \text{where } P \in \mathcal{X}^\lambda,$$

assuming a sorting $(f_1, \cdots, f_\lambda)$ of the fitnesses of P in descending order. In essence, $\beta(\gamma, P)$ is the probability of selecting an individual at least as good as the $\lceil \gamma \lambda \rceil$-ranked individual of P,

When sampling λ times with select(P_t) and recording the outcomes as vector $I_t \in [\lambda]^\lambda$, the *reproductive rate* of $P_t(i)$ is

$$\alpha_t(i) := \mathbf{E}\left[R_t(i) \mid P_t \right] \quad \text{where } R_t(i) := \sum_{j=1}^{\lambda} [I_t(j) = i].$$

Thus $\alpha_t(i)$ is the expected number of times that $P(i)$ is selected. The reproductive rate α_0 of Algorithm 1 is defined as $\alpha_0 := \sup_{t \geq 0} \max_{i \in [\lambda]} \{\alpha_t(i)\}$.

The operator mutate is represented by a transition matrix $p_{\text{mut}} \colon \mathcal{X} \times \mathcal{X} \to [0, 1]$, and we use $p_{\text{mut}}(y \mid x)$ to denote the probability to mutate an individual x into y.

In this paper, we consider the *unbiased* mutation operators [17]. This means that the probability distribution $p_{\text{mut}}(y \mid x)$ is invariant under bijection transformations of the Boolean cube $\{0, 1\}^n$, preserving the Hamming distance between any pair of bitstrings x, y. This invariance may be regarded as invariance under systematic flipping of arbitrary but fixed set of bit positions, and invariance under systematically applying an arbitrary but fixed permutation to all the bits.

One of the most frequently used unbiased mutation operators, the *bitwise mutation* (also known as the *standard bit mutation*), changes each bit of a given solution with a fixed mutation probability p_{mut}. Usually it is assumed that $p_{mut} = \chi/n$ for some parameter $\chi > 0$. For the bitwise mutation with mutation probability χ/n we have

$$\forall x, y \in \{0,1\}^n : p_{\mathrm{mut}}(y \mid x) = \left(\frac{\chi}{n}\right)^{H(x,y)} \left(1 - \frac{\chi}{n}\right)^{n-H(x,y)}.$$

Another well-known mutation operator, the *point mutation operator,* chooses i randomly from $[n]$ and changes only the ith bit in the given solution. Note that both of these mutation operators treat the bit values 0 and 1 indifferently, as well as the bit positions, and therefore satisfy the conditions of unbiasedness.

3   Upper Bounds for Expected Runtime

3.1   Tournament and (μ, λ)-Selection

First of all, due to similarity of function $\mathrm{PLATEAU}_r$ with the well-known Jump function JUMP_r, analogously to the proof of Theorem 11 (its JUMP_r case) from [2] we get

Theorem 1. *The EA applied to* $\mathrm{PLATEAU}_r$, $r = \mathcal{O}(1)$, *using*

- *a bitwise mutation given a mutation rate χ/n for any fixed constant $\chi > 0$,*
- *k-tournament selection or (μ, λ)-selection with their parameters k or λ/μ (respectively) being set to no less than $(1 + \delta)e^\chi$, where $\delta \in (0, 1]$ being any constant, and*
- *population size $\lambda \geq c \ln n$, for a sufficiently large constant c*

has the expected runtime $\mathcal{O}(n^r + n\lambda)$.

Note that by a slight modification of the proof of Theorem 11 [2], one can also obtain the $\mathcal{O}(n^r + n\lambda)$ upper bound on the expected EA runtime in the case of Jump function.

In the general case of unbiased mutation we prove the following.

Theorem 2. *The EA applied to* $\mathrm{PLATEAU}_r$, $r = \mathcal{O}(1)$, *using*

- *an unbiased mutation with $\Pr(\xi = 0) \geq p_0 = \Omega(1)$ and $\Pr(\xi = 1) = \Omega(1)$, where ξ is the random variable equal to the number of bits flipped in mutation,*
- *k-tournament selection or (μ, λ)-selection with their parameters k or λ/μ (respectively) being set to no less than $(1 + \delta)/p_0$, where $\delta \in (0, 1]$ being any constant, and*
- *population size $\lambda \geq c \ln n$ for sufficiently large constant c, independent of r*

has the expected runtime $\mathcal{O}(\lambda n^{r+1})$.

Proof. Let us consider a partition of $\mathcal{X}$ into $m = n - r + 1$ subsets, where $A_i := \{x : |x| = i - 1\}$, $i \in [m - 1], A_m := \{x : |x| \geq n - r\}$. Then from Theorem 7, analogously to the proof of Theorem 11 in [2], it follows that in conditions formulated above, on average after at most Cn iterations the EA will produce a population with at least $\gamma_0 \lambda$ individuals on the plateau A_m, where C

and γ_0 are positive constants. By the Markov's inequality, this implies that with probability at least $1/2$, starting from any population, within $2Cn$ iterations, the EA produces a population with at least $\gamma_0\lambda$ individuals on the plateau.

Consider any iteration t, when population P_t contains at least $\gamma_0\lambda$ individuals on the plateau. For any offspring in the population P_{t+1}, the probability to have not less than $n - r + 1$ ones is $\Omega(1/n)$. Given an individual x with $n - r + i$ ones, the probability to produce $\texttt{mutate}(x)$ with at least $n-r+i+1$ ones is also $\Omega(1/n)$. By the EA outline, all individuals in each population P_{t+i}, $i = 1, \ldots, r$ are identically distributed, so, by the inductive argument, for any $i = 1, \ldots, r$, the probability that the first individual produced in population P_{t+i} will have at least $n - r + i$ ones is $\Omega(1/n^i)$. (Of course, an individual with any other index in population P_t may be fixed here.) On the iteration $t + r$, the individual number one is optimal with probability $\Omega(1/n^r)$.

Now we can consider a sequence of series of the EA iterations, where the length of each series is $2Cn + r = \mathcal{O}(n)$ iterations. Suppose, D_j, $j = 1, 2, \ldots$, denotes an event of absence of the optimal individuals in the population throughout the jth series. In view of the above consideration, the probability of each event D_j, $j = 1, 2, \ldots$, is $1 - \Omega(1/n^r)$, so the probability to reach the optimum in at most j series is lower bounded by $(1 - C'/n^r)^j$ for some constant C'.

Let Y denote the random variable equal to the number of the first series when the optimal solution is obtained. By the properties of expectation (see, e.g., [10]),

$$E[Y] = \sum_{j=0}^{\infty} \Pr(Y > j) = 1 + \sum_{j=1}^{\infty} \Pr(D_1 \& \ldots \& D_j) \leq 1 + \sum_{j=1}^{\infty}(1 - C'/n^r)^j = \mathcal{O}(n^r).$$

Consequently, the expected runtime is $\mathcal{O}(\lambda n^{r+1})$.   □

The requirement of a positive constant lower bound on probability to mutate none of the bits $\Pr(\xi = 0) = \Omega(1)$ may be avoided at the expence of very high selection pressure and a factor of λ longer runtime, using Theorem 9:

Theorem 3. *The EA applied to* PLATEAU$_r$, $r = \mathcal{O}(1)$, *using an unbiased mutation with* $\Pr(\xi = 1) = \Omega(1)$, *where* ξ *is the random variable equal to the number of bits flipped in mutation,*

- *and k-tournament selection, $k \geq n(1 + \ln n)e/\Pr(\xi = 1)$ with a population of size $\lambda \geq k$,*
- *or (μ, λ)-selection with $\lambda/\mu \geq n(1 + \ln n)/\Pr(\xi = 1)$*

has the expected runtime $\mathcal{O}(\lambda^2 n^{r+1})$.

The proof is analogous to that of Theorem 2, but now the probability to choose a parent with the fitness $n - r$ within $n - r$ iterations of each series is lower-bounded only by $1/\lambda$, rather than by the constant γ_0. So, the probability of each event D_j is $1 - \Omega(n^{-r}/\lambda)$ and $E[Y] = \mathcal{O}(n^r\lambda)$.

3.2   Fitness-Proportionate Selection and Low Mutation Rate

Theorem 4. *The expected runtime of the EA on* PLATEAU$_r$, $r = \mathcal{O}(1)$, *using*

- *fitness-proportionate selection,*
- *bitwise mutation with mutation probability* χ/n, $\chi = (1-c)/n$, *for any constant* $c \in (0,1)$
- *population size* $\lambda \geq c'n^2 \ln(n)$, $\lambda = \mathcal{O}(n^K)$, *where* c' *and* K *are positive sufficiently large constants*

is $O(\lambda n^2 \log n + n^{2r+1})$.

Proof. We will apply Theorem 8 as in the proof of Theorem 5 from [3]. To this end, we use a partition of $\mathcal{X}$ into $m = n+1$ subsets, where $A_i := \{x : |x| = i-1\}$, $i \in [m-1]$, $A_m := \{x : |x| = n\}$.

Given $x \in A_j$ for any $j < m-1$, among the first $j+1$ bits, there must be at least one 0-bit, thus it suffices to flip the first 0-bit on the left while keeping all the other bits unchanged to produce a search point at a higher level. The probability of such an event is $\frac{\chi}{n}\left(1 - \frac{\chi}{n}\right)^{n-1} > \frac{\chi}{n}\left(1 - \frac{1}{n}\right)^{n-1} \geq \frac{1-c}{en^2} =: s_j, j \in [m-1]$. For s_{m-1} we have $s_* := s_{m-1} = \Omega((\frac{\chi}{n})^r)$. This choice of s_j satisfies (M1). To satisfy (M2), we pick $p_0 := (1 - \chi/n)^n$, i.e., the probability of not flipping any bit position by mutation.

In (M3), we choose $\gamma_0 := c/4$ and for any $\gamma \leq \gamma_0$, let f_γ be the fitness of the $\lceil \gamma\lambda \rceil$-ranked individual of any given $P \in \mathcal{X}^\lambda$. Thus there are at least $k \geq \lceil \gamma\lambda \rceil \geq \gamma\lambda$ individuals with fitness at least f_γ and let $s \geq kf_\gamma \geq \gamma\lambda f_\gamma$ be their sum of fitness. We can pessimistically assume that individuals with fitness less than f_γ have fitness $f_\gamma - 1$, therefore

$$\beta(\gamma, P) \geq \frac{s}{s + (\lambda - k)(f_\gamma - 1)} \geq \frac{s}{s + (\lambda - \gamma\lambda)(f_\gamma - 1)}$$

$$\geq \frac{\gamma\lambda f_\gamma}{\gamma\lambda f_\gamma + (\lambda - \gamma\lambda)(f_\gamma - 1)} = \frac{\gamma}{1 - (1-\gamma)/f_\gamma}$$

$$\geq \frac{\gamma}{1 - (1-c/4)/f^*} \geq \gamma e^{(1-c/4)/f^*},$$

where $f^* := n$ and in the last line we apply the inequality $e^{-x} \geq 1 - x$. Note that $p_0 = (1 - \chi/n)^n \geq e^{-\chi/(1-\varepsilon)}$ for any constant $\varepsilon \in (0,1)$ and sufficiently large n. Indeed, by Taylor theorem, $e^{-z} = 1 - z + z\alpha(z)$, where $\alpha(z) \to 0$ as $z \to 0$. So given any $\varepsilon > 0$, for all sufficiently small $z > 0$ holds $e^{-z} \leq 1 - (1-\varepsilon)z$. For any $\varepsilon \in (0,1)$ we can assume that $z = \chi/(n(1-\varepsilon))$, then for all sufficiently large n it holds that $(1 - \chi/n)^n \geq e^{-zn} = e^{-\chi/(1-\varepsilon)}$. So we conclude that

$$\beta(\gamma, P)p_0 \geq \gamma e^{(1-c/4)/f^*} e^{-\chi/(1-\varepsilon)} \geq \gamma\left(1 + \frac{1 - c/4 - \chi f^*/(1-\varepsilon)}{f^*}\right).$$

Since $\chi f^* \leq \chi n = 1 - c$, choosing $\varepsilon := 1 - \frac{1-c}{1-c/2} \in (0,1)$ implies $\chi f^*/(1-\varepsilon) \leq 1 - c/2$. Condition (M3) then holds for $\delta := c/(4n)$ because

$$\beta(\gamma, P)p_0 \geq \gamma\left(1 + \frac{1 - c/4 - (1 - c/2)}{f^*}\right) \geq \gamma\left(1 + \frac{c}{4n}\right).$$

To verify condition (M4'), we assume $C = 1$ and note that

$$\frac{8}{\gamma_0 \delta^2} \log \left(\frac{Cm}{\delta} \left(\log \lambda + \frac{1}{\gamma_0 s_* \lambda} \right) \right) \leq c'' r n^2 \ln(n),$$

for some constant $c'' > 0$, since $\lambda \leq n^K$ and $s^* = \Omega(n^{-2r})$. So (M4') holds if c' is large enough. By Theorem 8, we conclude that on average after at most $O(\lambda n^2 \log n + n^{2r+1})$ fitness evaluations the EA will produce the optimum. $\square$

In the case of $r = 0$, the application of Theorem 4 for $\lambda = \Theta(n^2 \log n)$ gives $\mathbf{E}[T] = O(n^4 \log^2 n)$, the same as the upper bound in Theorem 4.1 in [7].

4   Inefficiency of Fitness-Proportionate Selection Given Standard Mutation Rate

In this section, we consider Algorithm 1 with fitness-proportionate selection and the bitwise mutation given a constant value of the parameter $\chi > \ln 2$. This algorithm turns out to be inefficient on PLATEAU_r for any constant r. For the proof we will use the same approach as suggested for lower bounding the EA runtime on the ONEMAX fitness function in [15]. In order to obtain an upper bound on the reproductive rate, we first show that, roughly speaking, it is unlikely that the average number of ones in the individuals of EA population becomes less than $n/2$ sometime during an exponential number of iterations.

Lemma 1. *Let r, $\varepsilon > 0$ and $\delta > 0$ be constants. Define T to be the smallest t such that Algorithm 1, applied to PLATEAU_r function, with population size $\lambda \geq n^{2+\delta}$, using an f-monotone selection mechanism, bitwise mutation with $\chi = \Omega(1)$, has a population P_t where $\sum_{j=1}^{\lambda} |P_t(j)| \leq \lambda(n/2)(1 - \varepsilon)$. Then there exists a constant $c > 0$ such that $\Pr(T \leq e^{cn}) = e^{-\Omega(n^\delta)}$.*

The proof of Lemma 1 is analogous to that of Lemma 9 from [14]. It is provided in the Appendix B for the sake of completeness.

The main result of this section is Theorem 5 which establishes an exponential lower bound for the expected time till finding an approximate solution with an approximation ratio $1+w$ for $w < (1-\ln(2)/\chi)$. The proof of this bound is based on the negative drift theorem for populations [14] (see Theorem 6) and Lemma 1. Since this proof is rather technical, we start with a more straightforward lower bound for the EA runtime.

Proposition 1. *Let $\delta > 0$ be a constant, then there exists a constant $c > 0$ such that during e^{cn} generations Algorithm 1 with population size $\lambda \geq n^{2+\delta}$, and $\lambda = \text{poly}(n)$, bitwise mutation with mutation probability χ/n for any constant $\chi > \ln 2$, and fitness-proportionate selection, obtains the optimum of $\text{PLATEAU}_r(x)$ with probability at most $e^{-\Omega(n^\delta)}$.*

The proof of Proposition 1 is based on the same ideas as Corollary 13 [15].

Proof. It follows by Lemma 1 that with probability at least $1 - e^{-\Omega(n^\delta)}$ for any constant $\varepsilon' \in (0,1)$ we have $\sum_{j=1}^{\lambda} \text{PLATEAU}_r(P_t(j)) \geq (1 - \varepsilon')\lambda(n - r)/2$ during $e^{c'n}$ iterations for some constant $c' > 0$. Otherwise, with probability $e^{-\Omega(n^\delta)}$ we can pessimistically assume that the optimum is found before iteration $e^{c'n}$.

With probability at least $1 - e^{-\Omega(n^\delta)}$ the reproductive rate α_0 satisfies

$$\alpha_0 \leq \frac{\lambda n}{(1 - \varepsilon')\lambda(n - r)/2} \leq \frac{2}{1 - \varepsilon''}, \tag{1}$$

for some $\varepsilon'' \in (0,1)$, assuming n to be sufficiently large.

Inequality (1) implies that for a sufficiently small ε'' it holds that $\alpha_0 < e^\chi$ and using Corollary 1, we conclude that the probability to optimise a function PLATEAU_r within $e^{c''n}$ generations is $\lambda e^{-\Omega(n)}$ for some constant $c'' > 0$. Therefore with $c = \min\{c', c''\}$, the proposition follows.   □

The inapproximability result is established in

Theorem 5. *Let $\delta > 0$ be a constant, then there exists a constant $c > 0$ such that during e^{cn} generations the EA with population size $\lambda \geq n^{2+\delta}$, and $\lambda = \text{poly}(n)$, bitwise mutation probability χ/n for any constant $\chi > \ln(2)$, and fitness-proportionate selection, applied to PLATEAU_r, will obtain a $(1 - w)$-approximate solution, $w < (1 - \ln(2)/\chi)^2/2$, with probability at most $e^{-\Omega(n^\delta)}$.*

Proof. As in the proof of Proposition 1 we claim that with probability at least $1 - e^{-\Omega(n^\delta)}$ the reproductive rate α_0 satisfies the inequality $\alpha_0 \leq \frac{2}{1-\varepsilon''}$, for any $\varepsilon'' \in (0,1)$ assuming n to be sufficiently large, and then the upper bound $\alpha := \frac{2}{1-\varepsilon''}$ satisfies condition 1 of Theorem 6 for any $a(n)$ and $b(n)$. Note that this α also satisfies the inequality $\ln(\alpha) = \ln(2) - \ln(1 - \varepsilon') < \ln(2) + \varepsilon'e$ for any $\varepsilon' \in (0, 1/e)$.

Condition 2 of Theorem 6 requires that $\ln(\alpha)/\chi + \delta' < 1$ for a constant $\delta' > 0$. This condition is satisfied because $\frac{\ln(\alpha)}{\chi} < \frac{\ln(2)+\varepsilon'e}{\chi} < 1$ for a sufficiently small ε'. Here we use the assumption that $\chi > \ln(2)$. It suffices to assume $\varepsilon' = \frac{\chi - \ln 2}{2e}$. Define $\psi := \frac{\ln(2)+\varepsilon'e}{\chi} = \frac{\ln(2)}{2\chi} + \frac{1}{2}$.

To ensure Condition 3 of Theorem 6, we denote $\rho := \ln(2)/\chi < 1$ and

$$M(\chi) := \frac{1 - \sqrt{\psi(2 - \psi)}}{2} = \frac{1 - \sqrt{\rho/2 - \rho^2/4 + 3/4}}{2}.$$

Note that $M(\chi)$ is decreasing in ρ and therefore increasing in χ, besides that $M(\chi)$ is independent of n and r. Now we define $a(n)$ and $b(n)$ so that $b(n) < M(\chi)n$ and $b(n) - a(n) = \omega(n)$. Assume that $a(n) := n(1 - \varepsilon)M(\chi)$ and $b(n) := n(1 - \varepsilon/2)M(\chi)$, where $\varepsilon > 0$ is a constant.

Application of Theorem 6 shows that with probability at most $e^{-\Omega(n^\delta)}$ the EA obtains a search point with less than

$$z(\varepsilon) := \frac{n(1 - \varepsilon)}{2} \cdot \left(1 - \sqrt{\frac{\rho}{2} - \left(\frac{\rho}{2}\right)^2 + \frac{3}{4}}\right) \tag{2}$$

zero-bits for any constant $\varepsilon \in (0,1)$. Finally, using the Taylor series for the square root, we note that for any positive constant $w < (1-\rho)^2/2$, there exists $\varepsilon \in (0,1)$, such that the number of zero-bits in any $(1+w)$-approximate solution to PLATEAU$_r$ is at most $z(\varepsilon)$.   □

5   Discussion

It is shown in [1] that under very general conditions we have mentioned in the introduction, the (1+1) EA easily (in expected $\mathcal{O}(n \log n)$ time) reaches the plateau and then performs a random walk on it, quickly approaching to a "nearly-uniform" distribution. A similar behaviour may be expected from the elitist EAs like $(\mu + \lambda)$ EA, where the best incumbent, once having reached the plateau, will travel on it, until the optimum is found. One can expect that in the case of non-elitist EAs, if the selection is strong enough, the population will stick to the plateau and spread on it as well. In the present paper, however, we have not identified such regimes yet.

In the case of bitwise mutation, our Theorem 1 relies on a scenario, where the EA quickly reaches the edge of the plateau and most of the remaining time (with seldom possible retreats from the plateau) spends on the attempts to hit the optimum by "large" mutations, inverting up to $n-r$ zero-bits. Theorem 2, applicable to a wider class of mutation operators, relies on a more graduate scenario, where the search may consist of multiple stages, each one starting from an "arbitrary bad" population, then reaches the edge of the plateau in expected $\mathcal{O}(n \log n)$ time and tries to hit the optimum by making r sequential single-bit mutations, reducing the Hamming distance to the optimum by 1 in each EA iteration. If such an attempt fails, then we consider the next stage, over-pessimistically assuming that the search starts from a population of all-zero strings. Theorem 3 is even less demanding to the properties of mutation operators, but demanding very high selection pressure. It is likely that the runtime bound in this case may be significantly improved, since with such a high selection pressure the non-elitist EA becomes so close to the (1+1) EA. While Theorems 1 – 3 deal with the tournament or (μ, λ)-selection, Theorem 4 shows that a similar situation may be observed in the case of fitness proportionate selection, although in this case we require that the mutation probability is reduced to $\Theta(1/n^2)$ because otherwise the EA is likely to spend exponential time on the way to the plateau, as it is shown in Theorem 5.

6   Conclusions

This paper demonstrates the results, which are accessible by the available tools of runtime analysis. It also naturally leads to several questions for further research, some of which may require to develop principially new tools for EA analysis:

- What are the leading constants in the obtained upper bounds?
- What lower bounds can complement the obtained upper bounds?

- Under what conditions on selection pressure is it possible to transfer the tight results on (1+1) EA from [1] to the non-elitist EAs?
- How to extend the detailed runtime analysis to the Royal Road and Royal Staircase fitness functions (see, e.g., [18]) which have multiple plateaus?
- Would the genetic algorithms, which use the crossover operators, have any advantage over the mutation-based EAs considered in this paper?

Acknowledgment. The work was funded by program of fundamental scientific research of the Russian Academy of Sciences, I.5.1., project 0314-2019-0019. The author is grateful to Duc-Cuong Dang for helpful comments on preliminary version of the paper.

Appendix A

This appendix contains the formulations of results employed from other works. Some of the formulations are given with slight modifications, which do not require a special proof.

Our lower bound is based on the *negative drift theorem for populations* [14].

Theorem 6. *Consider the EA on $\mathcal{X} = \{0,1\}^n$ with bitwise mutation rate χ/n and population size $\lambda = \text{poly}(n)$, let $a(n)$ and $b(n)$ be positive integers such that $b(n) \leq n/\chi$ and $d(n) = b(n) - a(n) = \omega(\ln n)$. Given $x^* \in \{0,1\}^n$, define $T(n) := \min\{t \mid |P_t \cap \{x \in \mathcal{X} \mid H(x,x^*) \leq a(n)\}| > 0\}$. If there exist constants $\alpha > 1$, $\delta > 0$ such that*

(1) $\forall t \geq 0$, $\forall i \in [\lambda]$: if $a(n) < H(P_t(i), x^*) < b(n)$ then $\alpha_t(i) \leq \alpha$,
(2) $\psi := \ln(\alpha)/\chi + \delta < 1$,
(3) $b(n)/n < \min \left\{ 1/5, 1/2 - \sqrt{\psi(2 - \psi)/4} \right\}$,

then $\Pr\left(T(n) \leq e^{cd(n)}\right) = e^{-\Omega(d(n))}$ for some constant $c > 0$.

We also use a corollary of this theorem (Corollary 1 from [14]):

Corollary 1. *The probability that a non-elitist EA with population size $\lambda = \text{poly}(n)$, bitwise mutation probability χ/n, and maximal reproductive rate bounded by $\alpha < e^\chi - \delta$, for a constant $\delta > 0$, optimises any function with a polynomial number of optima within e^{cn} generations is $e^{-\Omega(n)}$, for some constant $c > 0$.*

To bound the expected optimisation time of Algorithm 1 from above, we use the *level-based analysis* [2]. The following theorem is a re-formulation of Corollary 7 from [2], tailored to the case of no recombination.

Theorem 7. *Given a partition $(A_1, \ldots, A_m)$ of $\mathcal{X}$, if there exist $s_1, \ldots, s_{m-1}$, p_0, $\delta \in (0,1]$, $\gamma_0 \in (0,1)$ such that*

(M1) $\forall P \in \mathcal{X}^\lambda, \forall j \in [m-1]$: $p_{\text{mut}}(y \in A_{\geq j+1} \mid x \in A_j) \geq s_j$,
(M2) $\forall P \in \mathcal{X}^\lambda, \forall j \in [m-1]$: $p_{\text{mut}}(y \in A_{\geq j} \mid x \in A_j) \geq p_0$,

(M3) $\forall P \in (\mathcal{X} \setminus A_m)^\lambda, \forall \gamma \in (0, \gamma_0]: \ \beta(\gamma, P) \geq (1 + \delta)\gamma/p_0,$

(M4) population size $\lambda \geq \dfrac{4}{\gamma_0 \delta^2} \ln\left(\dfrac{128m}{\gamma_0 s_* \delta^2}\right),$ where $s_* := \min\limits_{j \in [m-1]} \{s_j\},$

then, assuming $T_0 := \min\{t \mid |P_t \cap A_m| \geq \gamma_0 \lambda\}$, we have

$$\mathbf{E}\,[T_0] < \left(\frac{8}{\delta^2}\right) \sum_{j=1}^{m-1} \left(\ln\left(\frac{6\delta\lambda}{4 + \gamma_0 s_j \delta\lambda}\right) + \frac{1}{\gamma_0 s_j \lambda}\right). \qquad (3)$$

Note that literally the formulation of Corollary 2 in [2] gives the bound (3) only for the expected runtime, but it is easy to see from the proof therein that the bound actually holds for the expected number T_0 of the first population that contains at least $\gamma_0 \lambda$ individuals in level A_m as we put it in Theorem 7. This slight improvement is important in Sect. 3.

As an alternative to Theorem 7 we use the new level-based theorem based on the *multiplicative up-drift* [7]. Theorem 3.2 from [7] implies the following.

Theorem 8. *Given a partition* $(A_1, \ldots, A_m)$ *of* $\mathcal{X}$, *define* $T := \min\{t\lambda \mid |P_t \cap A_m| > 0\}$. *If there exist* $s_1, \ldots, s_{m-1}, p_0, \delta \in (0, 1], \gamma_0 \in (0, 1)$, *such that conditions (M1)–(M3) of Theorem 7 hold and*

(M4') *for some constant* $C > 0$, *the population size* λ *satisfies*

$$\lambda \geq \frac{8}{\gamma_0 \delta^2} \log\left(\frac{Cm}{\delta}\left(\log \lambda + \frac{1}{\gamma_0 s_* \lambda}\right)\right), \ \text{where } s_* := \min_{j \in [m-1]}\{s_j\},$$

then $\mathbf{E}\,[T] = \mathcal{O}\left(\frac{\lambda m \log(\gamma_0 \lambda)}{\delta} + \frac{1}{\delta}\sum_{j=1}^{m-1} \frac{1}{\gamma_0 s_j}\right).$

Theorem 8 improves on Theorem 7 in terms of dependence of the runtime bound denominator on δ, but only gives an asymptotical bound. Its proof is analogous to that of Theorem 7 and may be found in [3].

Theorem 9. *Given an* f-*based partition* $A_1, \ldots, A_m$ *of* $\mathcal{X}$, *if the EA uses the mutation, such that* $\Pr(\mathtt{mutate}(x) \in A_{\geq j+1}) \geq s_*$ *for any* $x \in A_j, j \in [m-1]$

- *and a* k-*tournament selection,* $k \geq \frac{(1+\ln m)e}{s_*}$ *with a population of size* $\lambda \geq k,$
- *or* (μ, λ)-*selection and* $\lambda \geq \frac{\mu(1+\ln m)}{s_*}$

then an element from A_m *is found in expectation after at most* em *genetations.*

The proof is analogous to that of the main result in [9].

Appendix B

This appendix contains the proofs provided for the sake of completeness.

Proof of Lemma 1. For the initial population, it follows by a Chernoff bound that $\Pr(T = 1) = e^{-\Omega(n)}$. We then claim that for all $t \geq 0$, $\Pr(T = t + 1 \mid T > t) \leq e^{-c'n}$ for a constant $c' > 0$, which by the union bound implies that $\Pr(T < e^{cn}) \leq e^{cn - c'n} = e^{-\Omega(n)}$ for any constant $c < c'$.

In the initial population, the expected number of ones of a k-th individual, $k \in [\lambda]$ is $|P_0(k))| \leq n/2$. It will be more convenient here to consider the number of zeros, rather than the number of ones. We denote $Z_t^{(j)} := n - |P_t(j)|$, for $t \geq 0$, $j \in [\lambda]$, and $Z_t := \lambda n - \sum_{j=1}^{\lambda} |P_t(j)|$. Let p_j be the probability of selecting the j-th individual when producing the population in generation $t+1$. For f-monotone selection mechanisms, it holds that $\sum_{j=1}^{\lambda} p_j Z_t^{(j)} \leq Z_t/\lambda$.

Let $P = (x_1, \ldots, x_\lambda)$ be any deterministic population. Denote the i-th bit of its k-th individual by $x^{(k,i)}$, $z_k := n - |x^{(k)}|$, $1 \leq k \leq \lambda$, and $Z(P) := \sum_{k=1}^{\lambda} z_k$.

Let us consider the bitwise mutation first. The expected number of zero-bits in any offspring $j \in [\lambda]$ produced from population $P_t = P$ is

$$\mathbf{E}\left[|P_{t+1}(j)| \mid P_t = P\right] = \sum_{k=1}^{\lambda} p_k \left[\sum_{i=1}^{n} \left(x^{(k,i)}(1 - \chi/n) + (1 - x^{(k,i)})\chi/n\right)\right],$$

so the expected value of $Z_{t+1}^{(j)}$ for any offspring $j \in [\lambda]$ is

$$\mathbf{E}\left[Z_{t+1}^{(j)} \mid P_t = P\right] \leq n - \mathbf{E}\left[|P_{t+1}(j)| \mid P_t = P\right]$$

$$= \sum_{k=1}^{\lambda} p_k \left[\sum_{i=1}^{n} \left(x^{(k,i)}(1 - \chi/n) + (1 - x^{(k,i)})\chi/n\right)\right]$$

$$= n - \sum_{k=1}^{\lambda} p_k \left[\chi + (1 - 2\chi/n) \sum_{i=1}^{n} x^{(k,i)}\right]$$

$$\leq n - \chi - (1 - 2\chi/n) \sum_{k=1}^{\lambda} p_k(n - z_k)$$

$$= \chi + (1 - 2\chi/n) \sum_{k=1}^{\lambda} p_k z_k \leq \chi + (1 - 2\chi/n)Z(P)/\lambda.$$

If $T > t$ and $Z(P) < \lambda n(1 + \varepsilon)/2$, then

$$\mathbf{E}\left[Z_{t+1} \mid P_t = P\right] \leq \lambda \chi + Z(P)(1 - 2\chi/n)$$

$$< \lambda \chi + \frac{\lambda n}{2}(1 + \varepsilon)(1 - 2\chi/n) = \frac{\lambda n}{2}(1 + \varepsilon) - \varepsilon\lambda\chi.$$

Now $Z_{t+1}^{(1)}, Z_{t+1}^{(2)}, \ldots, Z_{t+1}^{(\lambda)}$ are non-negative independent random variables, each bounded from above by n, so using the Hoeffding's inequality [12] we obtain

$$\Pr\left(Z_{t+1} \geq \frac{\lambda n}{2}(1 + \varepsilon)\right) \leq \Pr(Z_{t+1} \geq \mathbf{E}[Z_{t+1}] + \varepsilon\lambda\chi) \leq \exp\left(-\frac{2(\varepsilon\lambda\chi)^2}{\lambda n^2}\right),$$

which is $e^{-\Omega(n^\delta)}$ since $\lambda \geq n^{2+\delta}$. $\qquad \square$

References

1. Antipov, D., Doerr, B.: Precise runtime analysis for plateaus. In: Auger, A., Fonseca, C.M., Lourenço, N., Machado, P., Paquete, L., Whitley, D. (eds.) PPSN 2018. LNCS, vol. 11102, pp. 117–128. Springer, Cham (2018). https://doi.org/10.1007/978-3-319-99259-4_10
2. Corus, D., Dang, D., Eremeev, A.V., Lehre, P.K.: Level-based analysis of genetic algorithms and other search processes. IEEE Trans. Evol. Comput. **22**(5), 707–719 (2018)
3. Dang, D.C., Eremeev, A., Lehre, P.K.: Runtime analysis of fitness-proportionate selection on linear functions. ArXiv 1908.08686 [cs.NE] (2019)
4. Dang, D.C., Lehre, P.K.: Runtime analysis of non-elitist populations: from classical optimisation to partial information. Algorithmica **75**(3), 428–461 (2016)
5. Dang, D.C., Lehre, P.K.: Self-adaptation of mutation rates in non-elitist populations. In: Proceedings of PPSN 2016, pp. 803–813 (2016)
6. Doerr, B., Johannsen, D., Kötzing, T., Lehre, P.K., Wagner, M., Winzen, C.: Faster black-box algorithms through higher arity operators. In: Proceedings FOGA 2011, pp. 163–172 (2011)
7. Doerr, B., Kötzing, T.: Multiplicative up-drift. In: Proceedings of the Genetic and Evolutionary Computation Conference, GECCO 2019, Prague, Czech Republic, 13–17 July 2019, pp. 1470–1478 (2019)
8. Droste, S., Jansen, T., Tinnefeld, K., Wegener, I.: A new framework for the valuation of algorithms for black-box optimization. In: FOGA-7, pp. 253–270. Morgan Kaufmann, San Francisco (2003)
9. Eremeev, A.: Hitting times of local and global optima in genetic algorithms with very high selection pressure. Yugosl. J. Oper. Res. **27**(3), 323–339 (2017)
10. Gnedenko, B.V.: Theory of Probability. Gordon and Breach, London (1997)
11. Hampson, S., Kibler, D.: Plateaus and plateau search in Boolean satisfiability problems: when to give up searching and start again. In: Proceedings of the second DIMACS Implementation Challenge Cliques, Coloring and Satisfiability, pp. 437–456. American Mathematical Society (1996)
12. Hoeffding, W.: Probability inequalities for sums of bounded random variables. J. Am. Stat. Assoc. **58**(301), 13–30 (1963)
13. Jansen, T., Wegener, I.: On the utility of populations in evolutionary algorithms. In: Proceedings of GECCO 2001, pp. 1034–1041 (2001)
14. Lehre, P.K.: Negative drift in populations. In: Proceedings of PPSN 2010, pp. 244–253 (2010)
15. Lehre, P.K.: Fitness-levels for non-elitist populations. In: Proceedings of GECCO 2011, pp. 2075–2082 (2011)
16. Lehre, P.K., Özcan, E.: A runtime analysis of simple hyper-heuristics: to mix or not to mix operators. In: Proceedings of FOGA 2013, pp. 97–104 (2013)
17. Lehre, P.K., Witt, C.: Black-box search by unbiased variation. Algorithmica **64**, 623–642 (2012)
18. van Nimwegen, E., Crutchfield, J.: Optimizing epochal evolutionary search population-size independent theory. Comput. Meth. Appl. Mech. Eng. **186**(2–4), 171–194 (2000)
19. Sutton, A.M., Howe, A.E., Whitley, L.D.: Directed plateau search for max-k-sat. In: Proceedings of the Third Annual Symposium on Combinatorial Search, SOCS 2010, Stone Mountain, Atlanta, Georgia, USA, 8–10 July 2010 (2010)
20. Witt, C.: Population size versus runtime of a simple evolutionary algorithm. Theor. Comput. Sci. **403**(1), 104–120 (2008)

A Matheuristic for the Drilling Rig Routing Problem

Igor Kulachenko[1] and Polina Kononova[1,2]

[1] Novosibirsk State University, Novosibirsk, Russia
soge.ink@gmail.com
[2] Sobolev Institute of Mathematics SB RAS, Novosibirsk, Russia
pkononova@math.nsc.ru

Abstract. In this paper, we discuss the real-world Split Delivery Vehicle Routing Problem with Time Windows (SDVRPTW) for drilling rig routing in Siberia and the Far East. There is a set of objects (exploration sites) requiring well-drilling work. Each object includes a known number of planned wells and needs to be served within a given time interval. Several drilling rigs can operate at the same object simultaneously, but their number must not exceed the number of wells planned for this object. A rig that has started the work on a well completes it to the end. The objective is to determine such a set of rig routes (including the number of assigned wells for each object) to perform all well-drilling requests, respecting the time windows, that minimizes the total traveling distance. The main difference with traditional SDVRP is that it is the service time that is split, not the demand.

We propose a mixed-integer linear programming (MILP) model for this problem. To find high-quality solutions, we design the Variable Neighborhood Search based matheuristic. Exact methods are incorporated into a local search to optimize the distribution of well work among the rigs. Time-window constraints are relaxed, allowing infeasible solutions during the search, and evaluation techniques are applied to treat them. Results of computational experiments for the algorithm and a state-of-the-art MILP solver are discussed.

Keywords: Logistics · Uncapacitated vehicles · Split delivery service · Time windows · Metaheuristics · Mathematical models · Optimization problems

1 Introduction

Vehicle Routing Problems (VRP) are well-known NP-hard combinatorial optimization problems with a large number of various real-world applications [7, 23]. We consider the Drilling Rig Routing Problem (DRRP) arising in the oil and gas industry. In this problem, we have a fleet of uncapacitated drilling rigs

The reported study was funded by RFBR and Novosibirsk region according to the research project No. 19-47-540005.

(in short, rigs) and a set of objects that are exploration sites. The objects contain a predetermined number of wells planned to be drilled, and rigs have to perform work on their drilling during a certain planning horizon. The initial location for each rig is known. A rig travels from one object to another, performing drilling until it drills all assigned wells, and it does not have to return to the initial location after that. Each object has a time window that is a period of time during which all work on this object has to be started and completed. That means each object has the earliest time when a rig can start drilling and the latest time when all wells have to be already drilled. Several rigs can operate at the same object simultaneously, but their number must not exceed the number of wells planned for this object. A rig that has started the work on a well completes it to the end. The objective is to determine such a set of rig routes (including the number of assigned wells for each object) to perform all well-drilling requests, respecting the time windows, that minimizes the total traveling distance.

The described problem is similar to the Split Delivery Vehicle Routing Problem with Time Windows (SDVRPTW), where a fleet of capacitated homogeneous vehicles has to serve a set of customers requiring a service within a specific time interval, but contrary to the classical VRP a customer can be visited by several vehicles [2,11]. But usually SDVRPs consider split deliveries with respect to customer demand for goods, not service time. The fact that vehicles may not return to their initial location classifies the DRRP as Open VRP [15]. In [25], an Uncapacitated Open SDVRP with splittable service demand was studied. In this problem, the maximum service completion time was minimized, and time windows were not considered.

Since the DRRP can also be applied, with some alterations, to other real-world problems, we will sometimes refer to drilling rigs as vehicles and objects as customers as in classical VRP formulation.

The DRRP can be decomposed into three separate subproblems: assignment of customers to vehicles, sequencing of customers in vehicle routes, and distribution of well-drilling work among vehicles. The objective of the first one is to assign customers to vehicles in such a way that, while respecting time-window constraints, an assignment cost, which approximates the traveling distance, is minimized [4]. In the DRRP, this subproblem should also address the work sharing. The second subproblem corresponds to the solution of the traveling salesman problem for each vehicle separately [8]. The last subproblem requires the solution of the previous two subproblems as input and seeks to redistribute well-drilling work to get time-window feasibility. Thus, different decomposition approaches can be applied to solve the overall problem [3]. In this paper, we refer to the assignment and sequencing subproblems as the routing part of the DRRP.

Despite the advances in exact solution methods and hardware technology, most large-scale VRPTWs are still not able to be solved to optimality, and the performance of exact algorithms strongly varies with the time-window characteristics [6,23,24]. Thus, heuristic and metaheuristic approaches are usually applied to these problems. Hybrid metaheuristics, and matheuristics in particular, have seen significant advancement in recent years due to promising results [3,16,22,23].

We present a matheuristic combining a Variable Neighborhood Search (VNS) with a Mixed-Integer Linear Programming (MILP) [9,10,17]. The VNS is used to solve the routing part whereas the MILP is used to solve the well-drilling work distribution part. We also relax time-window constraints, allowing infeasible solutions during the search, and use evaluation techniques from [18,24] to effectively handle them.

The rest of this paper is structured as follows. We first introduce the mathematical model for the overall problem and the work redistribution subproblem in Sect. 2. Neighborhood structures are presented in Sect. 3. The framework of the VNS heuristic is described in Sect. 4. Computational results are discussed in Sect. 5. The last Sect. 6 concludes the paper.

2 Mathematical Model

2.1 Mathematical Model for the DRRP

The problem is represented by a fleet of identical vehicles denoted by K and a complete directed graph $G = (V, A)$ with a set of vertices V and a set of arcs A. Each vehicle $k \in K$ is initially located at an individual depot v_k. The set V is the union of a set of all depots V^K and a set of objects I requiring well-drilling work. Each object $i \in I$ has a given total number of planned wells W_i and an interval of times allowable to service $[e_i, l_i]$ called time window. That specifies that work on object i cannot be started before e_i and finished later than l_i. If some vehicle reaches an object i before the time e_i, it has to wait. We assume that drilling a well requires D days. Arcs $(i, j) \in A$ represent the possibility to travel from i to j with a distance d_{ij} and a duration t_{ij} in days.

We introduce the following binary decision variables:

$$x_{ijk} = \begin{cases} 1, & \text{if arc } (i,j) \in A \text{ is traversed by vehicle } k \in K, \\ 0, & \text{otherwise,} \end{cases}$$

$$y_{ik} = \begin{cases} 1, & \text{if vehicle } k \in K \text{ visits object or depot } i \in V, \\ 0, & \text{otherwise.} \end{cases}$$

It should be noted that variables y_{ik} are non-essential and are required for a clearer understanding of the model.

The decision variables $s_{ik} \geq 0$, defined for each vertex $i \in V$, represent the time when vehicle k starts to serve object i. In case vehicle k does not serve object $i \in I$ or $i \in V^k, i \neq v_k$, variable s_{ik} does not mean anything.

The variable $w_{ik} \in \mathbb{Z}_{\geq 0}$ represents the total number of wells drilled by rig k at object $i \in I$ and 0 for all depots.

Now we present the drilling rig routing problem as the mixed-integer linear program:

$$\min \sum_{k \in K} \sum_{i \in V} \sum_{j \in V} d_{ij} x_{ijk} \tag{1}$$

subject to

$$y_{ik} = \begin{cases} 1, i = v_k, \\ 0, i \neq v_k, \end{cases} \quad i \in V^K, k \in K, \tag{2}$$

$$s_{ik} = 0, \quad i = v_k, k \in K, \tag{3}$$

$$w_{ik} = 0, \quad i = v_k, k \in K, \tag{4}$$

$$\sum_{j \in V} x_{ijk} = \sum_{j \in V} x_{jik} = y_{ik}, \quad i \in V, k \in K, \tag{5}$$

$$\sum_{k \in K} w_{ik} = W_i, \quad i \in I, \tag{6}$$

$$w_{ik} \geq y_{ik}, \quad i \in I, k \in K, \tag{7}$$

$$w_{ik} \leq W_i y_{ik}, \quad i \in I, k \in K, \tag{8}$$

$$e_i \leq s_{ik}, \quad i \in I, k \in K, \tag{9}$$

$$s_{ik} + D w_{ik} \leq l_i, \quad i \in I, k \in K, \tag{10}$$

$$s_{ik} + D w_{ik} + x_{ijk} t_{ij} - s_{jk} \leq M_{ij}(1 - x_{ijk}), \quad i \in V, j \in I, k \in K, \tag{11}$$

$$x_{ijk}, y_{ik} \in \{0,1\}, \; s_{ik} \geq 0, \; w_{ik} \in \mathbb{Z}_{\geq 0}, \; i,j \in V, k \in K. \tag{12}$$

The objective function (1) minimizes the total traveling distance for all vehicles over all days of the planning horizon. Equalities (2)–(4) set the distribution of vehicles by their depots. Constraints (5) make sure that each object has exactly one predecessor and one successor in the route, and each vehicle returns to its depot. Equalities (6) ensure that the required number of wells are drilled at each object. Inequalities (7) guarantee that if a vehicle visits an object, it drills there at least one well, and inequalities (8) make sure that a vehicle cannot serve an object without visiting it. Constraints (9)–(11) ensure schedule feasibility with respect to time windows. Inequalities (11) also prevent subtours, and M_{ij}, $(i,j) \in A$, are large constants that can be set to $\max\{l_i - e_j, 0\}$. Constraints (12) define the types of variables.

Note that the problem (1)–(12) can be infeasible because of the limited fleet of vehicles and the time-window constraints. To overcome this, we relax the constraints (10) and include them into the objective function with penalties $\gamma \geq 0$. As a result, we get a relaxation of the original problem (1)–(12) as follows:

$$L(x, \tau, \gamma) = \min \sum_{k \in K} \sum_{i \in V} (\sum_{j \in V} d_{ij} x_{ijk} + \gamma \tau_{ik}) \tag{13}$$

subject to (2)–(9), (11)–(12) and additional constraints for new variables $\tau_{ik} \geq 0$ indicating the tardiness for each pair (i, k), $i \in I, k \in K$:

$$\tau_{ik} \geq s_{ik} + Dw_{ik} - l_i, \quad i \in I, k \in K. \tag{14}$$

Besides, constants M_{ij} in (11) now need to be set to new large enough values, for instance, $\max\{(\max_{m \in I} l_m) + DW_i - e_j, 0\}$.

Now the relaxed problem (2)–(9), (11)–(14) is feasible even if there is just one vehicle, and we can solve it by local search metaheuristics [21].

2.2 Mathematical Model for the Work Redistribution Subproblem

Employing the decomposition approach to the solution of the overall problem, we consider the work redistribution subproblem separately. In this subproblem, to achieve time-window feasibility while given a fixed visiting order, it is determined for each object how many wells are served by which vehicles visiting this object. To our knowledge, this problem is new, and its NP-hardness can be proved.

Let x_{ijk} and y_{ik} be already known, and we want to know the work distribution. If $y_{ik} = 0$, then object i is not visited by vehicle k. But if $y_{ik} = 1$, object i is not necessarily visited by vehicle k, since we can redistribute wells w_{ik} to other vehicles visiting this object.

Thereby, subproblem for well work distribution and readjustment of the routes arises. Let all the initial "maximum" routes be set. For vehicle k, it is a total preorder $I_k = \{i_0^k, i_1^k, \ldots, i_{u_k}^k\}$ with the following precedence relation: $i_l^k \preceq_{I_k} i_m^k$ if $l \leq m$, or $l = 0$, or $m = 0$. Here i_0^k corresponds to the depot of vehicle k and $i_{u_k}^k$ corresponds to the last object served by vehicle k.

We introduce new binary decision variables: x'_{ijk} defined only for $i, j \in I_k$, $i \preceq_{I_k} j$, and y'_{ik} such that

$$y'_{ik} \leq y_{ik}, \quad i \in V, k \in K. \tag{15}$$

Now we present well work redistribution subproblem, which is the problem (2)–(9), (11)–(14) defined on a reduced domain, as the MILP problem:

$$\min \sum_{k \in K} \sum_{i \in I_k} (\sum_{j \in I_k, i \preceq_{I_k} j} d_{ij} x'_{ijk} + \gamma \tau_{ik}) \tag{16}$$

$$\sum_{k \in K} w_{ik} = W_i, \quad i \in I, \tag{17}$$

$$w_{ik} \geq y'_{ik}, \quad k \in K, i \in I, \tag{18}$$

$$w_{ik} \leq W_i y'_{ik}, \quad k \in K, i \in I, \tag{19}$$

$$e_i \leq s_{ik}, \quad k \in K, i \in I_k, \tag{20}$$

$$s_{ik} + Dw_{ik} \leq l_i + \tau_{ik}, \quad k \in K, i \in I_k, \tag{21}$$

$$s_{ik} + Dw_{ik} + t_{ij}x'_{ijk} - s_{jk} \leq M_{ij}(1 - x'_{ijk}), \\ i \in I_k, j \in I_k \setminus \{v_k\}, i \preccurlyeq_{I_k} j, k \in K, \tag{22}$$

$$\sum_{j \in I_k, j \succ_{I_k} i} x'_{ijk} = \sum_{j \in I_k, i \succ_{I_k} j} x'_{jik} = y'_{ik}, \quad i \in I_k, k \in K, \tag{23}$$

$$x'_{ijk} \in \{0,1\}, \; s_{ik} \geq 0, \; \tau_{ik} \geq 0, \; i,j \in I_k, i \preccurlyeq_{I_k} j, k \in K, \tag{24}$$

$$y'_{ik} \in \{0,1\}, \; w_{ik} \in \mathbb{Z}_{\geq 0}, \; i \in V, k \in K. \tag{25}$$

The number of remaining variables now is by far less than for the problem (2)–(9), (11)–(14). Thus, the subproblem can be solved in a reasonable time to include it in the local search.

3   Neighborhoods

In the last four decades, local search has grown from a simple heuristic idea into a mature field of research in combinatorial optimization [1,21]. Local search is often used to solve NP-hard problems, such as VRP, since it provides a reliable approach for obtaining high-quality solutions to realistic-size problems in a reasonable time. Many neighborhoods for VRPs are introduced and studied from a theoretical and an empirical point of view [5,12,19,23]. Below we present six neighborhoods for the DRRP.

Let $\mu_k = (\mu_0^k, \mu_1^k, \ldots, \mu_{u_k}^k)$ be a sequence of visits in the route of vehicle $k \in K$, where μ_u^k, $u \in \{1, \ldots, u_k\}$ is the uth visited object, $\mu_0^k = v_k$. And let $\widetilde{\mu}_k$ be a sequence of the same length, each element $\widetilde{\mu}_u^k$ of which is equal to the number of wells of object μ_u^k served by the vehicle k, $\widetilde{\mu}_0^k = 0$. Then, the route for each vehicle $k \in K$ can be represented as a sequence $\sigma_k = (\sigma_0^k, \sigma_1^k, \ldots, \sigma_{u_k}^k)$, where σ_u^k corresponds to the pair $(\mu_u^k, \widetilde{\mu}_u^k)$ (Fig. 1).

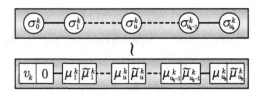

Fig. 1. The sequence corresponding to the route of vehicle k

A feasible solution σ can easily be associated with the corresponding set of sequences $\{\sigma_1, \ldots, \sigma_{|K|}\}$. Now we define the following neighborhood structures for solution σ. Neighborhoods N_1–N_3 and N_5–N_6 are focused on a route improvement and N_4 aims to improve well distribution.

- $N_1(\sigma)$ (Relocate): Relocate a subsequence $(\sigma_i^k,\ldots,\sigma_j^k)$ containing 1 or 2 customer visits. It can be relocated either to the same route or another.
- $N_2(\sigma)$ (Exchange): Exchange two disjoint subsequences $(\sigma_i^k,\ldots,\sigma_j^k)$ and $(\sigma_{i'}^{k'},\ldots,\sigma_{j'}^{k'})$ containing 1 or 2 customer visits (Fig. 2a).
- $N_3(\sigma)$ (Cross): Exchange two visit subsequences $(\sigma_i^k,\ldots,\sigma_{u_k}^k)$, $(\sigma_{i'}^{k'},\ldots,\sigma_{u_{k'}}^{k'})$ involving the extremities of two distinct routes (Fig. 2b). One of these subsequences can be empty.
- $N_4(\sigma)$ (Split): Split one customer visit to visit at the same position but with a smaller number of wells served (possibly zero) and a new visit by a different vehicle that was not assigned to this customer before (Fig. 2c).
- $N_5(\sigma)$ (Kernighan–Lin Cross): Kernighan–Lin version for $N_3(\sigma)$. The idea of this neighborhood structure is similar to the truncated Tabu Search method by one neighborhood, allowing ascents with respect to objective function within [13,14].
- $N_6(\sigma)$ (Kernighan–Lin Exchange and Relocate): Kernighan–Lin version for Exchange and Relocate neighborhood. In this type of neighborhood, we exchange two disjoint subsequences $(\sigma_i^k,\ldots,\sigma_j^k)$ and $(\sigma_{i'}^{k'},\ldots,\sigma_{j'}^{k'})$ containing r and r' customer visits, where $0 \le r, r' \le 12$, but at least one of them should be non-empty.

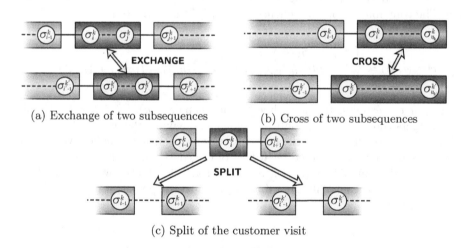

(a) Exchange of two subsequences   (b) Cross of two subsequences

(c) Split of the customer visit

Fig. 2. Neighborhood structures

For all neighborhoods, the resulting sequences are not allowed to contain a customer more than once. One or both subsequences in N_1–N_3 and N_5–N_6 can be reversed if they contain exactly two customer visits.

To reduce the neighborhoods we select a part of them by means of randomization and effective evaluation of the objective function. For the latter purpose, we use subsequence structures proposed in [24], which allow us for each subsequence $(\sigma_i^k, \ldots, \sigma_j^k)$ to get the value of the objective function

$$f(\sigma_k(i,j)) = d(\sigma_k(i,j)) + \gamma\tau(\sigma_k(i,j)) + \lambda\varepsilon(\sigma_k(i,j)).$$

Here d, τ, ε correspond to the distance traveled, the tardiness time, and the earliness time, related to visiting and serving the elements of the subsequence, and γ, λ correspond to the penalties for tardiness and earliness. The concatenation of two subsequences can be calculated in $\mathcal{O}(1)$ and retains all structure properties. This method uses the assumption that vehicles are able to go back in time to avoid late work completion [18]. So we get maximum tardiness instead of accumulated value for the whole route, and value $f(\sigma_k(i,j))$ is an estimation and not the exact value of the objective function. We use notation $L(\sigma)$ for the latter.

It can easily be seen that all moves for neighborhoods N_1–N_6 can be viewed as a separation of routes into subsequences, which are then concatenated into new routes [1,12]. So we can apply the above-mentioned subsequence structures to our neighborhoods. Since the place for subsequence insertion can be seen as an empty subsequence, two subsequences participate in moves for each of the neighborhoods. At least one of them is non-empty. For certainty, we refer to this subsequence as the first and the other one as the second.

To omit some unpromising candidates for the first subsequence we use the contribution of the subsequence to the total penalty for the corresponding vehicle. The contribution of subsequence $(\sigma_i^k, \ldots, \sigma_j^k)$ is estimated as

$$c(\sigma_k(i,j)) = \begin{cases} \sum_{l=i}^{l=j} \gamma\tau_{lk} + \lambda\varepsilon_{lk}, j = u_k, \\ \sum_{l=i}^{l=j} (\gamma\tau_{lk} + \lambda\varepsilon_{lk}) + f(\sigma_k(i,u_k)) - f(\sigma_k(j+1,u_k)) \\ \qquad\qquad - d(\sigma_k(i,j)), j \neq u_k, \end{cases}$$

where ε_{lk} corresponds to the earliness time for pair (l,k), $l \in I, k \in K$ (Fig. 3). For neighborhoods N_1–N_3, only subsequences having the value of contribution $c > 0$ are considered for the first subsequence. In neighborhoods N_5–N_6, the first subsequence is considered regardless of the contribution. For each customer in N_4, it is determined which visit to select from all potential with probabilities proportional to the contribution of these visits to the penalties.

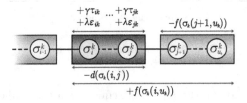

Fig. 3. Contribution of the subsequence to the objective function

If solution σ has no tardiness and waiting times, then, for all neighborhoods, the first subsequence is considered regardless of the contribution.

The insertion place (the second subsequence) is selected in the best way with respect to the objective function estimation from the places suitable for the time window of the first customer in the first subsequence. For neighborhood N_4, it also implies that all options for the split of the wells between two visits are considered.

To make the local search more diverse we use randomization inside neighborhood search. Thus, we consider neighborhoods $N_i^{q_i}(\sigma)$, $0 < q_i < 1$, $i \in \{1, 2, \ldots, 6\}$ which are random parts of the neighborhoods $N_i(\sigma)$. Each element of the set $N_i(\sigma)$ is included in the set $N_i^{q_i}(\sigma)$ with probability q_i independently of other elements. Randomization is applied not only when we select suitable first subsequence, but also when the insertion place is picked.

4 Optimization Method

The solution approach we propose consists in an interactive operation of VNS for the routing part of the problem and mathematical programming method for solving the work redistribution subproblem. Combination of a metaheuristic and mathematical programming techniques often leads to higher quality solutions, and, in recent years, interest on such hybrid metaheuristics has risen considerably in the field of optimization [3,9,16,22]. A similar approach was used in [10] for another routing problem.

4.1 Initial Solution

For a start, we need to create an initial solution σ. Although the VNS method can start from an arbitrary solution, we use a greedy heuristic that consists of three steps: assignment of customers to vehicles, generating a route permutation corresponding to this assignment, and distribution of well-drilling work among vehicles. First, each customer $i \in I$ is assigned to $\lceil 2DW_i/|l_i - e_i| \rceil$ nearest vehicles. Then, we construct routes using nearest-neighbor heuristic with respect to proximity measure (26) between two vertices $v_i, v_j \in V$ assigned to vehicle $k \in K$ [20].

$$\rho(v_i, v_j, k) = d_{ij} + \gamma^E \max\{e_j - s_{ik} - Dw_{ik} - t_{ij}, 0\} \\ + \gamma^T \max\{s_{ik} + Dw_{ik} + t_{ij} + Dw_{jk} - l_j, 0\} \quad (26)$$

This proximity measure considers not only distance but also the earliness and the tardiness time. Coefficients γ^E and γ^T are set to 1 and 0.05 respectively. We choose such values due to the excessive customer assignment and a rough estimation of work distribution. The number of assigned wells w_{ik} initially is set equal for each vehicle serving object i. After route construction, we distribute drilling work by solving approximately the model (15)–(25).

4.2   General Variable Neighborhood Search

Starting from an initial solution, the VNS consists of the following phases: shaking procedure, improvement procedure, and neighborhood change step [9]. In order to distinguish neighborhoods used in shaking and improvement procedures, we use two different notations $\mathcal{N}$ and N, respectively. The pseudo-code of the general VNS algorithm is presented below.

Algorithm 1. General Variable Neighborhood Search

1: **function** GENERAL_VNS(σ, k_{max}, l_{max}, $\mathcal{N}$, N)
2:    **while** stopping criterion is not reached **do**
3:       $k \leftarrow 1$
4:       **while** $k \leq k_{max}$ **do**
5:          $\sigma' \leftarrow$ Shake(σ, k, $\mathcal{N}$)
6:          $\sigma'' \leftarrow$ VND(σ', l_{max}, N)
7:          Neighborhood_change_sequential(σ, σ'', k)
8:    Solve to optimality the work redistribution subproblem w.r.t. the routes of σ
9:    **return** best found solution σ

As the stopping criterion (step 2) in our experiments, we use maximum CPU time allowed to be consumed by the method. In real applications, a more suitable criterion is achieving a predetermined number of iterations without improvement of the best found solution.

Shaking procedure (step 5) diversifies the search and tries to prevent getting stuck in a local optimum. Let $\mathcal{N} = \{\mathcal{N}_1, \ldots, \mathcal{N}_{k_{max}}\}$ be the set of neighborhoods used in the shaking procedure. Function Shake(σ, k, $\mathcal{N}$) consists in a selection of a random solution from $\mathcal{N}_k(\sigma)$. We use $k_{max} = 6$ in our experiments since neighborhoods $\mathcal{N}_k(\sigma)$, $k \geq 7$ are too diversified.

Algorithm 2. Shaking procedure

1: **function** SHAKE(σ, k, $\mathcal{N}$)
2:    **for** $i \in \{1, \ldots, k\}$ **do**
3:       Choose $j \in \{1, 2\}$ at random
4:       Choose $\sigma' \in \mathcal{N}_k^j(\sigma)$ at random
5:       $\sigma \leftarrow \sigma'$
6:    **return** σ

Neighborhood $\mathcal{N}_k^1(\sigma)$ in Algorithm 2 corresponds to $N_4(\sigma)$, and $\mathcal{N}_k^2(\sigma)$ corresponds to the cross-exchange operator, which exchanges two subsequences containing r and r' customer visits, where $1 \leq r \leq k$, $0 \leq r' \leq k$.

We use Variable Neighborhood Descent (VND) as an improvement procedure [9,17]. It also includes solving the subproblem (15)–(25).

Neighborhood change step makes a decision on which neighborhood will be explored next. We use sequential neighborhood change, steps of which are given

---

Algorithm 3. Variable Neighborhood Descent with work redistribution

---
1: **function** VND(σ, $l_{\max}$, N)
2: **repeat**
3:    $stop \leftarrow true$
4:    $l \leftarrow 1$
5:    $\sigma' \leftarrow \sigma$
6:    **repeat**
7:      $\sigma'' \leftarrow \arg\min_{x \in N_l(\sigma)} L(x)$
8:      Neighborhood_change_sequential(σ, σ'', l)
9:    **until** $l = l_{\max}$
10:   **if** $L(\sigma) \neq L(\sigma')$ **then**
11:     $\sigma' \leftarrow \sigma$
12:   Solve the work redistribution subproblem with regard to the routes of σ
13:   **if** $L(\sigma) < L(\sigma')$ **then**
14:     $stop \leftarrow false$
15: **until** $stop = true$
16: **return** σ

---

in Algorithm 4. At the final step of Algorithm 1, we solve the problem (15)–(25) to optimality for routes of the best found solution σ.

Algorithm 4. Sequential neighborhood change step

1: **procedure** NEIGHBORHOOD_CHANGE_SEQUENTIAL(σ, σ', k)
2: **if** $L(\sigma') < L(\sigma)$ **then**
3:    $\sigma \leftarrow \sigma'$
4:    $k \leftarrow 1$
5: **else**
6:    $k \leftarrow k + 1$

---

5   Computational Results

The described VNS algorithm is implemented in C++ with MSVC++ 14.16 compiler, using standard release options. All experiments are conducted on a computer with an AMD Ryzen 5 2600 3.4 GHz processor and 16 GB of RAM running under Microsoft Windows 10 (64-bit).

To test the algorithm we use location and service time data from the instance with 670 customers and 3 depots from our previous paper [14], but with different scaling coefficient for distances. We use this dataset only for the initial experiments presented in this paper. For further research, it is planned to generate other problem-specific datasets. We assign 30 wells to 1% of customers, 20 wells to 5% of customers, 10 wells to 20% of customers, and 5 wells to the rest. The number of days for drilling a well D is set to 2 days. The planning horizon is 365 days, and time windows for the customers are distributed uniformly through this period. A width of the time windows is varied between the value of the service time and this value plus a small random number. Random parts of the instance are selected to get smaller ones. As a result, we generate sets with 10 various instances for each of the following instance cardinalities: 20 customers and 3 vehicles, 50 customers and 6 vehicles, 150 customers and 12 vehicles. We also consider an alteration of these instances, in which time windows are halved through the reduction of the corresponding latest work completion time. Vehicles for such instances are doubled, so their number is 6, 12, and 24.

The matrices (d_{ij}) and (t_{ij}) are the same and consist of the non-negative integers indicating travel distances in days. Vehicles do not have to return to the depot. To take it into account, we modify the matrices (d_{ij}) and (t_{ij}) by the following rule:

$$d_{ij} = t_{ij} = \begin{cases} 0, & \text{if } i \in I, \ j \in V^K, \\ t_{ij}, & \text{otherwise.} \end{cases}$$

For Kernighan–Lin neighborhoods, we generate 25 neighboring solutions. The penalty γ is set to 10 since we aim to find feasible solutions with zero tardiness. The penalty λ used for waiting times in evaluations is set to 0.05. The fractions $q_1, \ldots, q_6$ are set in such a way that all neighborhoods $N_1^{q_1}(\sigma), \ldots, N_6^{q_6}(\sigma)$ have 200 neighbors unless it is less than 5% or more than 60% of the neighborhood.

MILP solver Gurobi (version 9.0) is used to solve the work redistribution subproblem during the search. We pass values corresponding to the incumbent solution for all the decision variables s_{ik}, w_{ik}, τ_{ik}, x'_{ijk}, and y'_{ij} to construct an initial MILP solution. The following Gurobi parameters are used for the subproblem: MIPFocus $= 1$, MIPGap $= 0.03$, TimeLimit $= 0.5$.

In Tables 1–3, we show results for instances of different sizes obtained by our algorithm and Gurobi. To compare the results, the objective function (13) with penalty $\gamma = 10$ is used. The columns τ indicate the related overall tardiness.

First, we test our algorithm on instances with 20 objects (Table 1). For this size, Gurobi is able to find an optimal solution for every instance within 20 min running with default parameters. The computation time of Gurobi varies for different instances due to the structural differences of them, including the number of wells and time windows. Each instance is run 100 times for 2 s for the VNS algorithm, and the columns n_{opt} indicate the number of optimal solutions found. For all the tables, the Gurobi results are underlined if the optimality of the solution is proved, and the best found solutions are shown in bold.

Next, we compare the algorithm with Gurobi on instances with 50 objects (Table 2). Each instance is run 10 times for 90 s for the VNS algorithm. For this size, Gurobi is unable to find optimal solutions in an hour for any instance. But it is able to prove optimality for a few instances if we extend calculation time to 5 h. Also, for instances 1–3 and 6–10 with usual time windows, Gurobi is run for an additional one hour with the passing of the best solution found in 10 runs for constructing an initial MILP solution. None of these best found solutions are improved. We use the following Gurobi parameters for these instances: ImproveStartTime $= 2520$ for initial an hour runs, and MIPFocus $= 1$, Presolve $= 2$, ImproveStartTime $= 1800$ for post-optimization an hour runs.

Table 1. The results for instances with 20 objects

#	Usual time windows, $\|K\|=3$				Halved time windows, $\|K\|=6$						
	Gurobi	VNS			Gurobi		VNS				
		Average	Best	n_{opt}			Average		Best		n_{opt}
					Distance	τ	Distance	τ	Distance	τ	
1	<u>**67**</u>	**67**	**67**	100	<u>**95**</u>	<u>**4**</u>	119.8	2.1	98	4	0
2	<u>**68**</u>	68.24	**68**	99	<u>**112**</u>	<u>**1**</u>	111.7	1.3	**112**	1	57
3	<u>**83**</u>	**83**	**83**	100	<u>**151**</u>	<u>**4**</u>	141	5.3	**151**	4	33
4	<u>**88**</u>	**88**	**88**	100	<u>**131**</u>	<u>**3**</u>	136.1	3.1	134	3	0
5	<u>**115**</u>	115.07	**115**	99	<u>**176**</u>	<u>**3**</u>	170.8	4.8	**176**	3	3
6	<u>**93**</u>	**93**	**93**	100	<u>**133**</u>	<u>**4**</u>	133.3	4	**133**	4	97
7	<u>**49**</u>	49.45	**49**	95	<u>**101**</u>	<u>**1**</u>	103.8	1.1	**101**	1	57
8	<u>**115**</u>	117.11	**115**	46	<u>**195**</u>	<u>**3**</u>	196.9	3.3	205	2	11
9	<u>**66**</u>	66.41	**66**	74	<u>**137**</u>	<u>**3**</u>	132.4	4.2	**137**	3	13
10	<u>**77**</u>	**77**	**77**	100	<u>**152**</u>	<u>**2**</u>	149.9	3.5	**152**	2	3
Gap (%)	**−0.00**	+0.38	**−0.00**	×	**−0.00**		+3.15		+0.41		×

Finally, we run the algorithm on instances with 150 objects (Table 3). Each instance is run 10 times for 300 s for the VNS algorithm. For this size, Gurobi shows poor results in an hour running for every instance if we run it without an initial solution. However, it is capable to improve solutions which we pass as initial. In Table 3, columns "Gurobi post opt" show the results for Gurobi post-optimization an hour runs when we pass the best solution found in 10 runs

Table 2. The results for instances with 50 objects

#	Usual time windows, $\|K\|=6$						Halved time windows, $\|K\|=12$					
	Gurobi		VNS				Gurobi		VNS			
			Average		Best				Average		Best	
	Distance	τ	Distance	τ	Distance	τ	Distance	τ	Distance	τ	Distance	τ
1	160	4	160	4	160	4	242	1	245.8	0.9	243	1
2	170	6	164	6	**159**	6	232	0	232.8	0.8	238	0
3	**73**	3	73.8	3	**73**	3	117	0	117.7	0	**117**	0
4	154	0	155.6	0.3	155	0	223	2	232.1	1.5	224	2
5	148	0	154.7	0	152	0	227	1	236.3	0.4	**227**	1
6	135	45	166.7	37.5	**164**	37	218	1	232.4	0	229	0
7	<u>**118**</u>	<u>**0**</u>	**118**	**0**	**118**	**0**	194	0	202.4	0.3	204	0
8	140	1	128	2.6	140	1	177	0	180.2	0	179	0
9	268	1	246.6	0.7	**237**	0	320	2	342.8	1.7	331	2
10	<u>**121**</u>	<u>**0**</u>	**121**	**0**	**121**	**0**	191	0	191.6	0	**191**	**0**
Gap (%)	−0.00		+0.25		−1.01		−0.00		+2.43		+1.33	

Table 3. The results for instances with 150 objects

	Usual time windows, $	K	= 12$					Halved time windows, $	K	= 24$					
	VNS				Gurobi		VNS				Gurobi	
	Average		Best		post opt		Average		Best		post opt	
#	Distance	τ	Distance	τ	Distance	τ	Distance	τ	Distance	τ	Distance	τ
1	318.5	0	**313**	**0**	**313**	**0**	479.2	1.3	460	1	**440**	**0**
2	306.5	0	**299**	**0**	**299**	**0**	500.7	0.3	484	0	**475**	**0**
3	235.8	0	232	0	**226**	**0**	360.3	0.2	357	0	**350**	**0**
4	355.1	1	**339**	**1**	**339**	**1**	444.6	0.4	439	0	**426**	**0**
5	282	0	**276**	**0**	**276**	**0**	428.1	1.5	**422**	**0**	**422**	**0**
6	330.5	0	**320**	**0**	**320**	**0**	500.7	0.6	500	0	**496**	**0**
7	272.2	3.8	**266**	**4**	**266**	**4**	395.1	1.7	393	1	**392**	**1**
8	272.6	12.1	**261**	**12**	**261**	**12**	408.9	5.1	404	5	**392**	**5**
9	369.7	1.8	**366**	**0**	**366**	**0**	472.5	0	462	0	**455**	**0**
10	377.1	2.3	**359**	**1**	**359**	**1**	514.4	1.2	511	1	**488**	**0**
Gap (%)	+3.50		−0.00		**−0.26**		+2.57		−0.00		**−2.47**	

for constructing an initial MILP solution. We use the same Gurobi parameters for these instances as in experiments for instances with 50 objects. It is worth noting that running the VNS algorithm for an hour improves all the results for the VNS presented in the table, but the results of Gurobi post-optimization runs on the instances with halved time windows are not always improved.

The tables show that our algorithm performs well and finds high-quality solutions, but the quality for the instances with tighter time windows and high sharing of the work can be improved.

6   Conclusion

In this paper, we have studied a new split delivery vehicle routing problem with time windows arising in a real-world context of mobile drilling rig routing. We formulated a well-drilling work distribution subproblem as a mixed-integer linear program (MILP) that turns out to be relatively easy to solve even for medium-sized instances. Thereby, we designed a matheuristic approach, where the solution of the work redistribution subproblem through the MILP is integrated within a VNS algorithm solving the routing part of the overall problem.

One of the new research directions is the optimization of robustness. Real-world routing problems often tend to have uncertainties in travel and service time, especially in long-term planning. In that case, it is important to build solutions that are insensitive to these uncertainties.

References

1. Aarts, E., Lenstra, J.: Local Search in Combinatorial Optimization. John Wiley & Sons, New York (1997)
2. Archetti, C., Speranza, M.G.: Vehicle routing problems with split deliveries. Int. Trans. Oper. Res. **19**(1–2), 3–22 (2012)
3. Archetti, C., Speranza, M.G.: A survey on matheuristics for routing problems. EURO J. Comput. Optim. **2**(4), 223–246 (2014). https://doi.org/10.1007/s13675-014-0030-7
4. Bramel, J., Simchi-Levi, D.: Probabilistic analyses and practical algorithms for the vehicle routing problem with time windows. Oper. Res. **44**(3), 501–509 (1996)
5. Bräysy, O., Gendreau, M.: Vehicle routing problem with time windows, Part I: route construction and local search algorithms. Transp. Sci. **39**(1), 104–118 (2005)
6. Gendreau, M., Tarantilis, C.D.: Solving large-scale vehicle routing problems with time windows: the state-of-the-art. In: CIRRELT-2010-04, Montreal (2010)
7. Golden, B.L., Raghavan, S., Wasil, E.A.: The Vehicle Routing Problem: Latest Advances and New Challenges. Springer, Boston (2008). https://doi.org/10.1007/978-0-387-77778-8
8. Gutin, G., Punnen, A.: The Traveling Salesman Problem and Its Variations. Springer, Boston (2002). https://doi.org/10.1007/b101971
9. Hansen, P., Mladenović, N., Todosijević, R., Hanafi, S.: Variable neighborhood search: basics and variants. EURO J. Comput. Optim. **5**(3), 423–454 (2016). https://doi.org/10.1007/s13675-016-0075-x
10. Hemmelmayr, V.C., Doerner, K.F., Hartl, R.F., Vigo, D.: Models and algorithms for the integrated planning of bin allocation and vehicle routing in solid waste management. Transp. Sci. **48**, 103–120 (2014)
11. Ho, S., Haugland, D.: A tabu search heuristic for the vehicle routing problem with time windows and split deliveries. Comput. Oper. Res. **31**(12), 1947–1964 (2004)
12. Irnich, S.: A unified modeling and solution framework for vehicle routing and local search-based metaheuristics. INFORMS J. Comput. **20**(2), 270–287 (2008)
13. Kernighan, B., Lin, S.: An efficient heuristic procedure for partitioning graphs. Bell Syst. Tech. J. **49**(2), 291–307 (1970)
14. Kulachenko, I., Kononova, P.: The VNS approach for a consistent capacitated vehicle routing problem under the shift length constraints. In: Bykadorov, I., Strusevich, V., Tchemisova, T. (eds.) MOTOR 2019. CCIS, vol. 1090, pp. 51–67. Springer, Cham (2019). https://doi.org/10.1007/978-3-030-33394-2_5
15. Li, F., Golden, B., Wasil, E.: The open vehicle routing problem: algorithms, large-scale test problems, and computational results. Comput. Oper. Res. **34**(10), 2918–2930 (2007)
16. Maniezzo, V., Stützle, T., Voß, S.: Matheuristics: Hybridizing Metaheuristics and Mathematical Programming. Springer, Boston (2009). https://doi.org/10.1007/978-1-4419-1306-7
17. Mladenovic, N., Hansen, P.: Variable neighborhood search. Comput. Oper. Res. **24**, 1097–1100 (1997)
18. Nagata, Y., Bräysy, O., Dullaert, W.: A penalty-based edge assembly memetic algorithm for the vehicle routing problem with time windows. Comput. Oper. Res. **37**(4), 724–737 (2010)
19. Savelsbergh, M.W.P.: The vehicle routing problem with time windows: minimizing route duration. INFORMS J. Comput. **4**, 146–154 (1992)

20. Solomon, M.M.: Algorithms for the vehicle routing and scheduling problems with time window constraints. Oper. Res. **35**, 254–265 (1985)
21. Talbi, E.G.: Metaheuristics: From Design to Implementation. Wiley, Hoboken (2009)
22. Talbi, E.G.: Hybrid Metaheuristics. Springer, Berlin (2013). https://doi.org/10.1007/978-3-642-30671-6
23. Toth, P., Vigo, D. (eds.): Vehicle Routing: Problems, Methods, and Applications, 2nd edn. Society for Industrial and Applied Mathematics, Philadelphia (2014)
24. Vidal, T., Crainic, T.G., Gendreau, M., Prins, C.: A hybrid genetic algorithm with adaptive diversity management for a large class of vehicle routing problems with time-windows. Comput. Oper. Res. **40**(1), 475–489 (2013)
25. Yakici, E., Karasakal, O.: A min-max vehicle routing problem with split delivery and heterogeneous demand. Optim. Lett. **7**, 1611–1625 (2013)

Locating Facilities Under Deliberate Disruptive Attacks

Anton V. Ushakov$^{(\boxtimes)}$ and Igor Vasilyev

Matrosov Institute for System Dynamics and Control Theory of SB RAS,
134 Lermontov str., 664033 Irkutsk, Russia
{aushakov,vil}@icc.ru
http://iv.icc.ru

Abstract. Facility disruptions or failures may occur due to natural disasters or a deliberate man-made attack. Such an attack is known as interdiction. Recently, facility location problems, addressing intentional strikes against operating facilities and strategies to reduce their impact, have received particular attention. In this paper, we present a new location-interdiction median problem aimed at designing a distribution network which is robust to the worst-case, long-term facility losses. We suppose that there are two players: defender (system designer) and attacker. The defender decides where to locate facilities to minimize the overall cost of supplying the demands of customers. The attacker determines which r facilities to interdict to maximize the cost of serving the customers from the remaining operational facilities. Note that we suppose that the facilities are attacked simultaneously and interdicted facilities become unavailable. We propose bilevel and single-level integer formulations of this problem. For a particular case when the attacker hits a single facility, we develop a fast local search procedure based on implicit enumeration of interdiction strategies. We test our approaches in a series of computational experiments on well-known test problems.

Keywords: Interdiction · Bilevel programming · p-median problem · Facility location · Disruption · Integer programming

1 Introduction

Facility location problems are ones of the most widely studied problems in combinatorial optimization and integer programming. Facilities and the corresponding connecting infrastructure are the main ingredients in any production, distribution, or service system and supply chains [29]. A traditional approach to the design of such systems assumes that all their components always remain operable. However, some accidents, like natural disasters or man-made attacks, can substantially reduce the system efficiency and even make it incapacitated. For example, the current coronavirus outbreak is disrupting the global supply chains from tech industry (production of cell phones and electronics) to fashion and car

© Springer Nature Switzerland AG 2020
A. Kononov et al. (Eds.): MOTOR 2020, LNCS 12095, pp. 359–372, 2020.
https://doi.org/10.1007/978-3-030-49988-4_25

manufacturing. Hyundai and Nissan announced that they would stop manufacturing in South Korea, Europe, and the US because of a lack of parts from China [5,8]. Meanwhile, Tesla considers to delay the production in Shanghai for at least one week. The disruptions can also be man-made. For example, the current French pension reform strike is the longest French strike over the last 50 years, which results in billion losses.

Given a service system, one may expect two main types of disruptions: facility disruptions (loss of production capabilities) and failures of links connecting facilities and customers. In the field of facility location, reliability and sustainability issues of service systems are extensively studied in the scope of the so-called *reliable facility location*. Such kind of models usually assume that the system components have some inherent probability of failure. Thus, the aim is to maximize the expected system efficiency, e.g. minimize the expected service cost in case of possible failures. The stochastic approach may be quite useful when modeling sudden losses of components due to natural disasters or some inherent system failures. However, disruptions may be caused not only by natural disasters (like earthquakes or fire) but also by disruptive man-made actions (sabotage, deliberate attacks). Intentional attack on a supply network is also called interdiction [11]. Interdiction models are aimed at identifying critical infrastructure elements such that the loss of them due to an attack causes the maximum possible harm and makes a system much less efficient. Originally, such models were mostly studied in military planning applications. Indeed, it is natural to identify the places for interdicting enemies supply lines in order to inflict the maximal damage. The first interdiction problem was considered in [22] where the aim of interdiction was to disrupt enemy supply flows with a limited budget available [35]. Note that most of the military-induced applications are focused on interdicting links of transportation networks (e.g. shortest paths) [18].

Recently, the interest has revived for interdiction problems due to fast growing threat of terrorist attacks (especially, after the September 11 attacks). Nowadays, we are witnessing how vulnerable may be facilities to such threats. International counter-terrorism efforts are forcing the terrorist groups, like al Qaeda, transfer their attention from so-called hard targets (embassies and military assets) to soft targets (civilian facilities). The main goal of such attacks is to engender the largest possible harm. For a service system, the harm is viewed as the loss of service coverage or efficiency (increase in the overall service costs). The first facility interdiction models were addressed in [11] where the r-interdiction median and covering problems were proposed. For example, the r-interdiction median problem is to find which r existing facilities to remove in order to decrease the efficiency of the existing supply system the most.

The next research in this field was focused on developing strategies of hedging against the worst-case attacks. Thus, in [10], the authors extend the r-interdiction problem by the possibility of fortifying (protecting) some existing facilities so that they become immune to attacks. The problem is then modeled as a bilevel integer program involving two players: attacker and defender [27]. This problem has received great attention and a number of effective solution algorithms has been developed. Among most effective are the implicit enumeration algorithm [27], an exact approach based on reformulating the problem as the maximal covering problem [28, 36], and a branch-and-cut algorithm from [26].

Numerous papers addressing various bilevel facility location protection-interdiction problems have been published since then, e.g. hub interdiction median problems [15, 24], the stochastic r-interdiction median problem with fortification [19], a r-interdiction median problem involving a defender's budget for fortification and penalties for serving the demands of customers from interdicted facilities [4], a partial interdiction model where some demand after interdiction can be outsourced at some cost [1], an interdiction problem with hierarchical system of facilities [7], a more general median protection/interdiction problem that involves outsourcing, capacitated nested hierarchical facilities, and the budget of interdiction [14], etc.

The aforementioned fortification-interdiction models assume that there is an existing service system and the goal is to thwart intentional attacks by protecting some facilities. Another approach consists in taking into account possible worst-case facility losses in the initial location of facilities. Such problems, known as location-interdiction models [29], are not widely addressed in the literature. For example, in [23], the authors extend the r-interdiction covering problem from [11] and introduce a maximal covering location-interdiction problem aimed at maximizing the coverage of customers if the most critical facilities are failed. Note that location-interdiction problems are more difficult to solve than the protection-interdiction ones and most of effective algorithms developed for protection problems are not applicable to them [29]. Other closely related problems are aimed at combining both location and protection decisions. An example is the model proposed in [2], where capacitated facilities may be located either protected or unprotected. In [3] the problem is to first locate a given number of facility followed by protecting some of them with a limited protection budget.

In this paper we propose a new facility location-interdiction model by extending the r-interdiction median problem in the way closely related to [23]. We formulate this problem as a bilevel integer linear program where the upper-level player (defender) finds p sites for locating facilities in order to minimize the overall cost of serving customers, taking into account that the lower-level player (attacker) hits some r facilities in order to inflict the maximal possible damage. We demonstrate that this problem can be reformulated as a single-level integer program so that small- and medium size problem instances can be solved with a general purpose commercial solver. For a particular case when the attacker hits a single open facility, we develop a fast local search heuristic.

2 Bilevel Facility Location-Interdiction Problem

Our model relies upon the uncapacitated p-median problem. It is one of the basic and widely studied facility location problems. It has been applied in designing service and distribution systems in private and public sectors [20]. It is also a powerful modeling tool with applications to production [9, 21] and machine learning [16, 30].

We suppose that there is a set I ($|I| = m$) of potential facility location sites and a set J ($|J| = n$) of customers. We denote as d_{ij} the shortest distance (service cost) between a customer j and the facility located at cite i. As for the p-median problem, each customer is served by the closest open facility. If for a customer, the closest open facility is interdicted, she/he is reassigned to the closest open non-interdicted facility. The interdicted facilities become completely unavailable. As the attacker has some limited resources for interdiction, we assume that only r out of p open facilities can be attacked. Note that we suppose that the attacker has perfect information about where the facilities are located, hence he/she always hits only open facilities. We also suppose that all r facilities chosen by the interdictor are attacked simultaneously, thus the defender has no way to thwart any losses.

Our problem can be viewed as a static Stackelberg game where two players make sequential decisions. The first player (leader or defender) first decides where to locate p facilities in order to minimize the overall sum of distances between customers and their closest open facilities, taking into account that some facilities may be interdicted; the second player (follower or attacker), knowing where the first player has located the facilities, tries to inflict the maximal possible harm to the service system by attacking r of them. Thus, some customers have to be reassigned to more distant facilities, which reduces the system efficiency.

Let us introduce the following decision variables:

$$y_i = \begin{cases} 1, & \text{if a facility is located at site i;} \\ 0, & \text{otherwise.} \end{cases}$$

$$x_{ij} = \begin{cases} 1, & \text{if a customer j is assigned to the facility located at site i;} \\ 0, & \text{otherwise,} \end{cases}$$

$$s_i = \begin{cases} 1, & \text{if a facility located at site i is interdicted;} \\ 0, & \text{otherwise,} \end{cases}$$

Furthermore, to ensure the assignment of customers to the closest facilities, we define the set $W_{ij} = \{k \in I : d_{kj} > d_{ij}\}$, i.e. the set of potential sites which are farther from customer j than site i. The bilevel interdiction-location median problem can be written as follows:

$$\min \sum_{i \in I} \sum_{j \in J} d_{ij} x_{ij}^*(y), \tag{1}$$

$$\sum_{i \in I} y_i = p, \tag{2}$$

$$y_i \in \{0, 1\} \qquad\qquad \forall\, i \in I, \tag{3}$$

where $x_{ij}^*(y)$ is a solution of the attacker's problem:

$$\max \sum_{i \in I} \sum_{j \in J} d_{ij} x_{ij}, \tag{4}$$

$$\sum_{i \in I} x_{ij} = 1 \qquad\qquad \forall\, j \in J, \tag{5}$$

$$x_{ij} \leq y_i \qquad\qquad \forall\, i \in I,\, j \in J, \tag{6}$$

$$\sum_{i \in I} s_i = r, \tag{7}$$

$$s_i \leq y_i \qquad\qquad \forall\, i \in I, \tag{8}$$

$$\sum_{k \in W_{ij}} x_{kj} \leq 1 + s_i - y_i \qquad\qquad \forall\, i \in I,\, j \in J, \tag{9}$$

$$s_i, x_{ij} \in \{0,\, 1\} \qquad\qquad \forall\, i \in I,\, j \in J. \tag{10}$$

Note that the defender and the attacker has the opposite objectives. The system planner attempts to locate p facilities to minimize the overall sum of distances between customers and the facilities remained after eliminating r of them. The interdictor determines which open facilities to hit in order to maximize the cost of serving customers. Constraints (2) set the number of facilities opened by the defender. Constraints (5) and (6) are the standard facility location constraints guaranteeing that each customer is served by exactly one open facility. Constraints (7) and (8) impose that the attacker can hit only r out of p open facilities. Finally, constraints (9) ensure that each customer is served from the closest open facility after attack occurs. Indeed, if a facility is located at site i and not interdicted, a customer j cannot be served by a more distant facility than i. As was noted in [27], these constraints can also be adapted to settings where customers are not supposed to be served from the closest open facilities. Instead, they may have some facilities preference ordering known for both the defender and the attacker (e.g. see [31–33]).

Bilevel programming problems are usually very hard to solve, especially when the problems on both levels are mixed-integer programs. In facility location, bilevel problems usually arise in competitive location when there are two players that compete for serving the demand of customers from a market [17]. A good example of competitive facility location problems is the well-known $(r|p)$-centroid problem [6,12]. This problem is known to be Σ_2^P hard [13].

As was noted above, the location-interdiction problems are much more challenging than widely studied fortification-interdiction problems. Their structure hinders from application some very efficient exact approaches based on implicit enumeration and decomposition techniques.

3   Integer Programming Formulation

In this section we demonstrate that the introduced location-interdiction median problem can be formulated as a single-level integer program. Thus, it can be solved by a commercial branch-and-bound solver.

For our formulation, we denote by H the set of all interdiction patterns, i.e. it consists of all possible strategies of interdicting r out of m potential location sites. We also denote by h the index corresponding to a specific interdiction strategy and by I_h the set of interdicted sites in pattern h. Our formulation uses the variables y_i that have the same definition as those in formulation (1)–(10). We also introduce the following binary variables:

$$x_{ij}^h = \begin{cases} 1, & \text{if a customer } j \text{ is assigned to the facility located at site } i \\ & \text{when interdiction pattern h occurs;} \\ 0, & \text{otherwise,} \end{cases}$$

Thus, we can formulate our problem as follows:

$$\min z \tag{11}$$

$$z \geq \sum_{i \in I} \sum_{j \in J} d_{ij} x_{ij}^h \qquad\qquad h \in H, \tag{12}$$

$$\sum_{i \in I} x_{ij}^h = 1 \qquad\qquad \forall j \in J,\ h \in H, \tag{13}$$

$$x_{ij}^h \leq y_i \qquad\qquad \forall i \in I,\ j \in J,\ h \in H, \tag{14}$$

$$\sum_{i \in I} y_i = p, \tag{15}$$

$$\sum_{i \in I_h} \sum_{j \in J} x_{ij}^h = 0, \qquad\qquad \forall h \in H, \tag{16}$$

$$y_i \in \{0, 1\} \qquad\qquad \forall i \in I, \tag{17}$$

$$x_{ij}^h \in \{0, 1\} \qquad\qquad \forall i \in I,\ j \in J,\ h \in H. \tag{18}$$

The objective function (11) is to minimize the overall cost z of serving customers from operational facilities remained after the worst-case attack on r potential location sites. Constraints (13)–(14) have the same definition as those in formulation (1)–(10), i.e. they ensure that each customer j must be assigned only to one open facility when interdiction pattern h occurs. Constraints (12) define the cost of serving customers from a given set of facilities if r sites corresponding to pattern h are lost. Note that given a set of facilities, if none of them are located at the sites from interdiction pattern h, then the cost z is equal to the p-median objective value. Finally, constraints (16) ensure that customers cannot be served by facilities located at the sites interdicted in pattern h.

Note that the model actually involves the exponential number of variables and constraints. However, the common assumption is that terrorist groups usually have limited resources and are able to simultaneously hit only a small number of facilities [27]. In this case, one can determine all possible $\binom{m}{r}$ interdiction patterns in advance. In particular, if only one facility is supposed to be attacked, the model contains $\mathcal{O}(m^2 n)$ decision variables and constraints.

4   Local Search

Bilevel integer problems are quite challenging for exact methods, especially in case of real-life instances of large size. Thus, the development of heuristics to find quality feasible solutions in reasonable time is quite natural. The main idea is to use a local search procedure which attempts to improve an initial solution by searching its neighbors. The location-interdiction median problem is based on the well-known p-median problem, hence the simplest and natural strategy is to develop a local search procedure over the SWAP neighborhood. It consists of all defender's solutions which are obtained from a current incumbent by closing one facility and opening another one in a different place.

In order to compute the defender's objective value for a given set of open facilities (the set of variables y_i), we actually have to find an optimal solution of the attacker's problem:

$$\max \sum_{i: y_i=1} \sum_{j \in J} d_{ij} x_{ij}, \tag{19}$$

$$\sum_{i:\ y_i=1} x_{ij} = 1, \qquad\qquad j \in J, \tag{20}$$

$$\sum_{k \in T_{ij}(y)} x_{kj} \leq s_i, \qquad\qquad i \in I:\ y_i = 1,\ j \in J, \tag{21}$$

$$\sum_{i:\ y_i=1} s_i = r, \tag{22}$$

$$s_i,\ x_{ij} \in \{0,\ 1\}, \qquad\qquad i \in I:\ y_i = 1, j \in J, \tag{23}$$

where $T_{ij}(y)$ is the set of open defender's facilities which are farther from customer j than facility i. This problem can be solved by a commercial branch-and-bound solver. On the other hand, in some applications where the numbers of both open and interdicted facilities are small, the attacker's problem can effectively be solved by enumerating all $\binom{p}{r}$ possible facility losses.

In this section we focus on a special case of the attacker's problem where $r = 1$. Our main goal is to develop a local search procedure which is faster than the naive implementation based on explicit enumeration.

Throughout this section we use the following notations. We denote as $d(i, j)$ the distance (service cost) between a customer j and a location i. We suppose that S is a solution of the defender's problem, i.e. S is any subset of I consisting of p elements. The closest open facility in S for customer j is denoted as $c_1(j)$. To perform a fast interchange, our algorithm needs the second and the third open facilities to j that we denote as $c_2(j)$ and $c_3(j)$, respectively. Searching the swap neighborhood, the algorithm picks one candidate facility to add to the current solution and one to drop from it that we refer to as f_i and f_r, respectively. Following [25], we assume that the distance between any facility site and any customer can be computed in $\mathcal{O}(1)$ time. For example, this is the case when the distance matrix is already calculated.

Note that for the p-median problem, there are several fast implementations of the swap-based local search, including prominent Whitaker's implementation [34] and a fast implementation from [25].

Given a solution S, we need to find a neighbor $\hat{S}$ providing a better value of the defender's objective or to check that such an improving neighbor does not exist. Thus, we have p sites for dropping and $m - p$ sites for insertion. Given a pair (f_i, f_r), we have to compute the gain of swapping f_r with f_i. A straightforward implementation assumes that, for each possible swap (f_i, f_r), we have to solve the attacker's problem, i.e. to enumerate all possible strategies of interdicting facilities in the solution obtained after swapping. If $r = 1$, the number of interdiction patterns is equal to p. If the three closest open facilities are not pre-computed and the calculation of distances takes $\mathcal{O}(1)$ time, determining the assignment of all customers under each interdiction pattern takes $\mathcal{O}(pn)$ time. Since the total number of swaps is equal to $p(m - p) = \mathcal{O}(pm)$ and the number of interdiction patterns for each possible swap is p, one iteration of local search takes $\mathcal{O}(p^3 mn)$ time. Suppose now that the three closest facilities to each customer j are found: this takes $\mathcal{O}(pn)$ time. In this case, one iteration of local search can be performed in $\mathcal{O}(p^2 mn)$ time.

However, the computation time can be further reduced by avoiding the explicit enumeration needed to find an optimal solution of the attacker's problem. Indeed, the value of the defender's objective function z^{lead} for a solution S can be written as follows:

$$z^{lead} = z^{med} + \max_{s \in S} \sum_{j : c_1(j) = s} d(c_2(j), j) - d(c_1(j), j), \tag{24}$$

where z^{med} is the value of the p-median objective. In other words, the cost that the defender will get after a disruptive strike can be computed as the sum of costs for serving customers from the closest open facilities plus the cost induced by the customers whose closest facility was lost due to attack. Thus, for each pair (f_i, f_r), we must compute the gain from swapping f_i and f_r according to both the p-median objective and the value induced by attack:

$$cost(f_i, f_r) = medcost(f_i, f_r) + losscost(f_i, f_r), \tag{25}$$

where $medcost(f_i, f_r)$ is computed as follows:

$$
\begin{aligned}
medcost(f_i, f_r) = &\sum_{j : c_1(j) \neq f_r} \min\{0, d(f_i, j) - d(c_1(j), j)\} \\
&+ \sum_{j : c_1(j) = f_r} (\min\{d(c_2(j), j), d(f_i, j)\} - d(c_1(j), j)).
\end{aligned}
\tag{26}
$$

In other words, if the closest facility of a customer j is not replaced (it is not f_r), the customer either reassigns to f_i, if it is closer, or stays assigned to the current facility. If customer j is served from f_r, then he/she assigns to the closest open facility: f_i or $c_2(j)$. If $medcost(f_i, f_r) < 0$, then we can gain from such a swap with respect to the p-median objective.

The value of $losscost(f_i, f_r)$ can be computed in a similar way (see Algorithm 1).

Algorithm 1 Finding an optimal solution of the attacker's problem when assessing the candidate pair (f_i, f_r)

1: **for all** $(i \in S)$ $closed(i) := 0$
2: **for all** $j \in J$ **do**
3:      **if** $c_1(j) = f_r$ **then**
4:          **if** $d(f_i, j) < d(c_2(j), j)$ **then** $closed(f_i) + = d(c_2(j), j) - d(f_i, j)$
5:          **else** $closed(c_2(j)) + = \min\{d(c_3(j), j), d(f_i, j)\} - d(c_2(j), j)$
6:          **end if**
7:      **else**
8:          **if** $d(f_i, j) < d(c_1(j), j)$ **then** $closed(f_i) + = d(c_1(j), j) - d(f_i, j)$
9:          **else if** $d(f_i, j) > d(c_1(j), j)$ and $d(f_i, j) < d(c_2(j), j)$ **then**
10:              $closed(c_1(j)) + = d(f_i, j) - d(c_1(j), j)$
11:          **else**
12:              **if** $c_2(j) = f_r$ **then** $closed(c_1(j)) + = \min\{d(f_i, j), d(c_3(j), j)\} - d(c_1(j), j)$
13:                  **else** $closed(c_1(j)) + = d(c_2(j), j) - d(c_1(j), j)$
14:              **end if**
15:          **end if**
16:      **end if**
17: **end for**
18: **return** $s^* = \operatorname{argmax}_{i \in S \cup \{f_i\} \setminus \{f_r\}} closed(i)$,
       $losscost(f_i, f_r) := closed(s^*) - \max_{s \in S} \sum_{j : c_1(j) = s} d(c_2(j), j) - d(c_1(j), j)$

The main idea is to implicitly calculate the costs of all interdiction patterns, considering each customer j independently. Note that the attacker can cause a harm to customer j if he hits the closest open facility to j.

If the closest facility to j is not dropped, f_i may be either the closest to j or the second closest. In the first case, the attacker may inflict damage only by hitting f_i, hence j remains assigned to the same facility (she/he was assigned before swapping). In the second case, the attacker striking the closest facility $c_1(j)$ forces the customer be served from the next facility: f_i. There is also another option. The candidate f_i may also be farther than the second closest facility $c_2(j)$. In this case, if $c_2(j)$ is dropped, the attacker engenders a harm by hitting the closest facility $c_1(j)$ and forcing the customer assign to the third closest facility $c_3(j)$. Otherwise, if $c_2(j)$ is not removed, the customer assigns to it in case of attack.

If the closest facility $c_1(j)$ is dropped, there are two options. The new facility f_i is closer than $c_2(j)$, thus an attack on f_i results in assigning the customer to the second closest facility $c_2(j)$. If the new facility is farther than $c_2(j)$, the customer in case of attack on $c_2(j)$ would be served from the third closest facility: $c_3(j)$ or f_i.

In all cases described above, for each customer j, only an attack on the closest facility (after swapping) can leads to reassigning the customer to the next closest facility. Thus, *closed* in Algorithm 1 accumulates the contribution of each particular customer to the cost of all possible interdiction patterns.

The values of $medcost(f_i, f_r)$ and $losscost(f_i, f_r)$ can obviously be computed in $\mathcal{O}(n)$ time for each candidate pair. As the total number of swaps is $p(m - p)$, each iteration of local search takes $\mathcal{O}(pmn)$ time. Note that the standard implementation of swap-based local search for the p-median problem requires the same time per iteration. After a swap is accepted, updating the closest facilities for each customer requires $\mathcal{O}(pn)$ time in the worst case.

5   Computational Experiments

In this section we report some computational experiments to test the proposed integer linear program as well as the local search procedure devised. We utilized several widely used test problem instances from the well-known TSPLIB[1]. Our integer linear model was set up using IBM ILOG Concert technology and C++ programming language. It was then solved with IBM ILOG CPLEX 12.8.0[2], freely available for non-commercial research. We implemented the local search procedure using C++ programming language as well. Note that our local search is not parallel, while CPLEX was allowed to use four parallel threads. We suppose that the number of interdictions r is equal to 1. The initial solutions for the local search were chosen at random. We performed only one run of the local search on each particular test problem.

The results are presented in Table 1, where the first column shows the problem name, column p contains the number of defender's facilities to be located. In *Objective values*, we report three following values. Column *Pmed* shows the optimal value of the p-median objective, i.e. the service cost without the threat of possible attacks. Columns *IntrCPLX* and *LS* demonstrate the values of the defender's objective function found by CPLEX and the local search, respectively. Columns $Time_{cplx}$ and $Time_{LS}$ represent the running times of CPLEX and the local search.

We can observe that CPLEX requires very much time to find an optimal solution of the single-level location-interdiction median problem. For example, CPLEX spent more than almost two hours to solve *berlin52* instance involving 52 customers and 5 facilities. Recall that the single-level problem contains very large number of variables and constraints, which makes it much harder than the underlying p-median problem.

Observe that the found objective values of the location-interdiction problem are in general much larger than those of the p-median problem, especially when the number of facilities is small. For example, for the problem *berlin52*, incorporating risks of a possible deliberate attack when locating 3 facilities results in increasing the service cost for over 27% with respect to the cost of the location strategy that does not assume facility losses. Thus, if the system planner is risk-averse, she may attempt to mitigate the impact of an attack but it may be quite

---

[1] http://elib.zib.de/pub/mp-testdata/tsp/tsplib.
[2] https://www.ibm.com/products/ilog-cplex-optimization-studio.

Table 1. Results on the TSPLIB problem instances, where p is the number of facilities opened by the defender, the number of interdictions $r = 1$.

Problem	m	p	Pmed	IntrCPLX	LS	$Time_{cplx}$	$Time_{LS}$
			Objective values				
ulysses22	22	2	70.02	123.47	123.47	7.1	0
		3	51.55	109.74	109.74	12.5	0
		4	42.57	67.54	67.54	9.5	0
		5	35.30	54.13	54.13	4.92	0
bayg29	29	2	13601.85	19469.92	19469.92	43.5	0
		3	10446.36	15343.86	16731.25	57.0	0
		4	9013.30	12180.74	12708.10	57.5	0
		5	7735.65	10239.33	10239.33	55.7	0.001
att48	48	2	72921.90	113222.34	113222.34	1262.3	0.001
		3	54623.16	85741.06	86337.62	2641.2	0.001
		5	39679.31	53308.19	53308.19	3854.7	0.001
		8	29159.05	35615.89	35615.89	2864.9	0.002
berlin52	52	2	14816.78	20000.29	20000.29	2825.8	0.001
		3	12057.82	16604.35	17599.84	4081.5	0.001
		5	8888.74	12158.69	12199.40	7125.8	0.002
		8	6402.17	8081.50	8080.88	5182.7	0.002
fl1400	1400	50	29090.23	–	31627.20	–	80.6
		100	16552.22	–	17500.34	–	264.3
		200	9355.34	–	10230.72	–	654.6

costly. Note that in case when 8 facilities are located, the increase in service cost is about 20%. However, for the problem $fl1400$, the increase is only 9% if 200 facilities are located. Thus, when the number of open facilities is small, the system planner may prefer a risky strategy (locating facilities according to the p-median protocol) rather than the robust design of a service system resulting in increase in the service cost. On the other hand, when the number of facilities is relatively large, a robust system can be designed with small increase in the service cost.

We see that the local search procedure found the same solutions as CPLEX in most cases (especially for small problem instances). Recall that we ran the local search only once. For $berlin52$, $p = 8$, it found a solution that even a little better. Note that this may happen due to a relatively large duality gap inherent in the single-level integer program.

Observe that the proposed heuristic is very fast, especially in comparison with the commercial solver. It spent less than one second to find very close to optimal solutions.

6   Conclusion

In this paper we proposed a new location-interdiction median problem based on the well-known p-median problem. This problem can naturally be formulated as a bilevel integer linear program where the system designer makes decision about where to locate facilities; and the attacker interdicts some facilities to engender the maximal harm to the system efficiency. We demonstrated that the problem can be formulated as a single-level integer linear problem that can be solved with a commercial solver. Since it can be utilized to solving only small problem instances, even assuming that only one facility is interdicted, we developed a fast implementation of local search over the basic SWAP-neighborhood.

Our further research may be focused on developing alternative formulations of the problem as well as exact solution approaches (branch-and-cut methods). Our research will also aim at extending the proposed technique for larger values of r and developing new heuristics. Finally, the developed local search procedures may be employed to devise various metaheuristics.

Acknowledgement. The reported study was funded by RFBR according to the research project No. 18-07-01037.

References

1. Aksen, D., Akca, S.S., Aras, N.: A bilevel partial interdiction problem with capacitated facilities and demand outsourcing. Comput. Oper. Res. **41**, 346–358 (2014). https://doi.org/10.1016/j.cor.2012.08.013
2. Aksen, D., Aras, N.: A bilevel fixed charge location model for facilities under imminent attack. Comput. Oper. Res. **39**(7), 1364–1381 (2012). https://doi.org/10.1016/j.cor.2011.08.006
3. Aksen, D., Aras, N., Piyade, N.: A bilevel p-median model for the planning and protection of critical facilities. J. Heuristics **19**(2), 373–398 (2013). https://doi.org/10.1007/s10732-011-9163-5
4. Aksen, D., Piyade, N., Aras, N.: The budget constrained r-interdiction median problem with capacity expansion. Cent. Eur. J. Oper. Res. **18**, 269–291 (2010). https://doi.org/10.1007/s10100-009-0110-6
5. Al Jazeera: Weakest link: Global supply chains disrupted by coronavirus (2020). https://www.aljazeera.com/ajimpact/weakest-link-global-supply-chains-disrupted-due-coronavirus-200212033618174.html
6. Alekseeva, E., Kochetov, Y., Plyasunov, A.: An exact method for the discrete $(r|p)$-centroid problem. J. Glob. Optim. **63**(3), 445–460 (2013). https://doi.org/10.1007/s10898-013-0130-6
7. Aliakbarian, N., Dehghanian, F., Salari, M.: A bi-level programming model for protection of hierarchical facilities under imminent attacks. Comput. Oper. Res. **64**, 210–224 (2015). https://doi.org/10.1016/j.cor.2015.05.016
8. Automative News Europe: Nissan faces global parts shortage due to coronavirus, report says (2020). https://europe.autonews.com/automakers/nissan-faces-global-parts-shortage-due-coronavirus-report-says

9. Avella, P., Boccia, M., Martino, C.D., Oliviero, G., Sforza, A., Vasilyev, I.: A decomposition approach for a very large scale optimal diversity management problem. 4OR **3**(1), 23–37 (2005). https://doi.org/10.1007/s10288-004-0059-1

10. Church, R.L., Scaparra, M.P.: Protecting critical assets: the r-interdiction median problem with fortification. Geogr. Anal. **39**(2), 129–146 (2007). https://doi.org/10.1111/j.1538-4632.2007.00698.x

11. Church, R.L., Scaparra, M.P., Middleton, R.S.: Identifying critical infrastructure: the median and covering facility interdiction problems. Ann. Am. Assoc. Geogr. **94**(3), 491–502 (2004). https://doi.org/10.1111/j.1467-8306.2004.00410.x

12. Davydov, I., Kochetov, Y., Carrizosa, E.: A local search heuristic for the $(r|p)$-centroid problem in the plane. Comput. Oper. Res. **52**, 334–340 (2014). https://doi.org/10.1016/j.cor.2013.05.003. Recent advances in Variable neighborhood search

13. Davydov, I., Kochetov, Y., Plyasunov, A.: On the complexity of the $(r—p)$-centroid problem in the plane. TOP **22**(2), 614–623 (2013). https://doi.org/10.1007/s11750-013-0275-y

14. Forghani, A., Dehghanian, F., Salari, M., Ghiami, Y.: A bi-level model and solution methods for partial interdiction problem on capacitated hierarchical facilities. Comput. Oper. Res. **114**, 104831 (2020). https://doi.org/10.1016/j.cor.2019.104831

15. Ghaffarinasab, N., Atayi, R.: An implicit enumeration algorithm for the hub interdiction median problem with fortification. Eur. J. Oper. Res. **267**(1), 23–39 (2018). https://doi.org/10.1016/j.ejor.2017.11.035

16. Hansen, P., Jaumard, B.: Cluster analysis and mathematical programming. Math. Program. **79**(1–3), 191–215 (1997)

17. Iellamo, S., Alekseeva, E., Chen, L., Coupechoux, M., Kochetov, Y.: Competitive location in cognitive radio networks. 4OR **13**(1), 81–110 (2014). https://doi.org/10.1007/s10288-014-0268-1

18. Israeli, E., Wood, R.K.: Shortest-path network interdiction. Networks **40**(2), 97–111 (2002). https://doi.org/10.1002/net.10039

19. Liberatore, F., Scaparra, M.P., Daskin, M.S.: Analysis of facility protection strategies against an uncertain number of attacks: the stochastic r-interdiction median problem with fortification. Comput. Oper. Res. **38**(1), 357–366 (2011). https://doi.org/10.1016/j.cor.2010.06.002

20. Marianov, V., Serra, D.: Median problems in networks. In: Eiselt, H.A., Marianov, V. (eds.) Foundations of Location Analysis. ISORMS, vol. 155, pp. 39–59. Springer, New York (2011). https://doi.org/10.1007/978-1-4419-7572-0_3

21. Masone, A., Sterle, C., Vasilyev, I., Ushakov, A.: A three-stage p-median based exact method for the optimal diversity management problem. Networks **74**(2), 174–189 (2019). https://doi.org/10.1002/net.21821

22. McMasters, A.W., Mustin, T.M.: Optimal interdiction of a supply network. Nav. Res. Logist. **17**(3), 261–268 (1970). https://doi.org/10.1002/nav.3800170302

23. O'Hanley, J.R., Church, R.L.: Designing robust coverage networks to hedge against worst-case facility losses. Eur. J. Oper. Res. **209**(1), 23–36 (2011). https://doi.org/10.1016/j.ejor.2010.08.030

24. Quadros, H., Roboredo, M.C., Pessoa, A.A.: A branch-and-cut algorithm for the multiple allocation r-hub interdiction median problem with fortification. Expert Syst. Appl. **110**, 311–322 (2018). https://doi.org/10.1016/j.eswa.2018.05.036

25. Resende, M., Werneck, R.: A fast swap-based local search procedure for location problems. Ann. Oper. Res. **150**, 205–230 (2007). https://doi.org/10.1007/s10479-006-0154-0

26. Roboredo, M.C., Pessoa, A.A., Aizemberg, L.: An exact approach for the r-interdiction median problem with fortification. RAIRO-Oper. Res. **53**(2), 505–516 (2019). https://doi.org/10.1051/ro/2017060

27. Scaparra, M.P., Church, R.L.: A bilevel mixed-integer program for critical infrastructure protection planning. Comput. Oper. Res. **35**(6), 1905–1923 (2008). https://doi.org/10.1016/j.cor.2006.09.019

28. Scaparra, M.P., Church, R.L.: An exact solution approach for the interdiction median problem with fortification. Eur. J. Oper. Res. **189**(1), 76–92 (2008). https://doi.org/10.1016/j.ejor.2007.05.027

29. Scaparra, M.P., Church, R.L.: Location problems under disaster events. In: Laporte, G., Nickel, S., Saldanha da Gama, F. (eds.) Location Science, pp. 631–656. Springer, Cham (2019). https://doi.org/10.1007/978-3-030-32177-2_22

30. Sidorov, D., Wei, W.S., Vasil'ev, I., Salerno, S.: Automatic defects classification with p-median clustering technique. In: 2008 10th International Conference on Control, Automation, Robotics and Vision, pp. 775–780. IEEE, Piscataway (2008). https://doi.org/10.1109/ICARCV.2008.4795615

31. Vasil'ev, I.L., Klimentova, K.B., Kochetov, Y.A.: New lower bounds for the facility location problem with clients' preferences. Comput. Math. Math. Phys. **49**(6), 1010–1020 (2009). https://doi.org/10.1134/S0965542509060098

32. Vasilyev, I., Klimentova, X., Boccia, M.: Polyhedral study of simple plant location problem with order. Oper. Res. Lett. **41**(2), 153–158 (2013). https://doi.org/10.1016/j.orl.2012.12.006

33. Vasilyev, I.L., Klimentova, K.B.: The branch and cut method for the facility location problem with client's preferences. J. Appl. Ind. Math. **4**(3), 441–454 (2010). https://doi.org/10.1134/S1990478910030178

34. Whitaker, R.A.: A fast algorithm for the greedy interchange for large-scale clustering and median location problems. INFOR **21**, 95–108 (1983). https://doi.org/10.1080/03155986.1983.11731889

35. Zhang, P., Fan, N.: Analysis of budget for interdiction on multicommodity network flows. J. Glob. Optim. **67**(3), 495–525 (2016). https://doi.org/10.1007/s10898-016-0422-8

36. Zheng, K., Albert, L.A.: An exact algorithm for solving the bilevel facility interdiction and fortification problem. Oper. Res. Lett. **46**(6), 573–578 (2018). https://doi.org/10.1016/j.orl.2018.10.001

Improving Effectiveness
of Neighborhood-Based Algorithms
for Optimization of Costly
Pseudo-Boolean Black-Box Functions

Oleg Zaikin$^{(\boxtimes)}$ and Stepan Kochemazov

ISDCT SB RAS, Irkutsk, Russia
zaikin.icc@gmail.com, veinamond@gmail.com

Abstract. Optimization of costly black-box functions is hard. Not only we know next to nothing about their nature, we need to calculate their values in as small number of points as possible. The problem is even more pronounced for pseudo-Boolean black-box functions since it is harder to approximate them. For such functions the local search methods where a neighborhood of a point must be traversed are in a particular disadvantage compared to evolutionary strategies. In the paper we propose two heuristics that make use of the search history to prioritize the more promising points from a neighborhood to be processed first. In the experiments involving minimization of an extremely costly pseudo-Boolean black-box function we show that the proposed heuristics significantly improve the performance of a hill climbing algorithm, making it outperform (1+1)-EA with an additional benefit of being more stable.

Keywords: Pseudo-Boolean optimization · Black-box optimization · Local search · Costly function · Boolean satisfiability problem

1 Introduction

Black-box optimization studies objective functions whose analytical form is not available [1,22]. Usually, black-box functions represent the result of a simulation or the reaction of some environment to a specific input. Depending on the processes involved, functions can be cheap or costly, defined everywhere or not, be smooth or not, etc. It goes without saying, that as the different classes of functions require special treatment, the same goes for the black-box functions. For example, if a black-box function is costly, then the amount of calculations should be as small as possible.

During the recent three decades, a considerable effort has been devoted to development of effective algorithms aimed at optimizing both continuous and discrete costly black-box objective functions. Some of these algorithms are based on building a surrogate model (see, e.g., [8,11,14]) that serves as an approximation of a considered function and allows one to use gradient methods. The

© Springer Nature Switzerland AG 2020
A. Kononov et al. (Eds.): MOTOR 2020, LNCS 12095, pp. 373–388, 2020.
https://doi.org/10.1007/978-3-030-49988-4_26

other algorithms rely only on direct calculations of a function (see the survey [17]). However, very few optimization algorithms are able to deal with costly pseudo-Boolean black-box functions. One of them was described in [27] where it was proposed to build a surrogate model based on discrete Walsh functions.

The present study is aimed at development of optimization algorithms for a class of costly pseudo-Boolean black-box functions. One can find their examples in many areas, from bioinformatics to cryptanalysis. The defining characteristics of these functions prevent the researcher from either obtaining any information about their gradient or from constructing a surrogate model that is anywhere near precision.

We propose two relatively simple heuristics that use the information accumulated during processing of the search space to improve the effectiveness of algorithms that optimize costly pseudo-Boolean black-box functions. Both heuristics are designed with neighborhood-based methods in mind. In short, for every Boolean variable we remember separately for its 1-value and 0-value the fraction of a function value in a point, obtained by changing this variable value from 0 to 1 or from 1 to 0, to the function value in the original point. We show that if we use this information to alter the order in which the points of a neighborhood are processed, then an improvement is reached at a lower cost (in terms of the number of function calculations) compared to the standard approach with randomly ordered neighborhoods.

In our computational experiments we employed the proposed heuristics for minimization of costly pseudo-Boolean black-box functions that arise in SAT-based cryptanalysis. They were described, e.g., in [31,32]. Their peculiar feature is that since a value is obtained by a specific procedure implementing the Monte Carlo method with limitations, they are not total, i.e. there are points in which the functions' values can not be evaluated in any reasonable time. We compare a hill climbing algorithm augmented with our heuristics with (1+1)-EA algorithm that tends to work really well with these functions (see [20,21,32]) and show that the proposed algorithm is more stable and often finds better records.

Let us give a brief outline of the paper. Section 2 contains the detailed description of the heuristics that we propose. In Sect. 3, the considered objective function is described. In Sect. 4 we conduct the computational experiments and discuss their results. After that we present brief conclusions and outline the directions for future research.

2   Proposed Heuristics

For optimization problems it is common to attempt to reduce the number of function calculations required by an optimization algorithm. However, the situation is much more dire when dealing with costly functions. In their case, one needs to make a decision, similar to that between exploration and exploitation in machine learning, at each step of the search. For example, in local-search-like algorithms whether to continue processing points from a neighborhood after obtaining an improvement in some point, hoping to find even better improvement, or to jump to a new point straight away. Note, that computing a costly function's value in some point can take minutes or even hours. Thus, taking into account realistic time limits of at most several days, the whole optimization process should fit into thousands of function's calculations at most. Therefore, for practical problems every decision of the sort can greatly impact the result of the algorithm in general. In contrast, for cheap functions it is often possible to compute their values several thousand times per second on one processor core, thus several million points more or several million points less do not really matter that much.

The often used method for directing the search is to use a surrogate model that exploits the search history to approximate a function behavior and suggest the most promising directions to move at. When dealing with a costly pseudo-Boolean black-box function, it is possible to construct such a model [27] if the function's value can be evaluated in any given point. However, if its values in some points are undefined, such a construction is not feasible. Nevertheless, during the search the information about processed points and the corresponding function's values is inevitably accumulated. Thus, it is sensible to make use of this information in an attempt to reduce the number of points to be processed.

In this section we propose two heuristics aimed at using the information about the search history. They are designed with neighborhood-based methods in mind, but can be adapted to other approaches, albeit in a less natural setting.

2.1   On Neighborhoods over Boolean Hypercube

Both proposed heuristics attempt to reduce the amount of function calculations before obtaining a next improvement by establishing a specific order on a part of a neighborhood of the current record.

In the present paper, we consider the case when a neighborhood of a point $v = \{v_1, \ldots, v_n\}, v_i \in \{0, 1\}, i \in \{1, \ldots, n\}$ is formed by all points v' at Hamming distance 1 and also all points v'' at Hamming distance 2 but with the same Hamming weight as v. In other words, we consider the so-called *add-remove-replace* neighborhood.

Recall that since we study the pseudo-Boolean functions defined over points from a Boolean hypercube, it means that each point of a search space represents a subset of a set of all variables occurring in a function. Then, changing one component of vector v, specifying a point of a search space, from 0 to 1 means adding a single variable to a set corresponding to v. We refer to all points constructed in such a way as to *add-neighborhood* of v. Changing the value of vector's component from 1 to 0 corresponds to removal of an element from the set, and the corresponding points are referred to as *remove-neighborhood*. Finally, if we replace a set element from a subset corresponding to v by a set element that does not belong to this subset, then it means that the Hamming weight of a corresponding point remains the same, but the Hamming distance between a new point and the original one is equal to 2. Let us say that such points form a *replace-neighborhood*. The pseudocode for forming the corresponding neighborhoods is presented at Algorithm 1.

Function GenAddNeighborhood(v):

> $N^{Add}(v) \leftarrow \emptyset, k \leftarrow 0$
> **for** $j \leftarrow 1$ to n, **if** $v[j] = 0$ **do**
>> $k \leftarrow k + 1, u_k \leftarrow v, u_k[j] \leftarrow 1$
>> $N^{Add}(v).\text{Append}(u_k)$
>
> **return** $N^{Add}(v)$

Function GenRemoveNeighborhood(v):

> $N^{Remove}(v) \leftarrow \emptyset, k \leftarrow 0$
> **for** $j \leftarrow 1$ to n, **if** $v[j] = 1$ **do**
>> $k \leftarrow k + 1, u_k \leftarrow v, u_k[j] \leftarrow 0$
>> $N^{Remove}(v).\text{Append}(u_k)$
>
> **return** $N^{Remove}(v)$

Function GenReplaceNeighborhood(v):

> $N^{Replace}(v) \leftarrow \emptyset, g \leftarrow 0$
> **for** $j \leftarrow 1$ to $n - 1, k \leftarrow j + 1$ to n **if** $v[k] \neq v[j]$ **do**
>> $g \leftarrow g + 1, w_g \leftarrow v, w_g[j] \leftarrow 1 - v[j], w_g[k] \leftarrow 1 - v[k]$
>> $N^{Replace}(v).\text{Append}(w_g)$
>
> **return** $N^{Replace}(v)$

Algorithm 1: Functions for generating add-, remove-, and replace-neighborhoods

Function ProcessNeighborhood($N,v^i,i,f_{bkv},Updated$):

> **for each** $u \in N$ **do**
>> $f(u) = \text{ComputeFunction}(u)$
>> **if** $f(u) < f_{bkv}$ **then**
>>> $f_{bkv} \leftarrow f(u), i \leftarrow i + 1, v^i = u$
>>> Updated $\leftarrow$ true
>>> **break**

Function RndOrder(N):

> /* Assume that $N = (u_1, \dots, u_{|N|})$ */
> $(k_1, \dots, k_{|N|}) \leftarrow \text{RandomPermutation}(1, \dots, |N|)$
> $N_{rand} \leftarrow (u_{k_1}, \dots, u_{k_{|N|}})$
> **return** N_{rand}

Algorithm 2: Auxiliary local search functions

In practice for costly functions it makes sense to implement neighborhood-based methods in such a way that the algorithm processes the neighborhood of Hamming radius 1 (add-remove-neighborhood) first and then proceeds to the remaining points. For some functions, depending on their overall properties and the starting point, it might make sense to first traverse the remove-neighborhood or the add-neighborhood.

Input: v_{start}
/* Compute value in starting point   */
$v^0 \leftarrow v_{start}, i \leftarrow 0, f_{bkv} \leftarrow f(v^0)$
Updated $\leftarrow$ true
while $Updated = True$ **do**
   | Updated $\leftarrow$ false
   | $N_{rand}^{Remove}(v^i) \leftarrow$ RndOrder(GenRemoveNeighborhood(v^i))
   | ProcessNeighborhood($N_{rand}^{Remove}(v^i), v^i, i, f_{bkv}$, Updated)
   | **if** Updated = false **then**
   | | $N_{rand}^{Add}(v^i) \leftarrow$ RndOrder(GenAddNeighborhood(v^i))
   | | ProcessNeighborhood($N_{rand}^{Add}(v^i), v^i, i, f_{bkv}$, Updated)
   | **if** Updated = false **then**
   | | $N_{rand}^{Replace}(v^i) \leftarrow$ RndOrder(GenReplaceNeighborhood(v^i))
   | | ProcessNeighborhood($N_{rand}^{Replace}(v^i), v^i, i, f_{bkv}$, Updated)

Algorithm 3: Simple hill climbing algorithm for an add-remove-replace neighborhood

2.2 Add-Remove-Sorting and Replace-Sorting Heuristics

In this subsection two heuristics are proposed: *add-remove-sorting* and *replace-sorting*. Note, that *replace-sorting* must be combined with *add-remove-sorting* while the latter can be used independently.

Let us consider simple hill climbing (or first-choice hill climbing, see, e.g., [23]) as an example of a neighborhood-based optimization algorithm. Assume that simple hill climbing starts from a point $v_{start} \in \{0,1\}^n$ and attempts to minimize a costly pseudo-Boolean black-box function f. The pseudocode of auxiliary functions that describe how neighborhoods are processed is outlined at Algorithm 2. Simple hill climbing then works as presented at Algorithm 3. Here f_{bkv} stands for the best known value of the objective function f. The goal of random reordering of neighborhoods in Algorithm 2 is to avoid the situations when the variables with lesser numbers gain an unfair advantage compared to the variables with larger numbers.

We assume that simple hill climbing processes remove-neighborhood first, then add-neighborhood second and replace-neighborhood third because it better suits the objective function we use in the computational experiments. However, the proposed heuristics can be adapted with little to no change to the case where add-neighborhood is processed before remove-neighborhood. If it is not known beforehand which neighborhood should be processed first (add or remove), then it makes sense to alternate choice of variants.

Function ComputeFunctionWithMemory(u,Add,$Remove$,v^i,f_{bkv}):

    $f_v \leftarrow \infty$

    if *not* $u \in ComputedPoints$ **then**

        | $ComputedPoints[u] \leftarrow$ ComputeFunction(u)

    $j \leftarrow$ index of the position in which u differs from v^i

    if HammingWeight(u)>HammingWeight(v^i) **then**

        | $Add[j] \leftarrow ComputedPoints[u]/f_{bkv}$

    else

        | $Remove[j] \leftarrow ComputedPoints[u]/f_{bkv}$

    return $ComputedPoints[u]$

Function ProcessNeighborhoodMem(N,v^i,i,f_{bkv},$Updated$, Add, $Remove$):

    for each $u \in N$ **do**

        $f(u) =$ ComputeFunctionWithMemory(u,Add,$Remove$,v^i,f_{bkv})

        if $f(u) < f_{bkv}$ **then**

            $f_{bkv} \leftarrow f(u)$, $i \leftarrow i+1$, $v^i = u$

            $Updated \leftarrow$ true, **break**

Algorithm 4: Auxiliary functions for simple hill climbing with sorting

Function SortedAddNeighborhood(v,Add):

    $N_{Sorted}^{Add}(v) \leftarrow \emptyset$, $N_{unknown}^{Add}(v) \leftarrow \emptyset$, $N_{known}^{Add}(v) \leftarrow \emptyset$, $k = 0$

    for $j \leftarrow 1$ to n, if $v[j] = 0$ **do**

        $k \leftarrow k+1$, $u_k \leftarrow v$, $u_k[j] \leftarrow 1$

        if $Add[j] = \infty$ **then** $N_{unknown}^{Add}(v)$.Append(u_k)

        else $N_{known}^{Add}(v)$.Append(u_k)

    $N_{Sorted}^{Add}(v) \leftarrow$ SortAsc($N_{known}^{Add}(v)$, Add)

    $N_{Sorted}^{Add}(v)$.Append($N_{unknown}^{Add}(v)$)

    return $N_{Sorted}^{Add}(v)$

Function SortedRemoveNeighborhood(v, $Remove$):

    $N_{Sorted}^{Remove}(v) \leftarrow \emptyset$, $N_{unknown}^{Remove}(v) \leftarrow \emptyset$, $N_{known}^{Remove}(v) \leftarrow \emptyset$, $k = 0$

    for $j \leftarrow 1$ to n, if $v[j] = 0$ **do**

        $k \leftarrow k+1$, $u_k \leftarrow v$, $u_k[j] \leftarrow 1$

        if $Remove[j] = \infty$ **then** $N_{unknown}^{Remove}(v)$.Append(u_k)

        else $N_{known}^{Remove}(v)$.Append(u_k)

    $N_{Sorted}^{Remove}(v) \leftarrow N_{unknown}^{Add}(v)$

    $N_{Sorted}^{Remove}(v)$.Append(SortAsc($N_{known}^{Remove}(v)$, $Remove$))

    return $N_{Sorted}^{Remove}(v)$

Algorithm 5: Functions for generating sorted add- and remove-neighborhoods

In the *add-remove-sorting* heuristic and *replace-sorting* heuristic we use the function values at already processed points to sort add-, remove- and replace-neighborhoods. In particular, we introduce three additional structures. The first is the associative array (map) *ComputedPoints*, that contains the function values for the points of the search space where it has already been computed. The second and third are two arrays *Add* and *Remove* of size n. The *Add*[j] stores the relative decrease in the value of a point from a neighborhood of some point v obtained by adding the variable x_j to v. $Remove[j]$ stores the similar value obtained for a point produced from some u by removing x_j from u. Note, that the currently stored values in Add and $Remove$ arrays correspond to the most recent computations of a function where these values could have been updated. Here we need to take into account the fact that the value of an objective function is not defined in some points (for example, its computation can be interrupted due to exceeding some reasonable time limit). We assume that the value of a function in such points is equal to ∞.

The pseudocode that describes how Add and $Remove$ arrays are updated is shown in the function `ComputeFunctionWithMemory` of Algorithm 4. The functions denoted as `SortedAddNeighborhood` and `SortedRemoveNeighborhood` in Algorithm 5 describe how they are used to alter the handling of the add- and remove-neighborhoods. In case of the add-neigborhood, informally, we first pick the part of a neighborhood for which the values of Add are less than ∞, i.e. they correspond to recently computed function values. We rearrange the order of these points in the ascending order of corresponding Add values. The order of the remaining points is randomly rearranged. Then, we put in the resulting sorted neighborhood the points with known Add values first and the remaining ones last. For the sorted remove-neighborhood we put the sorted points with known $Remove$ after the randomly rearranged other points. The algorithm is allowed to traverse randomly ordered points with unknown values from the add-remove neighborhood in order to add bias to the search, that can be useful for escaping local minima. Here the goal is to balance the exploitation and exploration.

Function SortedReplaceNeighborhood(v,Add,$Remove$,k):

> $N_{Sorted}^{Replace}(v_i) \leftarrow \emptyset, N_{Best}^{Replace}(v_i) \leftarrow \emptyset, N_{Rest}^{Replace}(v_i) \leftarrow \emptyset, g \leftarrow 0$
>
> $N_{SortAscAdd}^{Add} \leftarrow$ SortAsc(GenAddNeighborhood(v), Add)
>
> $N_{SortAscRemove}^{Remove} \leftarrow$ SortAsc(GenRemoveNeighborhood(v), $Remove$)
>
> **for** $j \leftarrow 1$ *to* $k, h \leftarrow 1$ *to* k **do**
>
> > /* replace in v 'remove' variable by 'add' variable   */
> >
> > $N_{Best}^{Replace}$.Append(Combine(v, $N_{SortAscAdd}^{Add}[j]$, $N_{SortAscRemove}^{Remove}[h]$))
>
> **for** $j \leftarrow k+1$ *to* $|N_{SortAscAdd}^{Add}|, h \leftarrow k+1$ *to* $|N_{SortAscRemove}^{Remove}|$ **do**
>
> > $N_{Rest}^{Replace}$.Append(Combine(v, $N_{SortAscAdd}^{Add}[j]$, $N_{SortAscRemove}^{Remove}[h]$))
>
> $N_{Sorted}^{Replace}(v) \leftarrow N_{Best}^{Replace}$.Append(RndOrder($N_{Rest}^{Replace}$))
>
> **return** $N_{Sorted}^{Replace}(v)$

Algorithm 6: Function for generating sorted replace-neighborhoods

---

The replace-sorting heuristic is a little bit trickier compared to the first one. There are two main distinctions here. First, the replace-neighborhood is usually much larger than the add-remove neighborhood. Second, we start traversing it only after having refreshed every single Add and $Remove$ value. Thus, the information in *Add* and *Remove* is significantly more relevant and captures the specifics of the current point the neighborhoods of which we process.

Input: v_{start}
$v^0 \leftarrow v_{start}, i \leftarrow 0, f_{bkv} \leftarrow f(v^0)$
$Add.\text{SetSize}(N), Remove.\text{SetSize}(N)$
$Add \leftarrow Remove \leftarrow (\infty, \dots, \infty),$ **Updated** $\leftarrow$ **true**,
while Updated $\leftarrow$ **true do**
$\quad$ Updated $\leftarrow$ **false**
$\quad N_{Sorted}^{Remove}(v^i) \leftarrow$ SortedRemoveNeighborhood(v^i,*Remove*)
$\quad$ ProcessNeighborhoodMem($N_{Sorted}^{Remove}(v^i), v^i, i, f_{bkv}$, **Updated**, *Add*, *Remove*) if
$\quad$ Updated $=$ **false then**
$\quad\quad N_{Sorted}^{Add}(v^i) \leftarrow$ SortedAddNeighborhood(v^i, *Add*)
$\quad\quad$ ProcessNeighborhoodMem($N_{Sorted}^{Add}(v^i), v^i, i, f_{bkv}$, **Updated**, *Add*, *Remove*)
$\quad$ if Updated $=$ **false then**
$\quad\quad N_{Sorted}^{Replace}(v^i) \leftarrow$ SortedReplaceNeighborhood(v,*Add*,*Remove*,*k*)
$\quad\quad$ ProcessNeighborhood($N_{rand}^{Replace}(v^i), v^i, i, f_{bkv}$, **Updated**)

Algorithm 7: Sorted-Add-Remove-Replace Simple Hill Climbing algorithm

---

Also, since the size of the replace-neighborhood is so much larger, the potential gain and the potential loss of imposing a specific order on it are much higher. Thus we introduce the parameter k to only apply our heuristic to the top $k \times k$ points. It works as shown in **SortedReplaceNeighborhood** function inbreak Algorithm 6. Essentially, it picks the first k points from the sorted add-neighborhood and first k points from the sorted remove-neighborhood and combines them to form $k \times k$ first points in the sorted replace-neighborhood. The order of the remaining points is randomly rearranged similar to how it is normally done in simple hill climbing.

Note, that the algorithm stops processing a current neighborhood as soon as it finds an improvement. Thus, the replace-neighborhood traversal is started only if no improvement was found in the add-remove neighborhood. According to our experiments (see Sect. 4), the replace-neighborhood is reached quite rarely, but it significantly helps to escape local minima.

We further refer to the variant of simple hill climbing algorithm that employs add-remove-sorting and replace-sorting heuristics as *SHC-ARR-sorted*. Its pseudocode is shown in Algorithm 7. The random variant of simple hill climbing shown in Algorithm 3 is called *SHC-ARR-random*. Note, that *SHC-ARR-sorted* is in fact a variant of the variable neighborhood descent metaheuristic [5].

3   Objective Function

Costly pseudo-Boolean black-box objective functions are encountered quite often in today's world. One example of such a function is the computation of influence spread in Influence Maximization Problem [15]. Other examples include various formulations of location problems in economics, specific optimization problems arising in cryptanalysis, etc. In the present paper, we focus on costly pseudo-Boolean black-box objective functions aimed at estimating the hardness of instances of the Boolean satisfiability problem (SAT, [4]). These functions are stochastic since they use the Monte Carlo method [19] to evaluate the runtime of some algorithm on a specific hard SAT instance. Different formulations of such functions are known [25,26,32]. In prior literature these functions were optimized via tabu search algorithms [25,26,32], evolutionary algorithms [20,21,32], genetic algorithms [20,21], etc. In [32], an improved version of the function from [25] was proposed that usually allows one to obtain more accurate estimations. In the rest of the paper, we apply the heuristics proposed in Sect. 2 to minimization of this very function.

Let us briefly describe the function from [32]. We will denote it as G. Assume that C is a Boolean formula over a set X of Boolean variables. The objective function G takes as an input a subset S of variables from X. Hereinafter such a subset is called *decomposition set*. For given C and S the function's value is the estimation of the runtime (in seconds) required to solve all simplified formulas produced by assigning all possible different combinations of values to variables from S in C. The function implements the Monte-Carlo method [19]: an estimation is calculated based on solving a relatively small number of simplified formulas that form a random sample. This approach closely resembles the Cube-and-Conquer approach [13], but is built on slightly different principles.

The function G has an important feature: its values in some points can be undefined. It happens when SAT instances from some random samples can not be solved in any reasonable time. Following [32], the processing of such samples is interrupted with the objective function value set to plus infinity.

We used the objective function implementation from the ALIAS tool [16] that is aimed at solving hard SAT instances. This tool in turn uses an IPASIR-based [2] version of the SAT solver ROKK [29] to solve simplified formulas. The following parameters control the function's accuracy: number of intervals u; interval size v. Similarly to [32], in all further experiments the following parameters' values were used: 100 and 10000, respectively.

4   Computational Experiments

To minimize the objective function from Sect. 3, we employed two versions of simple hill climbing, SHC-ARR-random and SHC-ARR-sorted, described in Sect. 2. In the replace-sorting heuristic k was equal to 10. We compared the mentioned algorithms with (1+1)-EA [9] because in [32] it showed the best overall performance among different algorithms for minimization of the considered function.

All computational experiments were conducted with the time limit of 1 day on one node of the computing cluster "Academician V.M. Matrosov" [7], equipped with 2 × 18-core Intel Xeon E5-2695 CPUs and 128 Gb of RAM. Note, that the implementation of the objective function is multithreaded, so at any moment all 36 CPU cores were employed to calculate the function's value.

In the following subsections, we describe the chosen test problems, show results of the experimental evaluation of proposed heuristics, and discuss the obtained results.

4.1   Test Problems

In [32], four hard pseudo-Boolean black-box optimization problems were studied using the objective function briefly described in the previous section. Each of them is in fact a problem of finding a decomposition set with low runtime estimation for a given conjunctive normal form (CNF). Every such CNF in turn encodes a SAT-based cryptanalysis [3] problem for a certain stream cipher [18]. In particular, four stream ciphers were studied: Trivium, Grain_v1, Mickey, Rabbit. In [32], it was shown that Mickey and Rabbit do not suit well for this type of SAT-based cryptanalysis. That is why for the present study we considered only the optimization problems for Trivium [6] and Grain_v1 [12]. A stream cipher's state is stored in registers, and its output is called *keystream*. The total size of Trivium's registers is 288 bits, while for Grain_v1 it is 160 bits. Keystream sizes were equal to 300 and 200, respectively.

We additionally considered two optimization problems for SAT-based cryptanalysis problems of the following cryptographic keystream generators: the alternating step generator (ASG) [10] with 192-bit total size of registers (keystream size is 200); the Wolfram generator [28] with 256-bit total size of registers (keystream size is 512).

The following variant of cryptanalysis was considered: given a known keystream of a stream cipher (or a keystream generator), to find the initial state of registers that was used to produce this keystream. CNFs for all considered problems were constructed by the Transalg tool [24]. The CNFs for Trivium and Grain_v1 were taken from [32], while for ASG-192 it was taken from [30]. As for Wolfram-256, it was constructed specifically for the present study.

Similarly to [32], in our experiments we limited the search space for all considered problems to only include the subsets of the set containing the variables corresponding to initial register's state, thus it was of size 2^{288} for Trivium; 2^{160} for Grain_v1; 2^{192} for ASG-192; 2^{256} for Wolfram-256. Following [32], as a starting point for every considered optimization algorithm we used a set of all Boolean variables that encodes the initial states of the corresponding registers.

4.2   Improving (1+1)-EA by Memory Heuristic

(1+1)-EA can actually generate a new point that will coincide with the point at which an iteration started. In [32] we detected such a situation directly, i.e. without memory usage. In [21] it was proposed to permanently store all points processed by (1+1)-EA in order to not calculate the objective function from [26] more than once in any point. We improved (1+1)-EA in the same way in application to the objective function from [32]. It means that the direct detecting was augmented by the memory usage. The improved version is further called (1+1)-EA_memory. All processed points are stored in an associative array, where a point representation as a Boolean vector is an array key, while a corresponding array value is an objective function value in this point. Let us further refer to this simple heuristic as to *memory* heuristic. We implemented the associative array using the *std::map* class from the C++ language, for which both insertion and lookup operations have logarithmic complexity. It might be better to use *std::unordered_map* because for it both insertion and lookup operations have constant average case complexity (albeit it requires more memory). Nevertheless, even for *std::map* the required computational resources were negligible. Experiments show that if an optimization algorithm is run for 1 day, the maintaining of the *memory* heuristic takes less than 1 s. The memory heuristic was also used in both SHC-ARR-random and SHC-ARR-sorted (see Subsect. 2.2).

Since the objective function is stochastic [32], in all further experiments 3 runs of an optimization algorithm on a given problem are conducted. Table 1 presents the following data for (1+1)-EA_memory: (1) the number of points in which the objective function was computed; (2) the number of points skipped via the memory heuristic. From the table it follows that the memory heuristic makes it possible to skip about 50% of already processed points. Note that only those points whose repetitions were detected by an associative array are considered as skipped. If we had additionally counted points whose repetitions were detected directly (see the explanation above), then the number of skipped points would have greatly increased.

4.3   Analysis of Add-Remove-Sorting and Replace-Sorting Heuristics

We compared SHC-ARR-random, SHC-ARR-sorted (see Subsect. 2.2), (1+1)-EA (in its original form), and (1+1)-EA_memory on four considered optimization problems. The results are shown in Fig. 1. Three runs of every algorithm are marked with _0, _1, and _2. Here x-axis shows the number of the objective function's calculations elapsed from the start of a run, while y-axis shows updates of the current best known objective function value. In Table 2, for each pair (algorithm, problem) the final objective function value of the best run (out of 3) is shown.

Table 1. The number of calculated and skipped points for all runs of the (1+1)-EA_memory algorithm.

Problem	Run 1		Run 2		Run 3	
	Calculated	Skipped	Calculated	Skipped	Calculated	Skipped
Trivium	7545	7075	8342	6717	9486	8243
Grain_v1	7155	4446	6676	7165	6995	5438
Wolfram-256	8372	5362	9272	5875	9013	4347
ASG-192	14222	14928	13833	13526	16438	18288

Table 2. Final objective function values found by the optimization algorithms in their best runs. The best result (among all algorithms) for every problem is marked with bold.

Problem	(1+1)-EA	(1+1)-EA_memory	SHC-ARR-random	SHC-ARR-sorted
Trivium	5.99e+41	4.68e+41	3.35e+42	**4e+41**
Grain_v1	1.08e+31	2.19e+30	2.95e+30	**1.63e+30**
Wolfram-256	1.47e+14	3.78e+06	1.46e+07	**3.69e+06**
ASG-192	3.63e+15	4.54e+14	**8.83e+13**	8.97e+13

It is clear that SHC-ARR-sorted shows very good results. In particular, it outperforms the competition (comparing results of the best runs) on 3 problems out of 4. Also, it shows more stable results, i.e. its runs show less variation in results (compared to other algorithms) and reach very close records. This advantage allows one to safely run SHC-ARR-sorted only once on a considered problem, while (1+1)-EA usually requires several runs due to high variation in results.

It turned out that both SHC algorithms are much less affected by the memory heuristic compared to (1+1)-EA. In all runs, less than 1% of points were repeated. Thus it can be concluded, that SHC-ARR-sorted showed such good and stable results mainly due to the sorting heuristics.

As it was stated in Sect. 3, the function computation can be interrupted in some points due to the time limit. We analyzed how often the function was interrupted on the considered problems when SHC-ARR-sorted was run. It turned out, that on Grain_v1 the amount of interruptions is comparable to the amount of successful calculations; on Trivium the amount of successful calculations was 6x–10x less than the amount of interruptions; on Wolfram-256 it was 15x–22x less; on ASG-192 it was 30x–60x less.

We also compared SHC-ARR-random and SHC-ARR-sorted by indexes of those points in neighborhoods on which the best known value of the objective function was updated. For this purpose we used a cumulative function $Index\_cumul$, whose value for a given serial number x of an update is the summation of indexes of all updates with serial number $\leq x$. The comparison is presented in Fig. 2. The x-axis shows serial numbers of updates, the y-axis shows values of $Index\_cumul$. From this figure it is clear that the combination of the proposed add-remove-sorted and replace-sorted heuristics allows the simple hill climbing algorithm to find new good points in a given neighborhood faster.

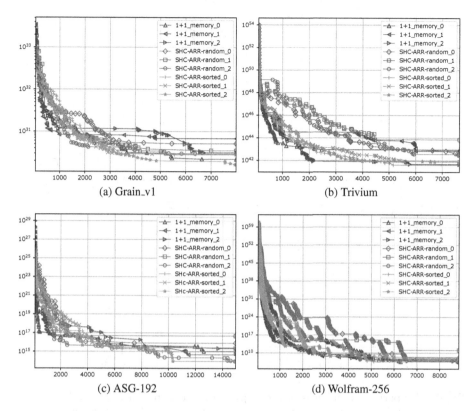

Fig. 1. Comparison of all considered optimization algorithms. The x-axis shows the number of the objective function's calculations elapsed from the start of a run, y-axis shows updates of the current best known objective function value.

4.4 Discussion

From the presented results it follows that, when improved by both proposed heuristics, simple hill climbing shows very good and stable results. This algorithm outperformed (1+1)-EA and a basic simple hill climbing on 3 considered problems out of 4. Note, that all analyzed algorithms were improved by the memory heuristic.

As a result of our study, for two optimization problems, ASG-192 and Grain_v1, better record values of the objective function were found compared to previously published results. In particular, in [30] for ASG-192 the best value was 2.64e+15, while in the present study it is 8.83e+13. For Grain_v1, we found the record point corresponding to 1.63e+30, while in [32] it was 2.96e+30. Our result for Trivium is slightly worse than the best published one. As for Wolfram-256, to the best of our knowledge the corresponding optimization problem is studied for the first time. Note, that earlier in [24] a much weaker variant of Wolfram (with 128-bit secret key, i.e. Wolfram-128) was studied by SAT-based cryptanalysis.

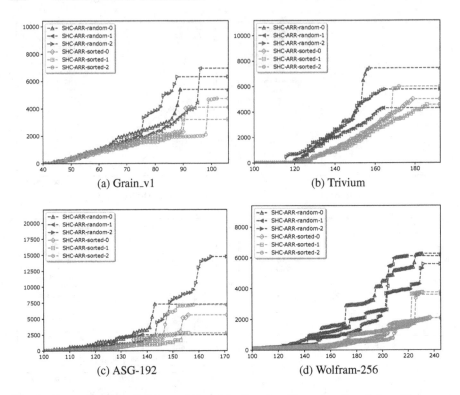

Fig. 2. Comparison by indexes of the objective function record updates. The x-axis shows the serial number of updates, the y-axis shows values of *Index_cumul*.

5   Conclusions

In the present study we proposed two heuristics for neighborhood-based pseudo-Boolean black-box optimization algorithms aimed at minimizing costly objective functions. These heuristics were used to improve the simple hill climbing algorithm. The improved version showed the expected increase in performance thanks to reducing the average number of function calculations between consecutive improvements of records. An additional benefit of the proposed heuristics in application to simple hill climbing consists in making the algorithm more stable.

In the future we are planning to apply the proposed heuristics to other neighborhood-based optimization algorithms, such as tabu search. Also we intend to study other costly pseudo-Boolean black-box objective functions.

Acknowledgements. The research was partially supported by Russian Foundation for Basic Research (grant no. 19-07-00746-a). Stepan Kochemazov is additionally supported by the Council for Grants of the President of Russia (stipend SP-2017.2019.5).

References

1. Audet, C., Hare, W.: Derivative-Free and Blackbox Optimization. Springer Series in Operations Research and Financial Engineering. Springer, Heidelberg (2017). https://doi.org/10.1007/978-3-319-68913-5
2. Balyo, T., Biere, A., Iser, M., Sinz, C.: SAT race 2015. Artif. Intell. **241**, 45–65 (2016)
3. Bard, G.V.: Algebraic Cryptanalysis, 1st edn. Springer, Heidelberg (2009). https://doi.org/10.1007/978-0-387-88757-9
4. Biere, A., Heule, M., van Maaren, H., Walsh, T. (eds.): Handbook of Satisfiability. Frontiers in Artificial Intelligence and Applications, vol. 185. IOS Press, Amsterdam (2009)
5. Brimberg, J., Hansen, P., Mladenovic, N., Taillard, É.D.: Improvement and comparison of heuristics for solving the uncapacitated multisource weber problem. Oper. Res. **48**(3), 444–460 (2000). https://doi.org/10.1287/opre.48.3.444.12431
6. De Cannière, C., Preneel, B.: TRIVIUM. In: Robshaw, M., Billet, O. (eds.) New Stream Cipher Designs. LNCS, vol. 4986, pp. 244–266. Springer, Heidelberg (2008). https://doi.org/10.1007/978-3-540-68351-3_18
7. Irkutsk supercomputer center of SB RAS. http://hpc.icc.ru
8. Costa, A., Nannicini, G.: RBFOpt: an open-source library for black-box optimization with costly function evaluations. Math. Program. Comput. **10**(4), 597–629 (2018). https://doi.org/10.1007/s12532-018-0144-7
9. Droste, S., Jansen, T., Wegener, I.: On the analysis of the (1+1) evolutionary algorithm. Theor. Comput. Sci. **276**(1–2), 51–81 (2002). https://doi.org/10.1016/S0304-3975(01)00182-7
10. Günther, C.G.: Alternating step generators controlled by de Bruijn sequences. In: Chaum, D., Price, W.L. (eds.) EUROCRYPT 1987. LNCS, vol. 304, pp. 5–14. Springer, Heidelberg (1988). https://doi.org/10.1007/3-540-39118-5_2
11. Gutmann, H.M.: A radial basis function method for global optimization. J. Glob. Opt. **19**(3), 201–227 (2001)
12. Hell, M., Johansson, T., Maximov, A., Meier, W.: The grain family of stream ciphers. In: Robshaw, M., Billet, O. (eds.) New Stream Cipher Designs. LNCS, vol. 4986, pp. 179–190. Springer, Heidelberg (2008). https://doi.org/10.1007/978-3-540-68351-3_14
13. Heule, M.J.H., Kullmann, O., Biere, A.: Cube-and-conquer for satisfiability. Handbook of Parallel Constraint Reasoning, pp. 31–59. Springer, Cham (2018). https://doi.org/10.1007/978-3-319-63516-3_2
14. Jones, D.R., Schonlau, M., Welch, W.J.: Efficient global optimization of expensive black-box functions. J. Glob. Opt. **13**(4), 455–492 (1998). https://doi.org/10.1023/A:1008306431147
15. Kempe, D., Kleinberg, J., Tardos, E.: Maximizing the spread of influence through a social network. In: KDD 2003, pp. 137–146. Association for Computing Machinery, New York (2003). https://doi.org/10.1145/956750.956769
16. Kochemazov, S., Zaikin, O.: ALIAS: a modular tool for finding backdoors for SAT. In: Beyersdorff, O., Wintersteiger, C.M. (eds.) SAT 2018. LNCS, vol. 10929, pp. 419–427. Springer, Cham (2018). https://doi.org/10.1007/978-3-319-94144-8_25
17. Kolda, T.G., Lewis, R.M., Torczon, V.: Optimization by direct search: new perspectives on some classical and modern methods. SIAM Rev. **45**(3), 385–482 (2003)
18. Menezes, A.J., Vanstone, S.A., Oorschot, P.C.V.: Handbook of Applied Cryptography, 1st edn. CRC Press Inc., Boca Raton (1996)

19. Metropolis, N., Ulam, S.: The Monte Carlo method. J. Am. Stat. Assoc. **44**(247), 335–341 (1949)
20. Pavlenko, A., Buzdalov, M., Ulyantsev, V.: Fitness comparison by statistical testing in construction of SAT-based guess-and-determine cryptographic attacks. In: GECCO 2019, pp. 312–320 (2019). https://doi.org/10.1145/3321707.3321847
21. Pavlenko, A., Semenov, A., Ulyantsev, V.: Evolutionary computation techniques for constructing SAT-based attacks in algebraic cryptanalysis. In: Kaufmann, P., Castillo, P.A. (eds.) EvoApplications 2019. LNCS, vol. 11454, pp. 237–253. Springer, Cham (2019). https://doi.org/10.1007/978-3-030-16692-2_16
22. Rios, L., Sahinidis, N.: Derivative-free optimization: a review of algorithms and comparison of software implementations. J. Glob. Opt. **56**, 1247–1293 (2013). https://doi.org/10.1007/s10898-012-9951-y
23. Russell, S., Norvig, P.: Artificial Intelligence: A Modern Approach, 3rd edn. Prentice Hall, Upper Saddle River (2009)
24. Semenov, A., Otpuschennikov, I., Gribanova, I., Zaikin, O., Kochemazov, S.: Translation of algorithmic descriptions of discrete functions to SAT with applications to cryptanalysis problems. Log. Methods Comput. Sci. **16**, 29:1–29:42 (2020)
25. Semenov, A., Zaikin, O.: Algorithm for finding partitionings of hard variants of Boolean satisfiability problem with application to inversion of some cryptographic functions. SpringerPlus **5**(1), 1–16 (2016)
26. Semenov, A., Zaikin, O., Otpuschennikov, I., Kochemazov, S., Ignatiev, A.: On cryptographic attacks using backdoors for SAT. In: AAAI 2018, pp. 6641–6648 (2018)
27. Verel, S., Derbel, B., Liefooghe, A., Aguirre, H., Tanaka, K.: A surrogate model based on walsh decomposition for pseudo-Boolean functions. In: Auger, A., Fonseca, C.M., Lourenço, N., Machado, P., Paquete, L., Whitley, D. (eds.) PPSN 2018. LNCS, vol. 11102, pp. 181–193. Springer, Cham (2018). https://doi.org/10.1007/978-3-319-99259-4_15
28. Wolfram, S.: Random sequence generation by cellular automata. Adv. Appl. Math. **7**(2), 123–169 (1986)
29. Yasumoto, T., Okuwaga, T.: Rokk 1.0.1. In: SAT Competition 2014: Solver and Benchmark Descriptions. Series of Publications B, vol. B-2017-1, p. 70. Department of Computer Science, University of Helsinki (2014)
30. Zaikin, O., Kochemazov, S.: An improved SAT-based guess-and-determine attack on the alternating step generator. In: Nguyen, P., Zhou, J. (eds.) ISC 2017. LNCS, vol. 10599, pp. 21–38. Springer, Cham (2017). https://doi.org/10.1007/978-3-319-69659-1_2
31. Zaikin, O., Kochemazov, S.: Black-box optimization in an extended search space for SAT solving. In: Khachay, M., Kochetov, Y., Pardalos, P. (eds.) MOTOR 2019. LNCS, vol. 11548, pp. 402–417. Springer, Cham (2019). https://doi.org/10.1007/978-3-030-22629-9_28
32. Zaikin, O., Kochemazov, S.: On black-box optimization in divide-and-conquer SAT solving. Opt. Methods Softw., 1–25 (2019). https://doi.org/10.1080/10556788.2019.1685993

Operational Research Applications

Securities and Cash Settlement Framework

Ekaterina Alekseeva[(⊠)], Sana Ghariani, and Nicolas Wolters

Banque de France, 48 rue Notre Dame des Victoires, 75002 Paris, France
{Ekaterina.ALEKSEEVA.external,Sana.GHARIANI,
Nicolas.WOLTERS.external}@banque-france.fr

Abstract. The securities settlement process consists in delivering securities from one financial actor to another in exchange for payment in currency. Each business day has a night-time settlement (NTS) period when transactions (exchange of cash and/or security for payment) are settled in batches. Banque de France is inter alia in charge of Mathematical Optimization Module (MOM) for the NTS period which is a component of a large European platform. To reduce the number of failed transactions some additional financial features can be triggered, such as partial settlement of eligible transactions and provision of credit (auto-collateralisation mechanism). MOM must settle as many transactions as possible respecting all business constraints and taking advantage of these financial features. Furthermore, MOM execution time is limited, the data size is large (several hundred thousands of transactions over a billion euro) and the number of transactions and their amounts require high numerical precision. In this work we introduce the necessary financial notions, explain the NTS process and formulate it as a discrete optimisation model. We expose heuristic, mixed integer and linear programming algorithmic approaches used to solve this large-scale problem. We present results obtained on production data and discuss some perspectives.

Keywords: Application of OR in finance · Discrete large-scale optimization · Mathematical programming · Heuristics

1 Introduction

From 2007 to 2015 four national central banks of Germany, France, Italy, and Spain developed Target2Securities (T2S) platform to facilitate cross-border settlement procedures among countries [10]. Currently, twenty-one Central Securities Depositories (CSD) - financial organizations holding securities - in twenty European countries use the platform, which is multi-currency since 2018. Banque de France is in charge of Mathematical Optimization Module (MOM) which makes fully automated decisions regarding settlement of transactions during the a night-time settlement (NTS) period by operations research (OR) algorithms. Since 2015, MOM is operating and meeting the needs of market. This paper addresses the optimisation problem related to the securities settlement process in order to maximize the volume and amounts of settlements in Europe.

© Springer Nature Switzerland AG 2020
A. Kononov et al. (Eds.): MOTOR 2020, LNCS 12095, pp. 391–405, 2020.
https://doi.org/10.1007/978-3-030-49988-4_27

Different challenges had to be faced. Firstly, NTS problem (NTSP) has stochastic nature, meaning that the input data: number and type of transactions, total amount to settle and available resources (securities, cash) vary significantly from one business day to another. Thus, it is a challenge to tune the optimisation algorithms and fix their settings to guaranty good quality settlement results for each business day.

Secondly, all amounts are converted into elementary units of their related currency (e.g., cents for euro) and expressed as large integer values (thousand billions) that result in numerical precision issues. For example, a solution with 9 digits after the decimal point (scale of 9) might cause the loss of an important amount of money in comparison with a solution with scale of 12. Even if the ready-to-use solvers try to handle these numerical difficulties [3], linear programming algorithms may return an optimal solution ineffective in reality.

Thirdly, since the problem size is large and MOM execution time is strictly limited, we neither can address a mixed integer mathematical formulation directly, nor use the exact optimization algorithms to get an optimal solution in a reasonable time. Moreover, the solution quality is a major issue. Therefore, the right equilibrium between execution time and optimality gap had to be found.

Finally, T2S platform is developed and run on different operating systems: Windows for development and z/OS for production. The choice of ready-to-use solvers and then the program behavior depend on the OS and computer hardware.

The contribution of this work is a presentation for the first time of a settlement framework that deals with heterogeneous financial assets (securities and currencies) using OR approaches. Section 2 presents state of the art of related applications. Section 3 introduces the main components of the NTS process, its objectives, and gives an example to explain the notions used in the paper. Section 4 gives a mathematical formulation of the NTSP. Section 5 introduces algorithms used by MOM. Sections 6 reports and discusses the computational results on real data: MOM's efficiency in terms of percentage of settled volume, settled amounts, and computational time. We conclude the paper with the future development directions.

2   Related Works

Since the mid-20th century, financial institutions have extensively used OR to define trading strategies, to value financial instruments, to solve the portfolio optimisation problem, to measure bank efficiency, and to solve other financial problems mentioned in [6,8]. Banking environment provides an appropriate context for application of OR techniques. The decision process in this environment is often fully automated, so that solutions obtained by algorithms are strictly implemented with low human intervention. However, the challenge is to solve the large associated optimisation problems handling large-amount transactions, a lot of constraints and variables and to overcome numerical issues in limited time. The aim is not necessarily to get an exact solution optimising some objective functions, but to produce feasible solutions of good quality quickly and consistently. In [3] authors give suggestions to tackle possible numerical and ill conditioning issues when ready-to-use solvers are used for linear and mixed integer problems.

To the best of our knowledge, there is no publication that exposes a solution of securities settlement process operating within a real-world platform. Several early works [1,4,5] explain organisation and operation of interbank payment and settlement systems presenting simulations results. In [4,5] authors describe a simulator implemented by Finnish central bank used for studying liquidity needs, system risks, pricing policies, and settlement modes. In [1] authors present a simulation-based approach to study relationships between settlement delays and intraday liquidity usage for Norwegian Bank's settlement system.

NTSP investigated in this paper may contain special cases of difficult combinatorial optimisation problems such as multidimensional knapsack problem and subset-sum problem [9]. These problems formalise so-called gridlock situation which is named bank clearing problem in financial context. This situation arises when there is no sufficient resources on a balance to settle a transaction alone but it might be possible to settle it simultaneously with a group of other ones, i.e., netting transactions, to bring the missing resources. So, the problem is to settle as many transactions between participants as possible applying netting mechanism. In [2] it was shown that the bank clearing problem is NP-complete when the number of participants is more than two. In comparison with knapsack problems, in [2,7] it was proved that there is no polynomial ϵ-approximative algorithm unless $P = NP$ for this problem even in case of two participants. In [2] authors discussed and developed several approximate efficient algorithms depending on the number of participants in netting. Algorithms were tested on random data generated in cooperation with a German bank. In [7] a new algorithm based on graph representation of the problem was implemented and compared with previous algorithm from [2] on data of Belarus interbank settlement center.

All works mentioned above do not consider securities-related transactions. In [2] authors mentioned extension of their research for securities settlement systems as a future perspective.

3   Main Components of NTSP

3.1   Securities and Cash Accounts

Each user of T2S platform has at least one securities account (SA) holding different securities emissions and one cash account (CA) holding money of the same currency. National Central Banks (NCBs) provide CAs to payment banks for securities settlement which in turn put CAs at disposal of other financial institutions. Legally, each SA and CA are located in accounting books of CSD and NCB, respectively.

Each SA and CA are divided in several compartments called "security positions" (SPs) and "cash balances" (CBs) depending on financial usage purpose of position and balance, respectively. For example, $SP1$ holds NOKIA bonds for trades, $SP2$ holds NOKIA shares for collateral operations, both within a single SA. We further suppose that each transaction debits or credits only one SP or CB.

Balance state of SPs belonging to issuers of securities and CBs belonging to central banks might be negative. In this case, transaction debiting the SP (CB) can always be fully settled.

3.2 Transaction Definition

Each user of T2S platform has a role of seller and/or buyer and may transfer only securities, or only cash, or securities against cash. A transaction is an agreement between a buyer and a seller to exchange resources (cash and/or security). Each transaction generates one or multiple movements of resources between balances and positions. Figure 1 presents T-account graphical representation that visualises movements for transactions of different type, as existing in accounting books, with debit side on the left, and credit side on the right. Figure 1 details the following transaction types considered in this paper:

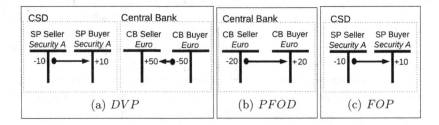

Fig. 1. Different types of business transactions.

- Delivery-Versus-Payment (*DVP*). Securities are transferred from seller's SP to buyer's SP and, simultaneously, cash is debited from buyer's CB and credited to seller's CB. In Fig. 1(a) 10 securities are debited from seller's SP and credited to buyer's SP against the payment of 50 euros debited from buyer's CB and credited to seller's CB.
- Payment-Free-Of-Delivery (*PFOD*). Cash is debited from seller's CB and credited to buyer's CB. In Fig. 1(b) 20 euros are debited from seller's CB and credited to buyer's CB.
- Free-Of-Payment (*FOP*). Securities are transferred from seller's SP to buyer's SP without compensation in cash. In Fig. 1(c) 10 securities are debited from seller's SP and are credited to buyer's SP.

Transactions might be linked together by two types of **links**:

- "with". Linked transactions must be settled together.
- "after". A transaction might be settled only if a linked transaction has already been settled.

One transaction may be involved in several links.

Each transaction has a **priority** for being settled. It is defined by two elements: a functional priority assigned by a T2S actor and time during which transaction stays unsettled in T2S platform. MOM has to favour the settlement of the oldest transactions with highest functional priority.

3.3   T2S Platform Features

T2S platform has two financial features intended to reduce the number of failed transactions.

Partial settlement functionality may be activated only for transactions of types DVP and FOP. Under this functionality transaction quantity and amount can be partially settled. This functionality is triggered if a transaction cannot be fully settled because of lack of securities available on the debited SP. Also, only transactions that meet thresholds criteria (minimum cash amount, minimum quantity) might be partially settled.

Auto-collateralisation (shortly, ACO) mechanism is a credit operation triggered when a participant (payment bank or its client) does not have enough cash to fully settle a transaction. The principle is to lend some cash against the pledge of some securities belonging to the participant to settle a transaction immediately. To manage the credit risk associated with this mechanism there are certain conditions:

- only certain securities may be eligible for collateralisation;
- amount of credit is calculated based on a daily price for each eligible security;
- maximum amount of credit might be limited;
- all credits must be reimbursed at the end of a business day.

To model these conditions there are two objects: **credit memorandum balance** (CMB) and **securities position valuation** (SPV). They allow establishing the links between the cash and securities accounts of participants providing and receiving credit to control the movements of resources on the related accounts. Depending on who owns and who uses a CA there are two types of CMB:

primary CMB establishes a link between CA and SA, both belonging to a payment bank. When there is not enough cash on the CA, then some credit might be provided from a central bank account. Central bank sets ACO credit limit for the payment bank. CMB logs amount of credit provided by central bank and headroom - remaining amount that a payment bank can still use without exceeding central bank ACO limit for current day.

secondary CMB establishes a link between CA of a payment bank and SA of a client of the payment bank. In this case the client uses the bank's CA for its transactions. The client has authorised usage (AU) limit. When the AU limit is reached, i.e., the client does not have anymore cash, then the bank provides a limited credit up to ACO limit. These limits vary according to the business needs, credit-worthiness of the client and legal rules of countries involved. Secondary CMB logs AU and ACO limits, and their headrooms.

We model CMB as a vector $(cmb^{type}, cb_{cmb}^P, cb_{cmb}^R, au_{cmb}^i, aco_{cmb}^i)$, where cmb^{type} is a primary or secondary CMB; cb_{cmb}^P, cb_{cmb}^R are the CBs providing and receiving the credit via ACO mechanism, respectively; au_{cmb}^i (only in case of secondary CMB) is the initial AU headroom; aco_{cmb}^i is the initial ACO headroom. These components are input data.

We model SPV as a vector $(s_{spv}, cmb_{spv}, sp_{spv}^P, sp_{spv}^R, p_{spv}, y_{spv}^q, y_{spv}^a)$, where s_{spv} is the security eligible for collateral; cmb_{spv} is CMB linking with a credit provider; sp_{spv}^P, sp_{spv}^R are the SPs providing and receiving the pledged securities, respectively; p_{spv} is a daily price of the security valuated for cmb_{spv} to calculate credit (or collateral amount); y_{spv}^q is the quantity of pledged securities (or collateral quantity); y_{spv}^a is the amount of cash lent (or collateral amount). Components $s_{spv}, cmb_{spv}, sp_{spv}^P, sp_{spv}^R$, and p_{spv} are input data, whereas values of y_{spv}^q and y_{spv}^a have to be found by MOM.

3.4 Assumptions

Without loss of generality, we exclude some business features from the paper, however, MOM handles them properly. Here are some of them:

- SA (CA) may have multiple SPs (CBs);
- order in which SPs (CBs) of the same SA (CA) are debited or credited;
- different restriction types for SPs and CBs (e.g., "reserved", "blocked" etc.);
- so-called "realignment chains" arising when there are cross-border transactions and resulting in additional technical transactions;
- some other credit limits in case of secondary CMB;
- some types of transactions which are in minority during NTS process;
- so-called "reverse transactions" reimbursing provided collateral amount.

3.5 Example

In this paragraph we demonstrate some notions introduced in the previous sections on a simple example.

Let payment bank $PB1$ sell 10 securities A. A buyer, payment bank $PB2$, possesses 50 securities A on its SP (securities on stock) and does not have any cash on its CB, Fig. 2. Seller $PB1$ has to deliver 10 securities to buyer $PB2$ against 20 euros. This is a DVP business transaction between two payment banks that generates two financial movements between CBs and SPs. Buyer's SP and CB are linked through a primary CMB cmb_1. Let initial available headroom aco_{cmb}^i be 100 euros. Providing account cb_{cmb}^P is a CB of central bank, that is its balance is not limited. The buyer's corresponding SPV object is composed of security A (s_{spv}); primary cmb_1 (cmb_{spv}); buyer's SP from where the pledged

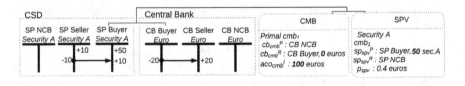

Fig. 2. DVP business transaction.

securities are taken (sp_{spv}^P); NCB's SP to receive the pledged securities (sp_{spv}^R); and security price (p_{spv}), let it be 0.4 euro.

Since buyer $PB2$ does not have enough cash to acquire securities but it has 50 securities on its SP, the ACO-mechanism is triggered. Figure 3 demonstrates the settlement process using the ACO-mechanism:

- collateral of 50 securities (y_{spv}^q) is taken from the buyer's SP and transferred to the NCB's SP linked via the CMB;
- ACO credit y_{spv}^a of 20 euros $(= 50 \cdot 0.4)$ is granted and transferred from NCB's CB to buyer's CB;
- ACO headroom is reduced to 80 euros $(= 100 - 20)$.

ACO-mechanism creates one technical DVP transaction t to move the collateral and the credit, respectively. This transaction must be settled together with the DVP business transaction.

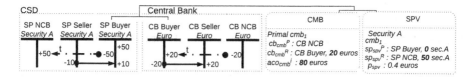

Fig. 3. Settlement using ACO-mechanism.

In real situation, the buyer may use the acquired securities as a collateral to obtain enough cash (securities on flow). Also, it is possible to use both collateral on flow and on stock composed of different securities. MOM manages both cases.

3.6   NTS Process

Figure 4 visualises NTS process which is a sequence of seven optimisation runs $R0-R6$ operating with batches of transactions of different types, financial nature and authorisation to be partially settled. Run $R0$ deals with only cash related transactions that "feed" CBs with resources and is solved by a particular algorithm out of MOM. Runs $R1-R6$ call MOM to settle cash and securities related transactions. Batch of transactions eligible for each run is composed of ones unsettled during a precedent run (dashed arrows in Fig. 4), new transactions specific to each run, and transactions arrived after the start of the precedent run. $R4$ is the largest in amount and volume, $R5$ solves newly entered and previously unsettled transactions. Run $R6$ authorises partial settlement of all transactions eligible for this functionality and failed to be fully settled before. Duration of each run varies and depends on batches sizes, transactions types and algorithms behaviors on given data. Execution time of each run is strictly limited to 15 min for runs $R1$, $R2$, and $R3$; and 45 min for runs $R4$, $R5$, and $R6$; that is in total 2 h 30 min for all runs of MOM. Taking into account the specific characteristics of runs, MOM uses different optimisation algorithms presented in Sect. 5.

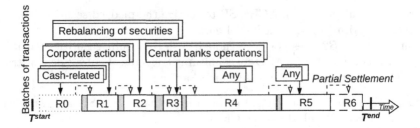

Fig. 4. Sequence of MOM runs during the NTS period.

4 Problem Formulation

In this section we introduce notations, variables and constraints to formulate NTSP as a mixed integer program (MIP) taking into account the assumptions listed in Sect. 3.4.

Input Data. Table 1 introduces the mathematical notations of input sets and data used in NTSP.

Variables. Let us introduce the following

binary variables $x_t = \begin{cases} 1, & \text{if transaction } t \text{ is settled, } \forall t \in \mathbb{T} \backslash \mathbb{T}^p \\ 0, & \text{otherwise;} \end{cases}$ continuous

variables x_t the settlement ratio of transaction t that can be partially settled, $x_t \in [0,1], \forall t \in \mathbb{T}^p$; and nonnegative integer variables

x_{spv}^{nbL} the number of lots of securities s_{spv} for $spv, \forall spv \in \mathbb{SPV}$;
x_t^{nbL} the number of lots of securities s_t for transaction $t, \forall t \in \mathbb{T}^p$;
y_{spv}^a the collateral amount associated with $spv, \forall spv \in \mathbb{SPV}$;
y_{spv}^q the quantity of pledged securities associated with $spv, \forall spv \in \mathbb{SPV}$;
z_{cmb}^a the collateral amount for $cmb, \forall cmb \in \mathbb{CMB}^1 \cup \mathbb{CMB}^2$;
aco_{cmb}^{Need} the necessary ACO amount related to $cmb, \forall cmb \in \mathbb{CMB}^1 \cup \mathbb{CMB}^2$.

Objective Function. Let R_k^A and R_k^V be the settlement amount and volume ratios for transactions of priority k, respectively:

$$R_k^A = \frac{\sum_{t \in \mathbb{T}_k} ra_t \cdot x_t}{\sum_{t \in \mathbb{T}_k} ra_t}, \quad R_k^V = \frac{\sum_{t \in \mathbb{T}_k} x_t}{|\mathbb{T}_k|} \tag{1}$$

MOM's has a twofold objective: to maximise the total settled amount and to maximise the total settled volume. Thus, the objective function is a sum of weighted ratios with respect to priority and with equal weights regarding settled amount and volume as follows:

$$\max \sum_{k \in \mathbb{K}} (R_k^A + R_k^V) \cdot w_k. \tag{2}$$

Table 1. Input sets and data notations.

Input sets	
CB^+, CB^-	Sets of CBs not allowed and allowed to be negative, resp.
SP^+, SP^-	Set of SPs not allowed and allowed to be negative, resp.
S	Set of all securities
T	Set of all transactions to settle
T^p	Set of all transactions that might be partially settled, $T^p \subseteq T$
CMB^1, CMB^2	Sets of primary and secondary CMBs, resp.
SPV	Set of SPVs
$\mathbb{K} = \{1, \ldots, 16\}$	Set of possible transaction priorities; "1" is the highest priority
T_k, $\forall k \in \mathbb{K}$	Set of transactions having priority k, $T = \bigcup_{k \in \mathbb{K}} T_k$
$\{U, C\}$	Set of modes to partially settle transactions
$\forall b \in CB^+ \cup CB^-$	
T_b^c, T_b^d	Set of transactions crediting and debiting CB b, resp., $T_b^c, T_b^d \subseteq T$
$\forall cmb \in CMB^1 \cup CMB^2$	
T_{cmb}^c, T_{cmb}^d	Set of transactions crediting and debiting cb_{cmb}^R, resp., $T_{cmb}^c, T_{cmb}^d \subseteq T$
SPV_{cmb}	Set of SPVs linked to cmb, $SPV_{cmb} \subseteq SPV$
$\forall p \in SP$	
T_p^c, T_p^d	Set of transactions crediting and debiting SP p, resp., $T_p^c, T_p^d \subseteq T$
SPV_p^c	Set of SPVs where SP p is sp_{spv}^R, $\forall spv \in SPV$
SPV_p^d	Set of SPVs where SP p is sp_{spv}^P, $\forall spv \in SPV$
Input data	
$w_k > 0$, $\forall k \in \mathbb{K}$	Weight associated with priority k and used in the objective function
q_p, $\forall p \in SP^+ \cup SP^-$	Initial quantity of securities available on SP p
$q_p \geq 0$, $\forall p \in SP^+$	
a_b, $\forall b \in CB^+ \cup CB^-$	Initial amount of cash available on CB b
$a_b \geq 0$, $\forall b \in CB^+$	
cmb_b^1, $\forall b \in CB$	Primary CMB, where CB b is cb_{cmb}^R
$\forall t \in T$	
$ra_t \geq 0$	Remaining amount of cash to settle for transaction t
$rq_t \geq 0$	Remaining quantity of securities to settle for transaction t
s_t, $\forall t \in T$	Security used in transaction t
$k_t \in \mathbb{K}$	priority of transaction t
$\forall t \in T^p$ in case of partial settlement of transaction t	
$m_t^p \in \{U, C\}$	Partial settlement mode for transaction t
$\forall s \in S$	
$q_s^{min} \geq 0$	Minimum quantity of securities s to settle
$a_s^{min} \geq 0$	Minimum amount valued for security s to settle
n_s^{min}	Minimum number of securities s that can be pledged
n_s^{lot}	Number of securities s per lot
$l_{t_1 t_2}^W = \begin{cases} 1, & \text{if transaction } t_1 \text{ must be settled with transaction } t_2, \forall t_1, t_2 \in T \setminus T^p \\ 0, & \text{otherwise} \end{cases}$	
$l_{t_a t_b}^A = \begin{cases} 1, & \text{if transaction } t_a \text{ must be settled after transaction } t_b, \forall t_a, t_b \in T \setminus T^p \\ 0, & \text{otherwise} \end{cases}$	
$\forall spv \in SPV$	
cmb_{spv}	CMB associated with SPV spv, $cmb_{spv} \in CMB^1 \cup CMB^2$
s_{spv}	Security eligible for pledge under SPV spv
$p_{spv} > 0$	Daily price of eligible security to calculate the credit amount
$au_{cmb}^i \geq 0$, $\forall cmb \in CMB^2$	initial headroom of authorised usage of balance cb_{cmb}^R
aco_{cmb}^i, $\forall cmb \in CMB^1 \cup CMB^2$	initial ACO headroom associated with balance cb_{cmb}^R
$aco_{cmb}^{min} \geq 0$, $\forall cmb \in CMB^1 \cup CMB^2$	minimum allowed amount of ACO related to cmb

Notice that, the objective function does not take into account the quantity of securities settled within each transaction. For each priority level, only transactions with payment (type DVP, $PFOD$) may contribute to both ratios in (2), whereas each transaction of type FOP may contribute only to volume ratio.

Constraints.

$$x_{t_1} \leq x_{t_2} + 1 - l^A_{t_1 t_2}, \quad x_{t_1} \leq x_{t_2} + 1 - l^W_{t_1 t_2}, \quad x_{t_2} \leq x_{t_1} + 1 - l^W_{t_1 t_2}, \quad \forall t_1, t_2 \in \mathbb{T} \backslash \mathbb{T}^p \quad (3)$$

Constraints (3) guaranty the right settlement of linked transactions not authorised for partial settlement.

$$\sum_{t \in \mathbb{T}^d_b} ra_t \cdot x_t - \sum_{t \in \mathbb{T}^c_b} ra_t \cdot x_t - z^a_{cmb^1_b} \leq a_b, \quad \forall b \in \mathbb{CB}^+ \quad (4)$$

Constraints (4) guaranty the non-negative final cash balance.

$$\sum_{t \in \mathbb{T}^d_p} rq_t \cdot x_t + \sum_{spv \in \mathbb{SPV}^d_p} y^q_{spv} - \sum_{t \in \mathbb{T}^c_p} rq_t \cdot x_t - \sum_{spv \in \mathbb{SPV}^c_p} y^q_{spv} \leq q_p, \forall p \in \mathbb{SP}^+ \quad (5)$$

Constraints (5) guaranty the non-negative balance for each position p from $\mathbb{SP}^+$.

$$z^a_{cmb} = \sum_{spv \in \mathbb{SPV}_{cmb}} y^a_{spv}, \quad \forall cmb \in \mathbb{CMB}^1 \cup \mathbb{CMB}^2 \quad (6)$$

Constraints (6) calculate the total collateral amount for each cmb.

$$\text{if } aco^{Need}_{cmb} = 0 \text{ then } z^a_{cmb} = 0, \quad \forall cmb \in \mathbb{CMB}^1 \cup \mathbb{CMB}^2 \quad (7)$$

Constraints (7) establish a link between z^a_{cmb} and aco^{Need}_{cmb}: if there is no need for ACO in cmb, then the corresponding collateral amount must be zero.

$$aco^{Need}_{cmb} = \begin{cases} 0, \quad \forall cmb \in \mathbb{CMB}^1 \text{ such that } cb^R_{cmb} \in \mathbb{CB}^- \\ \max\{0, \sum_{t \in \mathbb{T}^d_{cmb}} ra_t \cdot x_t - \sum_{t \in \mathbb{T}^c_{cmb}} ra_t \cdot x_t - a_{cb^R_{cmb}}\}, \\ \quad \forall cmb \in \mathbb{CMB}^1 \text{ such that } cb^R_{cmb} \in \mathbb{CB}^+ \\ \max\{0, \sum_{t \in \mathbb{T}^d_{cmb}} ra_t \cdot x_t - \sum_{t \in \mathbb{T}^c_{cmb}} ra_t \cdot x_t - \max\{0, au^i_{cmb}\}\}, \\ \quad \forall cmb \in \mathbb{CMB}^2 \end{cases} \quad (8)$$

Constraints (8) calculate the lack of cash (aco^{Need}_{cmb}) on the cash balance cb^R_{cmb} receiving the credit.

$$z^a_{cmb} \leq \max\{0, aco^i_{cmb}\}, \quad \forall cmb \in \mathbb{CMB}^1 \cup \mathbb{CMB}^2 \quad (9)$$

Constraints (9) guaranty non-negative flow on the collateral headroom for each cmb. If the initial ACO headroom is negative, then there is no collateral for this cmb.

$$\sum_{t\in\mathbb{T}^d_{cmb}} ra_t \cdot x_t - \sum_{t\in\mathbb{T}^c_{cmb}} ra_t \cdot x_t - z^a_{cmb} \leq \max\{0, au^i_{cmb}\}, \forall cmb \in \mathbb{CMB}^2 \quad (10)$$

Constraints (10) guaranty non-negative headroom for each secondary cmb.

$$\text{if } y^q_{spv} > 0 \text{ then } y^q_{spv} = x^{nbL}_{spv} \cdot n^{lot}_{s_{spv}} \text{ and } y^q_{spv} \geq n^{min}_{s_{spv}}, \quad \forall spv \in \mathbb{SPV} \quad (11)$$

Constraints (11) respect the rules to pledge securities by lots and minimal collateral quantity.

$$\text{if } y^q_{spv} > 0 \text{ then } y^a_{spv} = p_{spv} \cdot y^q_{spv} \text{ and } y^a_{spv} \geq aco^{min}_{cmb_{spv}}, \quad \forall spv \in \mathbb{SPV} \quad (12)$$

Constraints (12) calculate the amount brought in by the ACO and check that it satisfies to the minimum threshold amount for cmb associated with spv.

$$\text{if } 0 < x_t < 1 \text{ then } rq_t \cdot x_t \geq q^{min}_{s_t}, \quad \forall t \in \mathbb{T}^p \text{ such that } m^p_t \in \{U, C\} \quad (13)$$

$$\text{if } 0 < x_t < 1 \text{ then } ra_t \cdot x_t \geq a^{min}_{s_t}, \quad \forall t \in \mathbb{T}^p \text{ such that } m^p_t = C \quad (14)$$

$$\text{if } 0 < x_t < 1 \text{ then } rq_t \cdot x_t = n^{lot}_{s_t} \cdot x^{nbL}_t, \quad \forall t \in \mathbb{T}^p \text{ such that } m^p_t = U \quad (15)$$

Constraints (13)–(15) check the thresholds related to the different modes for partial settlement. If transactions are of mode U or C, then the minimum quantity threshold $q^{min}_{s_t}$ must be respected in (13), and, also, the minimum amount threshold $a^{min}_{s_t}$ for mode C in (14) or settlements by lots for mode U in (15).

Note that in (7)–(15) we keep "max" and "$if\ldots then$" for two reasons. First, to avoid overloading of MIP of NTSP with auxiliary variables and constraints in the paper. Second, we use CPLEX to solve MIPs (and associated linear programming problems) where relations "max" and "$if\ldots then$" can be modeled in two ways: via additional constraints with auxiliary binary variables and big constants, or via "indicator" constraints introduced in CPLEX 10 [11]. In MOM we use both ways.

5   Solution Approaches

MOM is composed of three solution phases and multiple algorithms as it is presented in Fig. 5. The NTSP MIP is too large to be solved directly by a general purpose optimizer. Thus, the preparatory phase reduces the problem size. Filtering procedure analyses all transactions and eliminates those that cannot or can be obviously settled. E.g., if transaction t_1 debits more resources than a CB (or SP) has, including resources that might be brought by incoming transactions, then $x_{t_1} = 0$ and t_1 is eliminated from optimization. If a CB is debited by the only one transaction t_2 and has enough cash to settle t_2, then $x_{t_2} = 1$ and t_2 is eliminated from optimization. The filtering procedure may also eliminate some

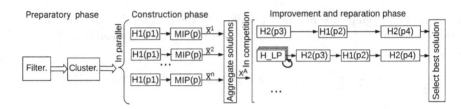

Fig. 5. MOM's algorithmic components

SPVs. E.g., if the quantity of all potentially pledged securities is low than $n_{s_{spv}}^{min}$ then this *spv* is eliminated.

Clusterisation step splits all remaining transactions into subsets to solve smaller problems. The subset size is limited, it is a parameter to tune. All linked transactions must be in the same subset. Clusterisation is based on ISIN code that uniquely identifies a specific security emission. On the other hand, each cluster corresponds to a connected component in a securities graph with SPs as nodes and transactions as arcs. We developed different strategies to allocate available resources (amount of cash on a balance and quantity of securities on a position) between subsets. One of them assigns the resources on a pro rata basis in accordance with the total amount of transactions from the same subset.

A construction phase launches a suite of algorithms $H1$ and MIP with parameters $p1$ and p, respectively, for each subset in parallel. Heuristic $H1(p1)$ considers unsettled transactions sorted in descending order according to priority and settlement ratios and tries to settle one after another. The attempt is successful if there are enough available resources to settle a transaction taking into account ACO credit. $MIP(p)$ solves MIP (2)–(10) by CPLEX with the subsolution obtained by $H1(p1)$ as initial solution. Parameter p allows (de)activating certain cuts (additional inequalities) in MIP. An aggregation step unifies the subsolutions $\bar{x}^1, \ldots, \bar{x}^n$ to each subset. The obtained solution x^A may still violate constraints (11)–(15).

A reparation phase launches several sequences of algorithms in competition to improve and get valid solution. Heuristic $H2$ verifies and makes all functional constraints satisfied. If a constraint is not satisfied then $H2$ deselects settled transactions one after another and propagates the impact on available resources and other transactions. Parameters $p3$ and $p4$ specify whether constraints (11)–(15) and others (mentioned in Sect. 3.4) must be respected or not. Heuristic $H1(p2)$ starts as $H1(p1)$ and if the attempt to settle transaction t is failed, $H1(p2)$ does not stop but looks for an adjacent (debiting the same balance or position) transaction to unsettle, bring some resource and improve the objective function value. The first found transaction resulting in successful attempt is selected. Parameter $p2$ regulates the number of adjacent transactions to explore. This algorithm reminds a stochastic local search.

H_{LP} iteratively solves the linear relaxation of (2)–(10) with additional constraints. Let x^i be solution obtained at iteration i, $k^H = \min\{k_t | t \in \mathbb{T}, x_t^i < 1\}$. Then, at iteration $i+1$, $\forall t \in \mathbb{T} \mid x_t^i < 1$ and $k_t = k^H$:

1. if $t \notin \mathbb{T}^p$ then variable x_t is fixed to 0;
2. otherwise, if $t \in \mathbb{T}^p$ constraint $x_t \leq \lfloor \frac{rq_t \cdot x_t^i}{n_{st}^{lot}} \rfloor \cdot \frac{n_{st}^{lot}}{rq_t}$ is added to respect (15).

The process stops when x_t is integer $\forall t \in \mathbb{T} \setminus \mathbb{T}^p$. Note that model (2)–(10) has always a trivial feasible solution (all variables are zero).

Several combinations of $H1$, $H2$, and H_{LP} with different parameters are used during the reparation phase. The best found solution is returned at the end.

6   Computational Results

MOM is coded in C++, uses CPLEX V.12.6.1 and runs on a mainframe computer IBM z15/OS with 16 dedicated processors. We present the computational results obtained on real data for 40 business days. In this data set, there were around 340 000 transactions for the total amount of around EUR 457 billion in average per day composed of 82% of type DVP, 16% of type FOP, and 2% of type $PFOD$. Figure 6 shows input data, the daily amount and volume to settle expressed as percentages over the considered period. Days 20 and 28 had the lowest amount and volume, respectively, and day 40 was the largest in both amount and volume to settle.

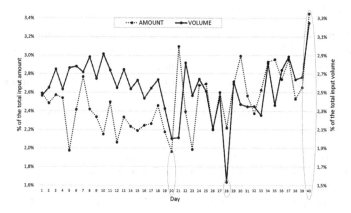

Fig. 6. Input data

We present here the final simulation results obtained with the configurations of algorithms and parameter settings in production conditions without detailing how they were tuned. Figure 7 shows the following indicators for each day d: the production computational time of MOM (right vertical axis), and the efficiency settlement ratios (left vertical axis) for amount and volume, respectively, calculated as follows:

$$SR_d^A(x) = \frac{\sum_{t \in \mathbb{T}_d} ra_t \cdot x_t}{\sum_{t \in \mathbb{T}_d} ra_t} \cdot 100\% \text{ and } SR_d^V(x) = \frac{\sum_{t \in \mathbb{T}_d} x_t}{|\mathbb{T}_d|} \cdot 100\%,$$

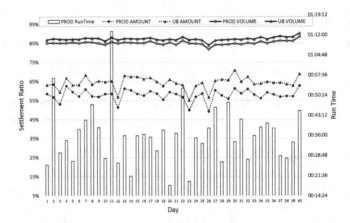

Fig. 7. Night-time settlement process results

where $\mathbb{T}_d$ is the total number of transactions during NTS period of business day d. As an optimal solution is unknown, we compare SR_d^A and SR_d^V for MOM solution x with their upper bounds in order to have an idea of the quality of obtained solutions. We calculated two upper bounds: on the total settled amount and volume separately, as follows:

– for each business day input data was gathered over all runs into one batch of transactions, and MOM ran only one time on the aggregated data;
– any transaction might have been partially settled;
– only the functional constraints related to the thresholds on minimum settled cash amount and lot size were omitted;
– two objective functions to maximise total settled amount and volume were optimised separately.

Let x^{UB_A} and x^{UB_V} be the solutions corresponding to the upper bounds on amount and volume, respectively. In Fig. 7 the ratios $SR_d^V(x^{UB_V})$, $SR_d^V(x)$, $SR_d^A(x^{UB_A})$, and $SR_d^A(x)$ are represented by curves UB VOLUME, PROD VOLUME, UB AMOUNT, and PROD AMOUNT for each day of the considered period.

Each run of MOM has to be finished as fast as possible and provide good quality solutions. The longest process took 1 h 13 min and the average computation time was 36 min long.

Note that reaching 100% ratios is unlikely to occur due to the links between transactions, their types and limited resources. MOM gives better results regarding the settled volume as curve PROD VOLUME is above PROD AMOUNT. In average, MOM is able to settle 85% of the total input volume versus 58% of the total input amount. For the considered data set, the largest difference between the upper bound and MOM settlement ratios was 10 points for amount and 3 points for volume. These gaps are not large, however, we believe that they depend on the topology of input data. As we do not know where the optima are located, it is difficult to say how much these gaps can be further reduced. Based on the results observed, it is difficult to conclude whether there is strong correlation between the size of instances (total input amount and volume) and the observed settlement indicators. This question is still open and under research.

7   Conclusions

The presented securities and cash settlement framework implemented in MOM and integrated in the innovational European platform allows managing hundred thousands of transactions over a billion euro every day. This is achieved by the combination of operations research optimisation approaches and state-of-the-art software. Results indicate the good solution efficiency and settlement ratios regarding amount and volume of transactions. However, when the new actors joint T2S platform, they might change the data topology (nature of transactions, links between them, etc.). Thus, to guarantee the robustness and high quality results of MOM, some future directions include the study of data topology, its impact on the calculation time and quality of the solutions obtained, and upgrade of the MOM's algorithms.

References

1. Asbjørn, E., Øverli, F.: Intraday liquidity and the settlement of large-value payments: a simulation-based analysis. Econ. Bull. 1(LXXVII), 41–48 (2006)
2. Güntzer, M.M., Jungnickel, D., Leclerc, M.: Efficient algorithms for the clearing of interbank payments. Eur. J. Oper. Res. 106(1), 212–219 (1998)
3. Klotz, E., Newman, A.M.: Practical guidelines for solving difficult linear programs. Surv. Oper. Res. Manag. Sci. 18, 18–32 (2013)
4. Leinonen, H.: Liquidity, risks and speed in payment and settlement systems: a simulation approach. Bank of Finland Studies (2005)
5. Leinonen, H., Soramaki, K.: Simulation: a powerful research tool in payment and settlement systems. Paym. Syst. Worldw. 15, 28–33 (2004)
6. Manfred, G., Maringer, D., Schumann, E.: Numerical Methods and Optimization in Finance. Academic Press, Cambridge (2019)
7. Shafransky, Y.M., Doudkin, A.A.: An optimization algorithm for the clearing of interbank payments. Eur. J. Oper. Res. 171(3), 743–749 (2006)
8. Sutcliffe, B.J., Charles, Z., William, T.: Applying operations research techniques to financial markets. Interfaces 33, 12–24 (2003)
9. Toth, P., Martello, S.: Knapsack Problems: Algorithms and Computer Implementations. Wiley, New York (1990)
10. ECB homepage. https://www.ecb.europa.eu/paym/target/target2/html/index-.en.html. Accessed 14 Feb 2020
11. IBM homepage. https://www.ibm.com/support/pages/difference-between-using-indicator-constraints-and-big-m-formulation. Accessed 14 Feb 2020

A Stable Alternative to Sinkhorn's Algorithm for Regularized Optimal Transport

Pavel Dvurechensky[1,3]($\boxtimes$) (iD), Alexander Gasnikov[2,3] (iD), Sergey Omelchenko[2] (iD), and Alexander Tiurin[4] (iD)

[1] Weierstrass Institute for Applied Analysis and Stochastics, Berlin, Germany
pavel.dvurechensky@gmail.com
[2] Moscow Institute of Physics and Technology, Moscow, Russia
[3] Institute for Information Transmission Problems RAS, Moscow, Russia
[4] National Research University Higher School of Economics, Moscow, Russia

Abstract. In this paper, we are motivated by two important applications: entropy-regularized optimal transport problem and road or IP traffic demand matrix estimation by entropy model. Both of them include solving a special type of optimization problem with linear equality constraints and objective given as a sum of an entropy regularizer and a linear function. It is known that the state-of-the-art solvers for this problem, which are based on Sinkhorn's method (also known as RSA or balancing method), can fail to work, when the entropy-regularization parameter is small. We consider the above optimization problem as a particular instance of a general strongly convex optimization problem with linear constraints. We propose a new algorithm to solve this general class of problems. Our approach is based on the transition to the dual problem. First, we introduce a new accelerated gradient method with adaptive choice of gradient's Lipschitz constant. Then, we apply this method to the dual problem and show, how to reconstruct an approximate solution to the primal problem with provable convergence rate. We prove the rate $O(1/k^2)$, k being the iteration counter, both for the absolute value of the primal objective residual and constraints infeasibility. Our method has similar to Sinkhorn's method complexity of each iteration, but is faster and more stable numerically, when the regularization parameter is small. We illustrate the advantage of our method by numerical experiments for the two mentioned applications. We show that there exists a threshold, such that, when the regularization parameter is smaller than this threshold, our method outperforms the Sinkhorn's method in terms of computation time.

Submitted to the editors DATE. This research was funded by Russian Science Foundation (project 18-71-10108).

Electronic supplementary material The online version of this chapter (https://doi.org/10.1007/978-3-030-49988-4_28) contains supplementary material, which is available to authorized users.

A. Kononov et al. (Eds.): MOTOR 2020, LNCS 12095, pp. 406–423, 2020.
https://doi.org/10.1007/978-3-030-49988-4_28

Keywords: Smooth convex optimization · Linear constraints ·
First-order methods · Accelerated gradient descent · Algorithm
complexity · Entropy-linear programming · Dual problem · Primal-dual
method · Sinkhorn's fixed point algorithm · Entropy-regularized
optimal transport · Traffic demand matrix estimation

1 Introduction

The main problem, we consider, is convex optimization problem of the following
form

$$(P_1) \qquad \min_{x \in Q \subseteq E} \{f(x) : A_1 x = b_1, A_2 x - b_2 \in -K\},$$

where E is a finite-dimensional real vector space, Q is a simple closed convex
set, A_1, A_2 are given linear operators from E to some finite-dimensional real
vector spaces H_1 and H_2 respectively, $b_1 \in H_1$, $b_2 \in H_2$ are given, $K \subseteq H_2$
is some cone, $f(x)$ is a γ-strongly convex function on Q with respect to some
chosen norm $\| \cdot \|_E$ on E. The last means that, for any $x, y \in Q$, $f(y) \geq f(x) +$
$\langle \nabla f(x), y - x \rangle + \frac{\gamma}{2} \|x - y\|_E^2$, where $\nabla f(x)$ is any subgradient of $f(x)$ at x and
hence is an element of the dual space E^*. Also we denote the value of a linear
function $\lambda \in E^*$ at $x \in E$ by $\langle \lambda, x \rangle$.

We are motivated to consider the described class of problems by two particular applications. The first one comes from transportation research and consists
in recovering a matrix of traffic demands between city districts from the information on population and workplace capacities of each district. As it is shown in
[23], a natural model of districts' population dynamics leads to an entropy-linear
programming optimization (see (9) below for the precise formulation) problem
for the traffic demand matrix estimation. In this case, the objective function in
(P_1) is a sum of an entropy function and a linear function. It is important to
note also that the entropy function is multiplied by a regularization parameter
γ and the model is close to reality, when the regularization parameter is small.
The same approach is used in IP traffic matrix estimation [54]. Close problems
arise also in more complicated congestion traffic modelling [4].

The second application is the calculation of regularized optimal transport
(ROT) between two probability measures introduced in [12]. The idea is to reg-
ularize the objective function in the classical optimal transport linear programming problem [31] by entropy of the transportation plan. This leads to the same
type of problem with a regularization parameter as in the traffic demands matrix
estimation. For the detailed problem statement, see (9). As it is argued in [13],
for the case of discretization of continuous probability measures, entropy regularization allows to obtain a better approximation for the optimal transportation
plan than the solution of the original linear programming problem. At the same
time, the regularization parameter γ should be small. Otherwise, the solution
of the regularized optimal transport problem will be a bad approximation for
the original optimal transport problem. To sum up, in both applications, it is
important to solve regularized problems with *small regularization parameter*.

The problem statement (P_1) covers many other applications besides mentioned above. For example, general entropy-linear programming (ELP) problem [20] arises in econometrics [24], modeling in science and engineering [32]. Such machine learning approaches as ridge regression [28] and elastic net [55] lead to the same type of problem.

1.1 Related Work

Sinkhorn's, RSA or Balancing Type Methods. Special types of Problem (P_1), such as traffic matrix estimation and regularized optimal transport, have efficient matrix-scaling-based solvers such as balancing algorithm, [8], Sinkhorn's method, [12,46], RAS algorithm [30]. Strong points of these algorithms are fast convergence in practice and easy parallel implementation. At the same time, these algorithms are suitable only for Problem (P_1) with special type of linear equality constraints. A generalization for a problem with a special type of linear inequalities constraints was suggested in [6], but without convergence rate estimates. Recently, [11] extended the approach of [12] for other special classes of entropy-minimization problems.

The problem of instability of the matrix-scaling approach for problems with small regularization parameter was addressed in [44], but the proposed techniques are less suitable for parallel computations than the initial algorithm. There is a proof of linear convergence of the Sinkhorn's method [21], but the theoretical bound is much worse than the rate in practice and theoretical rate is obtained in terms of convergence in a special metric, which is hard to interpret. The papers [2,18,27,35] analyse complexity of the Sinkhorn's algorithm to find an approximate solution to the regularized and non-regularized optimal transport problem. In particular, they show that the regularization parameter needs to be of the order of the desired accuracy, which can lead to the instability of the Sinkhorn's algorithm. An alternative matrix scaling algorithm was proposed in [1] together with theoretical analysis, but this method seems to be hard to implement in practice and no experimental results were reported.

In any case, all the mentioned algorithms are designed for a special instance of Problem (P_1).

First-Order Methods for Constrained Problems. We consider Problem (P_1) in large-scale setting, when the natural choice is some first-order method. Due to the presence of linear constraints, the applicability of projected-gradient-type methods to the primal problem is limited. Thus, the most common approach involves construction of the dual problem and primal-dual updates during the algorithm progress. There are many algorithms of this type like ADMM [7,25] and other primal-dual methods [5,9,19], see the extensive review in [48]. As it is pointed in [48], these methods have the following drawbacks. They need the tractability assumption of the proximal operator for the function f and some additional assumptions. These methods don't have appropriate convergence rate characterization: if any, the rates are non-optimal and are either only for the dual problem or for some weighted sum of primal objective residual and linear constraints infeasibility. In [48], the authors themselves develop a good alternative, based only on the assumption of proximal tractability of the function f, but only for problems with linear equality constraints. This approach was further developed in [53] for more general types of constraints. The key feature of the algorithm developed there is its adaptivity to the unknown level of smoothness in the dual problem. Nevertheless, the provided stopping criterion, which is based on the prescribed number of iterations, requires to know all the smoothness parameters. Further, in [49], the authors propose algorithms with optimal rates of convergence for a more general class of problems, but, for the case of strongly convex f, they assume that it is strongly convex with respect to a Euclidean-type norm. Thus, their approach is not applicable to entropy minimization problems, which are our main focus.

An advanced ADMM with provable convergence rate with appropriate convergence characterization was proposed in [41], but only for the case of equality constraints and Lipschitz-smooth f, which does not cover the case of entropy minimization. A general primal-dual framework for unconstrained problems was proposed in [14], but it is not applicable in our setting. An adaptive to unknown Lipschitz constant algorithm for primal-dual problems was developed in [36], but the authors work with a different from our problem statement and the case of strongly convex objective is considered only in Euclidean setting, which also does not cover the case of entropy minimization.

Several recent algorithms [10, 17, 22, 26, 34, 39, 42] are based on the application of accelerated gradient method [37,38] to the dual problem and have optimal rates. At the same time, these works do not consider general types of constraints as in Problem (P_1). Also the proposed algorithms use, as an input parameter, an estimate of the Lipschitz constant of the gradient in the dual problem, which can be very pessimistic and lead to slow convergence.

The idea of primal-dual accelerated gradient methods turned out to be quite fruitful in the context of distributed decentralized optimization and it application to Wasserstein baeycenter problem [15, 16, 29, 33, 43, 50, 51].

1.2 Contributions

1. In contrast to the existing methods for constrained problems in [3, 5, 7, 9, 10, 22, 25, 34, 36, 42, 48, 49, 53], we propose an algorithm simultaneously for Problem (P_1) with general linear equality and cone constraints; with optimal rate of convergence in terms of both primal objective residual and constraints infeasibility; with adaptivity to the Lipschitz constant of the objective's gradient; with online stopping criterion, which does not require the knowledge of this Lipschitz constant; with ability to work with entropy function as f. The main difference with [18] is that here we consider more general cone constraints and consider not only regularized optimal transport problems as application.

2. In contrast to existing Sinkhorn's-algorithm-based algorithms for solving entropy-regularized optimal transport problems [1, 2, 6, 8, 12, 30, 44, 46], we provide an algorithm simultaneously with provable convergence rate, easy implementability in practice and higher stability, when the regularization parameter is small.

3. In the experiments, we show that our algorithm is better than the Sinkhorn's method in situations of small regularization parameter in the primal problem, which means that the dual problem becomes less smooth problem.

The rest of the paper is organized as follows. In Sect. 2, we introduce notation, definition of approximate solution to Problem (P_1), main assumptions, and particular examples of (P_1) in applications. Section 3 is devoted to primal-dual algorithm for Problem (P_1) and its convergence analysis. Finally, in Sect. 4, we present the results of the numerical experiments for regularized optimal transport and traffic matrix estimation problems.

2 Preliminaries

For any finite-dimensional real vector space E, we denote by E^* its dual. We denote the value of a linear function $\lambda \in E^*$ at $x \in E$ by $\langle \lambda, x \rangle$. Let $\|\cdot\|_E$ denote some norm on E and $\|\cdot\|_{E,*}$ denote the norm on E^* which is dual to $\|\cdot\|_E$, i.e. $\|\lambda\|_{E,*} = \max_{\|x\|_E \leq 1} \langle \lambda, x \rangle$. In the special case, when E is a Euclidean space, we denote the standard Euclidean norm by $\|\cdot\|_2$. Note that, in this case, the dual norm is also Euclidean. For a cone $K \subseteq E$, the dual cone $K^* \subseteq E^*$ is defined as $K^* := \{\lambda \in E^* : \langle \lambda, x \rangle \geq 0 \;\; \forall x \in K\}$. By $\partial f(x)$ we denote the subdifferential of a function $f(x)$ at a point x. Let E_1, E_2 be two finite-dimensional real vector spaces. For a linear operator $A : E_1 \to E_2$, we define its norm as follows

$$\|A\|_{E_1 \to E_2} = \max_{x \in E_1, u \in E_2^*} \{\langle u, Ax \rangle : \|x\|_{E_1} = 1, \|u\|_{E_2,*} = 1\}.$$

For a linear operator $A : E_1 \to E_2$, we define the adjoint operator $A^T : E_2^* \to E_1^*$ in the following way $\langle u, Ax \rangle = \langle A^T u, x \rangle$, $\forall u \in E_2^*$, $x \in E_1$. We say that a function $f : E \to \mathbb{R}$ has a L-Lipschitz-continuous gradient if it is differentiable and its gradient satisfies Lipschitz condition $\|\nabla f(x) - \nabla f(y)\|_{E,*} \leq L\|x - y\|_E$, $\forall x, y \in E$. Note that, from this inequality, it follows that

$$f(y) \leq f(x) + \langle \nabla f(x), y - x \rangle + \frac{L}{2} \|x - y\|_E^2, \quad \forall x, y \in E. \tag{1}$$

Also, for any $t \in \mathbb{R}$, we denote by $\lceil t \rceil$ the smallest integer greater than or equal to t.

We characterize the quality of an approximate solution to Problem (P_1) by three quantities $\varepsilon_f, \varepsilon_{eq}, \varepsilon_{in} > 0$.

Definition 1. *We say that a point $\hat{x}$ is an $(\varepsilon_f, \varepsilon_{eq}, \varepsilon_{in})$-solution to Problem (P_1) iff the following inequalities hold*

$$|f(\hat{x}) - Opt[P_1]| \leq \varepsilon_f, \quad \|A_1 \hat{x} - b_1\|_2 \leq \varepsilon_{eq}, \quad \rho(A_2 \hat{x} - b_2, -K) \leq \varepsilon_{in}. \tag{2}$$

Here $Opt[P_1]$ denotes the optimal function value for Problem (P_1),

$$\rho(A_2 \hat{x} - b_2, -K) := \max_{\lambda^{(2)} \in K^*, \|\lambda^{(2)}\|_2 \leq 1} \langle \lambda^{(2)}, A_2 \hat{x} - b_2 \rangle.$$

Note that the last inequality in (2) is a natural generalization of linear constraints infeasibility measure $\|(A_2 x_k - b_2)_+\|_2$ for the case $K = \mathbb{R}_+^n$. Here the vector v_+ denotes the vector with components $[v_+]_i = (v_i)_+ = \max\{v_i, 0\}$.

The Lagrange dual problem to Problem (P_1) is

$$(D_1) \quad \max_{\lambda \in \Lambda} \left\{ -\langle \lambda^{(1)}, b_1 \rangle - \langle \lambda^{(2)}, b_2 \rangle + \min_{x \in Q} \left(f(x) + \langle A_1^T \lambda^{(1)} + A_2^T \lambda^{(2)}, x \rangle \right) \right\}.$$

Here we denote $\Lambda = \{\lambda = (\lambda^{(1)}, \lambda^{(2)})^T \in H_1^* \times H_2^* : \lambda^{(2)} \in K^*\}$. It is convenient to rewrite Problem (D_1) in the equivalent form of a minimization problem

$$(P_2) \quad \min_{\lambda \in \Lambda} \left\{ \langle \lambda^{(1)}, b_1 \rangle + \langle \lambda^{(2)}, b_2 \rangle + \max_{x \in Q} \left(-f(x) - \langle A_1^T \lambda^{(1)} + A_2^T \lambda^{(2)}, x \rangle \right) \right\}.$$

It is obvious that

$$Opt[D_1] = -Opt[P_2], \tag{3}$$

where $Opt[D_1]$, $Opt[P_2]$ are the optimal function value in Problem (D_1) and Problem (P_2) respectively. The following inequality follows from the weak duality

$$Opt[P_1] \geq Opt[D_1]. \tag{4}$$

We denote

$$\varphi(\lambda) = \varphi(\lambda^{(1)}, \lambda^{(2)}) = \langle \lambda^{(1)}, b_1 \rangle + \langle \lambda^{(2)}, b_2 \rangle$$
$$+ \max_{x \in Q} \left(-f(x) - \langle A_1^T \lambda^{(1)} + A_2^T \lambda^{(2)}, x \rangle \right). \tag{5}$$

Since f is strongly convex, $\varphi(\lambda)$ is a smooth function and its gradient is equal to (see e.g. [38])

$$\nabla \varphi(\lambda) = \begin{pmatrix} b_1 - A_1 x(\lambda) \\ b_2 - A_2 x(\lambda) \end{pmatrix}, \tag{6}$$

where $x(\lambda)$ is the unique solution of the strongly-convex problem

$$\max_{x \in Q} \left(-f(x) - \langle A_1^T \lambda^{(1)} + A_2^T \lambda^{(2)}, x \rangle \right). \tag{7}$$

Note that $\nabla \varphi(\lambda)$ is Lipschitz-continuous (see e.g. [38]) with constant

$$L \leq \frac{1}{\gamma} \left(\|A_1\|_{E \to H_1}^2 + \|A_2\|_{E \to H_2}^2 \right).$$

Previous works [10,22,34,42] rely on this quantity in the algorithm and use it to define the stepsize of the proposed algorithm. The drawback of this approach is that the above bound for the Lipschitz constant can be way too pessimistic. In this work, we propose an adaptive method, which has the same complexity bound, but is faster in practice due to the use of a "local" estimate for L in the stepsize definition.

We assume that the dual problem (D_1) has a solution $\lambda^* = (\lambda^{*(1)}, \lambda^{*(2)})^T$ and there exist some $R_1, R_2 > 0$ such that

$$\|\lambda^{*(1)}\|_2 \le R_1 < +\infty, \quad \|\lambda^{*(2)}\|_2 \le R_2 < +\infty. \tag{8}$$

It is worth noting that the quantities R_1, R_2 will be used only in the convergence analysis, but not in the algorithm itself.

To motivate the considered problem we describe two particular problems which can be written in the form of Problem (P_1).

Traffic demand matrix estimation, [52], and Regularized optimal transport problem, [12].

$$\min_{X \in \mathbb{R}_+^{p \times p}} \left\{ \gamma \sum_{i,j=1}^p x_{ij} \ln x_{ij} + \sum_{i,j=1}^p c_{ij} x_{ij} : Xe = \mu, X^T e = \nu \right\}, \tag{9}$$

where $e \in \mathbb{R}^p$ is the vector of all ones, $\mu, \nu \in S_p(1) := \{x \in \mathbb{R}^p : \sum_{i=1}^p x_i = 1, x_i \ge 0, i = 1, ..., p\}$, $c_{ij} \ge 0, i, j = 1, ..., p$ are given, $\gamma > 0$ is the regularization parameter, X^T is the transpose matrix of X, x_{ij} is the element of the matrix X in the i-th row and the j-th column. This problem with small value of γ is our primary focus in this paper.

General entropy-linear programming problem, [20].

$$\min_{x \in S_n(1)} \left\{ \sum_{i=1}^n x_i \ln (x_i/\xi_i) : Ax = b \right\}$$

for some given $\xi \in \mathbb{R}_{++}^n = \{x \in \mathbb{R}^n : x_i > 0, i = 1, ..., n\}$.

3   Primal-Dual Algorithm

In this section, we return to the primal-dual pair of problems (P_1)–(D_1). We apply Algorithm 1 in the supplementary of [18] to Problem (P_2) and incorporate in the algorithm a procedure, which allows to reconstruct also an approximate solution of Problem (P_1). The main novelty of this paper is the primal-dual analysis of this algorithm in the presence of inequality constraints. We choose Euclidean proximal setup, which means that we introduce euclidean norm $\|\cdot\|_2$ in the space of vectors λ and choose the prox-function $d(\lambda) = \frac{1}{2}\|\lambda\|_2^2$. Then, we have $V[\zeta](\lambda) = \frac{1}{2}\|\lambda - \zeta\|_2^2$.

Our primal-dual algorithm for Problem (P_1) is listed below as Algorithm 1. Note that, in this case, the set Λ has a special structure

$$\Lambda = \{\lambda = (\lambda^{(1)}, \lambda^{(2)})^T \in H_1^* \times H_2^* : \lambda^{(2)} \in K^*\}$$

as well as $\varphi(\lambda)$ and $\nabla\varphi(\lambda)$ are defined in (5) and (6) respectively. Thus, the step (10) of the algorithm can be written explicitly.

$$\zeta_{k+1}^{(1)} = \zeta_k^{(1)} + \alpha_{k+1}(A_1 x(\lambda_{k+1}) - b_1), \quad \zeta_{k+1}^{(2)} = \Pi_{K^*} \left(\zeta_k^{(2)} + \alpha_{k+1}(A_2 x(\lambda_{k+1}) - b_2) \right),$$

where $\Pi_{K^*}(\cdot)$ denotes euclidean projection on the cone K^*.

It is worth noting that, besides solution of the problem (7), the algorithm uses only matrix-vector multiplications and vector operations, which made it amenable for parallel implementation.

Algorithm 1. Primal-Dual Adaptive Similar Triangles Method (PDASTM)

Require: starting point $\lambda_0 = 0$, initial guess $L_0 > 0$, accuracy $\tilde{\varepsilon}_f, \tilde{\varepsilon}_{eq}, \tilde{\varepsilon}_{in} > 0$.
1: Set $k = 0$, $C_0 = \alpha_0 = 0$, $\eta_0 = \zeta_0 = \lambda_0 = 0$.
2: **repeat**
3:     Set $M_k = L_k/2$.
4:     **repeat**
5:         Set $M_k = 2M_k$, find α_{k+1} as the largest root of the equation $C_{k+1} := C_k + \alpha_{k+1} = M_k \alpha_{k+1}^2$.
6:         Calculate $\lambda_{k+1} = (\lambda_{k+1}^{(1)}, \lambda_{k+1}^{(2)})^T = (\alpha_{k+1}\zeta_k + C_k\eta_k)/C_{k+1}$.
7:         Calculate

$$\zeta_{k+1} = (\zeta_{k+1}^{(1)}, \zeta_{k+1}^{(2)})^T = \arg\min_{\lambda \in \Lambda}\left\{\frac{1}{2}\|\lambda - \zeta_k\|_2^2 + \alpha_{k+1}(\varphi(\lambda_{k+1}) + \langle\nabla\varphi(\lambda_{k+1}), \lambda - \lambda_{k+1}\rangle)\right\}. \tag{10}$$

8:         Calculate $\eta_{k+1} = (\eta_{k+1}^{(1)}, \eta_{k+1}^{(2)})^T = (\alpha_{k+1}\zeta_{k+1} + C_k\eta_k)/C_{k+1}$.
9:     **until**

$$\varphi(\eta_{k+1}) \leq \varphi(\lambda_{k+1}) + \langle\nabla\varphi(\lambda_{k+1}), \eta_{k+1} - \lambda_{k+1}\rangle + \frac{M_k}{2}\|\eta_{k+1} - \lambda_{k+1}\|_2^2. \tag{11}$$

10:    Set $\hat{x}_{k+1} = \frac{1}{C_{k+1}}\sum_{i=0}^{k+1}\alpha_i x(\lambda_i) = (\alpha_{k+1}x(\lambda_{k+1}) + C_k\hat{x}_k)/C_{k+1}$.
11:    Set $L_{k+1} = M_k/2$, $k = k + 1$.
12: **until** $|f(\hat{x}_{k+1}) + \varphi(\eta_{k+1})| \leq \tilde{\varepsilon}_f$, $\|A_1\hat{x}_{k+1} - b_1\|_2 \leq \tilde{\varepsilon}_{eq}$, $\rho(A_2\hat{x}_{k+1} - b_2, -K) \leq \tilde{\varepsilon}_{in}$.
Ensure: The points $\hat{x}_{k+1}, \eta_{k+1}$.

Theorem 1. *Let the main assumptions hold. Then Algorithm 1 will stop not later than k equals to*

$$\max\left\{\left\lceil\sqrt{\frac{16L(R_1^2 + R_2^2)}{\tilde{\varepsilon}_f}}\right\rceil, \left\lceil\sqrt{\frac{16L(R_1^2 + R_2^2)}{R_1\tilde{\varepsilon}_{eq}}}\right\rceil, \left\lceil\sqrt{\frac{16L(R_1^2 + R_2^2)}{R_2\tilde{\varepsilon}_{in}}}\right\rceil\right\}.$$

Moreover, no later than k equals to

$$\max\left\{\left\lceil\sqrt{\frac{32L(R_1^2 + R_2^2)}{\varepsilon_f}}\right\rceil, \left\lceil\sqrt{\frac{16L(R_1^2 + R_2^2)}{R_1\varepsilon_{eq}}}\right\rceil, \left\lceil\sqrt{\frac{16L(R_1^2 + R_2^2)}{R_2\varepsilon_{in}}}\right\rceil\right\},$$

the point $\hat{x}_{k+1}$ generated by Algorithm 1 is an approximate solution to Problem (P_1) in the sense of (2) and $\|\hat{x}_{k+1} - x^*\| \leq \sqrt{\frac{2\varepsilon_f}{\gamma}}$, where x^* is a solution to Problem (P_1).

Remark 1. Note that the result of Theorem 1 can be reformulated as follows. For any $k \geq 1$, the output $(\hat{x}_k, \eta_k)$ of Algorithm 1 satisfies

$$-\frac{16L(R_1^2 + R_2^2)}{(k+1)^2} \leq f(\hat{x}_k) - Opt[P_1] \leq f(\hat{x}_k) + \varphi(\eta_k) \leq \frac{16L(R_1^2 + R_2^2)}{(k+1)^2},$$

$$\|A_1\hat{x}_k - b_1\|_2 \leq \frac{16L(R_1^2 + R_2^2)}{R_1(k+1)^2}, \quad \rho(A_2\hat{x}_k - b_2, -K) \leq \frac{16L(R_1^2 + R_2^2)}{R_2(k+1)^2},$$

$$\|\hat{x}_k - x^*\|_E \leq \frac{8}{k+1}\sqrt{\frac{L(R_1^2 + R_2^2)}{\gamma}}.$$

4   Numerical Experiments

In this section, we focus on the problem (9), which is motivated by important applications to traffic demand matrix estimation, [52], and regularized optimal transport calculation, [12]. We provide the results of our numerical experiments, which were performed on a PC with processor Intel Core i5-2410 2.3 GHz and 4 GB of RAM using pure Python 2.7 (without C code) under managing OS Ubuntu 14.04 (64-bits). Numpy.float128 data type with precision 1e−18 and with max element ≈ 1.19e+4932 was used. No parallel computations were used. We compare the performance of our algorithm with Sinkhorn's-method-based approach of [12], which is the state-of-the art method for problem (9). We use two types of cost matrix C and three types of vectors μ and ν.

Cost Matrix C. The first type of the cost matrix C is usually used in optimal transport problems and corresponds to 2-Wasserstein distance. Assume that we need to calculate this distance between two discrete measures μ, ν with finite support of size p. Then, the element c_{ij} of the matrix C is equal to Euclidean distance between the i-th point in the support of the measure μ and j-th point in the support of the measure ν. We will refer to this choice of the cost matrix as *Euclidean cost*. The second type the cost matrix C comes from traffic matrix estimation problem. Let's consider a road network of Manhattan type, i.e. districts present a $m \times m$ grid. We build a m^2 by m^2 matrix D of pairwise Euclidian distances processing the grid rows one by one and calculating euclidean distances from the current grid element to all the others elements of the grid. Then, as it suggested in [45], we form the cost matrix C as $C = \exp(-0.065D)$, where the exponent is taken elementwise. We will refer to this choice of the cost matrix as *Exp-Euclidean cost*.

   To set a natural scale for the regularization parameter γ, we normalize in each case the matrix C dividing all its elements by the average of all elements.

Vectors μ and ν. The first type of vectors μ and ν is *normalized uniform random*. Each element of each vector is taken independently from the uniform distribution on $[0, 1]$ and then each vector is normalized so that each sums to 1, i.e.

The second type of vectors is *random images*. The first $p/2$ elements of μ are normalized uniform random and the second $p/2$ elements are zero. For ν the situation is the opposite, i.e. the first $p/2$ elements are zero, and the second $p/2$ elements are normalized uniform random. In our preliminary experiments we found that the methods behave strange on vectors representing pictures from MNIST dataset. We supposed that the reason is that these vectors have many zero elements and decided to include the described random images to the experiments setting. Finally, the third type are vectors of intensities of *images* of handwritten digits from MNIST dataset. The size of each image is 28 by 28 pixels. Each image is converted to gray scale from 0 to 1 where 0 corresponds to black color and 1 corresponds to white, then each image is reshaped to a vector of length 784. In our experiments, we normalize these vectors to sum to 1.

Accuracy. We slightly redefine the accuracy of the solution and use relative accuracy with respect to the starting point, i.e.

$$\tilde{\varepsilon}_f = [\text{Accuracy}] \cdot |f(x(\lambda_0)) + \varphi(\eta_0)|, \quad \tilde{\varepsilon}_{eq} = [\text{Accuracy}] \cdot \|A_1 x(\lambda_0) - b_1\|_2,$$

where we used the fact that $\lambda_0 = \eta_0 = 0$ and there are no cone constraints in (9).

Adaptive vs Non-adaptive Algorithm. First, we show that the adaptivity of our algorithm with respect to the Lipschitz constant of the gradient of φ leads to faster convergence in practice. For this purpose, we use normalized uniform random vectors μ and ν and both types of cost matrix C. We compare our new Algorithm 1 with non-adaptive Similar Triangles Method (STM), which has cheaper iteration than the existing non-adaptive methods [10,22,34,42]. We choose $m = 10$, and, hence, $p = 100$, Accuracy is 0.05. For the Exp-Euclidean cost matrix C, we use $\gamma \in \{0.1, 0.2, 0.3, 0.4, 0.5\}$, and, for Euclidean cost matrix C, we use $\gamma \in \{0.02, 0.1, 0.2, 0.3, 0.4, 0.5\}$. The results are shown in Fig. 1. In both cases our new Algorithm 1 is much faster than the STM. This effect was observed for other parameter values, so, in the following experiments, we consider PDASTM.

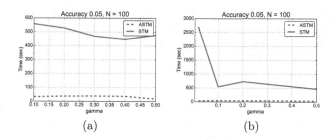

(a)   (b)

Fig. 1. The perfomance of PDASTM vs STM, Accuracy 0.05, Exp-Euclidean C (left) and Euclidean C (right).

Warm Start. During our experiments on the images from MNIST dataset PDASTM worked worse than on the normalized uniform random vectors. Possible reason is the large number of zero elements in the former vectors (a lot of black pixels). So we decided to test the performance of the algorithms on the random images vectors μ and ν. Also we decided to apply the idea of warm start to force PDASTM to converge faster. As we know, Sinkhorn's method works very fast when γ is relatively large. Thus, we use it in this regime to find a good starting point for the PDASTM for the problem with small γ. Notably, the running time of Sinkhorn's method is small in comparison with time of ASTM running. We test the performance of PDASTM versus PDASTM with warm start on problems with Exp-Euclidean matrix C and $\gamma \in \{0.001, 0.003, 0.005, 0.008, 0.01\}$ and on problems with Euclidean matrix C and $\gamma \in \{0.005, 0.01, 0.015, 0.02, 0.025\}$. The results are in Fig. 2. Other parameters are stated in the figure. The experiments were run 7 times, the results were averaged. As we can see, warm start accelerates the PDASTM. Similar results were observed in other experiments, so, we made the final comparison between the Sinkhorn's method and PDASTM with warm start.

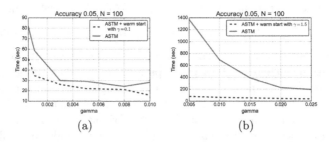

(a)   (b)

Fig. 2. The perfomance of PDASTM vs PDASTM with warm start, Accuracy 0.05, Exp-Euclidean C (left) and Euclidean C (right).

4.1   Sinkhorn's Method vs PDASTM with Warm Start

First we compare Sinkhorn's method and PDASTM with warm start on the problem with normalized uniform random vectors μ, ν and Euclidean cost matrix C with different values of $p \in \{100, 196, 289, 400\}$, Accuracy $\in \{0.01, 0.05, 0.1\}$, and $\gamma \in [0.005; 0.025]$. On each graph we point the value of γ used for generating a starting point for PDASTM with warm start by Sinkhorn's method. Each experiments was run 5 times and then the results were averaged. The results are shown on the Figs. 3, 4.

For the Exp-Euclidean cost matrix C, we performed the same experiments. For the space reasons, we provide the results on the Fig. 5 only for Accuracy 0.05. The results for other Accuracy values were similar.

In another series of experiments we compare the performance of PDASTM with warm start and Sinkhorn's method on the problem with images from MNIST dataset and Euclidean cost matrix C. We run both algorithms for the same set of γ values for 5 pairs of images. The results are aggregated by γ and the performance is averaged for each γ. We take three values of Accuracy, $\{0.01, 0.05, 0.1\}$. The results are shown on the Fig. 6.

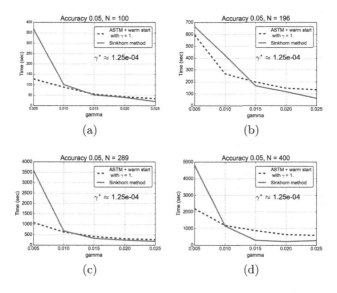

Fig. 3. The perfomance of PDASTM with warm start vs Sinkhorn's method, Accuracy 0.05, Euclidean cost matrix C.

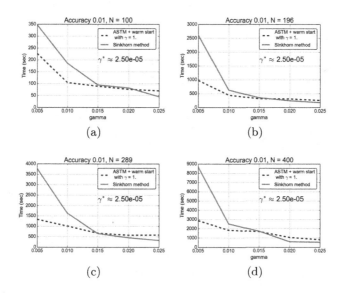

Fig. 4. The perfomance of PDASTM with warm start vs Sinkhorn's method, Accuracy 0.01, Euclidean cost matrix C.

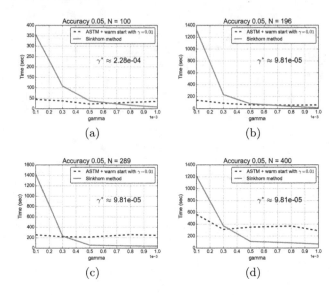

Fig. 5. The perfomance of PDASTM with warm start vs Sinkhorn's method, Accuracy 0.05, Exp-Euclidean cost matrix C.

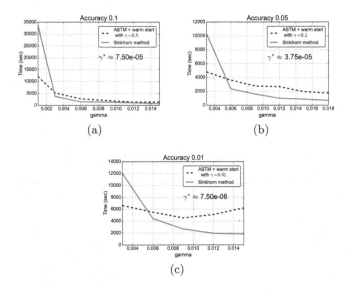

Fig. 6. The perfomance of PDASTM with warm start vs Sinkhorn's method, Euclidean cost matrix C, MNIST dataset.

As we can see on all graphs, for small values of γ, namely, smaller than some threshold γ_0, our method outperforms the state-of-the-art Sinkhorn's method. Note that, from [38], it follows that, for very small values of γ, less than some threshold $\gamma* = \frac{\varepsilon}{4\ln p}$, a good approximation of the solution to the problem (9) can be obtained by solution of the linear programming problem corresponding to $\gamma = 0$. We point these thresholds $\gamma*$ on the figures above. It should be noted that the threshold γ_0 is larger than $\gamma*$. This means that it is better to use our method, but not some method for linear programing problems.

Finally, we investigate the dependence of running time of PDASTM with warm start on the problem dimension p. As we can see from the Fig. 7, the dependence is close to quadratic, which was expected from the theoretical bounds. Also this dependence is close to that of the Sinkhorn's method.

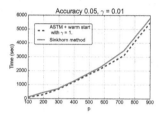

Fig. 7. Dependence of running time from the problem dimension p.

Conclusion

In this article, we propose a new adaptive accelerated gradient method for convex optimization problems and prove its convergence rate. We apply this method to a class of linearly constrained problems and show, how an approximate solution can be reconstructed. In the experiments, we consider two particular applied problems, namely, regularized optimal transport problem and traffic matrix estimation problem. The results of the experiments show that, in the regime of small regularization parameter, our algorithm outperforms the state-of-the-art Sinkhorn's-method-based approach. It would be interesting to extend the adaptive primal-dual methods for the stochastic setting [40] and for problems with inexact model of the objective [47].

References

1. Allen-Zhu, Z., Li, Y., Oliveira, R., Wigderson, A.: Much faster algorithms for matrix scaling. In: 2017 IEEE 58th Annual Symposium on Foundations of Computer Science (FOCS), pp. 890–901 (2017). arXiv:1704.02315
2. Altschuler, J., Weed, J., Rigollet, P.: Near-linear time approximation algorithms for optimal transport via Sinkhorn iteration. In: Guyon, I., et al. (eds.) Advances in Neural Information Processing Systems 30, pp. 1961–1971. Curran Associates, Inc. (2017). arXiv:1705.09634

3. Anikin, A.S., Gasnikov, A.V., Dvurechensky, P.E., Tyurin, A.I., Chernov, A.V.: Dual approaches to the minimization of strongly convex functionals with a simple structure under affine constraints. Comput. Math. Math. Phys. **57**(8), 1262–1276 (2017)

4. Baimurzina, D.R., et al.: Universal method of searching for equilibria and stochastic equilibria in transportation networks. Comput. Math. Math. Phys. **59**(1), 19–33 (2019). arXiv:1701.02473

5. Beck, A., Teboulle, M.: A fast dual proximal gradient algorithm for convex minimization and applications. Oper. Res. Lett. **42**(1), 1–6 (2014)

6. Benamou, J.D., Carlier, G., Cuturi, M., Nenna, L., Peyré, G.: Iterative Bregman projections for regularized transportation problems. SIAM J. Sci. Comput. **37**(2), A1111–A1138 (2015)

7. Boyd, S., Parikh, N., Chu, E., Peleato, B., Eckstein, J.: Distributed optimization and statistical learning via the alternating direction method of multipliers. Found. Trends Mach. Learn. **3**(1), 1–122 (2011)

8. Bregman, L.: Proof of the convergence of Sheleikhovskii's method for a problem with transportation constraints. USSR Comput. Math. Math. Phys. **7**(1), 191–204 (1967)

9. Chambolle, A., Pock, T.: A first-order primal-dual algorithm for convex problems with applications to imaging. J. Math. Imaging Vis. **40**(1), 120–145 (2011)

10. Chernov, A., Dvurechensky, P., Gasnikov, A.: Fast primal-dual gradient method for strongly convex minimization problems with linear constraints. In: Kochetov, Y., Khachay, M., Beresnev, V., Nurminski, E., Pardalos, P. (eds.) DOOR 2016. LNCS, vol. 9869, pp. 391–403. Springer, Cham (2016). https://doi.org/10.1007/978-3-319-44914-2_31

11. Chizat, L., Peyré, G., Schmitzer, B., Vialard, F.X.: Scaling algorithms for unbalanced optimal transport problems. Math. Comput. **87**(314), 2563–2609 (2018). arXiv:1607.05816

12. Cuturi, M.: Sinkhorn distances: lightspeed computation of optimal transport. In: Burges, C.J.C., Bottou, L., Welling, M., Ghahramani, Z., Weinberger, K.Q. (eds.) Advances in Neural Information Processing Systems 26, pp. 2292–2300. Curran Associates, Inc. (2013)

13. Cuturi, M., Peyré, G.: A smoothed dual approach for variational Wasserstein problems. SIAM J. Imaging Sci. **9**(1), 320–343 (2016)

14. Dünner, C., Forte, S., Takáč, M., Jaggi, M.: Primal-dual rates and certificates. In: Proceedings of the 33rd International Conference on International Conference on Machine Learning, ICML 2016, vol. 48. pp. 783–792. JMLR.org (2016)

15. Dvinskikh, D., Gorbunov, E., Gasnikov, A., Dvurechensky, P., Uribe, C.A.: On primal and dual approaches for distributed stochastic convex optimization over networks. In: 2019 IEEE 58th Conference on Decision and Control (CDC), pp. 7435–7440 (2019). https://doi.org/10.1109/CDC40024.2019.9029798. arXiv:1903.09844

16. Dvurechensky, P., Dvinskikh, D., Gasnikov, A., Uribe, C.A., Nedić, A.: Decentralize and randomize: faster algorithm for Wasserstein barycenters. In: Bengio, S., Wallach, H., Larochelle, H., Grauman, K., Cesa-Bianchi, N., Garnett, R. (eds.) Advances in Neural Information Processing Systems 31, NeurIPS 2018, pp. 10783–10793. Curran Associates, Inc. (2018). arXiv:1806.03915

17. Dvurechensky, P., Gasnikov, A., Gasnikova, E., Matsievsky, S., Rodomanov, A., Usik, I.: Primal-dual method for searching equilibrium in hierarchical congestion population games. In: Supplementary Proceedings of the 9th International Conference on Discrete Optimization and Operations Research and Scientific School (DOOR 2016) Vladivostok, Russia, 19–23 September 2016, pp. 584–595 (2016). arXiv:1606.08988

18. Dvurechensky, P., Gasnikov, A., Kroshnin, A.: Computational optimal transport: complexity by accelerated gradient descent is better than by Sinkhorn's algorithm. In: Dy, J., Krause, A. (eds.) Proceedings of the 35th International Conference on Machine Learning. Proceedings of Machine Learning Research, vol. 80, pp. 1367–1376 (2018). arXiv:1802.04367

19. Dvurechensky, P., Nesterov, Y., Spokoiny, V.: Primal-dual methods for solving infinite-dimensional games. J. Optim. Theory Appl. **166**(1), 23–51 (2015)

20. Fang, S.-C., Rajasekara, J. R., Tsao, H.-S. J.: Entropy Optimization and Mathematical Programming. Kluwer' International Series. Springer, Boston (1997)

21. Franklin, J., Lorenz, J.: On the scaling of multidimensional matrices. Linear Algebra Appl. **114**, 717–735 (1989). Special Issue Dedicated to Alan J. Hoffman

22. Gasnikov, A.V., Gasnikova, E.V., Nesterov, Y.E., Chernov, A.V.: Efficient numerical methods for entropy-linear programming problems. Comput. Math. Math. Phys. **56**(4), 514–524 (2016)

23. Gasnikov, A., Gasnikova, E., Mendel, M., Chepurchenko, K.: Evolutionary derivations of entropy model for traffic demand matrix calculation. Matematicheskoe Modelirovanie **28**(4), 111–124 (2016). (in Russian)

24. Golan, A., Judge, G., Miller, D.: Maximum Entropy Econometrics: Robust Estimation with Limited Data. Wiley, Chichester (1996)

25. Goldstein, T., O'Donoghue, B., Setzer, S., Baraniuk, R.: Fast alternating direction optimization methods. SIAM J. Imaging Sci. **7**(3), 1588–1623 (2014)

26. Guminov, S.V., Nesterov, Y.E., Dvurechensky, P.E., Gasnikov, A.V.: Accelerated primal-dual gradient descent with linesearch for convex, nonconvex, and nonsmooth optimization problems. Dokl. Math. **99**(2), 125–128 (2019)

27. Guminov, S., Dvurechensky, P., Tupitsa, N., Gasnikov, A.: Accelerated alternating minimization, accelerated Sinkhorn's algorithm and accelerated Iterative Bregman Projections (2019). arXiv:1906.03622

28. Hastie, T., Tibshirani, R., Friedman, J.: The Elements of Statistical Learning. Springer Series in Statistics. Springer, New York (2001). https://doi.org/10.1007/978-0-387-21606-5

29. Jakovetić, D., Xavier, J., Moura, J.M.F.: Fast distributed gradient methods. IEEE Trans. Autom. Control **59**(5), 1131–1146 (2014)

30. Kalantari, B., Khachiyan, L.: On the rate of convergence of deterministic and randomized RAS matrix scaling algorithms. Oper. Res. Lett. **14**(5), 237–244 (1993)

31. Kantorovich, L.: On the translocation of masses. Doklady Acad. Sci. USSR (N.S.) **37**, 199–201 (1942)

32. Kapur, J.: Maximum – Entropy Models in Science and Engineering. Wiley, New York (1989)

33. Kroshnin, A., Tupitsa, N., Dvinskikh, D., Dvurechensky, P., Gasnikov, A., Uribe, C.: On the complexity of approximating Wasserstein barycenters. In: Chaudhuri, K., Salakhutdinov, R. (eds.) Proceedings of the 36th International Conference on Machine Learning. Proceedings of Machine Learning Research, Long Beach, California, USA, 09–15 June 2019, vol. 97, pp. 3530–3540. PMLR (2019). arXiv:1901.08686

34. Li, J., Wu, Z., Wu, C., Long, Q., Wang, X.: An inexact dual fast gradient-projection method for separable convex optimization with linear coupled constraints. J. Optim. Theory Appl. **168**(1), 153–171 (2016)
35. Lin, T., Ho, N., Jordan, M.: On efficient optimal transport: an analysis of greedy and accelerated mirror descent algorithms. In: Chaudhuri, K., Salakhutdinov, R. (eds.) Proceedings of the 36th International Conference on Machine Learning. Proceedings of Machine Learning Research, Long Beach, California, USA, 09–15 June 2019, vol. 97, pp. 3982–3991. PMLR (2019)
36. Malitsky, Y., Pock, T.: A first-order primal-dual algorithm with linesearch. SIAM J. Optim. **28**(1), 411–432 (2018)
37. Nesterov, Y.: Introductory Lectures on Convex Optimization: A Basic Course. Kluwer Academic Publishers, Boston (2004)
38. Nesterov, Y.: Smooth minimization of non-smooth functions. Math. Program. **103**(1), 127–152 (2005)
39. Nesterov, Y., Gasnikov, A., Guminov, S., Dvurechensky, P.: Primal-dual accelerated gradient methods with small-dimensional relaxation oracle. Optim. Methods Softw., 1–28 (2020). https://doi.org/10.1080/10556788.2020.1731747. arXiv:1809.05895
40. Ogaltsov, A., Dvinskikh, D., Dvurechensky, P., Gasnikov, A., Spokoiny, V.: Adaptive gradient descent for convex and non-convex stochastic optimization (2019). arXiv:1911.08380
41. Ouyang, Y., Chen, Y., Lan, G., Eduardo Pasiliao, J.: An accelerated linearized alternating direction method of multipliers. SIAM J. Imaging Sci. **8**(1), 644–681 (2015)
42. Patrascu, A., Necoara, I., Findeisen, R.: Rate of convergence analysis of a dual fast gradient method for general convex optimization. In: 2015 54th IEEE Conference on Decision and Control (CDC), pp. 3311–3316 (2015)
43. Scaman, K., Bach, F., Bubeck, S., Lee, Y.T., Massoulié, L.: Optimal algorithms for smooth and strongly convex distributed optimization in networks. In: Precup, A., Teh, Y.W. (eds.) Proceedings of the 34th International Conference on Machine Learning. Proceedings of Machine Learning Research, vol. 70, International Convention Centre, Sydney, Australia, 06–11 August 2017, pp. 3027–3036. PMLR (2017)
44. Schmitzer, B.: Stabilized sparse scaling algorithms for entropy regularized transport problems. SIAM J. Sci. Comput. **41**(3), A1443–A1481 (2019). arXiv:1610.06519
45. Shvetsov, V.I.: Mathematical modeling of traffic flows. Autom. Remote Control **64**(11), 1651–1689 (2003)
46. Sinkhorn, R.: Diagonal equivalence to matrices with prescribed row and column sums. II. Proc. Am. Math. Soc. **45**, 195–198 (1974)
47. Stonyakin, F.S., et al.: Gradient methods for problems with inexact model of the objective. In: Khachay, M., Kochetov, Y., Pardalos, P. (eds.) MOTOR 2019. LNCS, vol. 11548, pp. 97–114. Springer, Cham (2019). https://doi.org/10.1007/978-3-030-22629-9_8. arXiv:1902.09001
48. Tran-Dinh, Q., Cevher, V.: Constrained convex minimization via model-based excessive gap. In: Proceedings of the 27th International Conference on Neural Information Processing Systems, NIPS 2014, pp. 721–729. MIT Press, Cambridge (2014)
49. Tran-Dinh, Q., Fercoq, O., Cevher, V.: A smooth primal-dual optimization framework for nonsmooth composite convex minimization. SIAM J. Optim. **28**(1), 96–134 (2018). arXiv:1507.06243

50. Tupitsa, N., Dvurechensky, P., Gasnikov, A., Uribe, C.A.: Multimarginal optimal transport by accelerated gradient descent (2020). arXiv:2004.02294
51. Uribe, C.A., Dvinskikh, D., Dvurechensky, P., Gasnikov, A., Nedić, A.: Distributed computation of Wasserstein barycenters over networks. In: 2018 IEEE Conference on Decision and Control (CDC), pp. 6544–6549 (2018). arXiv:1803.02933
52. Wilson, A.: Entropy in Urban and Regional Modelling. Monographs in Spatial and Environmental Systems Analysis. Routledge, Abingdon (2011)
53. Yurtsever, A., Tran-Dinh, Q., Cevher, V.: A universal primal-dual convex optimization framework. In: Proceedings of the 28th International Conference on Neural Information Processing Systems, NIPS 2015, pp. 3150–3158. MIT Press, Cambridge (2015)
54. Zhang, Y., Roughan, M., Lund, C., Donoho, D.L.: Estimating point-to-point and point-to-multipoint traffic matrices: an information-theoretic approach. IEEE/ACM Trans. Netw. **13**(5), 947–960 (2005)
55. Zou, H., Hastie, T.: Regularization and variable selection via the elastic net. J. Roy. Stat. Soc. B **67**(2), 301–320 (2005)

Most Favorable Russell Measures of Efficiency: Properties and Measurement

Chiang Kao[✉]

Department of Industrial and Information Management,
National Cheng Kung University, Tainan, Taiwan, Republic of China
ckao@mail.ncku.edu.tw

Abstract. Conventional radial efficiency measurement models in data envelopment analysis are unable to produce appropriate efficiency scores for production units lying outside the cone generated by the convex hull of the extreme efficient production units. In addition, in the case of production technologies with variable returns to scale, the efficiency scores measured from the input and output sides are usually different. To solve these problems, the Russell measure of efficiency, which takes both the inputs and outputs into account, has been proposed. However, the conventional Russell efficiency is measured under the least favorable conditions, rather than the general custom of measuring under the most favorable ones. This paper develops a model to measure Russell efficiency under the most favorable conditions in two forms, the average and the product. They can be transformed into a second-order cone program and a mixed integer linear program, respectively, so that the solution can be obtained efficiently. A case of Taiwanese commercial banks demonstrates that they are more reliable and representative than the radial measures. Since the most favorable measures are higher than the least favorable measures, and the targets for making improvements are the easiest to reach, they are more acceptable to the production units to be evaluated.

Keywords: Data envelopment analysis · Russell measure · Radial measure · Slacks-based measure

1 Introduction

Efficiency measurement is an important management task because it reveals the extent to which the performance of a production unit, or more generally, a decision making unit (DMU), has been unsatisfactory in the past and provides a direction for making improvements in the future. Many ideas for measuring efficiency have been proposed [14]. Since the seminal work of Charnes et al. [11], data envelopment analysis (DEA) has been considered an effective technique for measuring the relative efficiency of a set of DMUs that applies multiple inputs to produce multiple outputs.

Charnes et al.'s model [11], usually referred to as the CCR model, is applied to production technologies with constant returns to scale (CRS). Banker et al. [7] developed a modified model that allows for technologies with variable returns to scale (VRS). This model is commonly referred to as the BCC model in the literature. The efficiencies

© Springer Nature Switzerland AG 2020
A. Kononov et al. (Eds.): MOTOR 2020, LNCS 12095, pp. 424–439, 2020.
https://doi.org/10.1007/978-3-030-49988-4_29

measured from the CCR and BCC models are a form of radial measure. One weakness of this type of efficiency measure is that the efficiency scores of the DMUs lying outside the cone generated by the convex hull of the extreme efficient DMUs cannot be appropriately assigned. The radial efficiency can be measured from either the input or output side. In the case of the BCC model, there is another weakness. While the efficiencies measured from the input and output sides are the same for the CCR model, they are usually different for the BCC model. Which model should be used between the input and output sides depends on the purpose of the evaluation. When there is no specific purpose, there is no rule to follow in deciding which model to use.

One way to solve these problems is to use the Russell measure of efficiency [12, 13] to take all the inputs and outputs into account. The corresponding model is nonlinear. To obtain a linear model, Pastor et al. [20] proposed an enhanced Russell efficiency measure. Tone [23] termed this measure the slacks-based measure (SBM).

One feature of the DEA methodology is it allows the DMUs being evaluated to select the most favorable conditions by which to measure efficiency. This feature makes this methodology widely accepted for performance evaluation. While the Russell measures can solve the problems of inappropriate efficiency scores being assigned to certain DMUs and different results being obtained from the input and output models, they are calculated under the least favorable conditions for inefficient DMUs. In other words, the target on the production frontier selected for measuring efficiency is the farthest, rather than the general custom of being the closest, point to the DMU being evaluated. The results are thus unfair to inefficient DMUs.

Various approaches for measuring efficiency based on the closest targets have been proposed in the literature, starting with the works of Briec [8, 9]. The major differences of the approaches are the ways in which the production frontier and distance are defined. For example, Aparicio et al. [6] developed a mixed integer linear programming model to find the closest target in the conventional production possibility set. Aparicio and Pastor [4, 5] searched for the closest target in the extended facet production possibility set defined by Olesen and Petersen [19]. Fukuyama et al. [15] investigated the least-distance p-norm measures on an extended free disposable set based on the work of Ando et al. [1]. Petersen [21] developed a model to find the direction with the shortest distance to the production frontier. Aparicio [2] conducted a survey of the literature on this topic.

In this paper, we develop a model to measure the most favorable Russell efficiency based on the frontier used in the conventional way of measuring the least favorable Russell efficiency. Gonzaléz and Álvarez [16] initiated this study with an input-oriented Russell measure. Aparicio et al. [6] developed a model in the primal (envelopment form) and dual (multiplier form) combined spaces to measure the non-oriented Russell measure. However, Aparicio et al. [3] showed that, while this model works correctly for non-oriented measures, it cannot be successfully applied to input- or output-oriented measures, and they proposed a bilevel linear programming model. The model developed in the current study has two forms, the average and the product. The former can be transformed into a second-order cone program and the latter into a mixed integer linear program such that both can be solved efficiently. Since more favorable efficiency measures imply higher efficiency scores and closer targets for inefficient DMUs to reach with less effort, the results are more persuasive and acceptable to the DMUs being evaluated.

2 Conventional Efficiency Measures

Suppose a set of n DMUs that applies m inputs X_i, $i = 1, \ldots, m$ to produce s outputs Y_r, $r = 1, \ldots, s$. Let X_{ij} and Y_{rj} denote the ith input and rth output, respectively, of DMU j, $j = 1, \ldots, n$. The production possibility set constructed from these DMUs under variable returns to scale is $T = \{(x, y) \mid \sum_{j=1}^{n} \lambda_j X_{ij} \leq x_i, i = 1, \ldots, m, \sum_{j=1}^{n} \lambda_j Y_{rj} \geq y_r, r = 1, \ldots, s, \sum_{j=1}^{n} \lambda_j = 1, \lambda_j \geq 0, j = 1, \ldots, n\}$. The strongly efficient frontier of this set is $\partial^S(T) = \{(x, y) \in T \mid \hat{x} \leq x, \hat{y} \geq y, \text{ and } (\hat{x}, \hat{y}) \neq (x, y) \Rightarrow (\hat{x}, \hat{y}) \notin T\}$, which is the set of strongly efficient points of T. Theoretically, a DMU should select a point on the strongly efficient frontier to measure efficiency. The BCC model [7] for measuring the efficiency of DMU k in the envelopment form can be formulated from the input or output side, as follows:

Input-orientation

$$\theta_k^I = \min. \ \theta - \varepsilon \left(\sum_{i=1}^{m} s_i^- + \sum_{r=1}^{s} s_r^+ \right) \tag{1a}$$

$$\text{s.t.} \quad \sum_{j=1}^{n} \lambda_j X_{ij} + s_i^- = \theta X_{ik}, \quad i = 1, \ldots, m$$

$$\sum_{j=1}^{n} \lambda_j Y_{rj} - s_r^+ = Y_{rk}, \quad r = 1, \ldots, s$$

$$\sum_{j=1}^{n} \lambda_j = 1$$

$$\lambda_j, s_i^-, s_r^+ \geq 0, \quad j = 1, \ldots, n, \ i = 1, \ldots, m, \ r = 1, \ldots, s$$

$$\theta \text{ unrestricted in sign.}$$

Output-orientation

$$\frac{1}{\theta_k^O} = \max. \ \varphi + \varepsilon \left(\sum_{i=1}^{m} s_i^- + \sum_{r=1}^{s} s_r^+ \right) \tag{1b}$$

$$\text{s.t.} \quad \sum_{j=1}^{n} \lambda_j X_{ij} + s_i^- = X_{ik}, \quad i = 1, \ldots, m$$

$$\sum_{j=1}^{n} \lambda_j Y_{rj} - s_r^+ = \varphi Y_{rk}, \quad r = 1, \ldots, s$$

$$\sum_{j=1}^{n} \lambda_j = 1$$

$$\lambda_j, \; s_i^-, \; s_r^+ \geq 0, \quad j = 1,\ldots,n, \; i = 1,\ldots, m, \; r = 1, \ldots, s,$$

φ unrestricted in sign,

where ε is a small non-Archimedean number used to avoid ignoring the unfavorable factors when measuring efficiency. The input efficiency θ_k^I and output efficiency θ_k^O need not be the same. When the convexity constraint $\sum_{j=1}^{n} \lambda_j = 1$ is deleted, the BCC model becomes the CCR model [11]. In this case, the input and output models produce the same efficiency score, which is denoted as θ_k^{CCR} in this paper.

One way to solve the problems caused by the non-Archimedean number and input-output difference in efficiency measurement is to apply a non-radial measure, such as the Russell measure of efficiency. The Russell measure under variable returns to scale is calculated via the following model [12]:

$$R_k^{min} = \text{min.} \quad \frac{1}{m+s} \left(\sum_{i=1}^{m} \theta_i + \sum_{r=1}^{s} \frac{1}{\varphi_r} \right) \tag{2}$$

$$\text{s.t.} \sum_{j=1}^{n} \lambda_j X_{ij} \leq \theta_i X_{ik}, \; \theta_i \leq 1, \; i = 1,\ldots,m$$

$$\sum_{j=1}^{n} \lambda_j Y_{rj} \geq \varphi_r Y_{rk}, \; \varphi_r \geq 1, \; r = 1,\ldots,s$$

$$\sum_{j=1}^{n} \lambda_j = 1$$

$$\lambda_j \geq 0, \qquad j = 1,\ldots,n.$$

The efficiency is defined as the average of individual factor efficiencies. The constraints $\theta_i \leq 1$ and $\varphi_r \geq 1$ are imposed to restrict the target points for evaluating efficiency to those that dominate the DMU being evaluated. If an assumption of constant returns to scale is desired, then one simply deletes the convexity constraint $\sum_{j=1}^{n} \lambda_j = 1$.

The Russell measure defines efficiency as the average of the efficiencies of all input and output factors. Pastor et al. [20] and Tone [23] defined efficiency as the product of the arithmetic average of the efficiencies of the m inputs and the harmonic average of the efficiencies of the s outputs in the form of:

$$Q_k^{min} = \min. \frac{\frac{1}{m}\sum_{i=1}^{m}\theta_i}{\frac{1}{s}\sum_{r=1}^{s}\varphi_r},$$

subject to the same constraints as those in Model (2). Substituting θ_i with $(X_{ik} - s_i^-)/X_{ik}$ and φ_r with $(Y_{rk} + s_r^+)/Y_{rk}$, one obtains the following equivalent model:

$$Q_k^{min} = \min. \frac{1 - \frac{1}{m}\sum_{i=1}^{m} s_i^-/X_{ik}}{1 + \frac{1}{s}\sum_{r=1}^{s} s_r^+/Y_{rk}} \tag{3}$$

$$\text{s.t.} \sum_{j=1}^{n} \lambda_j X_{ij} + s_i^- = X_{ik}, \ i = 1, \ldots, m$$

$$\sum_{j=1}^{n} \lambda_j Y_{rj} - s_r^+ = Y_{rk}, \ r = 1, \ldots, s$$

$$\sum_{j=1}^{n} \lambda_j = 1$$

$$\lambda_j, \ s_i^-, \ s_r^+ \geq 0, \quad j = 1, \ldots, n, \ i = 1, \ldots, m, \ r = 1, \ldots, s.$$

This model is called the slacks-based measure (SBM) model in Tone [23]. The advantage of this model over Model (2) is that Model (2) is a nonlinear program, while this model is a fractional linear program, which can be linearized by applying a variable substitution technique proposed in Charnes and Cooper [10].

Different from the radial measure that requires either all inputs to be reduced in the same proportion θ as in Model (1a) or all outputs to be expanded in the same proportion φ as in Model (1b), the Russell measure takes the inputs and outputs into account at the same time, and the proportions θ_i and φ_r can be different for different factors. More importantly, the projection point used to measure efficiency is on the strongly efficient frontier. Pastor et al. [20] and Tone [23] proved that the Russell efficiency measure of the product form is less than or equal to both the input and output radial efficiency measures. In symbols, it is $Q_k^{min} \leq \theta_k^I$ and $Q_k^{min} \leq \theta_k^O$.

The objective of Model (2) or Model (3) is to find the greatest rates for reducing the inputs and expanding the outputs of the DMU being evaluated within the production possibility set at the same time. The purpose of the model is actually to identify the production frontier, rather than measuring efficiencies. The objective value, known as the efficiency of the DMU, is a by-product of this frontier identification process. However, since the objective function has a minimization direction, the efficiency measured from this model is the lowest among all possible measures, which contradicts the basic idea of DEA suggesting that efficiency is measured under the most favorable conditions.

3 Most Favorable Measures

The envelopment form of the BCC input model (1a) is intended to find the minimum value for θ to reduce the inputs of the DMU being evaluated such that the resulting point is still in the production possibility set. The purpose is to identify a frontier facet from the production possibility set based on which efficiency of this DMU is measured. If this DMU lies in the cone generated by the convex hull of the extreme efficient DMUs in the input space so that the slack variables are zero, then the target point $(\sum_{j=1}^{n} \lambda_j X_j, \sum_{j=1}^{n} \lambda_j Y_j) = (\theta_k^I X_k, Y_k)$, where $X_j = (X_{1j}, \ldots, X_{mj})$ and $Y_j = (Y_{1j}, \ldots, Y_{sj})$, reflects that its efficiency is θ_k^I, which is a by-product of this process. Conceptually, one should find the maximum value for θ to be the most favorable efficiency measure after all the frontier facets are identified. Due to the geometric property of the radial measures, the minimum and maximum values for θ are the same. Consider six DMUs, labelled as $A \sim F$ in Fig. 1, which apply different combinations of inputs X_1 and X_2 to produce one output Y. In measuring the efficiency of DMU D, the idea of the BCC input model is to identify a frontier facet by reducing X_{1D} and X_{2D} in the same proportion of θ along the ray $\overrightarrow{OD}$ until it reaches the boundary of the production possibility set at $\hat{D}$. The minimum value for θ, or the largest extent of contraction, is θ_k^I, which is the ratio of $O\hat{D}$ to OD. After all the frontier facets are identified, the strongly efficient frontier is then determined, and the efficiency is measured as the largest value for θ such that θD on the ray $O\hat{D}$ intercepts the strongly efficient frontier in the region of D' to D''. Since the intersection of the ray $\overrightarrow{OD}$ with the strongly efficient frontier in the region of D' to D'' is the unique point $\hat{D}$, the minimum and maximum values of θ are the same.

In measuring the Russell efficiency, all inputs and outputs are allowed to contract and expand in different proportions, respectively. The minimum and maximum values for the efficiency in this case may not be the same. More specifically, the target point found in the process of identifying the frontier facet via minimizing the distance parameters may not be the same as that found in the process of maximizing the parameters. For example, the efficiency of DMU D in Fig. 1 calculated from Model (2) is actually the lowest that can be obtained by using the points on the strongly efficient frontier in the region of D' to D'' as the target. The idea of the DEA technique, however, is to measure the efficiency under the most favorable conditions. Following this idea, one should search for a target in the region of D' to D'' that can produce the highest efficiency. The procedure for accomplishing this task can be separated into two phases, where Phase I is to identify the strongly efficient frontier and Phase II is to find a point on the strongly efficient frontier that will produce the highest efficiency.

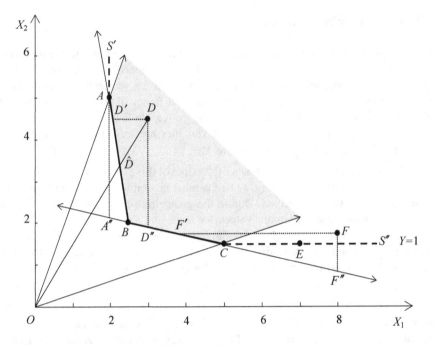

Fig. 1. Geometric interpretation of the efficiency measurement of various models.

To construct the strongly efficient frontier, all extreme efficient DMUs of the production possibility set that span the full dimensional efficient facets are identified first by applying any DEA model, e.g., Model (1a). Each strongly efficient frontier facet is the convex hull of a set of $m + s$ neighboring extreme efficient DMUs, provided the hyperplane extended from all sides of this facet envelops all DMUs. Let E_0 denote the set of the indices of $m + s$ extreme efficient DMUs such that the convex hull of these $m + s$ DMUs is a frontier facet F_0. The frontier facet F_0 can be expressed as $F_0 = \{(x, y) \mid \sum_{j \in E_0} \lambda_j X_{ij} = x_i, i = 1, \ldots, m, \sum_{j \in E_0} \lambda_j Y_{rj} = y_r, r = 1, \ldots, s, \sum_{j \in E_0} \lambda_j = 1, \lambda_j \geq 0, j \in E_0\}$. The frontier hyperplane extended from this frontier facet is $H_0 = \{(x, y) \mid \sum_{j \in E_0} \lambda_j X_{ij} = x_i, i = 1, \ldots, m, \sum_{j \in E_0} \lambda_j Y_{rj} = y_r, r = 1, \ldots, s, \sum_{j \in E_0} \lambda_j = 1, \lambda_j$ unrestricted in sign, $j \in E_0\}$. The mathematical expression of the frontier hyperplane H_0 differs from the frontier facet F_0 only in that the values of λ_j are allowed to be negative. Since F_0 is a frontier facet, the corresponding frontier hyperplane H_0 must envelop all n DMUs. In this case, every DMU d must have a projection point (target) on the hyperplane H_0 that dominates itself. The projection point for DMU d can be expressed as $(\sum_{j \in E_0} \lambda_j^{(d)} X_{ij}, \sum_{j \in E_0} \lambda_j^{(d)} Y_{rj})$, where $\sum_{j \in E_0} \lambda_j^{(d)} = 1$ and $\lambda_j^{(d)}$ are unrestricted in sign. Since every DMU d is dominated by its projection point $(\sum_{j \in E_0} \lambda_j^{(d)} X_{ij}, \sum_{j \in E_0} \lambda_j^{(d)} Y_{rj})$, we have $\sum_{j \in E_0} \lambda_j^{(d)} X_{ij} + s_i^{(d)-} = X_{id}, i = 1, \ldots, m, d = 1, \ldots, n$ and $\sum_{j \in E_0} \lambda_j^{(d)} Y_{rj} - s_r^{(d)+} = Y_{rd}, r = 1, \ldots, s, d = 1, \ldots, n$, where $s_i^{(d)-}, s_r^{(d)+} \geq 0$.

Using the DMUs in Fig. 1 to explain this, line segment $\overline{BC}$ is a frontier facet that can be expressed as $F = \{Z \mid Z = \lambda_B B + \lambda_C C, \lambda_B + \lambda_C = 1, \lambda_B, \lambda_C \geq 0\}$. The entire line $\overleftrightarrow{BC}$ is expressed as $H = \{Z \mid Z = \lambda_B B + \lambda_C C, \lambda_B + \lambda_C = 1, \lambda_B, \lambda_C$ unrestricted in sign$\}$. Since $\overline{BC}$ is a frontier facet, the corresponding line $\overleftrightarrow{BC}$ must envelop all DMUs by having nonnegative slacks. Consider four DMUs, A, D, C, and F, which can be projected to $A"$, $D"$, $C" = C$, and $F"$ on Line $\overleftrightarrow{BC}$, respectively, by fixing X_1 and Y at their current values. For DMU D, we have $D" = \lambda_B^{(D)} B + \lambda_C^{(D)} C = (3, 1.9; 1)^T$, with $\lambda_B^{(D)} = 0.8$ and $\lambda_C^{(D)} = 0.2$. Positive values for $\lambda_B^{(D)}$ and $\lambda_C^{(D)}$ indicate that $D"$ is located on line segment $\overline{BC}$. The corresponding slack variables have nonnegative values of $s_1^{(D)-} = 0$, $s_2^{(D)-} = 4.5 - 1.9 = 2.6$, and $s_1^{(D)+} = 0$. For DMU C, we have $C" = C$, with $\lambda_B^{(C)} = 0$ and $\lambda_C^{(C)} = 1$, and all the slacks are zero. For DMU A, we have $A" = \lambda_B^{(A)} B + \lambda_C^{(A)} C = (2, 2.1; 1)^T$, with $\lambda_B^{(A)} = 1.2$ and $\lambda_C^{(A)} = -0.2$, where $\lambda_C^{(A)}$ is negative. Positive $\lambda_B^{(A)}$ and negative $\lambda_C^{(A)}$ indicate that $A"$ is located to the left of DMU B on line $\overleftrightarrow{BC}$. The corresponding slacks are $s_1^{(A)-} = 0$, $s_2^{(A)-} = 5 - 2.1 = 2.9$, and $s_1^{(A)+} = 0$, which are nonnegative. Finally, for DMU F, we have $F" = \lambda_B^{(F)} B + \lambda_C^{(F)} C = (8, 0.9; 1)^T$, with $\lambda_B^{(F)} = -1.2$ and $\lambda_C^{(F)} = 2.2$, where $\lambda_B^{(F)}$ is negative. Negative $\lambda_B^{(F)}$ and positive $\lambda_C^{(F)}$ indicate that $F"$ is located to the right of DMU C on line $\overleftrightarrow{BC}$. The corresponding slacks are $s_1^{(F)-} = 0$, $s_2^{(F)-} = 1.75 - 0.9 = 0.85$, and $s_1^{(F)+} = 0$, which, again, are nonnegative.

Since it is not known beforehand which frontier facet of the strongly efficient frontier will be selected by DMU k to find the target to measure efficiency, all frontier facets must be considered. Let E denote the set of the indices of the extreme efficient DMUs. We use the binary variable B_j to indicate whether or not an extreme efficient DMU j is used to span the frontier facet. The conditions for DMU k to consider all possible frontier facets to measure efficiency can be expressed as:

$$\sum_{j \in E} \lambda_j^{(d)} X_{ij} + s_i^{(d)-} = X_{id}, \quad i = 1, \ldots, m, \quad d = 1, \ldots, n \tag{4.1}$$

$$\sum_{j \in E} \lambda_j^{(d)} Y_{rj} - s_r^{(d)+} = Y_{rd}, \quad r = 1, \ldots, s, \quad d = 1, \ldots, n \tag{4.2}$$

$$\sum_{j \in E} \lambda_j^{(d)} = 1, \quad d = 1, \ldots, n \tag{4.3}$$

$$s_i^{(d)-}, s_r^{(d)+} \geq 0, \quad i = 1, \ldots, m, \quad r = 1, \ldots, s, \quad d = 1, \ldots, n \tag{4.4}$$

$$\lambda_j^{(k)} \geq 0, \quad j \in E \tag{4.5}$$

$$\lambda_j^{(d)} \text{ unrestricted in sign}, \quad j \in E, \quad d = 1, \ldots, n \quad d \neq k \tag{4.6}$$

$$-MB_j \leq \lambda_j^{(d)} \leq MB_j, \quad j \in E, \quad d = 1, \ldots, n \tag{4.7}$$

$$\sum_{j \in E} B_j \leq m + s \tag{4.8}$$

$$B_j \in \{0, 1\}, \quad j \in E, \tag{4.9}$$

where M is a large number for allowing all possible $\lambda_j^{(d)}$ values to appear. The frontier facet spanned by the efficient DMUs corresponding to $B_j = 1$ is the facet for DMU k to measure efficiency. Constraints (4.1)–(4.4), for $d = k$, and (4.5) require the assessed DMU k to select a point on this facet to calculate efficiency. Constraints (4.1)–(4.4) and (4.6) ensure that all DMUs are enveloped by the hyperplane extended from the frontier facet.

The frontier hyperplane has a dimension of $m + s$. The sum of B_j is thus equal to $m + s$. However, to account for the degenerate case where the number of extreme efficient DMUs is less than $m + s$, we require $\sum_{j \in E} B_j \leq m + s$ in Constraint (4.8).

For cases of constant returns to scale, the convexity constraint $\sum_{j \in E} \lambda_j^{(d)} = 1$ is not needed. Moreover, since the frontier facets must pass through the origin, this implies that the origin must always be used with the $m + s - 1$ of other efficient DMUs to constitute the frontier facet. Thus, $m + s$ in constraint (4.8) is changed to $m + s - 1$.

To measure the Russell efficiency based on the closest target to the assessed DMU, one first applies Model (2), or any DEA model, to identify the extreme efficient DMUs, with their indices comprising the set E. One then uses the following mathematical program to calculate the efficiency of DMU k:

$$R_k^{\max} = \text{max.} \quad \frac{1}{m+s} \left[\sum_{i=1}^{m} \left(\frac{X_{ik} - s_i^{(k)-}}{X_{ik}} \right) + \sum_{r=1}^{s} \left(\frac{Y_{rk}}{Y_{rk} + s_r^{(k)+}} \right) \right] \tag{5}$$

s.t.   Constraint Set (4).

The objective function is nonlinear, and the constraints are linear, in that some binary variables are involved. This model can be solved efficiently by transforming it into a second-order cone program, as proposed in Sueyoshi and Sekitani [22] or a semidefinite program, as discussed in Halická and Trnovská [17].

Similarly, the Russell measure of the product form, i.e., the slacks-based measure, with the target closest to the assessed DMU can be calculated via the following model:

$$Q_k^{\max} = \text{max.} \quad \frac{1 - \frac{1}{m} \sum_{i=1}^{m} s_i^{(k)-} / X_{ik}}{1 + \frac{1}{s} \sum_{r=1}^{s} s_r^{(k)+} / Y_{rk}} \tag{6}$$

s.t.   Constraint Set (4).

This model is a fractional mixed integer program. By applying the variable substitution technique proposed in Charnes and Cooper [10], with $1/(1 + \frac{1}{s}\sum_{r=1}^{s} s_r^{(k)+}/Y_{rk}) = w$, $w\lambda_j^{(d)} = \mu_j^{(d)}$, $ws_i^{(d)-} = t_i^{(d)-}$, and $ws_r^{(d)+} = t_r^{(d)+}$, a linear mixed integer program for the VRS case is obtained as follows:

$$Q_k^{\max} = \text{max. } w - \frac{1}{m}\sum_{i=1}^{m}\frac{t_i^{(k)-}}{X_{ik}} \tag{7}$$

$$\text{s.t. } w + \frac{1}{s}\sum_{r=1}^{s}\frac{t_r^{(k)+}}{Y_{rk}} = 1$$

$$\sum_{j\in E}\mu_j^{(d)}X_{ij} + t_i^{(d)-} = wX_{id} \qquad i = 1,\ldots,m, \quad d = 1,\ldots,n$$

$$\sum_{j\in E}\mu_j^{(d)}Y_{rj} - t_r^{(d)+} = wY_{rd}, \quad r = 1,\ldots,s, \quad d = 1,\ldots,n$$

$$\sum_{j\in E}\mu_j^{(d)} = w, \quad d = 1,\ldots,n$$

$$t_i^{(d)-}, t_r^{(d)+} \geq 0, \quad i = 1,\ldots,m, \ r = 1,\ldots,s, \quad d = 1,\ldots,n$$

$$\mu_j^{(k)} \geq 0, \ j \in E$$

$$\mu_j^{(d)} \text{unrestricted in sign}, \quad j \in E, d = 1,\ldots,n, \quad d \neq k$$

$$-MB_j \leq \mu_j^{(d)} \leq MB_j, \quad j \in E, \quad d = 1,\ldots,n$$

$$\sum_{j\in E}B_j \leq m + s$$

$$w \geq 0$$

$$B_j \in \{0, 1\}, \qquad j \in E,$$

This model is much easier than Model (6) to solve.

4 Some Properties

The most favorable Russell measures of efficiency have several properties. First, it is noted that the conventional Russell measure of the product form, $R_k^{\min}$, is calculated based on the target that is the farthest to the assessed DMU. It can be calculated by changing the direction of optimization in Model (5) from maximization to minimization although the model in this case is more complicated than Model (2). This is also true for the product form of Model (6). We thus have the following theorem:

Theorem 1. The most favorable Russell measures of efficiency, both the average and product forms, are greater than or equal to the least favorable measures.

Second, every DMU uses a point on the strongly efficient frontier as the target to measure efficiency. The constraints (4.1)–(4.4) for $d = k$ and (4.5) in models (5) and (6), where all X_{ik} and Y_{rk} are positive, ensure that their most favorable Russell measures, both the average and product forms, are always positive. This leads to the following theorem:

Theorem 2. The most favorable Russell measures are always positive.

Third, models (2) and (3) use the average and product, respectively, of the input and output efficiencies as the DMU efficiency. The output efficiency in Model (3) is a harmonic average of the efficiencies of individual outputs, instead of the usual arithmetic average. To make the two measures comparable, the Russell efficiency of the average form can be defined as the average of the arithmetic average of the input efficiencies and the harmonic average of the output efficiencies, that is,

$$\hat{R}_k^{\max} = \frac{1}{2}[(\frac{1}{m}\sum_{i=1}^m \theta_i) + (\frac{1}{\frac{1}{s}\sum_{r=1}^s \varphi_r})].$$

Based on the arithmetic-geometric mean inequality stipulating that the arithmetic mean is greater than or equal to the geometric mean, we have the following relationship:

$$\hat{R}_k^{\max} = \frac{1}{2}[(\frac{1}{m}\sum_{i=1}^m \hat{\theta}_i) + (\frac{1}{\frac{1}{s}\sum_{r=1}^s \hat{\varphi}_r})] \geq \frac{1}{2}[(\frac{1}{m}\sum_{i=1}^m \theta_i^*) + (\frac{1}{\frac{1}{s}\sum_{r=1}^s \varphi_r^*})]$$

$$\geq \left(\frac{\frac{1}{m}\sum_{i=1}^m \theta_i^*}{\frac{1}{s}\sum_{r=1}^s \varphi_r^*}\right)^{1/2} \geq \frac{\frac{1}{m}\sum_{i=1}^m \theta_i^*}{\frac{1}{s}\sum_{r=1}^s \varphi_r^*} = Q_k^{\max},$$

where $(\hat{\theta}_i, \hat{\varphi}_r)$ and $(\theta_i^*, \varphi_r^*)$ are the optimal solutions corresponding to $\hat{R}_k^{\max}$ and $Q_k^{\max}$, respectively. The last inequality is obtained due to the fact that the value in the parentheses is less than or equal to one, and its square has a smaller value. This proves the following theorem:

Theorem 3. When the output efficiency in the Russell measure of the average and product forms is defined as the same, the Russell measure of the average form, $\hat{R}_k^{\max}$, is greater than or equal to that of the product form, $Q_k^{\max}$.

Finally, in radial measures, the efficiency scores are difficult to interpret when some slack variables have positive values. Geometrically, if a DMU lies in the cone generated by the extreme efficient DMUs, then all the slack variables will be zero when using the radial model to measure efficiency. In this case, the radial input efficiency of a DMU k can be measured via Model (5) with the constraints corresponding to DMU k replaced with $\sum_{j \in S} \lambda_j^{(k)} X_{ij} = \theta X_{ik}$, $i = 1, \ldots, m$ and $\sum_{j \in S} \lambda_j^{(k)} Y_{rj} = Y_{rk}$, $r = 1, \ldots, s$, and the objective function replaced with min θ. If we change the direction of optimization from minimization to maximization, we still obtain the same objective value because the ray emanating from the origin to DMU k intersects the frontier facet at only one point. We thus have $\theta^* = \min \theta = \max \theta$.

To compare the most favorable Russell measure with the radial input measure, we can formulate the constraints corresponding to DMU k in Model (5) as $\sum_{j \in S} \lambda_j^{(k)} X_{ij} = \theta_i X_{ik}$, $\theta_i \leq 1$, $i = 1,\ldots, m$ and $\sum_{j \in S} \lambda_j^{(k)} Y_{rj} = \varphi_r Y_{rk}$, $\varphi_r \geq 1$, $r = 1, \ldots,$ s, with the objective function of

$$R_k^{\max} = \max \frac{1}{m+s} \left(\sum_{i=1}^{m} \theta_i + \sum_{r=1}^{s} \frac{1}{\varphi_r}\right).$$

Since BCC input efficiency θ_k^I is a special case of the average form of the Russell measure for $\theta_i = \theta$ for all i, and $\varphi_r = 1$ for all r, we have

$$R_k^{\max} = \max \frac{1}{m+s} \left(\sum_{i=1}^{m} \theta_i + \sum_{r=1}^{s} \frac{1}{\varphi_r}\right) \geq \max \frac{1}{m+s} (m\theta + s) \geq \max \theta = \theta_k^I.$$

Similarly, since BCC output efficiency θ_k^O is a special case of the average form of the Russell measure for $\theta_i = 1$ for all i, and $\varphi_r = \varphi$ for all r, we have

$$R_k^{\max} = \max \frac{1}{m+s} \left(\sum_{i=1}^{m} \theta_i + \sum_{r=1}^{s} \frac{1}{\varphi_r}\right) \geq \max \frac{1}{m+s} (m + \frac{s}{\varphi}) \geq \max \frac{1}{\varphi} = \theta_k^O.$$

We thus have the following theorem:

Theorem 4. For DMUs lying in the cone generated by the convex hull of the extreme efficient DMUs, the most favorable Russell measure $R_k^{\max}$ is greater than or equal to both the radial input measure θ_k^I and output measure θ_k^O.

This theorem also holds for production technologies of constant returns to scale. Combined with the property where the conventional least favorable Russell measure $Q_k^{\min}$ is less than or equal to both BCC input efficiency θ_k^I and output efficiency θ_k^O, we have the following result for DMU k:

$$Q_k^{\min} \leq \left\{ \begin{matrix} \theta_k^I \\ \theta_k^O \end{matrix} \right\} \leq R_k^{\max}.$$

Note that the second inequality holds only for DMUs lying in the cone generated by the extreme efficient DMUs, while the first holds for all situations.

Table 1. Efficiencies measured from different models for the twelve inefficient banks.

| Bank | Radial efficiency | Russell efficiency | | | |
| | CCR | Average form | | Product form (SBM) | |
	θ_k^{CCR} (rank)	R_k^{min} (rank)	R_k^{max} (rank)	Q_k^{min} (rank)	Q_k^{max} (rank)
1	0.9960 (1)	0.9571 (1)	0.9960 (1)	0.8964 (2)	0.9920 (1)
2	0.9498 (5)	0.9182 (3)	0.9905 (2)	0.8388 (3)	0.9810 (2)
5	0.9933 (2)	0.8532 (6)	0.8634 (8)	0.6985 (5)	0.7143 (7)
7	0.8894 (8)	0.7662 (9)	0.8705 (7)	0.3389 (11)	0.6462 (9)
8	0.7328 (12)	0.6837 (11)	0.7577 (12)	0.2642 (12)	0.5721 (12)
9	0.9877 (4)	0.9477 (2)	0.9644 (3)	0.8971 (1)	0.9290 (3)
11	0.9379 (6)	0.8592 (5)	0.9558 (4)	0.7271 (4)	0.9122 (4)
12	0.9910 (3)	0.8900 (4)	0.9501 (5)	0.6252 (7)	0.8763 (5)
15	0.8607 (9)	0.7445 (10)	0.8233 (9)	0.4195 (9)	0.6654 (8)
17	0.9333 (7)	0.7764 (8)	0.8076 (10)	0.4536 (8)	0.6269 (10)
21	0.8548 (10)	0.8176 (7)	0.9131 (6)	0.6280 (6)	0.8279 (6)
23	0.7593 (11)	0.6835 (12)	0.7819 (11)	0.3979 (10)	0.5804 (11)
Ave.	0.9072	0.8248	0.8895	0.5988	0.7770

5   Taiwanese Commercial Banks

In a study predicting the performance of banks, Kao and Liu [18] measured the efficiencies of twenty-four Taiwanese commercial banks using total deposits, interest expenses, and non-interest expenses as the inputs and total loans, interest income, and non-interest income as the outputs.

By applying the conventional CCR model to the data in Kao and Liu [18], the efficiencies of the twenty-four banks under constant returns to scale are calculated. There are twelve banks that are efficient. Column two of Table 1 shows the results for the twelve inefficient banks. The numbers in parentheses are the ranks of the banks among those that are inefficient. In calculating the CCR efficiency, it is noted that only Bank No. 2 of these twelve inefficient banks lies in the cone generated by the extremely efficient banks. In other words, the other eleven banks have at least one slack variable with positive values. Their efficiency scores are dependent on the values assigned to the non-Archimedean number ε. The rankings obtained from the CCR efficiency scores are thus not reliable.

By applying Model (5) under constant returns to scale with the objectives of minimization and maximization, the least and most favorable Russell measures of efficiency of the average form for the twelve inefficient banks are calculated, respectively. The results are shown in columns three and four of Table 1. As expected, the most favorable measures are greater than the least favorable measures for all twelve banks. The average of the most favorable measures of 0.8895, as shown in the last row, is 7.84% higher than that of the least favorable measures of 0.8248. The rankings based on the two measures are slightly different, with a mean absolute difference of 1.17 ranks.

According to Theorem 4, the most favorable Russell measures of those DMUs lying in the cone generated by the extreme efficient DMUs are greater than or equal to their radial measures. This implies that $R_k^{\max}$ in column four of Table 1 must be greater than or equal to the corresponding θ_k^{CCR} in column two. However, since only Bank No. 2 is in the defined cone, this relationship only holds for six of the twelve inefficient banks. The rankings based on $R_k^{\max}$ are quite different from those based on θ_k^{CCR}. The largest difference between the two rankings occurs for Bank No. 5, with a difference of six ranks. The mean absolute difference between the two rankings is 1.83 ranks. Due to the effect of the positive slack values on the efficiency scores, the rankings based on $R_k^{\max}$ are more reliable than those based on θ_k^{CCR}.

The product form of the least and most favorable Russell measures for the twelve inefficient banks under CRS can be calculated via Model (7), with the objectives of minimization and maximization, respectively. The results are shown in the last two columns of Table 1. The latter is obviously greater than the former for every bank. The average scores shown in the last row of Table 1 indicate that the latter is 29.76% higher than the former. The rankings based on these two measures are also different, with a mean absolute difference of 1.16 ranks.

Based on the theorem proved in Pastor et al. [20] and Tone [23], the least favorable Russell measures of the DMUs must be less than or equal to their radial measures. By comparing the numbers in columns two and five, this property is confirmed, and their averages show that the latter is 34% lower than the former.

Another pair of measures worth comparing is the least favorable Russell measures of the product form, $Q_k^{\min}$, and the most favorable Russell measures of the average form, $R_k^{\max}$. The former is the conventional SBM, which has the lowest efficiency measures among all types of Russell measures, while the latter, in contrast, has the highest efficiency measure. This is actually a consequence of Theorem 4. The numbers in columns four and five show that $R_k^{\max}$ is indeed greater than $Q_k^{\min}$ for every bank, and the average of the former, 0.8895, is 48.5% higher than that of the latter, 0.5988. The rankings based on the two measures differ not by much, with a mean absolute difference of 1.33 ranks. Since the efficiency measure of the former is higher, and the corresponding target is closer to the assessed bank, making it easier to reach, it is more acceptable to the banks being evaluated.

All the discussions in this example are based on the assumption of constant returns to scale. Similar discussions can be made under the assumption of variable returns to scale.

6 Conclusion

The Russell measure of efficiency was proposed to solve the problems of radial measures of efficiency that cannot provide appropriate efficiency scores for inefficient DMUs lying outside the cone generated by the convex hull of the extreme efficient DMUs and different scores produced from the input and output models under variable returns to scale. While the conventional Russell measures can be used to solve these problems, they are the least favorable measures, which contradict the idea of DEA that efficiency should be measured under the most favorable conditions. Moreover, the targets associated with the measures are more difficult for inefficient DMUs to reach to become efficient. To amend this drawback, a model is developed in this paper to calculate Russell measures based on the target that is closest to the assessed DMU.

Two forms of the Russell measure are considered, the average and the product. It is proven that, first, the average form produces higher efficiency scores than the product form when the output efficiency is defined in the same way. Second, the most favorable Russell measures of the average form are greater than or equal to the radial measures for DMUs lying in the cone generated by the extreme efficient DMUs. A case of Taiwanese commercial banks confirms these findings. In real world applications, the most favorable efficiency measures produce a target that requires least effort for an inefficient DMU to reach to become efficient. The corresponding rankings provide better information for the top management to make appropriate decisions. For these reasons, the most favorable Russell measures are more reliable and representative, and are more acceptable to the DMUs to be evaluated as their efficiency scores.

Acknowledgment. This research was partially supported by the Ministry of Science and Technology of the Republic of China (Taiwan), under grant MOST108-2410-H-006-102-MY3.

References

1. Ando, K., Kai, A., Maeda, Y., Sekitani, K.: Least distance based inefficiency measures on the Pareto-efficient frontier in DEA. J. Oper. Res. Soc. Jpn. **55**, 73–91 (2012)
2. Aparicio, J.: A survey on measuring efficiency through the determination of the least distance in data envelopment analysis. J. Cent. Cathedra **9**, 143–167 (2016)
3. Aparicio, J., Cordero, J.M., Pastor, J.T.: The determination of the least distance to the strongly efficient frontier in Data Envelopment Analysis oriented models: modelling and computational aspects. Omega **71**, 1–10 (2017)
4. Aparicio, J., Pastor, J.T.: A well-defined efficiency measure for dealing with closest target in DEA. Appl. Math. Comput. **219**(17), 9142–9154 (2013)
5. Aparicio, J., Pastor, J.T.: Closest targets and strong monotonicity on the strongly efficient frontier in DEA. Omega **44**, 51–57 (2014)
6. Aparicio, J., Ruiz, J.L., Sirvent, I.: Closest targets and minimum distance to the Pareto-efficient frontier in DEA. J. Prod. Anal. **28**(3), 209–218 (2007). https://doi.org/10.1007/s11123-007-0039-5
7. Banker, R.D., Charnes, A., Cooper, W.W.: Some models for estimating technical and scale efficiencies in data envelopment analysis. Manage. Sci. **30**, 1078–1092 (1984)
8. Briec, W.: Minimum distance to the complement of a convex set: duality result. J. Optim. Theory Appl. **93**(2), 301–319 (1997). https://doi.org/10.1023/A:1022697822407
9. Briec, W.: Hölder distance function and measurement of technical efficiency. J. Prod. Anal. **11**(2), 111–131 (1999). https://doi.org/10.1023/A:1007764912174
10. Charnes, A., Cooper, W.W.: Programming with linear fractional functionals. Nav. Res. Logist. Q. **9**, 181–186 (1962)
11. Charnes, A., Cooper, W.W., Rhodes, E.: Measuring the efficiency of decision making units. Eur. J. Oper. Res. **2**, 429–444 (1978)
12. Färe, R., Grosskopf, S., Lovell, C.A.K.: The Measurement of Efficiency of Production. Kluwer-Nijhoff, Dordrecht (1985)

13. Färe, R., Lovell, C.A.K.: Measuring the technical efficiency of production. J. Econ. Theory **19**(1), 150–162 (1978)
14. Førsund, F.R., Lovell, C.A.K., Schmidt, P.: A survey of frontier production functions and of their relationship to efficiency measurement. J. Econom. **13**, 5–25 (1980)
15. Fukuyama, H., Maeda, Y., Sekitani, K., Shi, J.: Input-output substitutability and strongly monotonic p-norm least distance DEA measures. Eur. J. Oper. Res. **237**, 997–1007 (2014)
16. Gonzaléz, E., Álvarez, A.: From efficiency measurement to efficiency improvement: the choice of a relevant benchmark. Eur. J. Oper. Res. **133**, 512–520 (2001)
17. Halická, M., Trnovská, M.: The Russell measure model: Computational aspects, duality, and profit efficiency. Eur. J. Oper. Res. **268**, 386–397 (2018)
18. Kao, C., Liu, S.T.: Predicting bank performance with financial forecasts: a case of Taiwan commercial banks. J. Bank. Finance **28**, 2353–2368 (2004)
19. Olesen, O.B., Petersen, N.C.: Indicators of ill-conditioned data sets and model misspecification in data envelopment analysis: an extended facet approach. Manage. Sci. **42**(2), 205–219 (1996)
20. Pastor, J.T., Ruiz, J.L., Sirvent, I.: An enhanced DEA Russell graph efficiency measure. Eur. J. Oper. Res. **115**, 596–607 (1999)
21. Petersen, N.C.: Directional distance functions in DEA with optimal endogenous directions. Oper. Res. **66**(4), 1068–1085 (2018)
22. Sueyoshi, T., Sekitani, K.: Computational strategy for Russell measure in DEA: second-order cone programming. Eur. J. Oper. Res. **180**, 459–471 (2007)
23. Tone, K.: A slacks-based measure of efficiency in data envelopment analysis. Eur. J. Oper. Res. **130**, 498–509 (2001)

Optimization of Gain in Symmetrized Itakura-Saito Discrimination for Pronunciation Learning

Andrey V. Savchenko[1(✉)], Vladimir V. Savchenko[2],
and Lyudmila V. Savchenko[3]

[1] Laboratory of Algorithms and Technologies for Network Analysis,
National Research University Higher School of Economics, Nizhny Novgorod, Russia
avsavchenko@hse.ru

[2] Nizhny Novgorod State Linguistic University, Nizhny Novgorod, Russia

[3] Department of Information Systems and Technologies, National Research
University Higher School of Economics, Nizhny Novgorod, Russia

Abstract. This paper considers an assessment and evaluation of the pronunciation quality in computer-aided language learning systems. We propose the novel distortion measure for speech processing by using the gain optimization of the symmetrized Itakura-Saito divergence. This dissimilarity is implemented in a complete algorithm for pronunciation learning and improvement. At its first stage, a user has to achieve a stable pronunciation of all sounds by matching them with sounds of an ideal speaker. At the second stage, the recognition of sounds and their short sequences is carried out to guarantee the distinguishability of learned sounds. The training set may contain not only ideal sounds but the best utterances of a user obtained at the previous step. Finally, the word recognition accuracy is estimated by using deep neural networks fine-tuned on the best words from a user. Experimental study shows that the proposed procedure makes it possible to achieve high efficiency for learning of sounds and their sequences even in the presence of noise in an observed utterance.

Keywords: Signal processing · Itakura-Saito divergence · Gain optimization · Computer-aided language learning · Speech quality assessment · Convolutional neural networks (CNN)

1 Introduction

The problem of pronunciation training and improvement appears in many practical tasks including foreign language learning [1], teaching to hearing-impaired patients [2,3] and many other applications of computer-aided language learning (CALL) [4,5]. A typical CALL tool records speech of a user, detects and diagnoses mispronunciations in it, and suggests a way for correcting them [6]. The most important task is a pronunciation quality evaluation. It is usually solved by using modern speech recognition techniques [7,8]. Pronunciation scoring makes it possible to automatically provide feedback on the overall pronunciation quality

© Springer Nature Switzerland AG 2020
A. Kononov et al. (Eds.): MOTOR 2020, LNCS 12095, pp. 440–454, 2020.
https://doi.org/10.1007/978-3-030-49988-4_30

and to point to specific production problems [9]. For example, a goodness of pronunciation measure [10] took into account sub-phonemic (senone) posteriors at the output of deep neural network (DNN) together with the state transition probabilities of hidden Markov model (HMM). The human scores of pronunciation's goodness at word or sentence levels correlate significantly with averaged frame-level posteriors of senones estimated by multi-layer, stacked Restricted Boltzman Machines [4]. The problem of pronunciation quality evaluation from the intonation point of view in second language learning is studied in [11]. Speaker adaptive training and a grammar-based decoding graph limited the search space to the frequent errors types of the hybrid DNN-HMM in teaching the Arabic pronunciation [12]. The paper [13] separates recognition from assessment stage by using two different acoustic models. The paired phone-posteriors are incorporated as input features into a neural net model for assessing learner's pronunciation quality in [14]. Quality assessment is posed as a classification problem in [15], and a joint model was proposed by exploring interdependencies of pronunciation and its dependent factors using DNN and LSTM (Long-Short Term Memory).

Unfortunately, the above-mentioned methods cannot be applied if the sounds produced by a user should be matched with sounds of ideal speakers. Indeed, HMM and DNN models are trained on large speech corpora, which pronunciation quality may be low, especially for initial quality assessment on phonetic level [16]. In such case it is typical to directly compare speech signals using appropriate dissimilarity measures between their spectrums [17]. Unfortunately, the dissimilarity between the same sounds may be too high due to the known speech variability [18] and the presence of noise [19–21] if the CALL system is launched on a laptop so utterances are recorded with a built-in microphone.

Though conventional spectral normalization techniques [22,23] are useful, they are still sensible to the level of noise in the input or reference signal. In order to deal with this issue, in this paper, we improve the quality and noise robustness of spectral distortions by using the gain optimization techniques [8,24,25]. In particular, we propose the novel spectral distortion by optimizing the gain in the symmetric Itakura-Saito divergence [17,26]. This distortion is implemented in a demo CALL application, which can be used for pronunciation learning of Russian and English sounds.

The rest of the paper is organized as follows. In Sect. 2 we describe several spectral distortion measures for assessment of sound pronunciation using an autoregression (AR) model [18,27]. In Sect. 3 we introduce the proposed approach based on gain optimization [17]. Section 4 contains experimental study of our approach for Russian and English sounds. Concluding comments are given in Sect. 5.

2   Sound Pronunciation Evaluation

The first task in most CALL systems is to learn correct pronunciation of $C \geq 1$ phonemes or short sounds corresponded to letters [5]. We assume that a dataset of $R \geq C$ signals $\{\mathbf{x}_r\}, r \in \{1, ..., R\}$ with known labels $c(r) \in \{1, ..., C\}$ of every reference sound $\mathbf{x}_r$ is available. Each signal in this dataset should represent the utterance produced by an ideal speaker. A user learns to produce every cth sound to be as close as possible to one of ideal signals $\mathbf{x}_r$, where r is chosen so that $c(r) = c$. Hence, the following condition is tested:

$$\min_{r \in \{1,...,R | c(r)=c\}} \rho(\mathbf{x}, \mathbf{x}_r) < \rho_c, \tag{1}$$

where $\rho(\mathbf{x}, \mathbf{x}_r)$ is an arbitrary dissimilarity measure between utterances $\mathbf{x}$ and $\mathbf{x}_r$, and ρ_c is a fixed threshold that may depend on the class c of the learned sound. If criterion (1) is satisfied for several attempts to produce the c-th sound, one can assume that its pronunciation quality is appropriate. This procedure is repeated until condition (1) holds for all C sounds.

The discrimination $\rho(\mathbf{x}, \mathbf{x}_r)$ in speech processing is typically computed using power spectral densities (PSD) $\hat{G}_\mathbf{x}(f)$ and $\hat{G}_r(f)$ of the input signal $\mathbf{x}$ and $\mathbf{x}_r$. Here $f \in \{1, ..., F\}$ is the discrete frequency, and F is the sample rate. These PSDs may be estimated by assuming that the speech signals $\mathbf{x}$ and $\mathbf{x}_r$ for every sound can be represented as stationary AR ergodic Gaussian processes with zero mean [27,28]:

$$\hat{G}_\mathbf{x}(f) = \frac{\hat{\sigma}_\mathbf{x}^2}{2F} \left| 1 + \sum_{m=1}^{p} a_\mathbf{x}(m) e^{-i\pi m f / F} \right|^{-2}, \tag{2}$$

$$\hat{G}_r(f) = \frac{\hat{\sigma}_r^2}{2F} \left| 1 + \sum_{m=1}^{p} a_r(m) e^{-i\pi m f / F} \right|^{-2}, \tag{3}$$

where $i = \sqrt{-1}$ is the imaginary unit, p is the order of AR model, $a_\mathbf{x}(m)$ and $a_r(m), m \in \{1, ..., p\}$ are the AR or LPC (linear prediction coding) coefficients and $\hat{\sigma}_\mathbf{x}^2$ and $\hat{\sigma}_r^2$ are the gains or one-step prediction errors [17] that are equal to the variance of generative white noise. These parameters of the AR model can be estimated with the Levinson-Durbin algorithm and, e.g., the Burg method [27].

It is known [17,28] that the maximal likelihood solution for testing hypothesis about covariance matrix of the Gaussian signal $\mathbf{x}$ is achieved by using the Kullback-Leibler (KL) divergence [29] between the zero-mean Gaussian distributions. The latter can be computed as the Itakura-Saito (IS) distance [26] between PSDs $\hat{G}_\mathbf{x}(f)$ and $\hat{G}_r(f)$:

$$\rho_{IS}(\hat{G}_\mathbf{x}, \hat{G}_r) = \frac{1}{F} \sum_{f=1}^{F} \left(\frac{\hat{G}_\mathbf{x}(f)}{\hat{G}_r(f)} - \ln \frac{\hat{G}_\mathbf{x}(f)}{\hat{G}_r(f)} - 1 \right). \tag{4}$$

The IS divergence between PSDs (4) is well known in speech processing due to its strong correlation with the subjective MOS (mean opinion score) estimate of speech closeness [18]. Except the IS divergence (4), its symmetrized version, namely, COSH distance [17,30]) is widely used in practice:

$$\rho_{COSH}(\hat{G}_\mathbf{x}, \hat{G}_r) = \frac{2}{F} \sum_{f=1}^{F} \frac{(\hat{G}_\mathbf{x}(f) - \hat{G}_r(f))^2}{\hat{G}_\mathbf{x}(f)\hat{G}_r(f)}. \tag{5}$$

Unfortunately, the gains $\hat{\sigma}_{\mathbf{x}}^2$ and $\hat{\sigma}_r^2$ in the PSD estimates (2), (3) depend on the scale of signals $\mathbf{x}$ and $\mathbf{x}_r$. For example, if a user speaks two-times louder, the variance $\hat{\sigma}_{\mathbf{x}}^2$ will become twice higher. Thus, the gain normalization techniques are traditionally used in practice [17]:

$$\rho_{gn-IS}(\hat{G}_{\mathbf{x}}, \hat{G}_r) = \rho_{IS}(\hat{G}_{\mathbf{x}}/\hat{\sigma}_{\mathbf{x}}^2, \hat{G}_r/\hat{\sigma}_r^2). \tag{6}$$

The same normalization procedure is used to compute the gain-normalized version of the COSH distance (5):

$$\rho_{gn-COSH}(\hat{G}_{\mathbf{x}}, \hat{G}_r) = \rho_{COSH}(\hat{G}_{\mathbf{x}}/\hat{\sigma}_{\mathbf{x}}^2, \hat{G}_r/\hat{\sigma}_r^2). \tag{7}$$

However, even such normalization does not provide scale independence, which is especially crucial if the input signal contains noise, i.e. when a user's microphone is imperfect. It is known that the noise influence can be significantly reduced by using gain optimization of spectral distortions [17], in which the reference PSD is scaled in order to minimize the distance to the input signal. The first attempts of such optimization lead to the well-known Itakura distance [31]:

$$\rho_I(\hat{G}_{\mathbf{x}}, \hat{G}_r) = \ln\left(\frac{1}{F}\sum_{f=1}^{F}\frac{\hat{G}_{\mathbf{x}}(f)/\hat{\sigma}_{\mathbf{x}}^2}{\hat{G}_r(f)/\hat{\sigma}_r^2}\right). \tag{8}$$

Let us describe the usage of such approach for the COSH distance (5).

3 Proposed Approach

3.1 Gain-Optimized COSH Discrimination

In this paper we automatically scale each r-th reference instance to be as close as possible to the input signal $\mathbf{x}$. In particular, the gains in the COSH distance (5) are optimized as follows [8]:

$$\rho_{go-COSH}(\hat{G}_{\mathbf{x}}, \hat{G}_r) = \min_{\lambda>0} \rho_{COSH}(\hat{G}_{\mathbf{x}}, \lambda\hat{G}_r), \tag{9}$$

where

$$\rho_{COSH}(\hat{G}_{\mathbf{x}}, \lambda\hat{G}_r) = \frac{2}{F}\sum_{f=1}^{F}\frac{(\hat{G}_{\mathbf{x}}(f) - \lambda\hat{G}_r(f))^2}{\hat{G}_{\mathbf{x}}(f)\cdot\lambda\hat{G}_r(f)}. \tag{10}$$

Let us directly compute the minimum:

$$\frac{d\rho_{COSH}(\hat{G}_{\mathbf{x}}, \lambda\hat{G}_r)}{d\lambda} = \frac{2}{F}\frac{d}{d\lambda}\sum_{f=1}^{F}\left(\frac{\hat{G}_{\mathbf{x}}(f)}{\lambda\hat{G}_r(f)} + \frac{\lambda\hat{G}_r(f)}{\hat{G}_{\mathbf{x}}(f)} - 2\right) \tag{11}$$

$$= -\frac{2}{F\lambda^2}\sum_{f=1}^{F}\frac{\hat{G}_{\mathbf{x}}(f)}{\hat{G}_r(f)} + \frac{2}{F}\sum_{f=1}^{F}\frac{\hat{G}_r(f)}{\hat{G}_{\mathbf{x}}(f)} = 0.$$

Hence, the optimal value of the scaling factor is equal to

$$\lambda^* = \sqrt{\frac{\sum\limits_{f=1}^{F} \frac{\hat{G}_{\mathbf{x}}(f)}{\hat{G}_r(f)}}{\sum\limits_{f=1}^{F} \frac{\hat{G}_r(f)}{\hat{G}_{\mathbf{x}}(f)}}}. \tag{12}$$

By substituting this value into Eq. 10 and dividing it by 4, we obtain the final value of the gain-optimized COSH distance (9):

$$\rho_{go-COSH}(\hat{G}_{\mathbf{x}}, \hat{G}_r) = \frac{1}{F}\sqrt{\left(\sum\limits_{f=1}^{F} \frac{\hat{G}_{\mathbf{x}}(f)}{\hat{G}_r(f)}\right)\left(\sum\limits_{f=1}^{F} \frac{\hat{G}_r(f)}{\hat{G}_{\mathbf{x}}(f)}\right) - 1}. \tag{13}$$

This dissimilarity has many important properties, such as non-negativity, symmetry and dependence on the ratio of PSDs only. Despite the gain-normalized version, the proposed distortion (13) does not depend on scale: every PSD may be scaled without any affect. Moreover, it can be computed as efficiently as the original divergences from Sect. 2.

Finally, due to the above-mention equivalence of the IS and the KL divergences, one can use the known asymptotic distribution of the KL divergence between samples from the same distribution [29]. In fact, $(n(\mathbf{x}) - p)$-times symmetrized KL divergence has the chi-squared distribution with $p(p+1)/2$ degrees of freedom, where $n(\mathbf{x})$ is a duration (number of samples) of the input signal. As the condition (1) is tested by assuming that the input signal represents the c-th sound (correct null hypothesis), the threshold ρ_c can be set to $\frac{\chi^2_{\alpha,p(p+1)/2}}{4(n(\mathbf{x})-p)}$. Here $\chi^2_{\alpha,p(p+1)/2}$ is the α-quantile of the chi-squared distribution with $p(p+1)/2$ degrees of freedom and we take into account that (13) is 4-times lower than the original minimal COSH distance.

3.2   Pronunciation Learning Algorithm

Complete data flow of the proposed pronunciation learning procedure for CALL systems is presented in Algorithm 1. It is divided into three stages. At first, a user learns to pronounce each sound to be as close to the corresponding sound of a reference speaker as possible. The proposed gain-optimized COSH dissimilarity (13) is used in (1). Thresholds $\rho_c, c \in \{1, ..., C\}$ can be tuned in such a way that unexperienced non-native speaker has an ability to satisfy this criterion. However, type I error rate α and, as a consequence, these thresholds, should be adaptively made lower while a user reaches a certain level of progress.

As it is necessary to control the stability of correct pronunciation, there are three additional parameters, namely, a minimal number of trials $N_{\min} > 0$, a minimal ratio of close sounds $\delta_{\min} \in [0, 1]$ and a minimal "radius" $\bar{r}_0$. The latter is matched with the "radius" of a set of better pronounced sounds X_c:

$$\bar{r}(X_c) = \frac{1}{|X_c|} \sum_{\mathbf{x} \in X_c} \rho_{go-COSH}(\hat{G}_{\mathbf{x}}, \hat{G}_{\mathbf{x}_c^*}), \tag{14}$$

Algorithm 1. Proposed Approach for Pronunciation Learning

1: **for** $c \in \{1, ..., C\}$ **do** ▷ Learn isolated sounds
2:    $N_{reliable} := 0, N_{attempts} = 0, X_c := \{\}$
3:    **while** $N_{reliable} < N_{\min}$ AND $\frac{N_{reliable}}{N_{attempts}} < \delta_{\min}$ AND $\bar{r}(X_c) > \bar{r}_0$ (14) **do**
4:       $N_{attempts} := N_{attempts} + 1$
5:       Record speech signal $\mathbf{x}$ for the c-th sound
6:       Compute PSD $\hat{G}_{\mathbf{x}}$ (2) of signal $\mathbf{x}$
7:       **for** $r \in \{1, ..., R | c(r) = c\}$ **do**
8:          Compute distance $\rho_{go-COSH}(\hat{G}_{\mathbf{x}}, \hat{G}_r)$ (13)
9:          **if** $\rho_{go-COSH}(\hat{G}_{\mathbf{x}}, \hat{G}_r) < \rho_c$ **then**
10:             $N_{reliable} := N_{reliable} + 1$
11:             Append signal $\mathbf{x}$ to the set X_c
12:             Break
13:          **end if**
14:       **end for**
15:    **end while**
16:    (Optional) add the best utterance (17) to the dataset of reference sounds $\{\mathbf{x}_r\}$
17: **end for**
18: **repeat** ▷ Quality control: recognize isolated sounds
19:    $A := 0$
20:    **for** $c \in \{1, ..., C\}$ **do**
21:       **for** $n \in \{1, ..., N\}$ **do**
22:          Record speech signal $\mathbf{x}$ for the c-th sound
23:          Compute PSD $\hat{G}_{\mathbf{x}}$ (2) of the signal $\mathbf{x}$
24:          Obtain the nearest neighbor r^* (16)
25:          **if** $c(r^*) = c$ **then**
26:             $A := A + 1$
27:          **end if**
28:       **end for**
29:    **end for**
30:    Compute accuracy $A := A/(CN)$
31: **until** $A > A_0$
32: (Optional) Repeat quality control (Steps 18-31) for a sequence of isolated syllables
33: (Optional) Repeat quality control for DNN-based recognition of words

where

$$\mathbf{x}_c^* = \underset{\mathbf{x}^* \in X_c}{\operatorname{argmin}} \sum_{\mathbf{x} \in X_c} \rho_{go-COSH}(\hat{G}_{\mathbf{x}}, \hat{G}_{\mathbf{x}^*}). \qquad (15)$$

At the second stage it is necessary to control speech intelligibility [32]. We propose to recognize the user's sounds in order to verify that they are distinguishable from each other. We record $N \geq 1$ utterances for each sound and use the nearest neighbor rule for that purpose:

$$r^* = \underset{r \in \{1, ..., R\}}{\operatorname{argmin}} \rho_{go-COSH}(\hat{G}_{\mathbf{x}}, \hat{G}_r). \qquad (16)$$

The pronunciation is assumed to be correct only if estimated accuracy A is higher than a fixed threshold A_0, which can be increased over time. Unfortunately, the recognition accuracy would be decreased greatly for non-native speakers, compared with native ones if only an acoustic model from ideal speakers is used in this stage [13]. Hence, we propose to let a user add the best pronounced sound

$$\underset{\substack{\mathbf{x} \in X_c \\ }}{\text{argmin}} \; \underset{r \in \{1, \dots, R | c(r) = c\}}{\min} \rho_{go-COSH}(\hat{G}_\mathbf{x}, \hat{G}_r) \tag{17}$$

to the dataset of reference sounds during the first stage in order to memorize the best attempts and use them in subsequent stages.

At this point a learning procedure may be continued with optional steps. One can control the pronunciation quality for a sequence of sounds or isolated syllables similarly to how children learn to read. At first, the input signal is preprocessed in order to decrease its variability, detect voice activity regions, etc. [18]. Next, the largest piecewise quasi-stationary speech segments are detected and all the steps of our algorithm are repeated with only one exception, namely, the replacement of our dissimilarity measure (13) to the sum of these distances between corresponding syllables.

The final optional step consists in learning of words using existing automatic speech recognition techniques [18]. If the words produced by ideal speaker are available, conventional dynamic programming techniques, e.g., Dynamic Time Warping or the Viterbi algorithm in the HMM, are used to dynamically align the speech frames [11]. The quality control is implemented by measuring the distinguishability of words with conventional speech recognition based on DNNs.

The most computationally expensive steps in the described procedure have the following run-time complexities:

1. estimation of AR coefficients with the Levinson-Durbin method requires $O(p(n(\mathbf{x}) + p))$ operations;
2. estimation of the PSD (2) of the input signal needs $O(Fp)$ operations;
3. computing distortion (13) has complexity $O(F)$;
4. finding the nearest neighbor (16) linearly depends on the number of reference signals: $O(FR)$.

In practice, computational complexity can be significantly reduced by wrapping the PSD into the Mel-frequency scale $Mel(f) = 1125 \ln(1 + f/100)$ and computing the weighted sum of PSD samples at regular intervals. As a result, the number of spectral values will be significantly reduced. For example, if speech ranges ($f \in [200, 3400]$) are analyzed and a duration of a regular interval is equal to 55 Mels, such procedure will output only 31 samples of smoothed spectrum, which is much lower than 4000 samples for telephone speech ($F = 8000$ Hz).

The proposed approach has been implemented in .Net Framework 4.5 using C#. We developed the publicly available demo application (https://sites.google.com/site/frompldcreators/PhonemeTraining.zip) for learning of Russian and English sounds (Fig. 1). This application makes it possible to perform learning and recognition of isolated sounds, automatic segmentation and phoneme recognition in spontaneous speech. Moreover, it is possible to extend the set $\{\mathbf{x}_r\}$ by the sounds that are rather close to ideal signals to let a user reach not only the ideal speech but his or her best attempts.

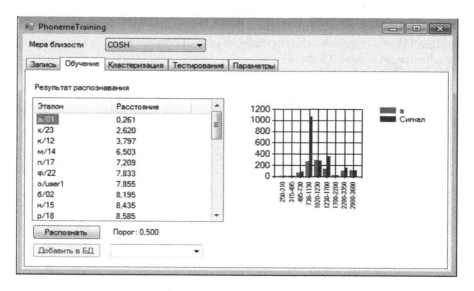

Fig. 1. Sample GUI of developed demo application "PhonemeTraining".

4   Experimental Results

In this section the proposed approach (Algorithm 1) is experimentally studied in pronunciation learning for English and Russian languages. In the former case, the reference dataset $\{\mathbf{x}_r\}$ contains sounds of $C = 10$ English letters ("a", "e", "i", "j", "o", "r", "u", "w", "x", "y") pronounced by ideal English native speaker from BBC. These sounds are used in the CALL software "Professor Higgins: English without accent" [5]. In the latter case, the set of reference sounds is filled by $C = 6$ Russian vowels (/aa/, /ee/, /oo/, /ii/, /y/, /uu/) pronounced by ideal native speaker for Russian version of "Professor Higgins" [5]. The following parameters are chosen: single reference signal per class ($R = C, c(r) = r$), sample rate $F = 8000$ Hz, AR-model order $p = 20$, the estimates of PSDs are smoothed at regular Mel-frequency intervals as described in the previous section.

In the first experiment, each of 5 Russian speakers (3 men and 2 women) produced 1200 isolated vowels (200 for each sound). An artificially generated white noise was added to each test utterance using the following procedure. At first, the signal-to-noise ratio (SNR) is fixed. Next, the initial and final pauses are detected in each utterance using simple energy thresholding, and the standard deviation of the remaining part with high energy is estimated. Finally, these standard deviation was corrected using given SNR, and uncorrelated normal random numbers with zero mean and the resulted standard deviation was added to each value of the speech signal.

The proposed gain-optimized dissimilarity measure (13) is compared with sounds' matching from "Professor Higgins" software [5] and conventional spectral distortions, namely, IS (4), Gain-normalized IS (6), Itakura (8) and original COSH (5) with its gain-normalized version (7). In order to test the quality for criterion (1), we computed the dependence of AUC (area under ROC curve) on the additive noise level (Table 1).

Table 1. Dependence of AUC (%) on SNR (dB)

Language	Distance	Signal-to-noise ratio, dB					
		26	20	16	14	12	10
Russian	"Professor Higgins" [5]	88.0	87.1	86.5	86.1	85.3	84.2
	IS (4)	64.1	62.8	61.3	59.3	57.1	55.4
	COSH (5)	79.5	76.6	74.2	71.9	71.4	70.6
	Gain-normalized IS (6)	92.7	92.4	91.1	89.0	86.4	84.0
	Itakura (8)	91.6	90.6	89.5	88.4	87.0	85.4
	Gain-normalized COSH (7)	94.7	94.3	93.2	91.9	90.2	88.7
	Proposed optimized COSH (13)	94.8	94.4	93.4	92.3	90.9	89.4
English	"Professor Higgins" [5]	76.7	77.1	76.8	77.3	75.3	72.8
	IS (4)	73.2	73.1	70.9	68.2	65.2	64.6
	COSH (5)	74.7	76.7	73.9	68.9	67.5	67.9
	Gain-normalized IS (6)	75.6	79.6	77.8	77.8	76.5	76.0
	Itakura (8)	78.2	79.1	77.8	78.1	76.9	76.4
	Gain-normalized COSH (7)	80.8	79.3	79.2	78.9	77.4	76.8
	Proposed optimized COSH (13)	80.7	79.4	79.3	79.0	77.8	77.4

Here, AUC of the testing of English sounds is much lower when compared to Russian sounds because non-native speech is recognized worth. The usage of unnormalized PSDs (2), (3) in IS (4) and COSH (5) distances is inappropriate as their AUCs are 13–30% lower when compared to gain normalization and optimization. The proposed optimization (13) leads to one of the highest values of AUC, which is 2–5% and 3–4% higher than conventional gain-normalized IS divergence (6) and its gain-optimized version (8).

AUC of our approach is slightly better than AUC of the original gain-normalized COSH distance (7) especially if the noise level becomes higher. The next experiment makes the advantages of our dissimilarity more noticeable. We examine the second stage of the proposed approach, namely, the quality control of sound pronunciation. The nearest neighbor rule (16) is used with the training and testing sets from the first experiment. The recognition accuracy is presented in Table 2.

Table 2. Dependence of accuracy (%) on SNR (dB)

Language	Distance	Signal-to-noise ratio, dB					
		26	20	16	14	12	10
Russian	"Professor Higgins" [5]	72.2	71.6	70.2	67.2	64.8	62.3
	IS (4)	32.1	26.4	19.9	17.5	17.4	17.1
	COSH (5)	32.8	33.0	33.0	32.4	31.0	30.0
	Gain-normalized IS (6)	85.7	81.7	75.7	66.9	59.3	55.5
	Itakura (8)	80.4	79.1	77.9	75.6	69.5	62.0
	Gain-normalized COSH (7)	86.3	84.3	79.4	74.3	65.2	58.3
	Proposed optimized COSH (13)	87.0	85.4	81.7	77.1	70.6	62.9
English	"Professor Higgins" [5]	47.5	45.0	47.5	50.0	47.5	35.0
	IS (4)	42.5	42.5	40.0	37.5	30.0	25.0
	COSH (5)	40.0	42.5	42.5	40.0	42.5	35.0
	Gain-normalized IS (6)	60.0	55.0	55.0	52.5	47.5	37.5
	Itakura (8)	55.0	55.0	52.5	50.5	40.0	37.5
	Gain-normalized COSH (7)	62.5	55.5	55.5	45.0	45.0	37.5
	Proposed optimized COSH (13)	62.5	57.5	57.5	52.5	52.5	45.0

As one can notice, the proposed approach is 0.7–4.5% more accurate when compared to the baseline (7). Gain normalization in the Itakura distance (8) also helps to achieve slow accuracy degradation with a decrease of SNR. It is necessary to emphasize that conventional matching of sounds in the "Professor Higgins" learning system is not as accurate as normalized and optimized spectral distortions. However, it is an order of magnitude faster. Moreover, its quality practically does not depend on the noise level, so that this software is the best one for $SNR = 10$ dB with only one exception that our dissimilarity (13) is still 0.6% more accurate.

An optional step 32 of the proposed Algorithm 1 is studied in the next experiment for Russian language. Two vocabularies are used, namely, 1) the list of 1832 Russian cities with corresponding regions; and 2) the list of 1913 drugs from a pharmacy of Nizhny Novgorod. All speakers pronounced every word from all vocabularies in isolated syllable mode, so that every vowel in the syllable is made stressed. The part of speech data suitable to reproduce our experiments is available for free download (https://sites.google.com/site/andreyvsavchenko/SpeechDataIsolatedSyllables.zip). Observed utterances are divided into 30 ms frames with 10 ms overlap. The syllables in the test signals are extracted with the amplitude detector and the vowels are recognized in each syllable by simple voting based on the results obtained using vowel recognition [20].

The dependence of the words recognition error rates on the SNR is shown in Fig. 2 and Fig. 3 for cities and drugs vocabularies, respectively. Here, the proposed optimization procedure leads to the dissimilarity (13) that is significantly (1–10%) more accurate than the gain-normalized spectral distortions. Hence, the

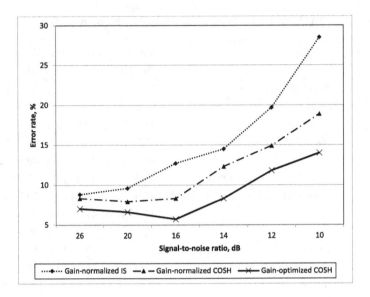

Fig. 2. Error rate (%), cities vocabulary.

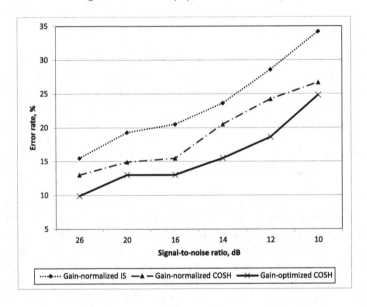

Fig. 3. Error rate (%), drugs vocabulary.

usage of our approach makes it possible to reduce the training time by minimizing the mistakes made by the quality control procedure.

In the last experiment we provide preliminarily results for the last step of our algorithm for recognition of the following English words: "down", "go", "left", "no", "off", "on", "right", "stop", "up", "yes". We use several existing convolutional neural networks (CNN) pre-trained on Google Speech Commands dataset, namely, 1) "conv" model based on CNN-trad-fpool3 [33]; 2) "low_latency_conv" model based on CNN-one-fstride4 [34]; 3) "low_latency_svdf" model with rank-constrained compression [35]; and 4) "tiny_conv" model with one fully connected layer. We compared the recognition accuracy for the original speaker independent model and their fine-tuned versions on the best recognized words from each speaker. Fine-tuning is implemented by using scripts from Simple Audio Recognition example of TensorFlow framework. Each speaker pronounced at least $N = 10$ examples of each word. In addition, inspired by our criterion (1), we chose the best examples by matching the confidence of speaker-independent CNN model with fixed threshold 0.9.

Table 3. Average accuracy (%) for CNN-based isolated word recognition

	conv	low_latency_conv	low_latency_svdf	tiny_conv
Pre-trained	72	27	46	20
Fine-tuned (all words)	91	92	55	50
Fine-tuned (best words)	94	96	60	65

The average accuracies for extra validation set of 100 commands for each speaker are shown in Table 3. Here, "conv" model is the most accurate one for conventional speaker-independent quality control. However, the usage of the speaker data in fine-tuned models drastically improves the overall accuracy on 19–65%. It is important to notice that the proposed selection of the best examples during initial training makes it possible to further improve accuracy on 3–15%.

5   Conclusion

To sum it up, this article introduced an approach to develop CALL systems (Algorithm 1), which most important part is a novel spectral distortion based on optimization of the symmetrized Itakura-Saito, i.e., COSH, divergence. It was experimentally shown that the proposed discrimination (13) makes it possible to achieve high AUC for pronunciation learning and accuracy for quality control even in the presence of noise in the input utterance.

In future, the last step of Algorithm 1 should be examined more thoroughly by using more complex DNN and LSTM models for speech recognition [36] to control the quality of word's pronunciation. Our preliminary study demonstrated (Table 3) that the overall accuracy becomes much higher if the best utterances of a learner are used similarly to our modification of the training set (17). Hence, it is necessary to study speaker adaptation techniques [12] in order to fine-tune the contemporary neural models. Finally, it is important to elaborate the possibility to use the proposed algorithm on extra datasets of other languages to analyze its performance and robustness thoroughly.

Acknowledgements. The work was prepared within the framework of the Basic Research Program at the National Research University Higher School of Economics (HSE).

References

1. Golonka, E.M., Bowles, A.R., Frank, V.M., Richardson, D.L., Freynik, S.: Technologies for foreign language learning: a review of technology types and their effectiveness. Comput. Assist. Lang. Learn. **27**(1), 70–105 (2014)
2. Sztahó, D., Kiss, G., Vicsi, K.: Computer based speech prosody teaching system. Comput. Speech Lang. **50**, 126–140 (2018)
3. Han, K.I., Park, H.J., Lee, K.M.: Speech recognition and lip shape feature extraction for English vowel pronunciation of the hearing-impaired based on SVM technique. In: Proceedings of the International Conference on Big Data and Smart Computing (BigComp), pp. 293–296. IEEE (2016)
4. Hu, W., Qian, Y., Soong, F.K.: A new DNN-based high quality pronunciation evaluation for computer-aided language learning (CALL). In: Proceedings of Interspeech, pp. 1886–1890 (2013)
5. Kneller, E., Karaulnyh, D.: System and method of converting voice signal into transcript presentation with metadata. RU Patent 2589851 C2, 10 July 2016
6. Agarwal, C., Chakraborty, P.: A review of tools and techniques for computer aided pronunciation training (CAPT) in English. Educ. Inf. Technol. **24**(6), 3731–3743 (2019). https://doi.org/10.1007/s10639-019-09955-7
7. Haikun, T., Shiying, W., Xinsheng, L., Yue, X.G.: Speech recognition model based on deep learning and application in pronunciation quality evaluation system. In: Proceedings of the International Conference on Data Mining and Machine Learning, pp. 1–5 (2019)
8. Savchenko, V.V.: Minimum of information divergence criterion for signals with tuning to speaker voice in automatic speech recognition. Radioelectron. Commun. Syst. **63**(1), 42–54 (2020). https://doi.org/10.3103/S0735272720010045
9. Franco, H., Bratt, H., Rossier, R., Rao Gadde, V., Shriberg, E., Abrash, V., Precoda, K.: Eduspeak®: a speech recognition and pronunciation scoring toolkit for computer-aided language learning applications. Lang. Test. **27**(3), 401–418 (2010)
10. Sudhakara, S., Ramanathi, M.K., Yarra, C., Ghosh, P.K.: An improved goodness of pronunciation (GoP) measure for pronunciation evaluation with DNN-HMM system considering hmm transition probabilities. In: Proceedings of Interspeech, pp. 954–958 (2019)
11. Arias, J.P., Yoma, N.B., Vivanco, H.: Automatic intonation assessment for computer aided language learning. Speech Commun. **52**(3), 254–267 (2010)
12. Elaraby, M.S., Abdallah, M., Abdou, S., Rashwan, M.: A deep neural networks (DNN) based models for a computer aided pronunciation learning system. In: Ronzhin, A., Potapova, R., Németh, G. (eds.) SPECOM 2016. LNCS (LNAI), vol. 9811, pp. 51–58. Springer, Cham (2016). https://doi.org/10.1007/978-3-319-43958-7_5
13. Huang, G., Ye, J., Shen, Y., Zhou, Y.: A evaluating model of English pronunciation for Chinese students. In: Proceedings of the 9th International Conference on Communication Software and Networks (ICCSN), pp. 1062–1065. IEEE (2017)
14. Xiao, Y., Soong, F., Hu, W.: Paired phone-posteriors approach to ESL pronunciation quality assessment. In: Proceedings of Interspeech, pp. 1631–1635 (2018)

15. Srinivasan, A., Yarra, C., Ghosh, P.K.: Automatic assessment of pronunciation and its dependent factors by exploring their interdependencies using DNN and LSTM. In: Proceedings of the 8th ISCA Workshop on Speech and Language Technology in Education (SLaTE), pp. 30–34 (2019)
16. Gu, L., Harris, J.G.: SLAP: a system for the detection and correction of pronunciation for second language acquisition. In: Proceedings of the International Symposium on Circuits and Systems (ISCAS), vol. 2, p. II. IEEE (2003)
17. Gray, R., Buzo, A., Gray, A., Matsuyama, Y.: Distortion measures for speech processing. IEEE Trans. Acoust. Speech Signal Process. **28**(4), 367–376 (1980)
18. Benesty, J., Sondhi, M.M., Huang, Y.A. (eds.): Springer Handbook of Speech Processing. SH. Springer, Heidelberg (2008). https://doi.org/10.1007/978-3-540-49127-9
19. Mošner, L., et al.: Improving noise robustness of automatic speech recognition via parallel data and teacher-student learning. In: Proceedings of the International Conference on Acoustics, Speech and Signal Processing (ICASSP), pp. 6475–6479. IEEE (2019)
20. Savchenko, A.V., Savchenko, L.V.: Towards the creation of reliable voice control system based on a fuzzy approach. Pattern Recogn. Lett. **65**, 145–151 (2015)
21. Savchenko, L.V., Savchenko, A.V.: Fuzzy phonetic decoding method in a phoneme recognition problem. In: Drugman, T., Dutoit, T. (eds.) NOLISP 2013. LNCS (LNAI), vol. 7911, pp. 176–183. Springer, Heidelberg (2013). https://doi.org/10.1007/978-3-642-38847-7_23
22. Su, H.Y., Gao, Y.: Adaptive gain reduction for encoding a speech signal. US Patent 9,269,365, 23 February 2016
23. Dionelis, N., Brookes, M.: Speech enhancement using modulation-domain Kalman filtering with active speech level normalized log-spectrum global priors. In: Proceedings of the 25th European Signal Processing Conference (EUSIPCO), pp. 2309–2313. IEEE (2017)
24. Erkelens, J., Jensen, J., Heusdens, R.: A data-driven approach to optimizing spectral speech enhancement methods for various error criteria. Speech Commun. **49**(7–8), 530–541 (2007)
25. Bastos, I., Oliveira, L.B., Goes, J., Silva, M.: MOSFET-only wideband LNA with noise cancelling and gain optimization. In: Proceedings of the 17th International Conference Mixed Design of Integrated Circuits and Systems (MIXDES), pp. 306–311. IEEE (2010)
26. Itakura, F., Saito, S.: Analysis synthesis telephony based on the maximum likelihood method. In: Proceedings of the 6th International Congress on Acoustics, pp. 17–20 (1968)
27. Marple Jr., S.L.: Digital Spectral Analysis with Applications, 2nd edn. Dover Publications, Mineola, New York (2019). 432 p.
28. Savchenko, V.V.: Itakura–Saito divergence as an element of the information theory of speech perception. J. Commun. Technol. Electron. **64**(6), 590–596 (2019). https://doi.org/10.1134/S1064226919060093
29. Kullback, S.: Information Theory and Statistics. Dover Publications, New York (1997)
30. Savchenko, A.V., Belova, N.S.: Statistical testing of segment homogeneity in classification of piecewise-regular objects. Int. J. Appl. Math. Comput. Sci. **25**(4), 915–925 (2015)
31. Itakura, F.: Minimum prediction residual principle applied to speech recognition. IEEE Trans. Acoust. Speech Signal Process. **23**(1), 67–72 (1975)

32. Savchenko, V.V., Savchenko, L.V.: Method for measuring the intelligibility of speech signals in the Kullback–Leibler information metric. Meas. Tech. **62**(9), 832–839 (2019). https://doi.org/10.1007/s11018-019-01702-1

33. Sainath, T.N., Parada, C.: Convolutional neural networks for small-footprint keyword spotting. In: Proceedings of the Sixteenth Annual Conference of the International Speech Communication Association, pp. 1478–1482 (2015)

34. Zhang, Y., Pezeshki, M., Brakel, P., Zhang, S., Bengio, C.L.Y., Courville, A.: Towards end-to-end speech recognition with deep convolutional neural networks. arXiv preprint arXiv:1701.02720 (2017)

35. Nakkiran, P., Alvarez, R., Prabhavalkar, R., Parada, C.: Compressing deep neural networks using a rank-constrained topology. In: Proceedings of the Sixteenth Annual Conference of the International Speech Communication Association, pp. 1473–1477 (2015)

36. Kuchaiev, O., et al.: Nemo: a toolkit for building AI applications using neural modules. arXiv preprint arXiv:1909.09577 (2019)

Integer Programming Approach to the Data Traffic Paths Recovering Problem

Igor Vasilyev[1,3]([ID]), Dong Zhang[2], and Jie Ren[3]

[1] Matrosov Institute for System Dynamics and Control Theory of Siberian Branch of Russian Academy of Sciences, Irkutsk, Russia
vil@icc.ru

[2] Algorithm and Technology Development Department, Global Technical Service Department, Huawei Technologies, Co., Ltd., Dongguan, China
zhangdong48@huawei.com

[3] Moscow Advanced Software Technology Lab, Huawei Russian Research Institute, Moscow, Russia
{vasilyev.igor,renjie21}@huawei.com

Abstract. In this paper, we propose a novel approach to recovering path relationships in communication networks. The path relationship is one of the key input data which is necessary for network operation and maintenance. We have a continuous network transformation, upgrades, expansions, service allocations, thus the network physical topology and paths relationship are permanently changing with high frequency. Our approach is aimed at recovering the path relationships through flow information of each arc in the network. Getting the flow information is not a big technical problem and its control is included in the basic toolbox for network monitoring. We consider two scenarios which lead us to integer linear programs. The both of them minimize the flow deviation, where in the first one we look for a directed spanning tree (r-arborescence) and, in the second one—more general origin/destination paths (OD-paths). We propose mixed integer linear programming formulations for both problems. Their feature is that they contain the non-polynomial number of constraints which are considered implicitly by the cutting planes approach. The preliminary computation results showed that the large-scale instances of the first scenario can easily be solved. At the same time, the optimal solutions of second scenario problems can be found only on small- and medium-size instances, which inspires for the further research.

Keywords: Communication network · Data flow · Mixed integer linear programming · Branch-and-cut algorithm

1 Introduction

In telecommunication networks, the physical topology of a network and the path relationships of services are the key ingredient for network operation and maintenance (including network reconstruction and expansion, service distribution

© Springer Nature Switzerland AG 2020
A. Kononov et al. (Eds.): MOTOR 2020, LNCS 12095, pp. 455–469, 2020.
https://doi.org/10.1007/978-3-030-49988-4_31

and change, delineation of network problems, etc.). For example, the emergence of 5G technologies results in merging of 3/4/5G networks and substantial development of a number of new services, such as the Internet of Things and Vehicles. Telecommunication networks are subject to constant transformations, upgrades, expansions, service allocations. Thus, the network physical topology and paths relationships are permanently changing with high frequency.

It causes that the problems of accurate network topology tracking and service path recovering are very important and challenging. This paper focuses on the latter, i.e. the path recovery problem, supposing that the network physical topology is already known. Traditionally, a telecommunication company attempts to recover path relationships by analyzing the complex network configuration files. However, this method has the following main disadvantages. First, it is very time consuming because each network involves a huge number of configuration files have to be collected and analyzed. Second, network usually do not assume centralized management and their various parts are controlled by different maintainers. Thus, there is no the common standard of such kind of files. It means that, for each network, we need to put a lot of efforts every time the network is upgraded and the recovering process need to be done again.

In this paper, we propose a novel approach to recovering network path relationships. It is aimed at recovering the path relationships by analyzing the information on data flows through the network arcs. Note that getting the flow information is not a challenging technical problem, since the control of this information is included in the basic toolbox for network monitoring. We consider two scenarios (problems) which lead us to integer linear programs. The both of them minimize the flow deviation: In the first one we look for a directed spanning tree (so called r-arborescence), while in the second one—more general origin/destination paths (OD-paths).

The integer programming is a common approach in the field of telecommunication networks and is widely used in a range of communication network problems (see [18,21] for a survey). Steiner tree problem on graphs and its version for directed graphs—r-arborescence problem—are well-known combinatorial optimization problems [7,11,13] with many applications to computer networks [9,10,16,17,23]. Different integer linear formulations were proposed for this problem [12,15,19,22], which allow one to solve large-scale instances to optimality. More general structures on networks, like finding OD-paths, are also common, especially in closely related network design and loading problems [2,3,6,8].

All the references mentioned above are about design, load and routing of data traffic. As far as we know, the problem of recovering and predicting paths is not addressed in the literature on communication networks. We could find only a few related papers, addressing transportation networks [1,14,20], where the variation between the historical time link flows and the simulated ones is minimized. Note that the problems considered in this paper have some key features that do not allow applying the same modelling and solution approaches.

The remainder of the paper is structured as follows. The statement of the problem and its integer linear formulations are introduced in Sects. 2 and 3,

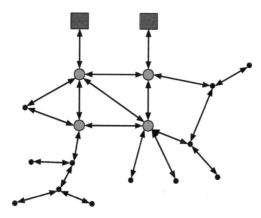

Fig. 1. Network graph $G(V, A)$, HC nodes are red boxes, HB—green circles, HA—black dots. (Color figure online)

respectively. Section 4 describes test instances and some pre-solving techniques to reduce the problem size. The approach based on branch-and-bound algorithm is outlined in Sect. 5. Finally, Sect. 6 gives preliminary computation results and concluding remarks.

2   Problem Statement

We have two scenarios for the path recovering problem. Let us describe some features related to both of them. Given a communication network which is described as follows:

$G(A, V)$—a simple directed graph which defines the network topology;
V—a set of vertices (nodes) consisting of three types of nodes:
V_A—a set of nodes of the base layer (HA nodes). We can think about these nodes as users.
V_B—a set of nodes of the intermediate layer (HB nodes). These nodes are supposed to be hubs or transmission nodes, i.e. they do not generate nor absorb flows.
V_C—a set of nodes of the core layer (HC nodes); These nodes can be considered as routers, i.e. they connect our network with other "outside" networks.
A—a set of arcs. The arcs are thought as data channels which connect nodes;
l_{ij}—the length of arc ij;
f_{ij}—an amount of data traversing arc $ij \in A$;

An example of the network is illustrated in Fig. 1.

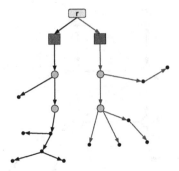

Fig. 2. r-arborescences from the dummy node r.

There are several main conditions to be taken into account:

1. HC nodes are not connected with each others directly.
2. There are no direct arcs between HC and HA nodes (between users and routers).
3. We are given with the data traffic d_t from one of HC node $s \in V_C$ to a base layer node $t \in V_A$. This traffic is unsplittable, i.e. it goes along one path p_{st} from s to t. The path p_{ts} from $t \in V_A$ to a HC node, which is also unsplittable, goes in opposite direction of p_{st}.
4. An additional condition is that no path can go backward from HA nodes to HB or HC ones and from HB nodes to HC ones.

Since path p_{st} is different from path p_{ts} solely in direction, we can only consider paths which go from HC nodes. These paths form a so-called *road map*. Our problem consists in recovering a road map so as the transfer of traffics d_t and q_t through the network corresponds as close as possible to the given values of f_{ij}, for arcs $ij \in A$, while the paths lengths are minimized.

There are additional conditions on paths, depending on the scenarios described in the following subsections.

2.1   First Scenario

In this scenario the main feature is that there is only one unique path from one HC node to any HA and HB nodes which the data flow can go along. Let us consider a dummy node r and arcs rs $\forall s \in V_C$. In this case, the road map consists of Steiner arborescence rooted at r with HA nodes as terminals, i.e. we have an r-arborescence (directed tree) with the root in r and all the HA nodes are reachable form r. An example of this r-arborescences is given in Fig. 2.

Given such r-arborescence, there is only a unique path p_{st} to any $t \in V_A$ from one $s \in V_C$.

Our problem consists in finding an r-arborescence such that transferring the traffics d_t and q_t is as close as possible to the given f_{ij} for each arc $ij \in A$ and the r-arborescence is of the minimal length.

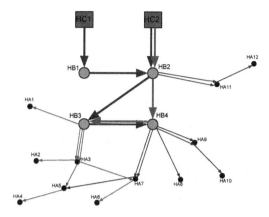

Fig. 3. Road map of the second scenario.

2.2 Second Scenario

The required road map has a different structure than that from the first scenario. Any path p_{st} now consists of two parts. The first one is a subpath which goes from a HC node to a HB node, and the second one is a subpath which starts from the last HB node and goes to its destination at a HA node. Note that the paths with a common HB node as a starting node of the second part have the common first parts, as depicted in Fig. 3. For this example, the road map consists of the following paths:

$$
\begin{aligned}
&\text{Path to } HA1\text{:} &&HC2 \rightarrow HB2 \rightarrow HB4 \rightarrow HB3 \rightarrow HA1 \\
&\text{Path to } HA2\text{:} &&HC2 \rightarrow HB2 \rightarrow HB4 \rightarrow HB3 \rightarrow HA3 \rightarrow HA2 \\
&\text{Path to } HA3\text{:} &&HC2 \rightarrow HB2 \rightarrow HB4 \rightarrow HB3 \rightarrow HA3 \\
&\text{Path to } HA4\text{:} &&HC2 \rightarrow HB2 \rightarrow HB4 \rightarrow HB3 \rightarrow HA3 \rightarrow HA5 \rightarrow HA4 \\
&\text{Path to } HA5\text{:} &&HC2 \rightarrow HB2 \rightarrow HB3 \rightarrow HB4 \rightarrow HA7 \rightarrow HA5 \\
&\text{Path to } HA6\text{:} &&HC2 \rightarrow HB2 \rightarrow HB4 \rightarrow HB3 \rightarrow HA3 \rightarrow HA7 \rightarrow HA6 \\
&\text{Path to } HA7\text{:} &&HC2 \rightarrow HB2 \rightarrow HB3 \rightarrow HB4 \rightarrow HA7 \\
&\text{Path to } HA8\text{:} &&HC2 \rightarrow HB2 \rightarrow HB3 \rightarrow HB4 \rightarrow HA8 \\
&\text{Path to } HA9\text{:} &&HC2 \rightarrow HB2 \rightarrow HB3 \rightarrow HB4 \rightarrow HA9 \\
&\text{Path to } HA10\text{:} &&HC2 \rightarrow HB2 \rightarrow HB3 \rightarrow HB4 \rightarrow HA9 \rightarrow HA10 \\
&\text{Path to } HA11\text{:} &&HC1 \rightarrow HB1 \rightarrow HB2 \rightarrow HA11 \\
&\text{Path to } HA12\text{:} &&HC1 \rightarrow HB1 \rightarrow HB2 \rightarrow HA11 \rightarrow HA12
\end{aligned}
$$

3 MILP Formulation

3.1 First Scenario

The problem can be viewed as a combination of two commodity flow problems, where the first one goes from the root and another goes to the root. Let $K = \{1, 2\}$ be the set of commodities, $f_{ij}^1 = f_{ij}$ be the flow of first commodity which goes through arc ij and $f_{ij}^2 = f_{ji}$—the flows of second commodity which goes through arc ij in opposite direction, $d_t^1 = d_t$ and $d_t^2 = q_t$—the corresponding demand $\forall t \in V_A$.

In order to find a r-arborescence, we add the dummy node and arcs to the graph:

$$V := V \cup \{r\}, \quad A := A \cup \{rs : \ s \in V_C\}.$$

The conditions imposing that any path cannot go upward from HA nodes to HB ones and from HB nodes to HC ones can be taken into account by setting

$$A := A \setminus \{ij : \ (i \in V_B \land j \in V_C) \lor (i \in V_A \land j \in V_B)\}.$$

We use the following notations. Let $W \subset V$, then

$$\delta^-(W) = \{ij \in A : \ i \in V \setminus W, \ j \in W\},$$

$$\delta^+(W) = \{ij \in A : \ i \in W, \ j \in V \setminus W\},$$

$$\delta^-(v) = \delta^-(\{v\}), \quad \delta^+(v) = \delta^+(\{v\}),$$

$$\overleftrightarrow{A} = \{ij \in A \setminus \delta^+(r) : \ ij \in A \land ji \in A\},$$

$$\overrightarrow{A} = \{ij \in A \setminus \delta^+(r) : \ ij \in A \land ji \notin A\},$$

$$\kappa(k) = \begin{cases} 1 \text{ if } k = 2, \\ 2 \text{ if } k = 1. \end{cases}$$

Let us introduce the variables y_{ij}:

$$y_{ij} = \begin{cases} 1 \text{ if arc } ij \text{ belongs to the } r\text{-arborescence,} \\ 0 \text{ otherwise,} \end{cases} \quad \forall ij \in A,$$

the variables x_{ij}^k defining the flow through arc $ij \in A$ of demand $k \in K$; and the variables $z_{ij}^k \ \forall ij \in A \setminus \delta^+(r)$ defining the deviations from the given f_{ij}^k.

With these variables, the MILP can be formulated as follows:

$$\min \ \alpha \cdot \sum_{k \in K} \sum_{ij \in A \setminus \delta^+(r)} z_{ij}^k + \beta \cdot \sum_{ij \in A} l_{ij} y_{ij} \tag{1}$$

subject to

$$z_{ij}^k \geq f_{ij}^k - x_{ij}^k - x_{ji}^{\kappa(k)} \qquad \forall k \in K, \ \forall ij \in \overleftrightarrow{A} \tag{2}$$

$$z_{ij}^k \geq x_{ij}^k + x_{ji}^{\kappa(k)} - f_{ij}^k \qquad \forall k \in K, \ \forall ij \in \overleftrightarrow{A} \tag{3}$$

$$z_{ij}^k \geq f_{ij}^k - x_{ij}^k \qquad \forall k \in K, \ \forall ij \in \overrightarrow{A} \tag{4}$$

$$z_{ij}^k \geq x_{ij}^k - f_{ij}^k \qquad \forall k \in K, \ \forall ij \in \overrightarrow{A} \tag{5}$$

$$\sum_{ij \in \delta^-(v)} y_{ij} \leq 1 \qquad \forall v \in V_B \cup V_C \tag{6}$$

$$\sum_{ij\in\delta^-(t)} y_{ij} = 1 \qquad\qquad \forall t \in V_A \qquad (7)$$

$$\sum_{ij\in\delta^+(W)} y_{ij} \geq 1 \qquad\qquad \forall W \subset V : r \in W,\ V_A \setminus W \neq \varnothing \qquad (8)$$

$$\sum_{ij\in\delta^+(r)} x_{ij}^k = \sum_{t\in V_A} d_t^k \qquad\qquad \forall k \in K \qquad (9)$$

$$\sum_{ij\in\delta^-(v)} x_{ij}^k - \sum_{ij\in\delta^+(v)} x_{ij}^k = 0 \qquad\qquad \forall k \in K,\ \forall v \in V_B \cup V_C \qquad (10)$$

$$\sum_{ij\in\delta^-(t)} x_{ij}^k - \sum_{ij\in\delta^+(t)} x_{ij}^k = d_t^k \qquad\qquad \forall k \in K,\ \forall t \in V_A \qquad (11)$$

$$x_{ij}^k \leq M \cdot y_{ij} \qquad\qquad \forall k \in K,\ \forall ij \in A \qquad (12)$$

$$y_{ij} \in \mathbb{B} \qquad\qquad \forall ij \in A \qquad (13)$$

$$x_{ij}^k \geq 0 \qquad\qquad \forall k \in K,\ \forall ij \in A \qquad (14)$$

$$z_{ij}^k \geq 0 \qquad\qquad \forall k \in K,\ \forall ij \in A \setminus \delta^+(r) \qquad (15)$$

The objective function (1) minimizes the flow deviation with respect to both Manhattan distance and the arborescence length. The parameters α and β define the weights of the corresponding objectives and allow us to find different Pareto optimal solutions of initial bi-objective optimization problem. z_{ij}^k is the flow deviation on arc ij defined by (2)–(5). Constraints (7) guarantee that all HA nodes are reachable once occurred in the r-arborescence and constraints (6) ensure that some HB nodes can be reached as well. Constraints (8) are the well know directed Steiner tree cut constraints and their number is non-polynomial [5, 12]. Constraints (9)–(11) are the flow conservation constraints. Inequality (12) bind x and y variables, i.e. the traffic can go only along the arborescence.

3.2   Second Scenario

To formulate the integer program for the second scenario, we also consider a dummy node r and arcs to each HC node, but also arcs to HB nodes, i.e.

$$V := V \cup \{r\}, \quad A := A \cup \{ri : i \in V_C \cup V_B\}.$$

Paths which go upward from HA nodes to HB or HC nodes and from HB nodes to HC ones are forbidden by

$$A := A \setminus \{ij : (i \in V_B \wedge j \in V_C) \vee (i \in V_A \wedge j \in V_B) \vee (i \in V_A \wedge j \in V_C)\}.$$

In addition to the notation of the first scenario, for some $W^1 \subseteq V$ and $W^2 \subseteq V$ let as denote

$$(W^1 : W^2) = \{ij \in A : i \in W^1,\ j \in W^2\}.$$

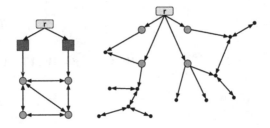

Fig. 4. $G(V^1, A^1)$ is on the left and $G(V^2, A^2)$ is on the right

The problem is divided into two parts. The first one is to find paths from HC to HB nodes, and the second one is to find paths from HB to HA nodes. Thus, the graph is divided into two corresponding subgraphs: $G^1(V^1, A^1)$ and $G^2(V^2, A^2)$, where

$$V^1 = \{r\} \cup V_C \cup V_B, \; A^1 = (\{r\} : V_C) \cup (V_C : V_B) \cup (V_B : V_B),$$

$$V^2 = \{r\} \cup V_B \cup V_A, \; A^2 = (\{r\} : V_B) \cup (V_B : V_A) \cup (V_A : V_A),$$

The example corresponding to Fig. 1 is depicted in Fig. 4.

Let us also introduce the following notations:

$$\overleftrightarrow{A^1} = \{ij \in A^1 \setminus \delta^+(r) : \; ij \in A^1 \wedge ji \in A^1\},$$

$$\overrightarrow{A^1} = \{ij \in A^1 \setminus \delta^+(r) : \; ij \in A^1 \wedge ji \notin A^1\},$$

$$\overleftrightarrow{A^2} = \{ij \in A^2 \setminus \delta^+(r) : \; ij \in A^2 \wedge ji \in A^2\},$$

$$\overrightarrow{A^2} = \{ij \in A^2 \setminus \delta^+(r) : \; ij \in A^2 \wedge ji \notin A^2\}.$$

The problem of finding the first part of paths defined on the graph $G(V^1, A^1)$ can be considered as a multi-commodity flow problem with $|K| \cdot |V_B|$ commodities. To define required paths and flows, the following variables are introduced:

$$y_{ij}^w = \begin{cases} 1 \text{ if arc } ij \text{ belongs to the path from } r \text{ to } w, \\ 0 \text{ otherwise,} \end{cases} \forall ij \in A^1, \; \forall w \in V_B,$$

x_{ij}^{kw}—the flow through arc $ij \in A^1$ of commodity $(k, w) \in K \times V_B$.

On the graph $G(V^2, A^2)$, the second part of paths can be considered as a multi-commodity flow problem with 2 commodities $k \in K = \{1, 2\}$, using the following variables:

$$y_{ij}^t = \begin{cases} 1 \text{ if arc } ij \text{ belongs to the path from } r \text{ to } t, \\ 0 \text{ otherwise,} \end{cases} \forall ij \in A^2, \; \forall t \in V_A.$$

For both $G(V^1, A^1)$ and $G(V^2, A^2)$, the variables z_{ij}^k are the deviations from the given $f_{ij}^k \; \forall ij \in A \setminus \delta^+(r)$.

With this basis, we can now formulate the following MILLP:

$$\min \ \alpha \cdot \sum_{k \in K} \sum_{ij \in A \setminus \delta^+(r)} z_{ij}^k + \beta \cdot \left(\sum_{w \in V_B} \sum_{ij \in A^1} l_{ij} y_{ij}^w + \sum_{t \in V_A} \sum_{ij \in A^2} l_{ij} y_{ij}^t \right) \quad (16)$$

subject to

$$z_{ij}^k \geq f_{ij}^k - \sum_{w \in V_B} x_{ij}^{kw} - \sum_{w \in V_B} x_{ji}^{\kappa(k)w} \qquad \forall k \in K, \ \forall ij \in \overleftrightarrow{A^1} \quad (17)$$

$$z_{ij}^k \geq \sum_{w \in V_B} x_{ij}^{kw} + \sum_{w \in V_B} x_{ji}^{\kappa(k)w} - f_{ij}^k \qquad \forall k \in K, \ \forall ij \in \overleftrightarrow{A^1} \quad (18)$$

$$z_{ij}^k \geq f_{ij}^k - \sum_{w \in V_B} x_{ij}^{kw} \qquad \forall k \in K, \ \forall ij \in \overrightarrow{A^1} \quad (19)$$

$$z_{ij}^k \geq \sum_{w \in V_B} x_{ij}^{kw} - f_{ij}^k \qquad \forall k \in K, \ \forall ij \in \overrightarrow{A^1} \quad (20)$$

$$z_{ij}^k \geq f_{ij}^k - \sum_{t \in V_A} d_t^k y_{ij}^t - \sum_{t \in V_A} d_t^{\kappa(k)} y_{ji}^t \qquad \forall k \in K, \ \forall ij \in \overleftrightarrow{A^2} \quad (21)$$

$$z_{ij}^k \geq \sum_{t \in V_A} d_t^k y_{ij}^t + \sum_{t \in V_A} d_t^{\kappa(k)} y_{ji}^t - f_{ij}^k \qquad \forall k \in K, \ \forall ij \in \overleftrightarrow{A^2} \quad (22)$$

$$z_{ij}^k \geq f_{ij}^k - \sum_{t \in V_A} d_t^k y_{ij}^t \qquad \forall k \in K, \ \forall ij \in \overrightarrow{A^2} \quad (23)$$

$$z_{ij}^k \geq \sum_{t \in V_A} d_t^k y_{ij}^t - f_{ij}^k \qquad \forall k \in K, \ \forall ij \in \overrightarrow{A^2} \quad (24)$$

$$\sum_{ij \in \delta^-(v)} y_{ij}^w - \sum_{ij \in \delta^+(v)} y_{ij}^w = 0 \qquad \forall w \in V_B, \ \forall v \in V^1 \setminus \{r, w\} \quad (25)$$

$$\sum_{ij \in \delta^-(v)} x_{ij}^{wk} - \sum_{ij \in \delta^+(v)} x_{ij}^{wk} = 0 \qquad \begin{matrix} \forall k \in K, \forall w \in V_B, \\ \forall v \in V^1 \setminus \{r, w\} \end{matrix} \quad (26)$$

$$\sum_{ij \in \delta^+(r)} y_{ij}^w - \sum_{ij \in \delta^-(w)} y_{ij}^w = 0 \qquad \forall w \in V_B \quad (27)$$

$$\sum_{ij \in \delta^+(r)} x_{ij}^{wk} - \sum_{ij \in \delta^-(w)} x_{ij}^{wk} = 0 \qquad \forall k \in K, \ \forall w \in V_B \quad (28)$$

$$x_{ij}^{kw} \leq M \cdot y_{ij}^w \qquad \forall k \in K, \ \forall w \in V_B, \ \forall ij \in A^1 \quad (29)$$

$$\sum_{ij \in \delta^-(v)} y_{ij}^t - \sum_{ij \in \delta^+(v)} y_{ij}^t = 0 \qquad \forall t \in V_A, \ \forall v \in V^2 \setminus \{r, t\} \quad (30)$$

$$\sum_{ij \in \delta^+(r)} y_{ij}^t = 1 \qquad \forall t \in V_A \quad (31)$$

$$\sum_{ij \in \delta^-(t)} y_{ij}^t = 1 \qquad \forall t \in V_A \quad (32)$$

$$\sum_{ij \in (W:W)} y_{ij}^w \leq |W| - 1 \qquad \forall w \in V_B, \ \forall W \subseteq V^1 \setminus \{r, w\} \qquad (33)$$

$$\sum_{ij \in (W:W)} y_{ij}^t \leq |W| - 1 \qquad \forall t \in V_A, \ \forall W \subseteq V^2 \setminus \{r, t\} \qquad (34)$$

$$\sum_{ij \in \delta^-(w)} x_{ij}^{wk} = \sum_{t \in V_A} d_t^k y_{ij}^t \qquad \forall k \in K, \ \forall w \in V_B \qquad (35)$$

$$y_{ij}^w \in \mathbb{B} \qquad \forall w \in V_B, \ \forall ij \in A^1 \qquad (36)$$

$$y_{ij}^t \in \mathbb{B} \qquad \forall t \in V_A, \ \forall ij \in A^2 \qquad (37)$$

$$x_{ij}^{kw} \geq 0 \qquad \forall k \in K, \forall w \in V_B, \quad \forall ij \in A^1 \qquad (38)$$

$$z_{ij}^k \geq 0 \qquad \forall k \in K, \ \forall ij \in A \setminus \delta^+(r) \qquad (39)$$

Objective function (16) minimizes both the total flow deviation and paths length. The flow deviations z_{ij}^k on the arc ij are defined by (17)–(24). The flow conservation constraints on $G^1(V^1, A^1)$ are met due to (25)–(28), while the flows and paths are binded by (29). Constraints (30)–(32) are the flow conservation on $G^2(V^2, A^2)$. Constraints (33) and (34) are the well-know subtour elimination constraints for $G^1(V^1, A^1)$ and $G^2(V^2, A^2)$ correspondingly and their number is non-polynomial [4]. The flows between these graphs are linked by (35).

4   Test Networks and Presolving

We were provided with data of many networks. Analyzed their structure, we proposed two techniques to reduce the problem size.

4.1   Problem Decomposition

Relying on networks analysis and our own knowledge about real telecommunication networks, we can conclude that network has a structure with the following properties:

1. HC nodes are not connected with each other.
2. HA and HB nodes are divided into several clusters which are not connected. Each cluster is quite sparse. HA nodes are connected with HB nodes either directly or by chains or circles with rare branching. HB nodes are connected by circle with some chords.

Therefore, we can conclude that any required path in both of the scenarios goes only within one cluster. Using these network properties, the problem can be decomposed into subproblems related to each cluster. Let us describe how the decomposition can be done. We are given with a network defined by the digraph $G(V, A)$. The clusters in $G(V, A)$ can be found by eliminating HC nodes. Thus, we get a disconnected graph, where each weakly connected component is a cluster. The components can be easily identified, for example, by the Depth First Search (DFS).

Table 1. Details on problem decomposition

Name	$|V_C|$	$|V_B|$	$|V_A|$	$|V|$	$|A|$
t_04_0579_c1	2	10	185	197	373
t_04_0579_c2	2	9	126	137	248
t_04_0579_c3	2	6	96	104	195
t_04_0579_c4	2	8	137	147	278
t_04_0579	**2**	**33**	**544**	**579**	**1300**

Name	$|V_C|$	$|V_B|$	$|V_A|$	$|V|$	$|A|$
t_04_0579_c1	6	51	1380	1437	2614
t_04_0579_c2	6	6	221	233	442
t_04_0579_c3	6	5	193	204	363
t_04_0579_c4	6	20	469	495	934
t_04_0579_c5	6	5	128	139	261
t_04_0579_c6	6	5	197	208	391
t_04_0579_c7	6	5	158	169	290
t_07_2849	**6**	**97**	**2746**	**2849**	**6216**

To meet the specific requirements for paper length, we consider only two networks: the smallest (t_04_0579) and the largest (t_07_2849) ones. The details of problem decomposition are given in Table 1, where $Name$ is the network name (in appended _cX, X means the cluster index). $|V_C|$ is the number of HC nodes, $|V_B|$ is the number of HB nodes, $|V_A|$ is the number of HA nodes, $|V|$ is the total number of nodes, $|A|$ is the number of arcs. It is easily seen that the problem is decomposed in several subproblems of much smaller dimension. The impact of this decomposition for different scenarios will be studied below.

4.2 Variable Fixing for the Second Scenario

In the second scenario, formulation (16)–(39) contains much more variables than that in the first scenario. Note that the problem decomposition described above is also valid for it. Furthermore, the following preprocessing, which allows us to drastically reduce the formulation size, plays a very important role.

Actually, due to the particular graph structure and sparsity, a node $t \in V_A$ can be reached only from a subset of nodes of $G(V^2, A^2)$. Let $V_t^2 \subseteq V^2$ be a subset of these nodes together with t, i.e. it contains t and the nodes for which there is a path to node t. It is quite evident that the required path from r to t goes along the arcs between these nodes. Thus, the variable defined on other arcs can be fixed to zero, i.e.

$$y_{ij}^t = 0 \ \ \forall ij \notin A_t^2, \text{ where } A_2^t = (V_t^2 : V_t^2). \tag{40}$$

V_t^2 and A_t^2 can easily be found, for example, by the DFS.

This preprocessing can significantly reduce the problem dimension. Figure 5 illustrates how sets A_1^2 look like for the example presented in Fig. 4. Table 2 shows the impact of this procedure in sense of the number of variables and constraints in the formulation (16)–(39). $Name$ is the network name (in appended _cX, X means the cluster index), $nvar$ is the number of variables, $ncon$ is the number of constraints without subtour elimination constraints (33) for the original formulation and after the variable fixing is performed, respectively. The problem dimension is reduced drastically, though it cannot actually be solved without decomposition and variable fixing even in case of the smallest instances t_04_0579_c3.

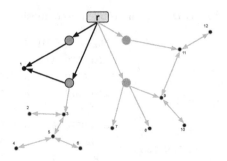

Fig. 5. Arc subset A_1^2 (black arcs)

Table 2. Number of variables and constraints in the second scenario

Name	Original		After fixing	
	nvar	ncon	nvar	ncon
t_04_0579_c1	67 991	111 414	5 544	8 568
t_04_0579_c2	30 986	53 299	3 656	5 689
t_04_0579_c3	18 820	30 866	2 578	4 055
t_04_0579_c4	37 796	61 858	3 894	6 058
t_04_0579	576 760	958 923	21 672	32 418
t_07_2849_c1	3 483 117	5 970 059	69 416	99 968
t_07_2849_c2	95 470	153 552	8 578	13 254
t_07_2849_c3	68 896	117 058	5 823	9 064
t_07_2849_c4	424 938	698 246	19 605	29 672
t_07_2849_c5	33 042	52 925	3 660	5 780
t_07_2849_c6	76 284	121 946	6 886	10 757
t_07_2849_c7	45 470	79 243	5 575	8 722
t_07_2849	14 009 099	23 556 017	173 795	244 793

5   Solution Approach

For initial testing of proposed formulations, a simple approach based on the branch-and-cut algorithm was implemented. We use a MIP solver as the branch-and cut framework. Steiner tree cut constraints (8) and subtour elimination constraints (33), (34) are considered as *lazy constraints*, i.e. they are added to the formulation in the case of violation by an integer solution.

In the first scenario, let us denote by $\mathcal{W}$ the set of all possible subsets in (8) i.e.

$$\mathcal{W} = \{W \subset V : r \in W, \ V_A \setminus W \neq \varnothing\},$$

$\overline{\mathcal{W}}$ be a subset of $\mathcal{W}$ and $P(\overline{\mathcal{W}})$ be problem (1)–(15) with (8) defined only on $\overline{\mathcal{W}}$. Our approach is outlined in Algorithm 1. It is necessary to solve a separation problem on Step 2. In our simplified approach $\bar{y}$ is integer, so this problem can

Algorithm 1. Implementation of formulation (1)-(15)

Step 0. Define initial $\overline{\mathcal{W}}$.

Step 1. Solve $P(\overline{\mathcal{W}})$, let $(\bar{y}, \bar{x}, \bar{z})$ be the solution found.

Step 2. Find $\overline{W} \in \mathcal{W} : \sum\limits_{ij \in \delta^+(\overline{W})} \bar{y}_{ij} = \min\limits_{W \in \mathcal{W}} \sum\limits_{ij \in \delta^+(W)} \bar{y}_{ij}$.

Step 3. If $\sum\limits_{ij \in \delta^+(\overline{W})} \bar{y}_{ij} >= 1$ then goto Stop.

Step 4. If $\sum\limits_{ij \in \delta^+(\overline{W})} \bar{y}_{ij} = 0$ then $\overline{\mathcal{W}} := \overline{\mathcal{W}} \cup \overline{W}$ and goto Step 1.

Stop. $(\bar{y}, \bar{x}, \bar{z})$ is a optimal solution of problem (1)-(15).

Table 3. First scenario results

Name	Time	
	Original	Decomp
t_04_0579_v01	1.19	1.20
t_07_2849_v01	10.14	8.83

easily be solved. Let us consider arcs $\bar{A} = \{ij \in A : \bar{y}_{ij} = 1\}$. $\overline{W}$ consists of r and the other nodes from V which are reachable from r over $\overline{A}$. It forms an arborescence, which can be easily found by DFS. If $V_A \subset \overline{W}$, the problem is solved, otherwise the corresponding constraint (8) is added. Initially $\mathcal{W} = \{r\}$.

As in the first scenario, sets W, on which the corresponding constraints (34) and (33) are violated, can be easily found by scanning the arcs corresponding to the integer solutions.

6   Computation Results and Concluding Remarks

For both scenarios, the approach has been implemented and tested on workstation with Intel Core i7-3770 CPU 3.4 GHz and 32 GB RAM. To solve the MILP problems, we use MIPCL[1] solver. We consider a case $\alpha = 1$ and $\beta = 0$. In our preliminary results we consider only the so-called *steady* case, i.e. we suppose that the flow data is exact, without uncertainty and noise, which results in that the optimal value of objective function equals to zero.

The results for the first scenario are given in Table 3, where running time in seconds is given for the original and decomposed (Decomp) networks. As we can see, all the instances are very simple and can be solved very fast even without problem decomposition.

The second scenario seems to be much more challenging. The results are given in Table 4, where *ncuts* is the number of generated subtour elimination constraints. Using the decomposition and the variable fixing procedure, the instances on network t_4_579 can be solved to optimality quite fast. The largest one was not solved, because two subproblems were not solved within a time limit of one hour.

---

[1] http://www.mipcl-cpp.appspot.com/index.html.

Table 4. Second scenario results

Name	ncuts	Time		Name	ncuts	Time
t_4_579_c1	13	22.7		t_7_2849_c1	0	*3600.0*
t_4_579_c2	0	14.4		t_7_2849_c2	2	381.6
t_4_579_c3	2	9.4		t_7_2849_c3	2	52.6
t_4_579_c4	0	21.1		t_7_2849_c4	0	*3600.0*
Total	**15**	**67.6**		t_7_2849_c5	0	12.1
				t_7_2849_c6	5	952.1
				t_7_2849_c7	5	1501.7

As concluding remarks, we can say that the problem on the first scenario is simple and can efficiently be solved by MIP solvers. The problem of second scenario is much harder and there is room for further research. It is worth mentioning that the LP relaxation can be solved fast, providing us with a good lower bound for the objective value, but general MIP solvers cannot find a good feasible solution. Therefore, it is necessary to develop a heuristic for searching for upper bounds in order to tackle large instances. The further research have to be also moved toward considering a more real scenario when the flow data is uncertain and contains noise.

References

1. Abadi, A., Rajabioun, T., Ioannou, P.A.: Traffic flow prediction for road transportation networks with limited traffic data. IEEE Trans. Intell. Transp. Syst. **16**(2), 653–662 (2015). https://doi.org/10.1109/TITS.2014.2337238
2. Akyildiz, I.F., Lee, A., Wang, P., Luo, M., Chou, W.: A roadmap for traffic engineering in SDN-openflow networks. Comput. Netw. **71**, 1–30 (2014). https://doi.org/10.1016/j.comnet.2014.06.002
3. Altın, A., Yaman, H., Pınar, M.C.: The robust network loading problem under hose demand uncertainty: formulation, polyhedral analysis, and computations. INFORMS J. Comput. **23**(1), 75–89 (2011). https://doi.org/10.1287/ijoc.1100.0380
4. Applegate, D.L., Bixby, R.E., Chvátal, V., Cook, W.J.: The Traveling Salesman Problem: A Computational Study. Princeton University Press, Princeton (2006). http://www.jstor.org/stable/j.ctt7s8xg
5. Avella, P., Boccia, M., Sforza, A., Vasil'ev, I.: A branch-and-cut algorithm for the median-path problem. Comput. Optim. Appl. **32**(3), 215–230 (2005). https://doi.org/10.1007/s10589-005-4800-2
6. Avella, P., Mattia, S., Sassano, A.: Metric inequalities and the network loading problem. Discrete Optim. **4**(1), 103–114 (2007). https://doi.org/10.1016/j.disopt.2006.10.002. Mixed Integer Programming
7. Beasley, J.E.: An algorithm for the steiner problem in graphs. Networks **14**(1), 147–159 (1984). https://doi.org/10.1002/net.3230140112

8. Benhamiche, A., Mahjoub, A.R., Perrot, N., Uchoa, E.: Unsplittable non-additive capacitated network design using set functions polyhedra. Comput. Oper. Res. **66**, 105–115 (2016). https://doi.org/10.1016/j.cor.2015.08.009

9. Bharath-Kumar, K., Jaffe, J.: Routing to multiple destinations in computer networks. IEEE Trans. Commun. **31**(3), 343–351 (1983). https://doi.org/10.1109/TCOM.1983.1095818

10. Cheng, X., Li, Y., Du, D.Z., Ngo, H.Q.: Steiner trees in industry. In: Du, D.Z., Pardalos, P.M. (eds.) Handbook of Combinatorial Optimization, pp. 193–216. Springer, Boston (2005). https://doi.org/10.1007/0-387-23830-1_4

11. Dreyfus, S.E., Wagner, R.A.: The steiner problem in graphs. Networks **1**(3), 195–207 (1971). https://doi.org/10.1002/net.3230010302

12. Goemans, M.X., Myung, Y.S.: A catalog of steiner tree formulations. Networks **23**(1), 19–28 (1993). https://doi.org/10.1002/net.3230230104

13. Hakimi, S.L.: Steiner's problem in graphs and its implications. Networks **1**(2), 113–133 (1971). https://doi.org/10.1002/net.3230010203

14. Karbassi, A., Barth, M.: Vehicle route prediction and time of arrival estimation techniques for improved transportation system management. In: IEEE IV2003 Intelligent Vehicles Symposium, Proceedings (Cat. No. 03TH8683), pp. 511–516, June 2003. https://doi.org/10.1109/IVS.2003.1212964

15. Könemann, J., Pritchard, D., Tan, K.: A partition-based relaxation for steiner trees. Math. Program. **127**(2), 345–370 (2011). https://doi.org/10.1007/s10107-009-0289-2

16. Novak, R., Kandus, G.: Adaptive steiner tree balancing in distributed algorithm for multicast connection setup. Microprocess. Microprogr. **40**(10), 795–798 (1994). https://doi.org/10.1016/0165-6074(94)90042-6

17. Novak, R., Rugelj, J., Kandus, G.: Steiner Tree Based Distributed Multicast Routing in Networks, pp. 327–351. Springer, Boston (2001). https://doi.org/10.1007/978-1-4613-0255-1_10

18. Oliveira, C.A., Pardalos, P.M.: Mathematical Aspects of Network Routing Optimization. SOIA, vol. 53. Springer, New York (2011). https://doi.org/10.1007/978-1-4614-0311-1

19. Polzin, T., Daneshmand, S.V.: On steiner trees and minimum spanning trees in hypergraphs. Oper. Res. Lett. **31**(1), 12–20 (2003). https://doi.org/10.1016/S0167-6377(02)00185-2

20. Rathore, P., Kumar, D., Rajasegarar, S., Palaniswami, M., Bezdek, J.C.: A scalable framework for trajectory prediction (2018). https://arxiv.org/abs/1806.03582v3

21. Resende, M., Pardalos, P.: Handbook of Optimization in Telecommunications. Springer, Heidelberg (2008). https://doi.org/10.1007/978-0-387-30165-5

22. Siebert, M., Ahmed, S., Nemhauser, G.: A linear programming based approach to the steiner tree problem with a fixed number of terminals. Networks **75**(2), 124–136 (2020). https://doi.org/10.1002/net.21913

23. Voß, S.: Steiner tree problems in telecommunications. In: Resende, M.G.C., Pardalos, P.M. (eds.) Handbook of Optimization in Telecommunications, pp. 459–492. Springer, Boston (2006). https://doi.org/10.1007/978-0-387-30165-5_18

Author Index